EXPLORATIONS
IN BASIC BIOLOGY

EXPLORATIONS IN BASIC BIOLOGY

Eleventh Edition

Stanley E. Gunstream

Pasadena City College

San Francisco   Boston   New York
Cape Town   Hong Kong   London   Madrid   Mexico City
Montreal   Munich   Paris   Singapore   Sydney   Tokyo   Toronto

Acquisitions Editor: Jeff Howard
Associate Editor: Andrew Sobel
Assistant Editor: Jessica Berta
Executive Managing Editor: Kathleen Schiaparelli
Assistant Managing Editor: Beth Sweeten
Production Services: Progressive Publishing Alternatives
Composition: Progressive Information Technologies
Senior Managing Editor, Art Production and
    Management: Patricia Burns
Manager, Production Technologies: Matthew Haas
Managing Editor, Art Management: Abigail Bass
Art Production Editor: Rhonda Aversa

Illustrations: Argosy Publishing
Art Director: John Christiana
Cover Design: Yellow Dog Designs
Interior Design: Yellow Dog Designs
Manufacturing Manager: Alexis Heydt-Long
Manufacturing Buyer: Alan Fischer
Cover Image: Leopard (Panthera pardus). South Africa.
    SuperStock, Inc.
Photographer: Philip Van Den Berg
Text Printer: Courier-Kendallville
Cover Printer: Lehigh

Library of Congress Cataloging-in-Publication Data
Gunstream, Stanley E.
    Explorations in basic biology / Stanley E. Gunstream.—11th ed.
        p.   cm.
    Includes bibliographical references.
    ISBN 0-13-222913-7
    1. Biology—Laboratory manuals.   I. Title.

QH317.G82 2008
570—dc22   2006032636

ISBN-13 978-0-13-222913-5
ISBN-10   0-13-222913-7

CONTENTS

PREFACE

The eleventh edition of *Explorations in Basic Biology*, like earlier editions, is designed for use in the laboratory component of introductory general biology courses. It is compatible with any modern biology textbook. The exercises provide a variety of options for one- or two-semester courses and one-, two-, or three-quarter courses. The exercises are appropriate for three-hour laboratory sessions, but they are also adaptable to a two-hour laboratory format.

Explorations in Basic Biology is designed to enhance learning by students and to simplify the work of instructors.

MAJOR FEATURES

1. The forty-one exercises provide a *wide range of options* for the instructor, and the range of activities within an exercise further increases the available options. Several exercises contain investigative portions that ask students to design and conduct experiments on their own, at the discretion of the instructor.

2. Each exercise is basically *self-directing*, which allows students to work independently without continuous assistance by the instructor.

3. Each exercise and its major subunits are *self-contained* so that the instructor may arrange the sequence of exercises, or the activities within an exercise, to suit his or her preferences. In addition, portions of an exercise may be deleted without negatively affecting the continuity of the exercise.

4. Over 250 **illustrations** are provided to enhance students' understanding of both background information and laboratory procedures.

5. New **key terms** are in bold print for easy recognition by students.

6. Each exercise begins with a list of **Objectives** that outlines the minimum learning responsibilities of the student.

7. The text of each exercise starts with a discussion of **background information** that is necessary to (a) understand the subject of the exercise and (b) prepare the student for the activities that follow. The inclusion of the background information minimizes the need for introductory explanations and ensures that all lab sections receive the same background information. Background information always precedes the activity that students are to perform.

8. Before beginning the laboratory activities, students are asked to demonstrate their understanding of the background information by labeling illustrations and completing the portion of the laboratory report that covers this material. Students are asked to color-code selected illustrations to enhance their learning of anatomical features.

9. The required **Materials** (equipment and supplies) are listed for each activity in the exercise. This list helps the student to obtain the needed materials and guides the laboratory technician in setting up the laboratory. The exercises use standard equipment and materials that are available in most biology departments.

10. Activities to be performed by students are identified by an **Assignment** heading and icon for easy recognition. This heading clearly distinguishes activities to be performed from the preceding background information. The assignment sections are numbered sequentially within each exercise and on the laboratory report to facilitate identification and discussion. Each assignment begins with a list of required materials (when needed) followed by specific directions for the laboratory activity.

11. A **Laboratory Report** is provided for each exercise to guide and reinforce students' learning. The laboratory reports not only provide a place for students to record observations, impressions, collected data, and conclusions, but they also provide a convenient means of assessing student understanding.

Each of the diversity exercises (Exercises 10 through 16) includes a minipracticum section on the laboratory report. This challenges students to use knowledge gained in the laboratory session to identify organisms on the basis of their recognition characteristics, and it gives them a taste of a laboratory practicum.

MAJOR IMPROVEMENTS IN THE ELEVENTH EDITION

Information has been updated and procedures have been rewritten throughout the manual wherever needed to facilitate student learning. There are over twenty new or modified figures. Exercises with significant improvements include the following:

Exercise 1. Orientation

Exercise 7. Photosynthesis

Exercise 19. Blood and Circulation

Exercise 29. Fertilization and Development

Exercise 32. Transport in Plants

Exercise 37. Evolution

Exercise 40. Population Growth

ACKNOWLEDGMENTS

The suggestions by adopters continue to be especially helpful in improving the manual. Special recognition is due those whose critical reviews provided the basis for the improvements in the eleventh edition:

James Anderson, *De Anza College*

Linda Barham, *Meridian Community College*

Ralston Bartholomew, *Warren County Community College*

Diane B. Beechinor, *Palo Alto College*

Michael Boyle, *Seattle Central Community College*

Geralyn M. Caplan, *Owensboro Community & Technical College*

Danette Young Carlton, *Virginia State University*

Brad Chandler, *Palo Alto College*

Bruce R. Chorba, *Mercer County Community College*

Patricia Della Penta, *Warren County Community College*

Diane Doidge, *Grand View College*

Howard Duncan, *Norfolk State University*

Aaron Fried, *SUNY Cortland*

Bob Futrell, *Rockingham Community College*

Carla Gardner, *Coastal Carolina Community College*

L. A. Green, *Florence-Darlington Technical College*

Richard D. Harrington, *Rivier College*

Leland N. Holland Jr., *Pasco-Hernando Community College*

Kathleen P. Huffman, *Thomas Nelson Community College*

Judy Kaufman, *Monroe Community College*

Celeste A. Mazzacano, *Columbia College*

Thomas H. Milton, *Richard Bland College*

Barbara Newman, *Southwest Missouri State University*

Brian Palestis, *Wagner College*

Christina Pflugfelder, *Skagit Valley College, Whidbey Island*

Ed R. Pickering, *Wilberforce University*

O. Tacheeni Scott, *California State University*

Jacqueline L. Spencer, *Thomas Nelson Community College*

Sherri Townsend, *North Arkansas College*

Janet Westbrook, *Cerro Coso College*

Cherie L. R. Wetzel, *City College of San Francisco*

Edwin Wiles, *Surry Community College*

Lona E. Williams, *Clinch Valley College*

Karen R. Zagula, *Wake Technical Community College*

Martin Zahn, *Thomas Nelson Community College*

Jeffrey R. Zimdahl, *Centenary College*

The professional skills and helpfulness of the editorial and production personnel at Prentice Hall have eased the work of the revision. I especially want to thank Andrew Sobel, Associate Editor, for his support and contribution. It was also a pleasure to work with Heather Meledin of Progressive Publishing Alternatives, who skillfully guided the production process. Original line drawings and color illustrations by Mary Dersch have significantly enhanced the usefulness of the book. The contributions of these talented professionals, including all members of the book team, are gratefully acknowledged.

Photo Credits

Figure 1.1: Stanley E. Gunstream; **Figure 2.1:** Courtesy of Olympus America, Inc.; **Figure 9.6:** By permission of Turtox/Cambosco/Ward's Natural Science Establishment, Inc.; **Figure 9.7:** By permission of Ward's Natural Science Establishment, Inc.; **Figure 19.11:** Stanley E. Gunstream; **Figure 19.12a:** Stanley E. Gunstream; **Figure 19.12b:** Stanley E. Gunstream; **Figure 19.12c:** Stanley E. Gunstream; **Figure 20.7:** Stanley E. Gunstream; **Figure 23.10a:** Stanley E. Gunstream; **Figure 23.10b:** Stanley E. Gunstream; **Figure 24.6a:** Stanley E. Gunstream; **Figure 24.6b:** Stanley E. Gunstream; **Figure 27.7:** Stanley E. Gunstream; **Figure 27.8:** Stanley E. Gunstream; **Figure 29.1:** Stanley E. Gunstream; **Figure 29.2:** Stanley E. Gunstream; **Figure 29.3:** Stanley E. Gunstream; **Figure 35.6:** Used with permission of Carolina Biological Supply Company; **Figure 37.7a:** Stanley E. Gunstream; **Figure 37.7b:** Stanley E. Gunstream; **Plate 1:** P1.1A-© David M. Philips/Visuals Unlimited; P1.1B-© CNRI/Science Photo Library/Photo Researchers, Inc.; P1.1C-© Science Charles W. Stratton/Visuals Unlimited; P1.2-© Dr. Tony Brain/Science Photo Library/Custom Medical Stock Photo; P1.3-© John D. Cunningham/Visuals Unlimited; P1.4-© T.E. Adams/Visuals Unlimited; **Plate 2:** P2.1-© M.I. Walker/Photo Researchers, Inc.; P2.2-© Michael Abbey/Visuals Unlimited; P2.3-© M.I. Walker/Photo Researchers, Inc.; P2.4-© George J. Wilder/Visuals Unlimited; P2.5-© Michael Abbey/Visuals Unlimited; **Plate 3:** P3.1-© Manfred Kage/Peter Arnold, Inc.; P3.2-© Manfred Kage/Peter Arnold, Inc.; P3.3-© M.I. Walker/Photo Researchers, Inc.; P3.4-© Eric V. Grave/Photo Researchers, Inc.; P3.5-© G.W. Willis, M.D./Visuals Unlimited; **Plate 4:** P4.1-© T.E. Adams/Visuals Unlimited; P4.2-© Cabisco/Visuals Unlimited; P4.3-© David M. Phillips/Visuals Unlimited; P4.4-© PHOTOTAKE/Carolina Biological Supply Company; P4.5-© Eric Grave/Photo Researchers, Inc.; **Plate 5:** P5.1-© PHOTOTAKE/Carolina Biological Supply Company; P5.2-© PHOTOTAKE/Carolina Biological Supply Company; P5.3-© Runk/Schoenberger/Grant Heilman Photography;

EXPLORATIONS IN BASIC BIOLOGY

Part I

Fundamentals

ORIENTATION

OBJECTIVES

After completing the laboratory session, you should be able to:
1. Describe how to prepare for a laboratory session.
2. Describe laboratory safety procedures.
3. State the meaning of common prefixes and suffixes.
4. Convert from one metric unit to another and convert between metric and English units.
5. Make precise and detailed observations.
6. List the steps of the scientific method.
7. Define all terms in bold print.

Laboratory study is an important part of a course in biology. It provides opportunities for you to observe and study biological organisms and processes and to correlate your findings with the textbook and lectures. It allows the conduction of experiments, the collection of data, and the analysis of data to form conclusions. In this way, you experience the process of science, and it is this process that distinguishes science from other disciplines.

PROCEDURES TO FOLLOW

Your success in the laboratory depends on how you prepare for and carry out the laboratory activities. The procedures that follow are expected to be used by all students.

Preparation

Before coming to the laboratory, complete the following activities to prepare yourself for the laboratory session:
1. Read the assigned exercise to understand (a) the objectives, (b) the meaning and spelling of new terms, (c) the introductory background information, and (d) the procedures to be followed.
2. Label and color-code illustrations as directed in the manual, and complete the items on the laboratory report related to the background information.
3. Bring your textbook to the laboratory for use as a reference.

Working in the Laboratory

The following guidelines will save time and increase your chances of success in the laboratory:

1. Remove the laboratory report from the manual so you can complete it without flipping pages. Laboratory reports are three-hole-punched so you can keep completed reports in a binder.
2. Follow the directions explicitly and in sequence unless directed otherwise.
3. Work carefully and thoughtfully. You will not have to rush if you are well prepared.
4. Discuss your procedures and observations with other students. If you become confused, ask your instructor for help.
5. Answer the questions on the laboratory report thoughtfully and completely. They are provided to guide the learning process.

Laboratory Safety and Housekeeping

Exercises in this manual may involve unfamiliar procedures using hazardous chemicals and a variety of laboratory equipment. Therefore, you should follow a few simple safety procedures.

1. Come to class well prepared for the laboratory session. Understanding the background material and the procedures to be used is the first step toward a safe and successful experience.

2. At all times, follow the directions of your instructor.
3. Keep your work station clean and uncluttered. Unnecessary materials should be kept somewhere other than on your desktop.
4. Keep clothes and hair away from the open flame of a Bunsen burner.
5. Always use a mechanical pipetting device when pipetting fluid; never use your mouth.
6. Keep water away from electrical cords and electronic equipment. Electricity and water are not compatible.
7. Wear disposable protective gloves when working with hazardous chemicals or biologicals, such as blood products, DNA, or urine, to avoid contact with skin or clothes.
8. Inform your instructor immediately of any breakage, spills, or injuries, even minor ones.
9. Do not eat, drink, or apply makeup in the laboratory, and keep other materials away from your mouth. The laboratory is not germ free.
10. Be certain that you understand how to use a piece of equipment before trying to use it. When in doubt, ask your instructor.
11. At the end of the period:
    a. Return all equipment to the designated location.
    b. Wash all glassware and return it and other materials to their designated locations.
    c. Wash your desktop and your hands.

BIOLOGICAL TERMS

One of the major difficulties encountered by beginning students is learning biological terminology. Each exercise has new terms emphasized in bold print so that you do not overlook them. Be sure to know their meanings prior to the laboratory session.

Most biological terms are composed of a root word and either a prefix or a suffix, or both. The **root word** provides the main meaning of the term. It may occur at the beginning or end of the term, or it may be sandwiched between a **prefix** and a **suffix.** Both prefix and suffix modify the meaning of the root word. The parts of a term are often joined by adding *combining vowels* that make the term easier to pronounce. The following examples illustrate the structure of biological terms:

1. The term *endocranial* becomes *endo/crani/al* when separated into its components. *Endo-* is a prefix meaning "within"; *crani* is the root word meaning "skull"; *-al* is a suffix meaning "pertaining to." Therefore, the literal meaning of endocranial is "pertaining to within the skull."
2. The term *arthropod* becomes *arthr/o/pod* when separated into its components. *Arthr-* is a prefix

meaning "joint"; *o* is a combining vowel; *pod* is the root word meaning "foot." Therefore, the literal meaning of arthropod is "jointed foot."

Once you understand the structure of biological terms, learning the terminology becomes much easier. *Appendix A contains the meaning of common prefixes, suffixes, and root words.* Use it frequently to help master new terms.

 Assignment 1

Using Appendix A, **complete item 1 on Laboratory Report 1 that begins on page 11.**

UNITS OF MEASUREMENT

In making measurements, scientists use the **International System of Units (SI),** which is commonly called the **metric system.** The metric system is the only system of measurement used in many countries of the world. It is an easy and convenient system, once it is learned. Some of the common units are shown in Table 1.1. Study the table to become familiar with the names and values of the units. Note that within each category the units vary by powers of 10, which allows easy conversion from one unit to another. The English equivalents are given for some units for comparison and so you can convert values from one system to the other. If you study the table, you will be able to recognize the relative value of the prefixes in the names of the units, which are summarized here.

Prefix	Symbol	Meaning
mega-	M	$10^6 = 1,000,000$
kilo-	k	$10^3 = 1,000$
hecto-	h	$10^2 = 100$
deka-	da	$10^1 = 10$
		$10^0 = 1$
deci-	d	$10^{-1} = 0.1$
centi-	c	$10^{-2} = 0.01$
milli-	m	$10^{-3} = 0.001$
micro-	μ	$10^{-6} = 0.000001$

Length

Length is the measurement of a line, either real or imaginary, extending between two points. Your height, the distance between cities, and the size of a football field involve length. The basic unit of length is the **meter.**

TABLE 1.1	COMMON METRIC SYSTEM UNITS			
Category	Symbol	Unit	Value	English Equivalent
Length	km	kilometer	1,000 m	0.62 mi
	m	meter*	1 m	39.37 in.
	dm	decimeter	0.1 m	3.94 in.
	cm	centimeter	0.01 m	0.39 in.
	mm	millimeter	0.001 m	0.04 in.
	μm	micrometer	0.000001 m	0.00004 in.
Mass	kg	kilogram	1,000 g	2.2 lb
	g	gram*	1 g	0.04 oz
	dg	decigram	0.1 g	0.004 oz
	cg	centigram	0.01 g	0.0004 oz
	mg	milligram	0.001 g	
	μg	microgram	0.000001 g	
Volume	l	liter*	1 l	1.06 qt
	ml	milliliter	0.001 l	0.03 fl. oz
	μl	microliter	0.000001 l	

*Denotes the base unit.

Mass

Mass is the characteristic that gives an object inertia, the resistance to a change in motion. It is the quantity of matter in an object. Mass is not the same as weight because weight is dependent on the force of gravity acting on the object. The mass of a given object is the same on Earth as on the moon, but its weight is much less on the moon because the moon's force of gravity is less than Earth's. As a nonscientist, you probably can get by using mass and weight interchangeably, although this is technically incorrect.

In this section, you will use a triple-beam balance to measure the mass of objects. Obtain a balance and locate the parts labeled in Figure 1.1. Note that each beam is marked with graduations. The beam closest to you has 0.1-g and 1.0-g graduations, the middle beam has 100-g graduations, and the farthest beam has 10-g graduations. There is a movable mass attached to each beam. When the pan is empty and clean and the movable masses are moved to zero—as far to the left as possible—the balance mark on the right end of the

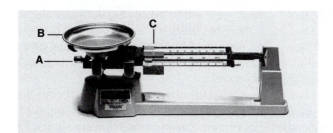

Figure 1.1 A triple-beam balance. **A.** Adjustment knob. **B.** Pan. **C.** Movable mass.

beam should align with the balance mark on the upright post. If not, rotate the adjustment knob under the pan at the left until it does align. Now the balance is ready to use.

Procedure for Measuring Mass

All three beams are used if you are measuring an object with a mass over 100 g; the first and third beams are used when measuring an object with a mass between 10 and 100 g; and the first beam only is used when measuring an object with a mass of 10 g or less. Here is how to do it.

Place the object to be measured in the center of the pan. Move the movable mass on the middle (100 g) beam to the right, one notch at a time, until the right end of the beam drops below the balance mark. Then, move the mass back to the left one notch. Now, slide the movable mass on the third (10 g) beam to the right, one notch at a time, until the right end of the beam drops below the balance mark. Next, move the mass back to the left one notch. Finally, slide the movable mass on the first (1 g) beam slowly to the right until the right end of the beam aligns with the balance mark. The mass of the object is then determined as the sum of the masses indicated on the three beams.

Volume

Volume is the space occupied by an object or a fluid. The liter is the basic unit of volume, but milliliters are the common units used for small volumes. Graduated cylinders and pipettes are used to measure fluids. Length may also be involved in volume determinations because 1 cubic centimeter (cm^3 or cc) equals 1 ml.

Conversion of Metric Units

A major advantage of the metric system is that units within a category may be converted from one to another by multiplying or dividing by the correct power of 10. The following examples show how this is done.

1. To convert 5.75 meters into millimeters, the first step is to determine how many millimeters are in a meter. Table 1.1 shows that 1 mm = 0.001 m, or 1/1,000 of a meter. Thus, there are 1,000 mm in 1 m. Because you are converting meters to millimeters, you multiply 5.75 m by a fraction expressing that there are 1,000 mm in 1 m and cancel like units. You use the unit that you want to convert *to* as the numerator of the fraction.

$$\frac{5.75 \text{ m}}{1} \times \frac{1,000 \text{ mm}}{1 \text{ m}} = 5,750 \text{ mm}$$

Note that this equation multiplies 5.75 by 1,000, which moves the decimal three places to the right and changes the units from meters to millimeters. Moving the decimal and changing the unit is a quick way to do such problems.

2. To convert 125 centimeters into meters, the first step is to determine how many centimeters are in a meter. Table 1.1 shows that 1 cm = 0.01 m, or 1/100 of a meter. Thus, there are 100 cm in 1 m. Because you are converting centimeters into meters, you multiply 125 cm by a fraction expressing that 1 m contains 100 cm and cancel like units. Do you know why 1 m is the numerator of the fraction?

$$\frac{125 \text{ cm}}{1} \times \frac{1 \text{ m}}{100 \text{ cm}} = 1.25 \text{ m}$$

Note that this equation divides 125 cm by 100, which moves the decimal two places to the left and changes the units from centimeters to meters.

Temperature

Scientists use the **degree Celsius (°C)** as the base unit for temperature measurements; most nonscientists in the United States use the **degree Fahrenheit (°F).** A simple comparison of the two systems is shown here.

	°C	°F
Boiling point of water	100	212
Freezing point of water	0	32

You can convert from one system to the other by using these formulas.

Celsius to Fahrenheit:

$$°F = \frac{9}{5}°C + 32$$

Fahrenheit to Celsius:

$$°C = \frac{5}{9}(°F - 32)$$

Assignment 2

Materials

Balance, triple beam
Beaker, 250 ml
Graduated cylinder, 25 ml
Millimeter ruler, clear plastic

1. **Complete item 2a on the laboratory report.**
2. Perform the procedures that follow and record your calculations and data in item 2 on the laboratory report.
   a. Determine the diameter of a penny in millimeters. Then convert your answer to centimeters, meters, and inches.
   b. Using a balance, determine the mass of a 250-ml beaker in grams. Then convert your answer to milligrams and ounces. If you need help in using the balance, see your instructor.
   c. Fill the beaker about half full with tap water. Pour a little water into the graduated cylinder. Look at the top of the water column from the side and note that it is curved rather than flat. This curvature is known as the **meniscus,** and it results because water molecules tend to "creep up" and stick to the side of the cylinder. When measuring the volume of fluid in a cylinder or pipette, you must read the volume at the *bottom of the meniscus,* as shown in Figure 1.2.
   d. While observing the meniscus from the side, add water to the graduated cylinder and fill it to the 15-ml mark. Repeat until you can do this with ease.
   e. Using a graduated cylinder, beaker, and balance, determine the weight (mass) of 20 ml of water.
3. **Complete item 2 on the laboratory report.**

OBSERVATIONS

In the laboratory, you will be asked to make many observations, and these observations require both seeing and thinking. The process of biology involves making careful observations to collect data, analyzing the data, and forming conclusions.

For example, careful observations are necessary to identify biological organisms. Each species (kind of organism) has certain characteristics that are similar to related organisms and characteristics that are different from related organisms. The greater the similarity

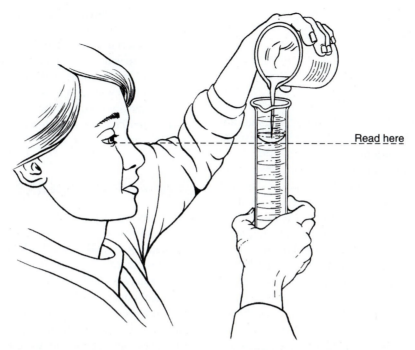

Figure 1.2 Reading the volume of water at the bottom of a meniscus.

between organisms, the closer is their relationship. Biologists use this principle of similarity to determine the closeness of relationship between organisms. You have subconsciously used this same principle in recognizing the various breeds of dogs as dogs and in distinguishing between dogs and cats.

When classifying organisms, biologists develop a **dichotomous key,** based on the characteristics of the organisms, to separate a variety of organisms according to species. A dichotomous key is constructed so a single characteristic is considered at each step, an either/or decision, to separate the organisms into *two groups.* A series of these either/or decisions ultimately separates each type of organism from other types. For example, if we were to develop a dichotomous key to distinguish rats, dogs, cats, monkeys, gorillas, and humans (while ignoring all other organisms), it might look like Figure 1.3. This dichotomous key is purely artificial, but it illustrates how one is constructed. *Note that only one distinguishing characteristic is considered at each step (branch).*

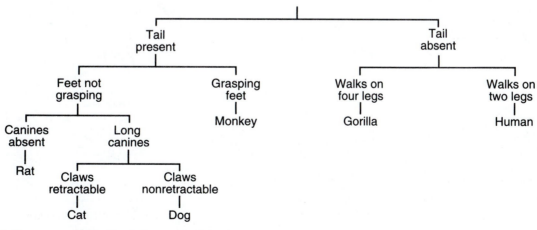

Figure 1.3 An example of a dichotomous key.

Assignment 3

Materials

Leaves of various types

1. Several sets of eight to ten different types of leaves are provided in the laboratory for your examination. Each leaf is numbered. Your objective is to develop a dichotomous key to categorize the leaves of one set according to their observable characteristics. To develop such a key, you must first observe the leaves carefully to determine their similarities and distinguishing characteristics. You are not trying to identify the names of the plants from which the leaves were obtained. Instead, you are to record the number of the leaf that fills the position at the end of each final branch of your key. At each branch of your key, record the distinguishing feature of leaves in that branch, as shown in the example in Figure 1.3.

   In making your key, consider leaf characteristics such as presence or absence of a petiole (leaf stalk), the arrangement of leaf veins (parallel or netlike), general shape (needlelike, scalelike, or flat, thin leaves), simple or compound leaves, shape of the leaf margin (e.g., smooth, toothed, indented), and so forth. Figure 1.4 illustrates certain characteristics of leaves to help you get started.

   There is no one correct way to make this dichotomous key. It can be made in several ways, but it will challenge your ability to recognize the distinguishing characteristics of the leaves.

2. ***Construct your dichotomous key in item 3 of the laboratory report.***

SCIENTIFIC INQUIRY

Biologists use a particular method to find answers to questions about life and living organisms. This method is called the **scientific method,** an orderly process that provides scientific evidence for explanations of natural phenomena. Unlike anecdotal evidence, scientific evidence consists of measurements or observations that are widely shared with other scientists and that are repeatable by anyone with the proper tools. Therefore, biologists can repeat the measurements or observations to either verify or reject the conclusions of other biologists. In this way, the scientific method makes biology, like other branches of science, an ongoing, self-correcting process of inquiry.

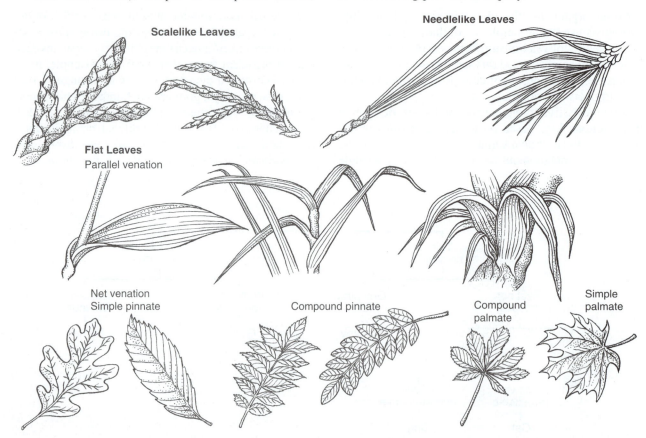

Figure 1.4 Examples of leaf types.

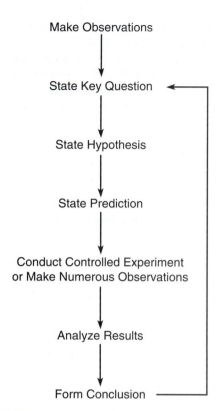

Figure 1.5 Steps of the scientific method.

You will participate in the scientific method as you perform the activities in this laboratory manual. The key steps in the scientific method are shown in Figure 1.5 and described below. Hypothetical examples are provided for each step.

1. The scientific method begins with careful, thoughtful **observations** of nature directly or indirectly.

   Example: General observations suggest that regular exercise may reduce the risk of heart attacks.

2. Observations raise questions in the biologist's mind that lead to a statement of the **key question** to be answered. This key question is sometimes called "the problem." This question is usually a "how," "what," or "why" question.

   Example: Does regular exercise reduce the risk of heart attacks and coronary artery disease?

3. A testable **hypothesis** is stated. This is a statement of the anticipated answer to the key question.

   Example: Regular exercise reduces the risk of heart attacks and coronary artery disease.

4. A **prediction** is made based on the hypothesis. It is usually phrased in an "if . . . then" manner. This tells the biologist what to expect if the results support the hypothesis.

   Example: If regular exercise reduces the risk of heart attacks and coronary artery disease, then mice on an exercise regimen will develop fewer heart attacks or less coronary heart disease than mice that do not exercise.

5. Either a **controlled experiment** is designed and conducted or **multiple observations** are made to test the hypothesis. A controlled experiment provides the most convincing evidence about a hypothesis, but some hypotheses cannot be tested this way. For example, hypotheses about the relationships of fossils in rock strata or the grouping of organisms into taxonomic categories cannot be experimentally tested and require numerous observations. In our example, a controlled experiment can be done and is preferred.

   In a controlled experiment, there are three kinds of variables (conditions). The **independent variable** (in our example, regular exercise) is the condition that is being evaluated for its effect on the **dependent variable** (in our example, heart attacks or coronary artery disease). **Controlled variables** are all other conditions that could affect the results but do not because they are kept constant. A controlled experiment consists of two groups of subjects. The *experimental group* is exposed to the independent variable, but the *control group* is not. All other variables are controlled (kept constant) in both groups.

   Example: A population of laboratory mice have been bred to have identical hereditary compositions that make them susceptible to coronary artery disease and heart attacks. Mice of similar age are selected from this population for the experiment. Fifty randomly selected mice are placed in the experimental group and 50 randomly selected mice are placed in the control group. Mice in the experimental group are placed on a regular exercise regimen (an exercise wheel is in each cage), but mice in the control group are placed in cages without exercise wheels. Data from the exercise wheels are tabulated to be certain that mice in the experimental group are exercising. All other variables are kept identical for each group (i.e., diet, temperature, humidity, light–dark cycles, water availability, etc.). The experiment runs for 120 days.

6. **Results** of the controlled experiment or multiple observations are collected and analyzed.

   Example: Among the experimental group, 18 mice had heart attacks or coronary artery disease. Among the control group, 25 mice had heart attacks or coronary artery disease. In addition, mice in the experimental group weighed 10% less, on average, than mice in the control group.

7. A **conclusion** regarding whether the hypothesis is accepted or rejected is made based on the results. A conclusion often leads to the formation of a new key question and hypothesis, which lead to additional experiments or observations.

Example: The hypothesis is accepted because the experimental group had fewer heart attacks or less coronary artery disease than the control group. Biologists use statistics to determine the significance of differences between results of experimental and control groups—but we won't be concerned about that here.

 Assignment 4

Complete item 4 on the laboratory report.

Laboratory Report 1

ORIENTATION

Student _____

Lab Instructor _____

1. LABORATORY PROCEDURES AND BIOLOGICAL TERMS

a. Describe how you are to prepare for each laboratory session.

_____

_____

_____

b. Using Appendix A, indicate the literal meaning of the following terms:

Biology _____

Morphology _____

Unicellular _____

Leukocyte _____

Organelle _____

Gastric _____

Pathology _____

Dermatitis _____

Osteocyte _____

Hypodermic _____

Pseudoscience _____

Intercellular _____

Extracellular _____

c. Using Appendix A, construct terms with the following literal meanings:

Study of tissues _____

Within a cell _____

Large molecule _____

2. MEASUREMENTS

a. Indicate the name and value of these metric symbols.

Symbol	Name of Unit	Value of Unit
km	_____	_____
ml	_____	_____
mg	_____	_____

cm   _____   _____

mm   _____   _____

μg   _____   _____

b. Indicate the diameter of a penny in the following units:

_____ mm   _____ cm   _____ m   _____ in.

c. Indicate the weight of a 250-ml beaker in the following units:

_____ g   _____ mg   _____ oz

d. List the steps that you used to determine the weight of 20 ml of water.

1. _____

2. _____

3. _____

4. _____

e. Indicate the weight of 20 ml of water.   _____ g

f. A physician directs a patient to take 1 g of vitamin C each day. How many 250-mg tablets must be taken each day?   _____

g. Normal body temperature is 37°C (98.6°F). If a patient's temperature is 38.5°C, what is the temperature in degrees Fahrenheit?   _____ °F

h. The U.S. National Research Council recommends that adults eat 0.8 g of protein daily per kilogram of body weight. What is the minimum number of grams of protein that should be eaten by a man weighing 185 pounds?

_____ g

What is your weight?   _____ lb;   _____ kg

What should be your minimum protein intake?   _____ g

i. If the ground beef in a $\frac{1}{4}$-lb hamburger contains 20% protein and 25% fat, indicate the grams of

fat _____ g

protein _____ g

3. OBSERVATIONS

a. Examine the leaves provided in the laboratory and construct a dichotomous key based on their characteristics. To help you get started, first separate the leaves into two groups: (1) needle- or scalelike leaves and (2) flat, thin leaves, as shown on the next page. Use the next page for the development of your key.

b. After you have completed your key, describe how you approached the problem and the steps or process that you used to construct your key.

_____

_____

_____

_____

_____

```
                            Leaves Provided
              ┌──────────────────┴──────────────────┐
      Needle- or scalelike leaves            Flat, thin leaves
```

4. SCIENTIFIC INQUIRY

a. Do you think the generalization that all organisms are composed of cells was established using controlled experiments or multiple observations? _____

_____

b. Why is a control group important in an experiment? _____

_____

c. What would you conclude if 23 mice in the experimental group had heart attacks or coronary artery disease? _____

_____

d. State a new key question that the result of the experiment triggers in your mind. _____

_____

e. State a testable hypothesis and prediction based on your new key question. _____

_____

  Hypothesis: _____

_____

  Prediction: _____

_____

f. Describe how you would test this hypothesis in a controlled experiment.

_____

_____

_____

_____

THE MICROSCOPE

OBJECTIVES

After completing the laboratory session, you should be able to:

1. Identify the parts of compound and stereo microscopes and state the function of each.
2. Describe and demonstrate the correct way to:
   - a. Carry a microscope.
   - b. Clean the lenses.
   - c. Prepare a wet-mount slide.
   - d. Focus with each objective.
   - e. Determine the total magnification.
   - f. Estimate the size of objects.
3. Define all terms in bold print.

A microscope is a precision instrument and an essential tool in the study of cells, tissues, and minute organisms. It must be handled and used carefully at all times. Most of the microscopic observations in this course will be made with a **compound microscope,** but a **stereo microscope** will be used occasionally. A microscope consists of a lens system, a controllable light source, and a mechanism for adjusting the distance between the lens system and the object to be observed.

To make the observations required in this course, you must know how to use a microscope effectively. This exercise provides an opportunity for you to develop skills in microscopy.

THE COMPOUND MICROSCOPE

The major parts of the compound microscope are shown in Figure 2.1. Your microscope may be somewhat different from the one illustrated. As you read the following descriptions, locate the parts in Figure 2.1.

The **base** rests on the table and, in most microscopes, contains a built-in **light source** and a **light switch.** Some microscopes have a **voltage control knob,** which regulates light intensity, located near the light switch. The **arm** rises from the base and supports the stage, lens system, and control mechanisms. The **stage** is the flat surface on which microscope slides are placed for viewing. **Stage clips,** or a **mechanical stage,** hold the slide in place. The **mechanical stage control** allows precise movement of the slide.

A **condenser** is located below the stage. It concentrates light on the object being observed. On some microscopes, it may be lowered or raised by a condenser control knob. Usually, the condenser should be raised to its highest position. It is fixed in this position on some microscopes. An **iris diaphragm** is built into the condenser, which enables you to control the amount of light entering the lens system. The size of the aperture in the iris diaphragm is usually controlled by a lever or a rotatable wheel.

The **ocular lens** (eyepiece) is located at the upper end of the **body tube** if the microscope is monocular. In binocular microscopes, as in Figure 2.1, the oculars are attached to a **rotatable head** attached to the body tube. The rotatable head allows a choice of viewing positions, and it is locked in place by a **lock screw.** The two oculars may be moved closer together or farther apart to adjust for the distance between the pupils of your eyes. The left ocular has an adjustment ring that may be used to accommodate any difference in visual acuity between your eyes. A **revolving nosepiece** with the attached **objective lenses** is located at the lower end of the body tube.

Student microscopes usually have three objectives. The shortest is the **scanning objective,** which has a magnification of 4×. The **low-power objective** with a 10× magnification is intermediate in length. The **high-power (high-dry) objective** is the longest and usually has a magnification of 40×, but may have a 43× or 45×

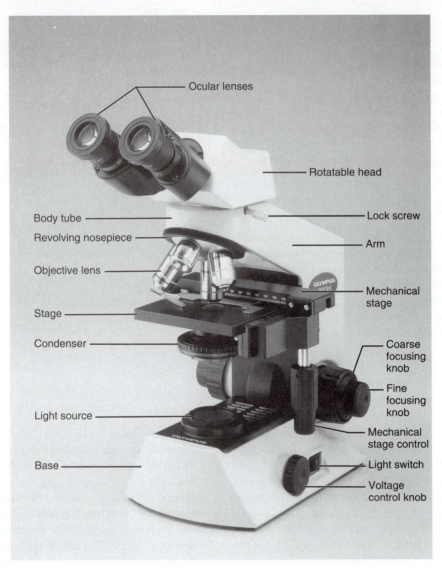

Figure 2.1 A binocular compound microscope (Courtesy of Olympus America Inc.).

magnification in some microscopes. In this manual, the high-power objective is often called the 40× objective. Some microscopes also have an **oil-immersion objective** (100× magnification) that is a bit longer than the high-power objective.

There are two focusing knobs. The **coarse-focusing knob** has the larger diameter and is used to bring objects into rough focus when using the 4× and 10× objectives. The **fine-focusing knob** has a smaller diameter and is used to bring objects into sharp focus. It is the *only* focusing knob used with the high-power and oil-immersion objectives.

Magnification

The magnification of each lens is fixed and inscribed on the lens. The ocular usually has a 10× magnification. The powers of the objectives may vary but usually are

4×, 10×, and 40×. The **total magnification** is calculated by multiplying the power of the ocular by the power of the objective.

Resolving Power

The quality of a microscope depends on its ability to **resolve** (distinguish) objects. Magnification without resolving power is of no value. Modern microscopes increase both magnification and resolution by a careful matching of light source and precision lenses. Most microscopes have a blue light filter located in either the condenser or the light source, because resolving power increases as the wavelength of light decreases. Blue light has a short wavelength. Student microscopes usually can resolve objects that are 0.5 μm or more apart. The best light microscopes can resolve objects that are 0.1 μm or more apart.

Contrast

Sufficient **contrast** must be present among the parts of an object for the parts to be distinguishable. Contrast results from the differential absorption of light by the parts of the object. Sometimes, stains must be added to a specimen to increase the contrast. A reduction in the amount of light improves contrast when viewing unstained specimens.

Focusing

A microscope is focused by increasing or decreasing the distance between the specimen on the slide and the objective lens. The focusing procedure used depends on whether your microscope has a **movable stage** or **movable body tube.** Both procedures are described here. Use the one appropriate for your microscope. As a general rule, you should start focusing with the low-power (10×) objective unless the large size of the object requires starting with the 4× objective.

Focusing with a Movable Stage

1. Rotate the 10× objective into viewing position.
2. While using the coarse-focusing knob and *looking from the side* (not through the ocular), raise the stage to its highest position or until the slide is about 3 mm from the objective.
3. While looking through the ocular, slowly lower the stage by turning the coarse-focusing knob away from you until the object comes into focus. Use the fine-focusing knob to bring the object into sharp focus.

Focusing with a Movable Body Tube

1. Rotate the 10× objective into viewing position.
2. While using the coarse-focusing knob and *looking from the side* (not through the ocular), lower the body tube until it stops or until the objective is about 3 mm from the slide.
3. While looking through the ocular, slowly raise the body tube by turning the coarse-focusing knob toward you until the object becomes visible. Use the fine-focusing knob to bring the object into sharp focus.

Switching Objectives

Your microscope is **parcentric** and **parfocal.** This means that if an object is centered and in sharp focus with one objective, it will be centered and in focus when another objective is rotated into the viewing position. However, slight adjustments to recenter and refocus (with the fine-focusing knob) may be necessary. As you switch objectives from 4× to 10× to 40× to increase magnification, the (1) working distance, (2)

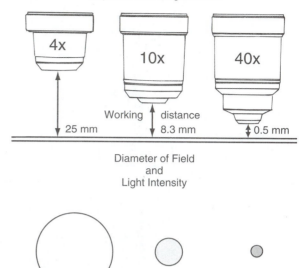

Objective lens magnification

Figure 2.2 Relationship of objective power to working distance, diameter of field, and light intensity.

diameter of the field, and (3) light intensity are *reduced* as magnification increases. Note this relationship in Figure 2.2.

Slide Preparation

Specimens to be viewed with a compound microscope are placed on a **microscope slide** and are usually covered with a **cover glass.** Specimens may be mounted on slides in two different ways. A **prepared slide** (permanent slide) has a permanently attached cover glass, and the specimen is usually stained. A **wet-mount slide** (temporary slide) has the specimen mounted in a liquid, usually water, and covered with a cover glass. In this course, you will use commercially prepared permanent slides and wet-mount slides that you will make. Wet-mount slides are prepared as shown in Figure 2.3.

Care of the Microscope

You should carry a microscope upright in front of you, not at your side. Use one hand to support the base and the other to grasp the arm. See Figure 2.4. Develop the habit of cleaning the lenses prior to using the microscope. Use only special lint-free lens paper. If the lens paper does not clean the lenses, inform your instructor. If any liquid gets on the lenses during use, wipe it off immediately and clean the lenses with lens paper.

When you are finished using the microscope, perform these steps:

1. Turn off the light switch.
2. Remove the slide. Clean and dry the stage.

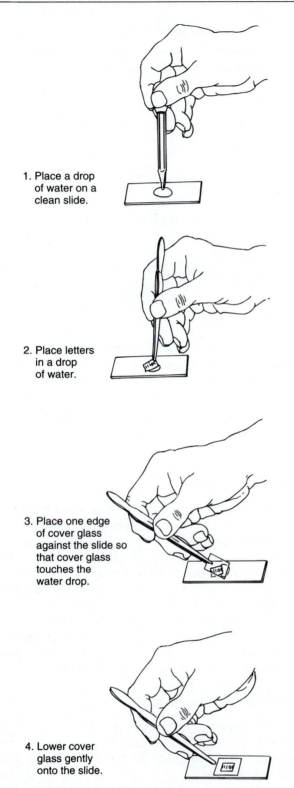

1. Place a drop of water on a clean slide.

2. Place letters in a drop of water.

3. Place one edge of cover glass against the slide so that cover glass touches the water drop.

4. Lower cover glass gently onto the slide.

Figure 2.3 Preparation of a wet-mount slide.

3. Clean the lenses with lens paper.
4. If your microscope has a rotatable head, rotate it so the ocular(s) extend over the arm and lock it in place with the lock screw.
5. Rotate the nosepiece so no objective extends beyond the front of the stage.

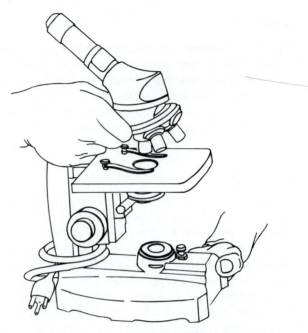

Figure 2.4 How to carry the microscope. Grasp the arm of the microscope with one hand and support the base with the other.

6. Raise the stage to its highest position *or* lower the body tube to its lowest position in accordance with the type of microscope you are using.
7. Unplug the light cord and loosely wrap it around the arm below the stage. Add a dustcover, if present.
8. Return the microscope to the correct cabinet cubicle.

Assignment 1

Materials

Compound microscope
Lens paper

1. Obtain the microscope assigned to you. Carry it as described above, and place it on the table in front of you. Locate the parts shown in Figure 2.1. Clean the lenses with lens paper. Try the knobs and levers to see how they work.
2. Raise the condenser to its highest position and keep it there. Plug in the light cord and turn on the light. If your microscope has a voltage control knob, adjust it to an intermediate position to prolong the life of the bulb.
3. Rotate the 4× objective into viewing position and look through the ocular(s). If your microscope is binocular, adjust the interocular distance to match your interpupillary distance (the distance between the pupils of your eyes) by pushing the oculars together or pulling them apart. The circle of light

that you see is called the **field of view** or simply the **field.**

4. While looking through the ocular, open and close the iris diaphragm and note the change in light intensity. Repeat for each objective and note that light intensity decreases as the power of the objective increases. Therefore, you will need to adjust the light intensity when you switch objectives. *Remember to use reduced light intensity when you are viewing unstained and rather transparent specimens.*

5. If your microscope has a voltage control knob, repeat item 4 while leaving the iris diaphragm open but changing the light intensity by altering the voltage.

6. ***Complete item 1 on Laboratory Report 2 that begins on page 25.***

Developing Microscopy Skills

The following microscopic observations are designed to help you develop skill in using a compound microscope.

Assignment 2

Materials

Compound microscope
Dissecting instruments
Kimwipes
Medicine droppers
Microscope slide and cover glass
Newspaper
Paper towel
Water in dropping bottle

1. Obtain a microscope slide and cover glass. If they are not clean, wash them with soap and water, rinse, and dry them. Use a paper towel to dry the slide, but use Kimwipes to blot the water from the fragile cover glass.

2. Use scissors to cut three sequential letters from a newspaper with the letter *i* as the middle letter.

3. Prepare a wet-mount slide of the letters as shown in Figure 2.3. Use a paper towel to soak up any excess water. If too little water is present, add a drop at the edge of the cover glass and it will flow under the cover glass.

4. Place the slide on the stage with the letters over the stage aperture, the circular opening in the stage. Secure it with either a mechanical stage or stage clamps. See Figure 2.5. The slide should be parallel to the edge of the stage nearest you with the letters oriented so they may be read with the naked eye.

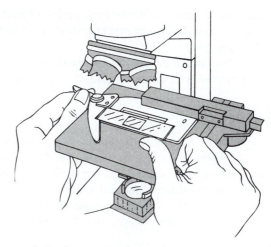

Figure 2.5 Placing the slide in the mechanical stage. The retainer lever is pulled back to place the slide against the stationary arm. Then the retainer lever is released to secure the slide.

5. Rotate the 4× objective into viewing position and bring the letters into focus using the focusing procedure previously described that is appropriate for your microscope.

   If your microscope is binocular, bring the object into focus first, viewing only with your right eye. Then, without moving the focusing knobs, turn the adjustment ring on the left ocular until the object is in sharp focus when viewing with your left eye. Now, you can view objects clearly with both eyes.

6. Center the letter *i* and bring it into sharp focus. Can you see all of the *i*? Can you see the other letters? What is different about the orientation of the letters when viewed with the microscope instead of the naked eye?

7. Move the slide to the left while looking through the ocular. Which way does the image move? Practice moving the slide while viewing through the ocular until you can quickly place a given letter in the center of the field.

8. Center the *i* and bring it into sharp focus. Rotate the 10× objective into position. Is the *i* centered and in focus? If not, center it and bring it into sharp focus. How much of the *i* can you see?

9. Rotate the 40× objective into position. Is the *i* centered and in focus? All that you can see at this magnification are "ink blotches" that compose the *i*. If you do not see this, center the *i* and bring it into focus with the *fine-focusing knob. Never use the coarse-focusing knob with the high-power objective.*

10. Practice steps 5 through 9 until you can quickly center the dot of letter *i* and bring it into focus with each objective. **Remember,** you are *never* to

start observations with the high-power objective. Instead, start at a lower power and work up to the 40× objective.

11. **Complete item 2 on the laboratory report.**
12. Remove the slide and set it aside for later use.

Depth of Field

When you view objects with a microscope, you obviously are viewing the objects from above. The vertical distance within which structures are in sharp focus is called the **depth of field,** and it decreases as magnification increases. You will learn more about depth of field by performing the observations that follow.

Assignment 3

Materials

Compound microscope
Prepared slide of fly wing

1. Obtain a prepared slide of a fly wing. Observe the tiny spines on the wing membrane with each objective, starting with the 4× objective. Can you see all of a spine at each magnification?
2. Using the 4× objective, locate a large spine at the base of the wing where the veins converge and center it in the field. Can you see all of it?
3. Rotate the 10× objective into position and observe the spine. Can you see all of it? Practice focusing up and down the length of the spine, and note that you can see only a portion of the spine at each focusing position.
4. Rotate the 40× objective into position and observe the spine. At each focusing position, you can see only a thin "slice" of the spine. To determine the spine's shape, you have to focus up and down the spine using the *fine-focusing knob.*

   The preceding observations demonstrate that when viewing objects with a depth (thickness) greater than the depth of field, you see only a two-dimensional plane "optically cut" through the object. To discern an object's three-dimensional shape, a series of these images must be "stacked up" in your mind as you focus through the depth of the object.
5. **Complete item 3 on the laboratory report.**

Diameter of Field

When using each objective, you must know the diameter of the field to estimate the size of observed objects. Estimate the diameter of field for each magnification of your microscope as described in the section that follows.

Assignment 4

Materials

Compound microscope
Metric ruler, clear plastic
Water-mount slide of newsprint

1. Place the clear plastic ruler on the microscope stage as shown in Figure 2.6. The edge of the ruler should extend across the diameter of the field. Focus on the metric scale with the 4× objective, and adjust the ruler so one of the millimeter marks is at the left edge of the field. Estimate and record the diameter of field at a total magnification of 40× by counting the spaces and portions thereof between the millimeter marks.
2. Use these equations to calculate the diameter of field at 100× and 400×.

$$\frac{\text{Diam. (mm)}}{\text{at } 100\times} = \frac{40\times}{100\times} \times \frac{\text{diam. (mm) at } 40\times}{1}$$

$$\frac{\text{Diam. (mm)}}{\text{at } 400\times} = \frac{40\times}{400\times} \times \frac{\text{diam. (mm) at } 40\times}{1}$$

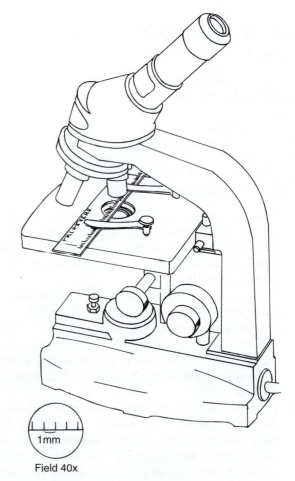

1mm

Field 40x

Figure 2.6 Estimating the diameter of field.

3. Return to your slide of newspaper letters, and estimate the diameter of the dot of the letter *i* and the length of the letter *i* including the dot.
4. ***Complete item 4 on the laboratory report.***

Application of Microscopy Skills

In this section, you will use the skills and knowledge gained in the preceding portions of the exercise.

Assignment 5

Materials

Compound microscope
Dissecting instruments
Hair segments, blond and brunette
Kimwipes
Medicine droppers
Microscope slide and cover glass
Water in dropping bottle
Pond water

1. Prepare a wet-mount slide of two crossed hairs, one blond and the other brunette. Obtain 1-cm lengths of hair from cooperative classmates.

2. Using the 4× objective, center the crossing point of the hairs in the field and observe. Are both hairs in sharp focus?
3. Examine the crossed hairs at 100× magnification. Are both hairs in sharp focus? Determine which hair is on top by using focusing technique. If you have trouble with this, see your instructor.
4. Examine the crossed hairs at 400× magnification. Are both hairs in focus? Move the crossing point to one side and focus on the blond hair. Using focusing technique, observe surface and optical midsection views of the hair as shown in Figure 2.7. Note that you cannot see both views at the same time. What does this tell you about the depth of field in relationship to the diameter of the hair?
5. ***Complete items 5a–5e on the laboratory report.***
6. Make slides of the pond water samples and examine them microscopically. Use reduced illumination when searching for organisms. The object is to sharpen your microscopy skills rather than to identify the organisms; however, you may see organisms like those shown in Figure 2.8. Note the size, color, shape, and motility of the organisms. ***Draw a few of the organisms in the space for item 5f on the laboratory report.***
7. Prepare your microscope for return to the cabinet as described previously, and return it.

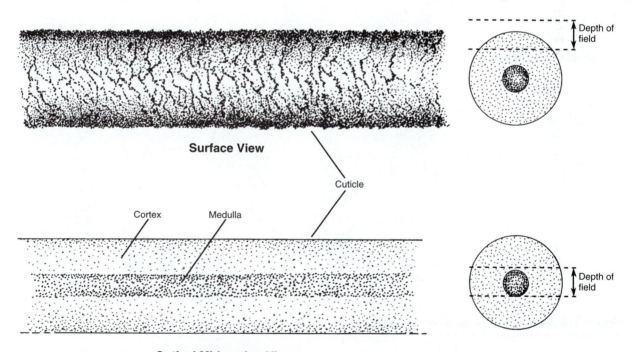

Surface View

Cuticle

Cortex   Medulla

Depth of field

Depth of field

Optical Midsection View

Figure 2.7 Surface and optical midsection views of a human hair, 400×.

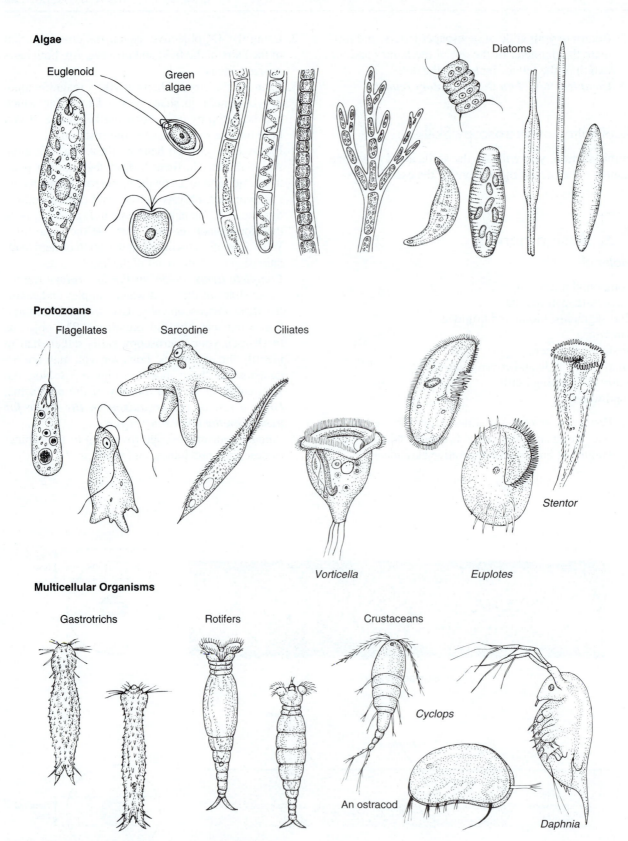

Algae

Euglenoid

Green algae

Diatoms

Protozoans

Flagellates

Sarcodine

Ciliates

Vorticella

Euplotes

Stentor

Multicellular Organisms

Gastrotrichs

Rotifers

Crustaceans

Cyclops

An ostracod

Daphnia

Figure 2.8 Representative pond water organisms.

THE STEREO MICROSCOPE

A **stereo microscope** is used to view objects that are too large or too opaque to observe with a compound microscope. The two oculars enable stereoscopic observations and usually are 10× in magnification. Most student models have two objectives that provide 2× and 4× magnification so that total magnification is 20× and 40×. Some models have a zoom feature that allows observations at intermediate magnifications. Objects are usually viewed with reflected light instead of transmitted light, although some stereo microscopes provide both types of light sources.

The parts of a stereo microscope are shown in Figure 2.9. Note the single focusing knob and the two oculars. The oculars may be moved inward or outward to adjust for the distance between the pupils of your eyes. One ocular has a focusing ring that may be adjusted to accommodate for differences in visual acuity in your eyes.

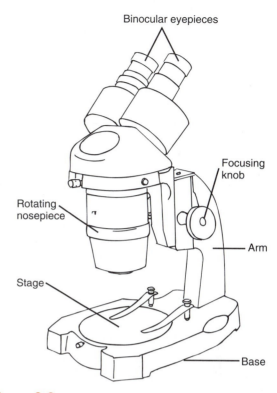

Figure 2.9 Stereo microscope.

 Assignment 6

Materials

Animal and plant specimens
Coin
Desk lamp
Stereo microscope
Metric ruler, clear plastic

1. Obtain a stereo microscope from the cabinet and locate the parts shown in Figure 2.9.
2. Place a coin or other object on the stage, illuminate it with a desk lamp, and examine it with both objectives. To accommodate for differences in acuity in your eyes, focus first with the focusing knob while viewing through the ocular that *cannot* be individually adjusted. Then use the **focusing ring** on the other ocular to bring the object into sharp focus for that eye.
3. Practice focusing until you are good at it. Move the coin on the stage, noting the direction in which the image moves. Examine the grooves and ridges of a finger tip that produce your unique fingerprint.
4. Determine the diameter of field at each magnification.
5. Examine the animal and plant specimens provided by your instructor.
6. *Complete items 6 and 7 on the laboratory report.*

THE MICROSCOPE

Student _____

Lab Instructor _____

1. THE COMPOUND MICROSCOPE

a. Write the term that matches each meaning.

1. Used as handle to carry microscope   _____

2. Lenses attached to the nosepiece   _____

3. Concentrates light on the object   _____

4. Lens you look through   _____

5. Platform on which slides are placed   _____

6. Rotates to change objectives   _____

7. The shortest objective   _____

8. The longest objective   _____

9. Control knob used for sharp focusing   _____

10. Control knob used for rough focusing   _____

11. Controls amount of light entering condenser   _____

b. List the power of the ocular and objectives on *your* microscope, and calculate the total magnification for the combinations noted.

Magnification		*Total*
Ocular(s)	*Objective*	*Magnification*
_____ ×	_____ ×	_____ ×
	_____ ×	_____ ×
	_____ ×	_____ ×
	_____ ×	_____ ×

c. Write the term that matches each meaning.

1. Tissue used to clean lenses   _____

2. Objective with the least working distance   _____

3. Slide with an attached cover glass   _____

4. Objective with the largest field   _____

2. INITIAL OBSERVATIONS

a. Indicate the direction the image moves when the slide is moved:

To the left _____   Away from you _____

b. Indicate the steps to be used in focusing on an object with the high-power objective. _____

_____

_____

c. Describe the best way to relocate an object that is "lost" while viewing with the high-power objective. _____

_____

d. Draw the letter *i* as it appears when observed with each of the following:

| **Unaided** | **4×** | **10×** | **40×** |
| **Eye** | **Objective** | **Objective** | **Objective** |

3. DEPTH OF FIELD

Based on your observations, draw a side view of a large spine from a fly wing at these *total magnifications*.

40×   **100×**

4. DIAMETER OF FIELD

a. Indicate the estimated diameter of field at each *total magnification* for your microscope.

Magnification	*Diameter of Field*
_____×	_____ mm
_____×	_____ mm
_____×	_____ mm

b. At these *total magnifications*, indicate the estimated lengths of objects that extend across (1 mm = 1,000 μm):

1. Two-thirds of the field at 40×   _____ mm   _____ μm

2. 25% of the field at 100×   _____ mm   _____ μm

3. One-half of the field at 400×   _____ mm   _____ μm

4. 80% of the field at 40×   _____ mm   _____ μm

c. Determine and record these measurements:

The length of the letter *i*, including the dot   _____ mm   _____ μm

The diameter of the dot   _____ mm   _____ μm

5. APPLICATIONS

a. When viewing the crossed hairs and the top hair is in sharp focus, is the other hair visible or in sharp focus at the following *total magnifications*?

Magnification	Visible	Sharp Focus
40×	_____	_____
100×	_____	_____
400×	_____	_____

b. At 400×, is the depth of field greater or less than the diameter of the blond hair? _____

c. Estimate the diameter of the blond hair. _____ μm

d. Describe the shape of the blond hair. _____

e. On the basis of your observations, are the following characteristics increased, decreased, or unchanged when magnification is increased?

Illumination _____   Depth of field _____

Working distance _____   Diameter of field _____

f. Draw five to six organisms observed in the pond water slides. Indicate the approximate size, color, and speed of motion of each. Make your drawings large enough to show details.

6. THE STEREO MICROSCOPE

a. When using the stereo microscope, indicate the direction the image moves when the coin is moved

To the left _____   Toward you _____

b. Measure and record the diameter of field at each *total magnification* of the stereo microscope.

Magnification	*Diameter of Field*
_____×	_____ mm
_____×	_____ mm

c. Draw two or three specimens as viewed with a stereo microscope.

7. REVIEW

a. Contrast a prepared slide and a wet-mount slide.

Prepared slide _____

_____

Wet-mount slide _____

_____

b. How should prepared slides be handled? _____

_____

c. Describe how to determine the total magnification when using any objective of your microscope. _____

_____

_____

d. Describe how to estimate the length of an object observed with your microscope. _____

_____

_____

e. How should light intensity be adjusted when viewing nearly transparent specimens? _____

_____

Part II

Cell Biology

THE CELL

OBJECTIVES

After completing the laboratory session, you should be able to:
1. Describe the basic characteristics of prokaryotic and eukaryotic cells.
2. Identify the basic parts of eukaryotic cells when viewed with a microscope.
3. Describe the major functions of the parts of a cell.
4. Distinguish between (a) prokaryotic, animal, and plant cells and (b) unicellular, colonial, and multicellular organisms.
5. Define all terms in bold print.

Living organisms exhibit two fundamental characteristics that are absent in nonliving things: **self-maintenance** and **self-replication.** The smallest unit of life that exhibits these characteristics is a single living cell. Thus, a single cell is the *structural and functional unit of life.* The *cell theory* states that:

1. All organisms are composed of cells.
2. All cells arise from preexisting cells.
3. All hereditary components of organisms occur in cells.

Two different types of cells occur in the biotic world: prokaryotic cells and eukaryotic cells. **Prokaryotic** (*pro* = "before," *karyon* = "nucleus") **cells** are more primitive, relatively simple, and very small (usually less than 10 μm in length), and they occur only in bacteria and archaeans. **Eukaryotic** (*eu* = "true") **cells** are more advanced, more complex, and much larger, and they compose all other organisms: protists, fungi, plants, and animals. A key distinguishing feature is that eukaryotic cells possess a **nucleus** and other membrane-bound **organelles,** which are tiny intracellular compartments specialized for specific functions. Prokaryotic cells lack a nucleus and other membrane-bound organelles.

PROKARYOTIC CELLS

The ultrastructure of a generalized bacterial cell is shown in Figure 3.1 as an example of a prokaryotic cell. The structures are shown as observed with an electron microscope. You cannot see this detail with your microscope. Locate the parts of the cell in Figure 3.1 as you read the following description.

A prokaryotic cell is surrounded by a **plasma membrane** (label 7), which controls the movement of materials into and out of the cell. The **cytoplasm** occupies the interior of the cell. Most of the cytoplasm consists of the **cytosol** (label 8), a semifluid solution of various chemicals in which intracellular particles are suspended. **Ribosomes** (label 2) are tiny particles of ribonucleic acid (RNA) and protein suspended in the cytosol. Ribosomes are sites of protein synthesis.

A single, highly coiled, circular molecule of deoxyribonucleic acid (DNA) constitutes the prokaryotic chromosome. It is concentrated in a region known as the **nucleoid.** There is no physical separation between the chromosome and the cytosol. Additional small circular DNA molecules called **plasmids** (label 9) may be located outside the nucleoid. Because DNA contains the coded hereditary information, cellular functions are controlled primarily by the chromosome and to a lesser extent by plasmids, which contain supplemental hereditary information.

A nonliving **cell wall** (label 6) lies external to the plasma membrane, and it gives shape and support to the cell. A protective **capsule** is secreted external to the cell wall. Tiny extensions of the plasma membrane and cytosol called **pili** (singular, *pilus*) extend through the cell wall and capsule. Pili attach the bacterial cell to surfaces and sometimes serve as channels for the

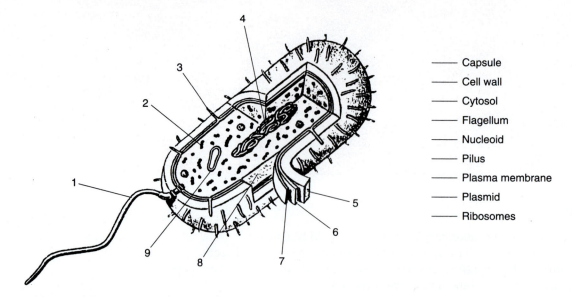

Figure 3.1 Ultrastructure of a prokaryotic cell.

exchange of DNA between two cells of the same species. Some bacteria possess one or more whiplike **flagella** (singular, *flagellum*) that rotate to propel the cell through its liquid environment. Prokaryotic flagella are distinctly different in structure from flagella found in eukaryotic cells.

Assignment 1
Materials

Compound microscope
Prepared slides of:
Pseudomonas aeruginosa
Staphylococcus aureus
Oscillatoria

1. Label Figure 3.1. Color-code the nucleoid and plasmid.
2. Examine prepared slides of *P. aeruginosa, S. aureus*, and *Oscillatoria* at 400× total magnification. The cells are smaller than you probably imagine, so use reduced light and focus carefully.
   Compare your observations with Figure 3.1 and Plate 1, which is located after page 115. Can you detect intracellular structures?
3. ***Complete item 1 on Laboratory Report 3 that begins on page 41.***

EUKARYOTIC CELLS

The presence of organelles in eukaryotic cells enables the compartmentalization of functions so that many functions can occur simultaneously. In this section,

you will study the ultrastructure of animal and plant cells as examples of eukaryotic cells. However, all eukaryotic cells do not have the same structure. For example, animal cells differ from plant cells, and an individual animal or plant is composed of many different types of cells. Cellular components are shown in the figures and described as they appear when viewed with an electron microscope. Most organelles are too small to be seen with your microscope.

The Animal Cell

As you read the following description of structures found in animal cells, locate and label the structures in the generalized animal cell shown in Figure 3.2. All these structures are not found in every animal cell.

Animal cells are surrounded by a **plasma membrane** (label 11), which controls the movement of materials into and out of the cell. The **cytoplasm** and its organelles occupy most of the cell volume. Within the cytoplasm, the **cytosol** (label 8) is the chemically rich solution that bathes and suspends the nucleus, cytoplasmic organelles, and other particles.

The **cytoskeleton** forms a complex framework that supports the cell. Three types of protein filaments compose the cytoskeleton: (1) very fine **microfilaments** (label 13), (2) slightly larger **intermediate filaments** (not labeled), and (3) the still larger **microtubules** (label 15). Shortening and lengthening of these protein fibers are responsible for the movement of organelles within the cell, and in some cases, they enable movement of the entire cell. A microtubule organizing center, the **centrosome** (not labeled), is located near the nucleus, and numerous microtubules radiate from it. In animal cells, two **centrioles** (label 16), short cylindrical rods formed of microtubules, are

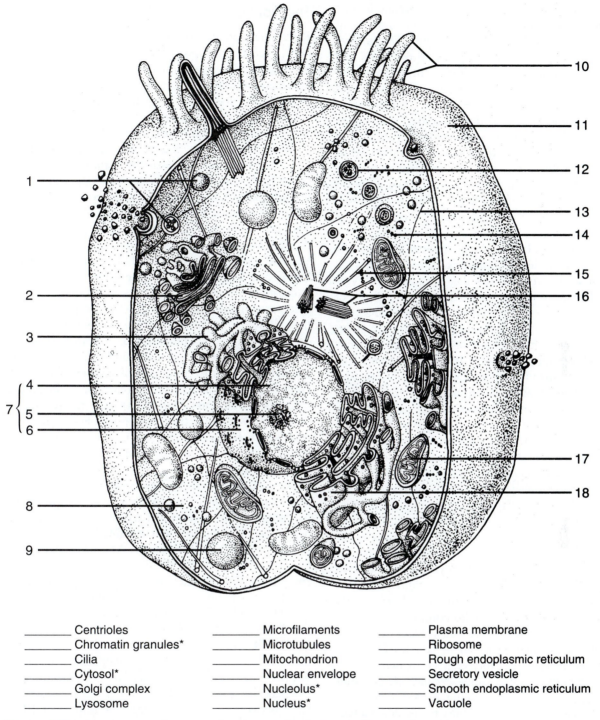

Figure 3.2 Ultrastructure of a generalized animal cell. Organelles are shown as seen with an electron microscope. Organelles visible with your microscope are noted with asterisks.

_____ Centrioles	_____ Microfilaments	_____ Plasma membrane
_____ Chromatin granules*	_____ Microtubules	_____ Ribosome
_____ Cilia	_____ Mitochondrion	_____ Rough endoplasmic reticulum
_____ Cytosol*	_____ Nuclear envelope	_____ Secretory vesicle
_____ Golgi complex	_____ Nucleolus*	_____ Smooth endoplasmic reticulum
_____ Lysosome	_____ Nucleus*	_____ Vacuole

oriented perpendicularly to each other within the centrosome. Some animal cells possess **cilia** (label 10), short hairlike projections whose wavelike motion moves fluid and particles across the cell surface. Sperm cells possess a long whiplike **flagellum** (not shown) that propels the sperm cell through its fluid environment. Both cilia and flagella contain microtubules that enable their movements.

The **nucleus** is a large spherical organelle that contains the **chromosomes,** which are composed of DNA and protein. In nondividing cells, the chromosomes are uncoiled and elongated so only bits and pieces of them can be seen as **chromatin granules** at each focal plane when the nucleus is viewed with a microscope. In dividing cells, the chromosomes coil tightly and appear as dark-staining, rod-shaped structures. The spherical, dark-staining structure in the nucleus is the **nucleolus,** which is composed of RNA and protein. The nucleus is surrounded by a **nuclear envelope** that is composed of two adjacent membranes perforated by pores. The pores allow movement of materials between the nucleus and cytoplasm.

The outer membrane of the nuclear envelope is continuous with the **endoplasmic reticulum (ER),** a series of folded membranes that permeate the cytoplasm. There are two types of ER. **Rough ER** (label 18) is confluent with the nuclear membrane and consists of folded membranes lying parallel to each other. It is called rough ER because its cytoplasmic surfaces are studded with ribosomes. **Ribosomes** (label 14) also occur suspended in the cytosol. They are tiny particles of RNA and protein, and they are the sites of protein synthesis. Ribosomes are assembled in the nucleolus and then move through pores in the nuclear envelope into the cytoplasm. The **smooth ER** (label 3) is so named because it lacks attached ribosomes. It extends outward from the rough ER and consists of interconnected tubules. Whereas the rough ER is involved in the synthesis and modification of proteins, the smooth ER is involved in synthesis of nonprotein molecules.

Both rough and smooth ER serve as channels for the movement of materials within the cytoplasm. They also package substances within tiny membranous sacs called **vesicles** for transport within the cell, especially to the Golgi complex. The **Golgi complex** (label 2) is a stack of membranes where substances are further modified, stored, and packaged in vesicles for transport. Lysosomes and secretory vesicles are two types of vesicles produced by the Golgi complex. **Lysosomes** (label 12) are vesicles containing powerful digestive enzymes that break down complex molecules, including worn-out or defective parts of a cell. **Secretory vesicles** (label 1) carry substances from the Golgi complex to the plasma membrane for export from the cell. Larger membranous sacs called **vacuoles** (label 9) are used to store substances.

Mitochondria are rather large organelles composed of a folded inner membrane surrounded by a smaller, nonfolded membrane. Mitochondria are sites of cellular respiration, a process that breaks down nutrient molecules to release energy for use by the cell.

The Plant Cell

Plant cells contain all the structures found in animal cells, except centrioles, lysosomes, and cilia, and only a few plants have reproductive cells with flagella. As you read the following description of structures found in plant cells, locate and label the structures in the generalized plant cell shown in Figure 3.3. All these structures are not found in every plant cell.

As in animal cells, the **plasma membrane** (label 10) is the outer boundary of the **cytoplasm,** where it controls the movement of materials into and out of the cell. A porous but rigid **cell wall,** composed of cellulose, is formed exterior to the plasma membrane. The cell wall provides support and protection for the cell. Most of the cell is occupied by a large **central vacuole** that is filled with fluid. The **vacuolar membrane** (not labeled) envelops the central vacuole. The hydrostatic pressure in the central vacuole presses the cytoplasm against the cell wall, which helps to support the cell by increasing the rigidity of the cell wall. The central vacuole also functions as a lysosome and a storage site. The cell is also supported by the **cytoskeleton,** which is composed of protein fibers. Two components of the cytoskeleton are shown in Figure 3.3: **microfilaments** (label 15) and **microtubules** (label 16). Although they lack centrioles, plant cells possess a **centrosome** (label 14), a microtubule organizing center. It is an indistinct area with microtubules radiating from it.

The spherical **nucleus** contains **chromosomes,** which are composed of DNA and protein. In nondividing cells, the chromosomes are uncoiled and extended so only tiny bits of them are visible as **chromatin granules** (label 3) when viewed with a microscope. Also in the nucleus is the dark-staining **nucleolus** that is composed of RNA and protein. The **nuclear envelope** consists of a double membrane with numerous pores that enable substances to move freely between the nucleus and the **cytosol** (label 9), the fluid within the cytoplasm.

The **ER** forms membranous channels for the movement of materials throughout the cytoplasm. The **rough ER** (label 1) has **ribosomes** attached to its membranes, whereas the **smooth ER** (label 12) lacks ribosomes. **Ribosomes** (label 13), tiny particles formed of RNA and protein, are sites of protein synthesis. Small groups or chains of free ribosomes occur in the cytoplasm, where they are suspended in the cytosol. The membranous stacks of the **Golgi complex** are also located near the nucleus. The Golgi complex processes and packages substances in **vesicles** (label 7) for transport.

Plastids, found only in plant cells, are large organelles enclosed in a double membrane. There are three types of plastids, which are classified according to the pigments that they contain. **Chloroplasts** (label 17) are large green organelles containing numerous stacks

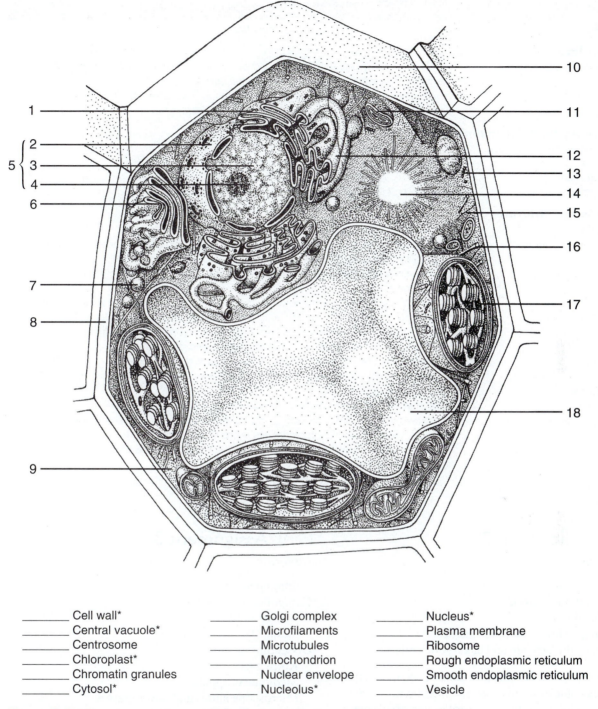

Figure 3.3 Ultrastructure of a generalized plant cell. Organelles are shown as seen with an electron microscope. Organelles visible with your microscope are noted with asterisks.

_____ Cell wall*
_____ Central vacuole*
_____ Centrosome
_____ Chloroplast*
_____ Chromatin granules
_____ Cytosol*

_____ Golgi complex
_____ Microfilaments
_____ Microtubules
_____ Mitochondrion
_____ Nuclear envelope
_____ Nucleolus*

_____ Nucleus*
_____ Plasma membrane
_____ Ribosome
_____ Rough endoplasmic reticulum
_____ Smooth endoplasmic reticulum
_____ Vesicle

of internal membranes that contain chlorophyll. Chloroplasts are the sites of photosynthesis, where light energy is converted into chemical energy. **Chromoplasts** (not shown) contain red to yellow pigments found in fruits and flowers. **Leukoplasts** (not shown) contain no pigments and are sites of starch storage.

As in animal cells, **mitochondria** (label 11) have an outer membrane enclosing a folded inner membrane. Mitochondria are sites of cellular respiration, which releases energy in nutrients for use by the cell.

The functions of cellular structures are summarized in Table 3.1.

TABLE 3.1   FUNCTION OF CELL STRUCTURES

Cell Structure	Function
Cell surface	
Plasma membrane	Controls the passage of materials into and out of the cell; maintains the integrity of the cell.
Cell wall	Protects and supports the cell. Absent in animal cells.
Cilia and flagella	Move fluid over cell surface or move the cell through fluid. Cilia absent in plant cells. Flagella present on reproductive cells of some plants.
Hereditary material organization	
Nuclear envelope	Controls the passage of materials into and out of the nucleus.
Nucleus	Acts as control center of the cell; contains the hereditary material.
Chromosomes	Contain DNA, the determiner of inheritance, which controls cellular processes.
Nucleolus	Assembles proteins and RNA that will form ribosomes.
Cytoplasmic structures	
Centrioles	Assist in the organization of microtubule systems. Absent in plant cells.
Endoplasmic reticulum	Membranous channels for the movement of materials from place to place within a cell; provide membranous surfaces for chemical reactions.
Golgi complex	Stores, modifies, and packages materials in vesicles for transport within the cell or for export from the cell.
Lysosomes	Contain enzymes to digest worn-out or damaged cells or cell parts. Absent in plant cells.
Microfilaments	Provide support for the cell; involved in movement of organelles and entire cells.
Microtubules	Provide support for the cell; involved in movement of organelles and entire cells.
Mitochondria	Sites of aerobic cellular respiration, which releases energy from nutrients and forms ATP.
Plastids	Chloroplasts are sites of photosynthesis. Chromoplasts contain pigments giving color to flowers and fruits. Leukoplasts are often sites of starch storage. Absent in animal cells.
Ribosomes	Sites of protein synthesis.
Vacuole	Contains water and solutes, wastes, or nutrients.
Vacuole, central	Contains water and solutes; provides hydrostatic pressure, which helps support cell. Provides lysosome function in plant cells. Absent in animal cells.
Vesicles	Carry substances from place to place within the cell. Secretory vesicles carry substances to the plasma membrane for export from the cell.

Assignment 2

Materials

Colored pencils
Compound microscope
Medicine droppers
Microscope slides and cover glasses
Mixture of prokaryotic and eukaryotic cells

1. Label Figures 3.2 and 3.3. Color-code these structures: nucleus, nucleolus, mitochondria, SER, RER, Golgi complex, centrioles, and chloroplasts.
2. ***Complete items 2a and 2b on the laboratory report.***
3. Prepare a wet-mount slide of the mixture of prokaryotic and eukaryotic cells. Examine the cells at 400× with reduced illumination and distinguish the cell types.
4. ***Complete item 2 on the laboratory report.***

Onion Epidermal Cells

Epidermal cells of an onion scale show many of the features found in nongreen plant cells and are excellent subjects to use in beginning your study of eukaryotic cells.

Assignment 3

Materials

Colored pencils
Compound microscope
Dropping bottles of:
    iodine ($I_2 + KI$) solution
Dissecting instruments
    Kimwipes
    Medicine droppers
    Microscope slides and cover glasses
    Onion bulb, red

1. Label Figure 3.4 using information from the previous section. Color-code the cytoplasm.
2. Prepare a wet-mount slide of the inner epidermis of an onion scale as shown in Figure 3.5. Use a drop of iodine solution as the mounting fluid and add a cover glass.
3. Examine the cells at 100× and 400× Note the arrangement of the cells. Locate the parts shown in Figure 3.4.
4. Prepare a wet-mount slide of the red epidermis from the outer surface of the onion scale. The color

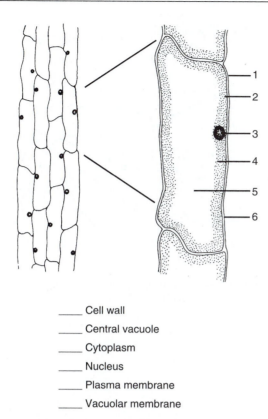

_____ Cell wall

_____ Central vacuole

_____ Cytoplasm

_____ Nucleus

_____ Plasma membrane

_____ Vacuolar membrane

Figure 3.4 Onion epidermal cell.

is due to **anthocyanin,** a water-soluble pigment in the central vacuole. Note the size of the vacuole and the location of the cytoplasm and nucleus.

5. **_Complete item 3 on the laboratory report._**

Elodea Leaf Cells

Elodea is a water plant with simple leaves composed of cells that are easy to observe and that exhibit the characteristics of green plant cells.

 Assignment 4

Materials

Compound microscope
Dissecting instruments
Kimwipes
Medicine droppers
Microscope slides and cover glasses
Elodea shoots

1. Label Figure 3.6.
2. Prepare a wet-mount slide of an _Elodea_ leaf in this manner.
   a. Use forceps to remove a young leaf from near the tip of the shoot and mount it in a drop of water.

1. Break a piece of onion scale.

2. Strip off a small piece of epidermis with your forceps.

3. Place a small piece of epidermis in a drop of iodine solution, flatten it to remove wrinkles, and apply a glass cover.

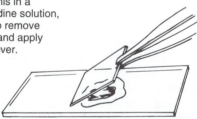

Figure 3.5 Preparation of a slide of onion epidermis.

   b. Add a cover glass and observe at 40× and 100×.
3. Note the arrangement of the cells and the "spine" cells along the edge of the leaf. Locate a light green area for study. The thickness of the leaf is composed of more than one layer of cells. Switch to the 40× objective and focus through the thickness of the leaf. Determine the number of cell layers present.
4. Using the 40× objective, focus through the depth of a cell and locate as many parts shown in Figure 3.6 as possible. The nucleus is spherical and slightly darker than the cytoplasm. Watch for moving chloroplasts. They are being carried by **cytoplasmic streaming,** a phenomenon in which the entire cytoplasm flows around the cell between the cell wall and central vacuole. The mechanism of cytoplasmic streaming is unknown, but microfilaments may be involved.

Surface View   **Optical Midsection**

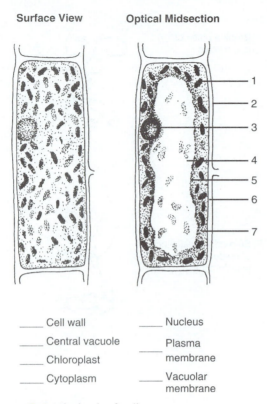

____ Cell wall	____ Nucleus
____ Central vacuole	____ Plasma membrane
____ Chloroplast	
____ Cytoplasm	____ Vacuolar membrane

Figure 3.6 *Elodea* leaf cells.

5. Focus carefully to observe surface and optical midsection views. See Figure 3.6. Note the basic shape of a cell.
6. Examine a spine cell at the edge of the leaf with reduced illumination to locate the nucleus, vacuole, and cytoplasm uncluttered with chloroplasts.
7. ***Complete item 4 on the laboratory report.***

Human Epithelial Cells

The epithelial cells lining the inside of your mouth are easily obtained for study, and they exhibit some of the characteristics of animal cells.

 Assignment 5

Materials

Compound microscope
Dropping bottles of:
   sodium chloride, 0.9%
Medicine droppers
Microscope slides and cover glasses
Toothpicks, flat

1. Label Figure 3.8.
2. Prepare a wet-mount slide of human epithelial cells as shown in Figure 3.7, using 0.9% sodium chloride (NaCl) as the mounting fluid.

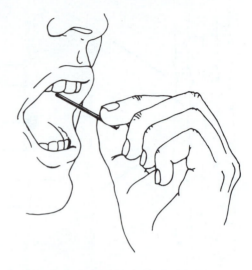

1. Gently scrape inside of mouth with a toothpick to obtain epithelial cells.

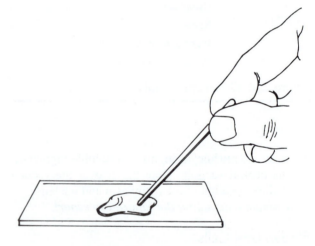

2. Swirl the toothpick in a drop of 0.9% NaCl on a clean slide.

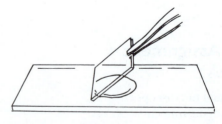

3. Gently apply a cover glass.

Figure 3.7 Preparation of a slide of human epithelial cells.

3. Add a cover glass and observe at 100× and 400×. Compare your cells with Figure 3.8. How do these cells differ from plant cells?
4. ***Complete item 5 on the laboratory report.***

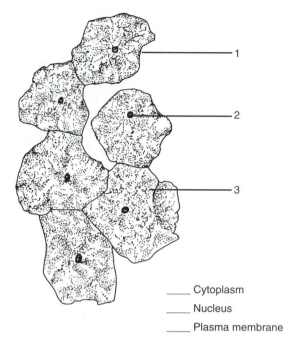

_____ Cytoplasm

_____ Nucleus

_____ Plasma membrane

Figure 3.8 Human epithelial cells.

The Amoeba

The common freshwater protozoan *Amoeba* exhibits many characteristics of animal cells, and it is large enough for you to see the cellular structure rather well. The nucleus, cytoplasm, vacuoles, and cytoplasmic granules are readily visible. The flowing movement of an amoeba allows it to capture minute organisms that are digested in **food vacuoles.** An amoeba also has **contractile vacuoles** that maintain its water balance by collecting and pumping out excess water.

Ectoplasm, the clear outer portion of the cytoplasm, is located just interior to the cell membrane. Most of the cytoplasm consists of the granular **endoplasm.** It contains the organelles and may be either gel-like, called the **plasmagel,** or fluid, called the **plasmasol.** Reversible changes between plasmagel and plasmasol result in the flowing **amoeboid movement** of an amoeba. Your white blood cells also exhibit amoeboid movement as they slip through capillary walls and wander among the body tissues, engulfing disease-causing organisms and cellular debris.

 Assignment 6

Materials

Compound microscope
Kimwipes
Medicine droppers
Microscope slides and cover glasses
Amoeba proteus culture

1. Use a medicine dropper to obtain fluid *from the bottom* of the culture jar and place a drop on a clean slide. Do *not* add a cover glass.
2. Use the 10× objective with reduced light to locate and observe an amoeba. Note the manner of movement and locate the nucleus, cytoplasm, and vacuoles. The nucleus is spherical and a bit darker than the granular cytoplasm. Contractile vacuoles appear as spherical bubbles in the cytoplasm, and food vacuoles contain darker food particles within the vacuolar fluid. Compare your specimen to that shown in Figure 3.9 and Plate 2.5. Observe the way the amoeba moves, while remembering that some of your white blood cells move in a similar way.
3. Note the appearance of the cytoplasm. Does the appearance of the inner and outer portions differ? Observe the nucleus and vacuoles. Gently add a cover glass to your slide and observe an amoeba at

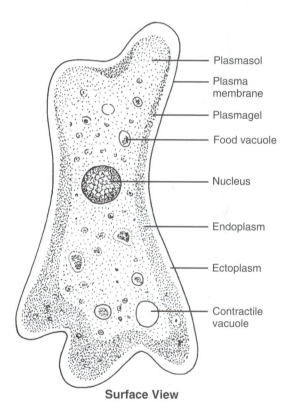

Plasmasol

Plasma membrane

Plasmagel

Food vacuole

Nucleus

Endoplasm

Ectoplasm

Contractile vacuole

Surface View

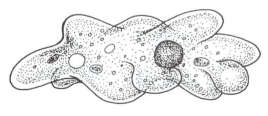

Side View

Figure 3.9 An amoeba.

400×. When the pressure of the cover glass becomes too great, the cell membrane will rupture and the contents of the cell will spill out, killing the amoeba.

4. ***Complete item 6 on the laboratory report.***

LEVELS OF ORGANIZATION

Organisms may exhibit unicellular, colonial, or multicellular forms of body organization. **Unicellular organisms** consist of a single cell that performs all the functions of the organism. **Colonial organisms** are composed of a group of similar cells, and each cell functions more or less independently. **Multicellular organisms** are usually composed of vast numbers of different types of cells, with each type performing a different function.

In most multicellular organisms, cells are arranged to form tissues, and different tissues are grouped together to form organs. A **tissue** is a group of similar cells performing a similar function. Muscle and nerve tissues are examples. An **organ** is composed of several different types of tissues that function together to carry out the particular functions of the organ. The heart and brain are examples. In most animals, coordinated groups of organs compose an **organ system** and work together to carry out the functions of the organ system. The circulatory and skeletal systems are examples.

In this section, you will examine prepared slides of multicellular organisms. There are three basic types of prepared slides: whole mounts (w.m.) contain the entire organism; longitudinal sections (l.s.) are thin sections cut through the organism along its longitudinal axis; and cross-sections or transverse sections (x.s.) are thin sections of the organism cut perpendicular to the longitudinal axis.

 Assignment 7

Materials

Compound microscope
Prepared slides of:
    earthworm (*Lumbricus*), x.s.
    buttercup (*Ranunculus*) root, x.s.

1. Examine the prepared slides of earthworm, x.s., and buttercup root, x.s. (Handle prepared slides by the edges or label to avoid getting fingerprints on the cover glass.) Each tissue is composed of similar cells that are characteristic for that tissue.

2. ***Complete the laboratory report.***

THE CELL

Student _____

Lab Instructor _____

1. PROKARYOTIC CELLS

a. List the labels for Figure 3.1.

1. _____   4. _____   7. _____

2. _____   5. _____   8. _____

3. _____   6. _____   9. _____

b. Sketch the appearance of the prokaryotic cells of bacteria and cyanobacteria at 400× as observed from your slides.

Pseudomonas aeruginosa   **Staphylococcus aureus**   **Oscillatoria**

2. EUKARYOTIC CELL STRUCTURE

a. Write the term that matches each statement.

1. Sites of photosynthesis   _____

2. Filaments of DNA and protein   _____

3. Control center of the cell   _____

4. Sites of aerobic cellular respiration   _____

5. Primary support for plant cells   _____

6. Separates the nucleus and cytoplasm   _____

7. Sites of protein synthesis   _____

8. Sacs of strong digestive enzymes   _____

9. Packages materials in vesicles for export from the cell   _____

10. Controls passage of materials into and out of the cell   _____

11. Semiliquid substance in which cellular organelles are embedded _____

12. Large fluid-filled organelle in mature plant cells   _____

13. Channels for the movement of materials within the cell   _____

14. Assembles precursors of ribosomes within the nucleus   _____

15. Nuclear components containing the genetic code
    controlling cellular processes   _____

b. List the labels for Figures 3.2 and 3.3.

Figure 3.2

1. _____   7. _____   13. _____

2. _____   8. _____   14. _____

3. _____   9. _____   15. _____

4. _____   10. _____   16. _____

5. _____   11. _____   17. _____

6. _____   12. _____   18. _____

Figure 3.3

1. _____   7. _____   13. _____

2. _____   8. _____   14. _____

3. _____   9. _____   15. _____

4. _____   10. _____   16. _____

5. _____   11. _____   17. _____

6. _____   12. _____   18. _____

c. Draw a few prokaryotic and eukaryotic cells from your observations of the cell mixture.

Prokaryotic Cells   **Eukaryotic Cells**

3. ONION EPIDERMAL CELLS

a. List the labels for Figure 3.4.

1. _____   3. _____   5. _____

2. _____   4. _____   6. _____

b. Diagram three adjacent onion epidermal cells as observed on your slide and label the parts that you see.

   c. Determine and record the average length and width of three adjacent cells.

      Width _____mm   Length _____mm

   d. Examination of your slide will show that a few nuclei do not appear next to a cell wall.

      Explain that observation. _____

      _____

4. *ELODEA* LEAF CELLS

   a. Are the cells flexible or rather rigid?_____ What causes this condition?

      _____

   b. What is the three-dimensional shape of a cell?_____

   c. How many cell layers compose the thickness of the leaf? _____

      Which layer has the largest cells? _____

   d. Draw a spine cell at the edge of the leaf at 100×. Label the observed parts.

   e. List the labels for Figure 3.6.

      1. _____   4. _____   6. _____

      2. _____   5. _____   7. _____

      3. _____

5. HUMAN EPITHELIAL CELLS

   a. What structures observed in onion epidermal cells are *not* present in human epithelial cells?

      _____

   b. Indicate the description that best describes the human epithelial cells:

      Thin and platelike _____   Thick and boxlike _____

   c. Are human epithelial cells thinner than *Elodea* cells? _____

   d. Diagram a few cells from your slides and label the parts observed.

   e. List the labels for Figure 3.8.

      1. _____   2. _____   3. _____

6. AMOEBA

a. **Diagram the amoeba on your slide and label the parts observed.**

b. **Is the cell membrane rigid or flexible?** _____

c. **Is the amoeba alive?** _____   **Why do you think so?** _____

_____

d. **Would you use this same evidence to determine if cells in your body are alive?** _____

Explain. _____

_____

7. LEVELS OF ORGANIZATION

a. **Define these terms:**

Unicellular organism _____

_____

Colonial organism _____

_____

Multicellular organism _____

_____

Tissue _____

_____

Organ _____

_____

Organ system _____

_____

b. Using the 4× objective, outline the general arrangement of tissues composing a buttercup root x.s. and an earthworm x.s. Using the 10× objective, draw a few cells of at least three tissues showing the basic shape and structure of the cells. Show where the cells were located on your outline sketch.

Buttercup Root x.s.

Earthworm x.s.

8. REVIEW

a. Name the two characteristics of living organisms that distinguish them from nonliving things.

_____   _____

b. Indicate the structural and functional unit of life. _____

c. Based on your study and observations, indicate the presence of the following structures in animal and plant cells by placing an X in the appropriate spaces.

Structure	Animal	Green Plant	Nongreen Plant
Cell wall	_____	_____	_____
Central vacuole	_____	_____	_____
Centrioles	_____	_____	_____
Chloroplasts	_____	_____	_____
Chromatin granules	_____	_____	_____
Cytoplasm	_____	_____	_____
Endoplasmic reticulum	_____	_____	_____
Lysosomes	_____	_____	_____
Mitochondria	_____	_____	_____
Microtubules	_____	_____	_____
Nucleolus	_____	_____	_____
Nucleus	_____	_____	_____
Plasma membrane	_____	_____	_____
Ribosomes	_____	_____	_____

CHEMICAL ASPECTS

OBJECTIVES

After completing the laboratory session, you should be able to:
1. Determine the number of protons, neutrons, and electrons in an atom of an element from data shown in a periodic table of the elements.
2. Diagram a shell model of a simple atom.
3. Describe the formation of ionic and covalent bonds in accordance with the octet rule.
4. Diagram or make ball-and-stick models of simple molecules.
5. Describe and perform simple tests for carbohydrates, fats, and proteins.
6. Define all terms in bold print.

Life at its most fundamental level consists of complex chemical reactions. Therefore, it is important to have a basic understanding of chemical reactions and the chemicals composing biological organisms.

Chemical substances are classified into two groups: elements and compounds. An **element** is a substance that cannot be broken down by chemical means into any simpler substance. Table 4.1 lists the most common elements found in living organisms. An **atom** is the smallest unit of an element that retains the properties (characteristics) of the element.

TABLE 4.1	COMMON ELEMENTS IN THE HUMAN BODY	
Element	Symbol	Percentage*
Oxygen	O	65
Carbon	C	18
Hydrogen	H	10
Nitrogen	N	3
Calcium	Ca	2
Phosphorus	P	1
Potassium	K	0.35
Sulfur	S	0.25
Sodium	Na	0.15
Chlorine	Cl	0.15
Magnesium	Mg	0.05
Iron	Fe	0.004
Iodine	I	0.0004

*By weight.

Two or more elements may combine to form a **compound.** Water (H_2O) and table salt (NaCl) are simple compounds; carbohydrates, fats, and proteins are complex compounds. The smallest unit of a compound that retains the properties of the compound is a **molecule.** It is formed of two or more atoms joined by **chemical bonds.**

ATOMIC STRUCTURE

Atoms are composed of three basic components: **protons, neutrons,** and **electrons.** Protons and neutrons have a mass of 1 atomic unit and are located in the central region of an atom, the nucleus. Protons possess a positive electrical charge of 1 (+1), but neutrons have no charge. Electrons have a negative electrical charge of 1 (−1), almost no mass, and orbit at near the speed of light around the atomic nucleus. The attraction between positive and negative charges is what keeps the electrons spinning about the nucleus.

Atoms of a given element differ from those of all other elements in the number and arrangement of protons, neutrons, and electrons, and these differences determine the properties of each element. Chemists have used these differences to construct a **periodic table of the elements.**

Table 4.2 is a simplified periodic table of the first 20 elements. Each element is identified by its **symbol.** Note the location of the **atomic number,** which indicates the number of protons in each atom. Because a

TABLE 4.2	SIMPLIFIED PERIODIC TABLE OF THE ELEMENTS—1 THROUGH 20*						
I	II	III	IV	V	VI	VII	VIII
1 H hydrogen 1.0							2 He helium 4.0
3 Li lithium 7.0	4 Be beryllium 9.0	5 B boron 11.0	6 C carbon 12.0	7 N nitrogen 14.0	8 O oxygen 16.0	9 F fluorine 19.0	10 Ne neon 20.2
11 Na sodium 23.0	12 Mg magnesium 24.3	13 Al aluminum 27.0	14 Si silicon 28.1	15 P phosphorus 31.0	16 S sulfur 32.1	17 Cl chlorine 35.5	18 Ar argon 40.0
19 K potassium 39.1	20 Ca calcium 40.1						

Atomic Number, Atomic Symbol, Atomic Mass labels shown within table.

*The dark line separates metals, on the left, from nonmetals.

neutral atom possesses the same number of protons as electrons, the atomic number also indicates the number of electrons in each atom of the element. Note that the elements are sequentially arranged in the table according to atomic number.

The atom's **mass number** (sometimes called atomic weight) equals the sum of protons plus neutrons in each atom. Because the number of neutrons may vary slightly in atoms of the same element, the atomic mass reflects the *average* mass of the atoms. Atoms of an element that vary slightly in mass due to a difference in the number of neutrons are called **isotopes.**

The **shell model** is a useful, although not quite accurate, representation of atomic structure. The energy levels occupied by electrons are represented by circles (shells) around the nucleus. Electrons are shown as dots on the circles.

Examine Figure 4.1. Note the arrangement of the electrons in sodium (Na) and chlorine (Cl). The innermost shell can contain a maximum of only two electrons. Atoms with two or more shells can contain a maximum of eight electrons in the outermost shell.

Assignment 1

1. Use the simplified periodic table of the elements (Table 4.2) to determine the number of protons, electrons, and neutrons in the first 20 elements.
2. Compare the number of electrons in the outer shell of the elements to their position in Table 4.2.
3. ***Complete item 1 on Laboratory Report 4 that begins on page 55.***

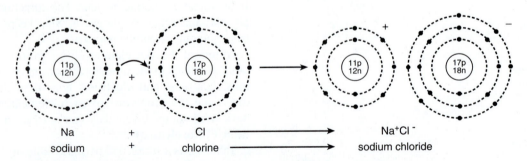

Figure 4.1 The formation of sodium chloride, an ionic reaction. Chlorine gains an electron from sodium, with the result that each atom has eight electrons in the outer shell.

REACTIONS BETWEEN ATOMS

The **octet rule** states that atoms react with each other to achieve eight electrons in their outermost shells. An exception is that, if the first shell is the outermost shell, it is filled by only two electrons.

The octet rule is achieved by (1) losing or gaining electrons or (2) sharing electrons. Atoms with one, two, or three electrons in the outer shell donate (lose) these electrons to other atoms so the adjacent shell, filled with electrons, becomes the outer shell. Atoms with six or seven electrons in the outer shell receive (gain) electrons from adjacent atoms to fill their outer shells. Other atoms usually attain an octet by sharing electrons.

Ionic Bonds

Typically, an atom has a net charge of zero because it has the same number of protons (positive charges) and electrons (negative charges). If an atom gains electrons, it becomes negatively charged because the number of protons is constant. Similarly, the loss of electrons causes an atom to be positively charged. An atom with an electrical charge is called an **ion.**

Ionic bonds are formed between two ions by the attraction of opposite charges that have resulted from gaining or losing electrons. Examine Figure 4.1. Sodium and chlorine unite by an ionic bond to form sodium chloride (NaCl). Note that, with the transfer of the single electron in the outer shell of sodium to chlorine, each ion has eight electrons in its outer shell. This transfer causes sodium to have a positive charge (+1) and chlorine to have a negative charge (−1). The opposite charges hold the ions together to form the molecule. This type of bonding occurs between **metals** (electron donors) and **nonmetals** (electron recipients).

Covalent Bonds

When nonmetals react with nonmetals, electrons are shared instead of being passed from one atom to another. This sharing of electrons usually forms a **covalent bond.** The electrons are shared in pairs, one from each atom. The electrons spend some time in the outer shell of each atom, so the outer shell of each atom meets the octet rule.

Examine Figure 4.2. Note that the oxygen atom, with six electrons in its outer shell, needs to obtain two additional electrons to complete the outer shell. Each hydrogen atom needs one additional electron to complete the first shell. Each hydrogen atom shares its electron with oxygen, and oxygen shares one of its electrons with each hydrogen atom. This sharing of the electrons forms a covalent bond between one oxygen atom and each of the two hydrogen atoms to form one water molecule (H_2O).

Table 4.3 shows the covalent bonding capacity of the four elements that form about 96% of the human body. The bonding pattern shows each potential covalent bond represented by a single line. In other words, each line represents one electron that may be shared with another atom that also has one electron to share to form a covalent bond. Note that an oxygen atom can form two covalent bonds and a hydrogen atom can form one. That is what is shown in the covalent reaction forming a water molecule in Figure 4.2 using shell models. By using a line to represent a covalent bond, chemists can quickly draw **structural formulas** without the hassle of drawing shell models. For example, the structural formula of a water molecule is H—O—H, an oxygen molecule is shown as O=O, and a hydrogen molecule is represented as H=H.

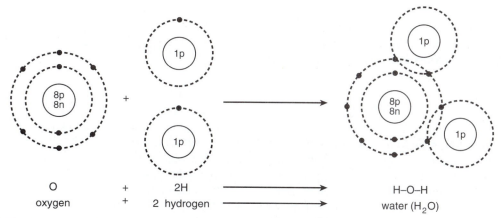

| O | + | 2H | | H–O–H |
| oxygen | + | 2 hydrogen | | water (H_2O) |

Figure 4.2 The formation of water, a covalent reaction. A pair of electrons is shared between each hydrogen atom and the oxygen atom. Shared electrons are counted as belonging to each atom.

TABLE 4.3	BONDING CAPACITIES	
Element	Number of Bonds	Potential Bonds
Carbon	4	—C—
Nitrogen	3	N
Oxygen	2	—O—
Hydrogen	1	H—

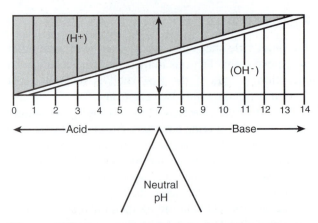

Figure 4.3 The pH scale. The diagonal line indicates the proportionate amount of hydrogen (H^+) and hydroxide (OH^-) ions at each pH value.

Assignment 2

1. Study Figures 4.1 and 4.2 to understand the characteristics of ionic and covalent bonds.
2. Using Figure 4.2 and Table 4.3, compare the shell model and structural formula of water.
3. ***Complete item 2 on the laboratory report.***

ACIDS, BASES, AND pH

An **acid** increases the concentration of **hydrogen ions** (H^+) in a solution by dissociating and releasing H^+. The greater the degree of dissociation, the greater is the strength of the acid. Hydrogen chloride is a strong acid.

$$HCl \longrightarrow H^+ + Cl^-$$

A **base** is a substance that decreases the H^+ concentration of a solution. Some bases, such as ammonia (NH_3), combine directly with H^+.

$$NH_3 + H^+ \longrightarrow NH_4^+$$

Other bases, such as sodium hydroxide (NaOH), indirectly decrease the H^+ concentration by dissociating to release **hydroxide ions** (OH^-), which then combine with H^+ to form water.

$$NaOH \longrightarrow Na^+ + OH^-$$

Chemists use a **pH scale** to indicate the strength of acids and bases. Figure 4.3 shows the scale, which ranges from 0 to 14, and indicates the proportionate concentration of hydrogen and hydroxide ions at the various pH values. When the concentration of hydrogen ions increases, the concentration of hydroxide ions decreases, and vice versa. A change of one (1.0) in the pH number indicates a ten-fold change in the hydrogen ion concentration because the pH scale is a logarithmic scale.

Pure water has a pH of 7, the point where the concentrations of hydrogen and hydroxide ions are equal. Observe this in Figure 4.3. Acids have a pH less than 7; bases have a pH greater than 7. The greater the concentration of hydrogen ions, the stronger is the acid and the *lower* the pH number. The greater the concentration of hydroxide ions, the stronger is the base and the *higher* the pH number.

Living organisms are sensitive to the concentrations of hydrogen and hydroxide ions and must maintain the pH of their cells within narrow limits. For example, your blood is kept very close to pH 7.4. Organisms control the pH of cellular and body fluids by buffers. A **buffer** is a compound or a combination of compounds that can either combine with or release hydrogen ions to keep the pH of a solution relatively constant.

Assignment 3

Materials

Beakers, 100 ml
Graduated cylinder, 50 ml
pH test papers, wide and narrow ranges
Buffer solution, pH 7, in 1,000-ml flask
Distilled water, pH 7, in 1,000-ml flask
Dropping bottles of:
    aftershave lotion
    Alka-Seltzer® solution
    bromthymol blue
    detergent solution
    hydrochloric acid, 1.0%

household ammonia
lemon juice
white vinegar

1. *Complete items 3a and 3b on the laboratory report.*
2. Use pH test papers to determine the pH of the various solutions provided. Place 1 drop of the solution to be tested on a small strip of pH test paper while holding it over a paper towel. Compare the color of the pH paper with the color code on the dispenser to determine the pH. Use the wide-range test paper first to determine the approximate pH. Then, use a narrow-range test paper for the final determination. *Record your pH determinations in item 3c on the laboratory report.*
3. Investigate the effect of a buffer on pH as follows:
   a. Place 25 ml of (i) distilled water, pH 7, and (ii) buffer solution, pH 7, in separate, labeled beakers. Measure and record the pH of each.
   b. Add 6 drops of bromthymol blue to each beaker. Mix by swirling the liquid. The fluid in each beaker should be a pale blue. Bromthymol blue is a pH indicator that is blue at pH 7.6 and yellow at pH 6.0.
   c. Place the beaker of distilled water and bromthymol solution on a piece of white paper. Add 1.0% HCl drop-by-drop to the solution and mix thoroughly by swirling the liquid between each drop. Count and record the number of drops required to turn the solution yellow, which indicates a pH of 6.

d. Use the preceding method to determine the number of drops of 1.0% HCl required to lower the buffer solution from pH 7 to pH 6.
4. *Complete item 3d on the laboratory report.*

IDENTIFYING BIOLOGICAL MOLECULES

Carbohydrates, lipids, and proteins are complex organic compounds found in living organisms. Carbon atoms form the basic framework of their molecules. In this section, you will analyze plant and animal materials for the presence of these compounds. Work in groups of two to four students. If time is limited, your instructor will assign certain portions to separate groups, with all groups sharing the data.

Carbohydrates

Carbohydrates are formed of carbon, hydrogen, and oxygen. Their molecules are characterized by the H—C—OH grouping in which the ratio of hydrogen to oxygen is 2 : 1. **Monosaccharides** are simple sugars containing three to seven carbon atoms, and they serve as building blocks for more complex carbohydrates. The combination of two monosaccharides forms a **disaccharide** sugar, and the union of many monosaccharides forms a **polysaccharide.** Six-carbon sugars, such as *glucose,* are primary energy sources for living organisms. *Starch* in plants and *glycogen* in animals are polysaccharides used for nutrient storage. See Figure 4.4.

Figure 4.4 Carbohydrates. **A.** Glucose is a 6-carbon monosaccharide. **B.** Maltose is a disaccharide composed of two glucose subunits. **C.** Starch is a polysaccharide composed of many glucose subunits.

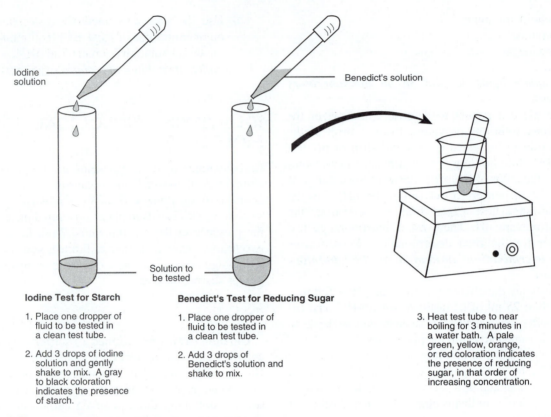

Iodine solution

Benedict's solution

Solution to be tested

Iodine Test for Starch

1. Place one dropper of fluid to be tested in a clean test tube.

2. Add 3 drops of iodine solution and gently shake to mix. A gray to black coloration indicates the presence of starch.

Benedict's Test for Reducing Sugar

1. Place one dropper of fluid to be tested in a clean test tube.

2. Add 3 drops of Benedict's solution and shake to mix.

3. Heat test tube to near boiling for 3 minutes in a water bath. A pale green, yellow, orange, or red coloration indicates the presence of reducing sugar, in that order of increasing concentration.

Figure 4.5 Testing for the presence of starch and sugar.

You will use the iodine test for starch and Benedict's test for reducing sugars (most 6-carbon and some 12-carbon sugars) to identify the presence of these carbohydrates. See Figure 4.5.

Iodine Test for Starch

1. Place 1 dropper (1 ml) of the liquid to be tested in a clean test tube. (A "dropper" means *one dropper full* of liquid.)
2. Add 3 drops (0.15 ml) of iodine solution to the liquid in the test tube and shake gently to mix.
3. A gray to blue-black coloration of the liquid indicates the presence of starch, in that order of increasing concentration.

Benedict's Test for Reducing Sugars

1. Place 1 dropper (1 ml) of the liquid to be tested in a clean test tube.
2. Add 3 drops (0.15 ml) of Benedict's solution to the liquid in the test tube and shake gently to mix.
3. Heat the tube to near boiling in a water bath for 2–3 min. (A water bath consists of a beaker that is half-full of water, heated on a hot plate or over a burner. The test tube is placed in the beaker so it is heated by the water in the beaker rather than directly by the heat source.)

4. A light green, yellow, orange, or brick-red coloration of the liquid indicates the presence of reducing sugars, in that order of increasing concentration.

Lipids

Lipids include an array of oily and waxy substances, but the most familiar are the neutral **fats** (triglycerides). They are composed of carbon, oxygen, and hydrogen and have fewer oxygen atoms than carbohydrates. A fat molecule consists of three long *fatty acid* molecules joined to a *glycerol* molecule. See Figure 4.6. Fats are not soluble in water. They serve as an important means of nutrient (energy) storage in organisms.

In a *saturated* fatty acid, all interior carbon atoms are joined to each other and to two hydrogen atoms by single bonds. In a *monounsaturated* fatty acid, two interior carbon atoms are joined by a double bond, and each of these carbon atoms also bonds to a single hydrogen atom. *Polyunsaturated* fatty acids contain two or more double carbon–carbon bonds. The presence of lipids may be determined by a paper spot test or a Sudan IV test. Sudan IV is a dye that selectively stains lipids.

Paper Spot Test

1. Place a drop of the liquid to be tested on a piece of paper. Blot it with a paper towel and let it dry. If

Figure 4.6 A triglyceride is composed of one glycerol molecule combined with three fatty acids. Note that the middle fatty acid is monounsaturated.

Figure 4.7 Proteins are formed of amino acids joined together by peptide bonds. **A.** Formation of a peptide bond. **B.** Amino acids composing a tiny part of a protein are arranged somewhat like beads on a string.

the substance to be tested is a solid, it can be rubbed on the paper. Remove the excess with a paper towel. Animal samples may need to be gently heated.

2. After the spot has dried, hold the paper up to the light and look for a permanent translucent spot on the paper indicating the presence of lipids.

Sudan IV Test

1. Place 1 dropper (1 ml) of distilled water in a clean test tube and add 3 drops (0.15 ml) of Sudan IV. Shake the tube from side to side to mix well.
2. Add 10 drops (0.5 ml) of the substance to be tested and mix again. Let the mixture stand for 6 min.
3. Lipids will be stained red and will rise to the top of the water or be suspended in tiny globules.

Proteins

Proteins are usually large, complex molecules composed of many **amino acids** joined together by **peptide bonds.** Their molecules contain nitrogen in addition to carbon, oxygen, and hydrogen. Proteins play important roles by forming structural components of cells and tissues and by serving as enzymes. See Figure 4.7. Proteins may be identified by a biuret test that is specific for peptide bonds.

Biuret Test

1. Place 1 dropper (1 ml) of the solution to be tested in a clean test tube.
2. Add 1 dropper (1 ml) of 10% sodium hydroxide (NaOH) and mix by shaking the tube from side to side. *Caution:* NaOH is a caustic substance that can cause burns. If you get it on your skin or clothing, wash it off *immediately* with lots of water.

3. Add 5 drops (0.25 ml) of 0.05% copper sulfate (CuSO₄) to the tube and shake from side to side to mix. Observe after 3 min.
4. A violet color indicates the presence of proteins.

Assignment 4

Materials

Hot plate
Beaker, 260 ml
Brown paper
Dissecting instruments
Medicine dropper
Mortar and pestle
Test tubes
Test-tube holder
Test-tube rack
Dropping bottles of:
albumin, 1.0%
Benedict's solution
copper sulfate, 0.05%
corn oil
egg white, 1.0%
glucose, 0.1%
iodine (I₂+KI) solution
sodium hydroxide, 10%
soluble starch, 0.1%
Sudan IV dye

1. To evaluate the results of the chemical tests that you will perform on the substances provided, it is necessary to prepare a **set of standards** that demonstrates both positive and negative results for each test. The standards are prepared by performing each test on (a) a substance giving a positive reaction and (b) a substance (water) giving a negative reaction.

2. Prepare a set of standards as shown on the chart in item 4a of the laboratory report, and record your results in the chart. The test tubes used must be clean; if in doubt, wash them. Follow these procedures.
   a. Label eight clean test tubes for the substances listed in the "Substance Tested" column of the chart. Note that the paper-spot test does not require test tubes.
   b. Add to the test tubes 1 dropper (about 1 ml) of the substances to be tested.
   c. Add the reagents for the tests to the appropriate test tubes and perform each test.
   d. Perform the paper-spot test on brown paper.
   e. *Save the standards for reference.* ***Record your results in the chart in item 4a on the laboratory report.***

3. ***Complete item 4b on the laboratory report.***
4. Perform each of the tests on substances (unknowns) listed on the chart in item 4c on the laboratory report or provided by your instructor. In some cases, it will be necessary to macerate the material to obtain liquid for testing. Use this procedure.
   a. Place a 1-cm cube of the substance to be tested in a mortar. Add 5 ml of distilled water. Use a pestle to crush and grind the cube into a pulpy mass.
   b. Pour off the liquid and use it as the test solution.
5. Compare your results with the standards, and ***record your results on the chart in item 4c and complete the laboratory report.***

CHEMICAL ASPECTS

Student _____

Lab Instructor _____

1. FUNDAMENTALS

a. Define these terms:

Element _____

_____

Compound _____

_____

Atom _____

_____

Molecule _____

_____

b. Consult Table 4.2 and indicate the (a) symbol and (b) number of protons, electrons, and neutrons in atoms of these elements.

Element	Symbol	Protons	Number of Electrons	Neutrons
Hydrogen	_____	_____	_____	_____
Oxygen	_____	_____	_____	_____
Nitrogen	_____	_____	_____	_____
Carbon	_____	_____	_____	_____
Chlorine	_____	_____	_____	_____
Calcium	_____	_____	_____	_____

2. REACTIONS BETWEEN ATOMS

a. Using Table 4.2 and Figure 4.1, draw shell models of an atom of these elements.

Calcium   **Carbon**

b. Define:

Ionic bond _____

Covalent bond _____

c. Indicate the number of electrons in the outer shell of these neutral atoms and ions:

Na _____   K$^+$ _____   Mg _____   H$^+$ _____   Ca _____   Cl _____

d. Using Figure 4.1 as a guide, draw a shell model of these molecules:

Potassium Chloride (KCl)   **Calcium Chloride (CaCl$_2$)**

e. Using Table 4.3 as a guide, draw the structural formula of these molecules:

Methane (CH$_4$)   **Ammonia (NH$_3$)**

Carbon Dioxide (CO$_2$)   **Formaldehyde (CH$_2$O)**

3. pH

a. Define:

Acid _____

Base _____

b. Identify these pH values as either acid or base. Rank the acids and bases from strongest (#1) to weakest (#4).

a. 10   b. 3   c. 2   d. 6   e. 8   f. 12   g. 1   h. 7.5

Acid: 1 _____ 2 _____ 3 _____ 4 _____ Base: 1 _____ 2 _____ 3 _____ 4 _____

c. Record the pH of these test solutions.

After-shave lotion _____   Vinegar _____   Household ammonia _____

Alka-Seltzer® _____   Detergent solution _____   Lemon juice _____

d. Indicate the number of drops of 1.0% HCl required to decrease the pH of these solutions from 7 to 6.

Distilled water _____   Buffer solution _____

Explain your results. _____

_____

_____

_____

_____

_____

4. BIOLOGICAL MOLECULES

a. Prepare a set of standards for the chemical tests and record your results in the following chart:

STANDARDS

Test	Organic Compound Testing for	Substance Tested	Results
Iodine	Starch	Starch	
		Water	
Benedict's	Reducing sugars	Glucose	
		Water	
Paper spot	Lipids	Corn oil	
		Water	
Sudan IV	Lipids	Corn oil	
		Water	
Biuret	Protein	Albumin	
		Water	

b. Why is the preparation of a set of standards important? _____

_____

_____

c. Perform the chemical tests on the substances provided. Compare your results with the standards and record your results in the following chart. Indicate your test results as:

Strongly positive   +++

Moderately positive   ++

Slightly positive   +

Negative   0

UNKNOWNS

Substance Tested	Test Results* Iodine	Benedict's	Paper spot	Sudan IV	Biuret	Organic Compounds Present
Onion						
Potato						
Apple						
Egg white						

*+ = positive; 0 = negative.

DIFFUSION AND OSMOSIS

OBJECTIVES

After completing the laboratory session, you should be able to:
1. Distinguish between passive and active transport through a plasma membrane.
2. Describe and explain the cause of Brownian movement, diffusion, and osmosis.
3. Describe the effects of temperature, molecular weight, and concentration gradient on the rate of diffusion and osmosis.
4. Use the scientific method in your laboratory investigations.
5. Define all terms in bold print.

Materials constantly move into and out of cells. The **selective permeability** of the plasma membrane permits some materials to pass through it, but prevents others. Furthermore, the materials that can pass through the membrane may change from moment to moment.

Materials move through a plasma membrane by two different processes: passive transport and active transport. **Passive transport,** commonly called **diffusion,** requires no energy expenditure by the cell. Molecules move from an area of their higher concentration to an area of their lower concentration (i.e., down a **concentration gradient**), due to their constant, random motion. In contrast, **active transport** requires the use of energy by the cell and is independent of molecular motion. Molecules may move either down or against a concentration gradient. In a similar way, materials that would normally diffuse out of a cell may be prevented from doing so by **active retention.**

In this exercise, you will use the scientific method of inquiry, including controlled experiments. See pages 8–10 in Exercise 1 to refresh your understanding of the steps in the scientific method and the nature of a controlled experiment, including controlled, independent, and dependent variables.

BROWNIAN MOVEMENT

Molecules of liquids and gases are in constant, random motion. A molecule moves in a straight-line path until it bumps into another molecule, and then it bounces off into a different straight-line course. Because this motion is temperature dependent, the higher the temperature, the greater the molecular movement is. This movement cannot be observed directly. However, tiny particles suspended in a liquid may be observed with a microscope as they are moved in a random fashion due to bombardment by molecules composing the liquid. This vibratory movement is indirect evidence of molecular motion and is called **Brownian movement** after Robert Brown, who first described it in 1827.

 Assignment 1

Materials

Compound microscope
Dissecting needle
Medicine dropper
Microscope slide and cover glass
Powdered carmine dye
Dropping bottle of detergent-water solution

1. Place 1 drop of detergent-water solution on a clean slide.
2. Dip a *dry* tip of a dissecting needle into powdered carmine and, while holding the tip above the drop of fluid, tap the needle with your finger to shake only a few dye particles into the drop. Add a cover glass.

3. Examine the dye particles at 400× to observe Brownian movement.
4. *Complete item 1 on Laboratory Report 5 that begins on page 65.*

DIFFUSION

The constant, random motion of molecules is what enables diffusion to occur. **Diffusion** is the net movement of the same kind of molecules from an area of their higher concentration to an area of their lower concentration. Thus, the molecules move down a **concentration gradient.** Molecules that initially are unequally distributed in a liquid or gas tend to move by diffusion throughout the medium until they are equally distributed. Diffusion is an important means of distributing materials within cells and of passively moving substances through cell membranes.

Diffusion and Temperature

In the first experiment, you will test this hypothesis: *Temperature has no effect on the rate of diffusion.* To test the hypothesis, you will perform a **controlled experiment** to determine whether the rate of diffusion is the same at two different temperatures. The water temperature in each beaker will serve as the control for the other. Note that there is only one independent variable—the difference in temperature—and that all other conditions are the same. Having only one independent variable in an experiment is desirable to obtain clear-cut results.

 Assignment 2

Materials

Hot plate
Beakers, 250 ml
Beaker tongs
Celsius thermometer
Forceps
Granules of:
    potassium permanganate
Ice water
Paper towels

1. *Complete item 2a on the laboratory report.*
2. Obtain two beakers. Use a glass-marking pen to label one "A" and the other "B." Fill Beaker A two-thirds full with water and heat it on a hot plate to about 50°C. Use beaker tongs to place the beaker on a pad of paper towels.
3. Place an equal amount of ice water (but no ice) in Beaker B. Record the temperature of the water in each beaker.

4. Wait 5 min for the movement of water currents to subside. *Keeping both beakers motionless so the water is also motionless*, add a granule of potassium permanganate to each. Record the time. Observe the rate of diffusion by the potassium permanganate molecules during a 15-min period. Is the rate identical in each beaker?
5. *Complete item 2 on the laboratory report.*

Diffusion and Molecular Weight

In this experiment, you will test the hypothesis: *Molecular weight has no effect on the rate of diffusion.* You are to do so by determining the rate of diffusion of two substances that have different molecular weights: methylene blue (mol. wt. 320) and potassium permanganate (mol. wt. 158). Both substances are soluble in water and readily diffuse through an agar gel that is about 98% water.

 Assignment 3

Materials

Forceps
Petri dish containing agar gel
Granules of:
    methylene blue
    potassium permanganate
Metric ruler

1. Place single, equal-size granules of potassium permanganate and methylene blue about 5 cm apart on an agar plate. Gently press the granules into the agar with forceps to ensure good contact. Record the time.
2. After 1 hr, determine the rate of diffusion by measuring the diameter of each colored circle on the agar plate. (Your instructor may want you to place the agar plate in an incubator at 37°C to speed up the process.)
3. *Complete item 3 on the laboratory report.*

Diffusion, Molecular Size, and Membrane Permeability

At times, small molecules are able to diffuse through plasma membranes, while larger molecules are prevented from passing through. This selectivity based on molecular size can be observed in an experiment in which a cellulose membrane simulates a plasma membrane. The cellulose membrane has many microscopic pores scattered over its surface, and molecules smaller than the pores are able to pass through the membrane.

You will use the iodine test for starch and Benedict's test for reducing sugars to determine the results of the experiment. See Figure 4.5 on page 52.

Starch Test

1. Place 1 dropper (1 ml) of the substance to be tested in a clean test tube.
2. Add 3 drops (0.15 ml) of iodine solution to the test tube and shake to mix. A gray to blue-black coloration indicates the presence of starch, in that order of increasing concentration.

Benedict's Test

1. Place 1 full dropper (1 ml) of the substance to be tested in a clean test tube.
2. Add 3 drops (0.15 ml) of Benedict's solution to the test tube, and mix by shaking.
3. Heat to near boiling for 2–3 min in a boiling water bath. A water bath is set up by placing the test tube in a beaker, which is about half full of water, and by heating the beaker on a hot plate. In this way the test tube is heated by the water in the beaker rather than directly by the hot plate. A light green, yellow, orange, or brick-red coloration of the liquid in the test tube indicates the presence of reducing sugars, in that order of increasing concentration.

 ## Assignment 4
Materials

Hot plate
Beakers, 250 ml
Cellulose tubing, 1-in. width
Rubber bands
Test tubes, large, 24 × 200 mm
Test tubes, 14 × 150 mm
Test-tube brush
Test-tube holders
Test-tube rack
Dropping bottles of:
  Benedict's solution
  iodine (I_2 + KI) solution
Squeeze bottles of:
  glucose, 20%
  soluble starch, 0.1%

1. Set up the experiment as shown in Figure 5.1. *Be certain* to rinse the outside of the sac before inserting it into the test tube.
2. Place the test tube in a beaker or rack for 20 min.
3. Examine your test tube and record any color change that has occurred.
4. Test the solution outside the sac to see if glucose diffused from the sac. Place 1 dropper of the solution in a clean test tube and perform Benedict's test.
5. **Complete item 4 on the laboratory report.**
6. Clean the test tubes thoroughly using a test-tube brush.

Betacyanin and Beet Root Cells

The deep red color of beet roots is due to the presence of a water-soluble pigment, betacyanin, within the central vacuole of beet root cells. When fresh beet roots are cut, betacyanin leaks out of the sectioned root cells, but it is retained within intact living beet root cells. How is betacyanin retained? Is it because the molecules are too large to pass through intact cell membranes? If so, it should be retained by intact, but nonliving, cell membranes. Investigate this problem by performing the experiment described in Assignment 5, which tests the hypothesis: *Betacyanin is not retained in beet root cells by living processes.*

 ## Assignment 5
Materials

Hot plate
Beakers, 250 ml
Dissecting instruments
Glass-marking pen
Test tubes, 14 × 150 mm
Isopropyl alcohol
Beet roots, fresh

1. Use a scalpel or sharp knife to cut a portion of a beet root into narrow pieces that will fit into a test tube. Pieces about 0.5 × 1.0 cm are fine.
2. Place the pieces in a beaker and rinse them with two washings of tap water to remove the betacyanin released by cutting the cells.
3. Place equal numbers of beet pieces into three numbered test tubes. Add sufficient water to cover the pieces in tubes 1 and 2, and add a similar amount of alcohol to tube 3.
4. Place tubes 1 and 3 in a test-tube rack. Heat the water in tube 2 to boiling for 3 min in a water bath, and then place tube 2 in the test-tube rack. Alcohol is a poison that kills the cells but leaves the membranes intact, whereas boiling kills the cells and ruptures the cell membranes.
5. After 30 min, pour the liquid from each tube into a similarly numbered, clean test tube. Compare the color of the liquids and the color of the beet pieces among test tubes 1, 2, and 3.
6. **Complete item 6 on the laboratory report.**

Osmosis

Water is essential for life. It is the **solvent** of living systems, and it is the aqueous medium in which the chemical reactions of life occur. Substances dissolved in a solvent are called **solutes.** Water is the most abundant substance in cells, and it contains both organic

1. Fill a large test tube two–thirds full with water and add 4 droppers (4 ml) of iodine solution. Place test tube in a beaker.

2. Soak a 20-cm length of cellulose tubing in water. Then tie a knot in one end to form a sac.

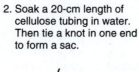

3. Fill the sac half full with 0.1% starch solution and add 5 droppers (5 ml) of 20% glucose solution.

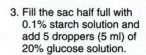
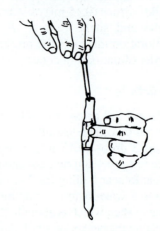

4. Hold sac closed and rinse outside of sac under the tap.

5. Insert the sac into the test tube.

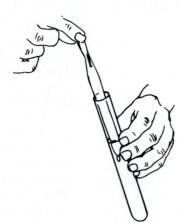

6. Bend top of sac over the lip of test tube and secure it with a rubber band. Let stand for 20 min.

Figure 5.1 Preparation of an experiment to study the effect of molecular size on the diffusion of molecules through a cellulose membrane.

and inorganic solutes. Water is a small molecule that diffuses freely in and out of cells in accordance with the concentration of water inside and outside the cells. The concentration of a substance is the relative quantity of the substance in a unit of the solution or mixture. Concentration is often expressed as a percentage. Thus, a 5% salt solution contains 5% salt and 95% water, and a 10% salt solution contains 10% salt and 90% water. In an aqueous solution, the concentration of water increases as the concentration of a solute decreases.

Osmosis is simply the diffusion of water through a semipermeable or selectively permeable membrane. Like all substances, water diffuses down a concentration gradient. It moves from an area of higher concentration of water to an area of lower concentration of water.

Osmosis and Concentration Gradients

In this experiment, you will test the hypothesis: *The magnitude of the concentration gradient does not affect the rate of osmosis.* Your instructor has set up two osmometers like those in Figure 5.2. The equal-size cellulose sacs are attached to rubber stoppers into which glass tubing has been inserted. The sacs have been filled with 10% and 20% sucrose, respectively, and immersed in tap water. The molecules of sucrose are too large to pass through the pores of the membrane; however, water will diffuse into the sacs, increasing the volume of the fluid, which generates pressure, forcing the fluid up the glass tubes.

When solute concentrations on opposite sides of a semipermeable membrane are unequal, water always

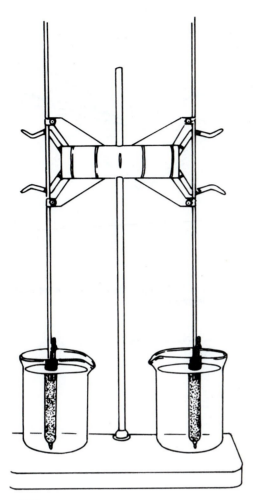

Figure 5.2 Osmometer setup.

moves from the **hypotonic solution** (lower solute concentration) into the **hypertonic solution** (higher solute concentration), that is, down the concentration gradient of water. If the concentrations of solutes on each side of the membrane are equal, the solutions are said to be **isotonic solutions.** They are at **osmotic equilibrium,** and no net movement of water occurs through the membrane.

 Assignment 6

1. Examine the osmometers set up by your instructor. Use a meterstick to determine the distance the fluid rises in each tube during a 30-min period.
2. *Complete item 6 on the laboratory report.*

Osmosis and Living Cells

It is common knowledge that plants wilt when deprived of water and "perk up" again when watered. Do you think that osmosis is involved? What do you think would happen if a plant is exposed to hypertonic

or hypotonic solutions? What happens at the cellular level? In Assignment 7, you will perform two experiments that will help you to answer these questions.

 Assignment 7

Materials

Beakers, 250 ml
Compound microscope
Microscope slides and cover glasses
Scalpel or sharp knife
Dropping bottles and flasks of:
  distilled water
  10% salt solution
Celery stalks
Elodea shoots

Experiment 1: Celery sticks are typically soaked in water for a time prior to serving so they will be crisp. How does this work?

1. Place about 200 ml of distilled water in one beaker and the same amount of salt solution in another beaker.
2. Cut four celery sticks (about 0.5 × 7 cm), and immerse two sticks in each beaker. Record the time.
3. Examine the crispness of the celery sticks after 1 hr.
4. *Complete items 7a–7c on the laboratory report.*

Experiment 2: To understand osmosis at the cellular level, perform the following experiment.

1. Prepare a water-mount slide of *Elodea* leaf and view it at 100× and 400× with your microscope. Note the size of the central vacuoles and the location of the protoplasm in the cells.
2. *In the space for item 7d on the laboratory report, draw an optical midsection view of a normal Elodea cell.* Show the distribution of the protoplasm and the central vacuole.
3. Place 1 drop of 10% salt solution at the edge of the cover glass, and draw it under the cover glass as shown in Figure 5.3. Repeat the process. Now, the leaf is mounted in salt solution. What do you think will happen?
4. Quickly examine the *Elodea* cells for several minutes with your microscope and watch what happens.
5. *Make a drawing of a cell mounted in salt solution in the space for item 7d on the laboratory report.*
6. Remove the leaf from the salt solution, rinse it, and remount it on a clean slide in tap water. Quickly observe it for several minutes with your microscope. What happens?
7. *Complete the laboratory report.*
8. Clean and dry the objectives and stage of your microscope to remove any salt water that may be present.

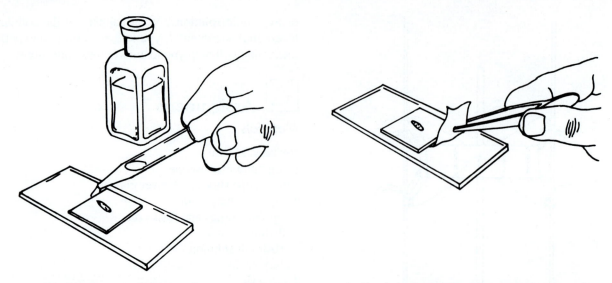

1. Add 1 to 2 drops of 10% salt solution to one edge of cover glass.

2. Touch the fluid at opposite edge with a piece of paper toweling.

Figure 5.3 Procedure to draw salt solution under a cover glass.

DIFFUSION AND OSMOSIS

Student _____

Lab Instructor _____

1. BROWNIAN MOVEMENT

Is Brownian movement a living process? _____ Describe the cause of Brownian movement.

_____

_____

_____

_____

_____

_____

2. DIFFUSION AND TEMPERATURE

a. Define diffusion. _____

_____

_____

What natural phenomenon enables diffusion to occur? _____

_____

_____

b. State the hypothesis to be tested. _____

_____

c. What is the independent variable? _____

d. What is the dependent variable? _____

e. Record the water temperature for each beaker.   A _____ °C   B _____ °C

f. Compare the relative rate (faster or slower) of diffusion in each beaker.

A _____   B _____

g. Do your results support the hypothesis? _____ What relationship seems to exist between temperature and the rate of diffusion? _____

_____

_____

h. Describe how you would test this new hypothesis. _____

_____

_____

_____

_____

_____

_____

3. DIFFUSION AND MOLECULAR WEIGHT

a. What is the hypothesis to be tested? _____

_____

What is the independent variable? _____

What is the dependent variable? _____

b. Record the diameter of the colored circles after 1 hr:

Methylene blue _____ mm   Potassium permanganate _____ mm

c. Do the results support the hypothesis? _____ What relationship seems to exist between molecular weight and
the rate of diffusion? _____

_____

_____

_____

_____

4. MOLECULAR SIZE AND DIFFUSION THROUGH A CELLULOSE MEMBRANE

a. Describe any color change in your test tube setup after 20 min. _____

_____

b. Explain any color change. _____

_____

_____

_____

_____

c. Did starch or glucose diffuse from the sac?

Starch _____   Glucose _____

d. Considering the molecules of starch, water, iodine, and glucose, which are small enough to diffuse through the
pores in the sac? _____

e. A 10% iodine solution fills a cellulose bag. How would you remove essentially all the iodine while leaving the water in the bag? Diagram your setup and describe your procedures.

5. BETACYANIN AND BEET ROOT CELLS

a. State the hypothesis to be tested.

_____

b. Indicate the presence (+) or absence (0) of betacyanin in the fluid of:

Tube 1 _____   Tube 2 _____   Tube 3 _____

Explain what happened to the cells to give these results.

Tube 1 _____

Tube 2 _____

Tube 3 _____

c. Do the results support the hypothesis? _____

d. State a conclusion from your results. _____

_____

e. Describe the meaning of "selective permeability" of plasma membranes. _____

_____

_____

6. OSMOSIS AND CONCENTRATION GRADIENTS

a. State the hypothesis to be tested. _____

_____

b. What is the independent variable? _____

c. Indicate the sucrose concentration and the distance the fluid rose in each osmometer tube during a 30-min interval.

Tube 1: % sucrose _____   Distance _____ mm

Tube 2: % sucrose _____   Distance _____ mm

d. Do the results support the hypothesis?_____ What seems to be the relationship between the magnitude of the concentration gradient and the rate of osmosis?

_____

_____

_____

e. Describe how you would test this new hypothesis. _____

_____

_____

7. OSMOSIS AND LIVING CELLS

a. Record the degree of crispness in the celery sticks in:

Distilled water _____   Salt water _____

b. In relation to the intracellular fluid of the celery cells, which solution is:

Hypotonic? _____   Hypertonic? _____

c. What can you conclude from the experiment? _____

_____

_____

_____

_____

d. Draw an optical midsection of an *Elodea* cell when mounted in:

Water   **Salt Solution**

e. Explain what happened to *Elodea* cells mounted in salt solution. _____

_____

_____

f. Plants wilt when deprived of water. What happens at the cellular level to cause the wilted appearance? _____

_____

_____

g. Why are plants killed when exposed to concentrated salt water? _____

_____

_____

h. Explain the basis of preserving food with salt or sugar. _____

_____

_____

i. Explain what happened to the *Elodea* cells when the leaf was remounted in tap water. _____

_____

_____

j. Based on what you have learned about osmosis, what movement of water would you expect if the extracellular fluid around your body cells, in comparison to the intracellular fluid, was:

Hypotonic? _____

_____

Hypertonic? _____

_____

What effect would the concentration gradient have on the rate of water movement through cell membranes?

_____

_____

_____

k. Considering osmosis at the cellular level, why is it necessary to drink adequate amounts of water each day?

_____

_____

_____

_____

ENZYMES

OBJECTIVES

After completing the laboratory session, you should be able to:
1. Explain the role of enzymes in living cells.
2. Describe how an enzyme functions in a chemical reaction.
3. Describe the results and state conclusions based on the experiments.
4. Define all terms in bold print.

Chemists often add inorganic **catalysts** to the reactants in order to increase the rate of some chemical reactions, but inorganic catalysts are not present in living organisms. Instead, living cells use organic catalysts called **enzymes** to speed up their chemical reactions. Thousands of different enzymes are known, and not all cells contain the same enzymes. Enzymes are vital to life. Cellular chemical reactions would not occur fast enough to support life without the aid of enzymes.

Both inorganic catalysts and enzymes act by stressing the chemical bonds of the reactants, which decreases the amount of **activation energy** required to start the reaction. Activation energy is the energy that must be added to reactants to start a chemical reaction. Using a match to start a fire is an example of applying activation energy (heat) to start a reaction. See Figure 6.1.

Enzymes are proteins, so each enzyme consists of a specific sequence of amino acids. Weak hydrogen bonds that form between some of the amino acids help to determine the three-dimensional shape of the enzyme, and it is this shape that allows the enzyme to fit onto a specific **substrate molecule** (substance acted upon). The enzyme and the substrate molecule must fit together like a lock and key.

The interaction of an enzyme and substrate is shown in Figure 6.2. In this reaction, the substrate is split into two products. Another way to express an enzymatic reaction is as follows:

$$E + S \longrightarrow ES \longrightarrow E + P$$

The *enzyme* (E) combines with the *substrate molecule* (S) to form a temporary *enzyme–substrate complex* (ES), where the specific reaction occurs. Then, the *product molecules* (P) separate from the enzyme, and the unchanged enzyme is recycled to combine with another substrate molecule. Note that the enzyme is not altered in the reaction, so a few enzyme molecules can catalyze a great number of reactions.

An enzyme is inactivated by a change in shape, and its shape is altered by anything that disrupts the pattern of hydrogen bonding. For example, many enzymes function best within rather narrow temperature and pH ranges because substantial changes in temperature or pH disrupt their hydrogen bonds and alter their shapes. However, some enzymes function well over rather broad temperature and pH ranges. It is the unique bonding pattern that determines the sensitivity of each enzyme to changes in temperature and pH.

ACTION OF CATALASE

All organisms using molecular oxygen produce hydrogen peroxide (H_2O_2) as a harmful by-product of some cellular reactions. Hydrogen peroxide is a strong oxidizing agent that can cause serious damage to cells. Fortunately, cells have an enzyme, **catalase,** which quickly breaks down hydrogen peroxide into water and oxygen, preventing cellular damage. Catalase has one of the highest turnover rates of any enzyme. Under optimum conditions, it can break down 40,000,000 molecules of H_2O_2 per second! Catalase is especially abundant in the liver cells of humans and other

71

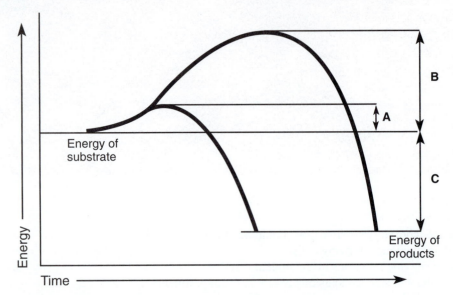

Figure 6.1 Enzymes and activation energy in an energy-releasing reaction. **A.** Activation energy required when an enzyme catalyzes the reaction. **B.** Activation energy required without the enzyme. **C.** Net energy released by the reaction.

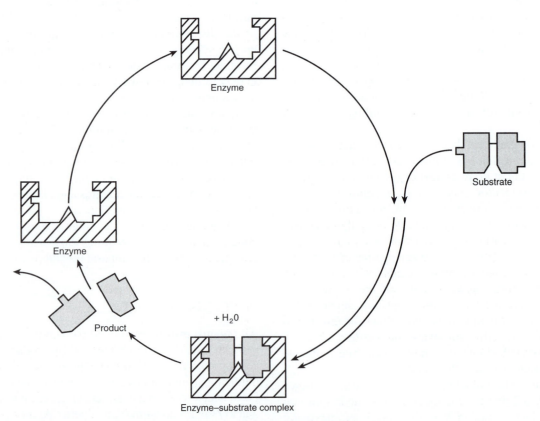

Figure 6.2 Interaction of enzyme and substrate. The substrate and enzyme fit together like a lock and key, forming the enzyme–substrate complex where the reaction occurs. Then, the products are released, and the unchanged enzyme is recycled.

vertebrates. The liver is an organ that detoxifies many harmful substances, including peroxides.

$$\text{Hydrogen Peroxide} \xrightarrow{\text{(catalase)}} \text{Water} + \text{Oxygen}$$
$$2H_2O_2 \qquad\qquad 2H_2O \qquad O_2$$

In a cell, oxygen released by the preceding reaction is used for other cellular processes; however, when the reaction occurs in a test tube, oxygen gas bubbles to the surface, producing a layer of foam on the surface of the peroxide. The amount of foam produced and the speed with which it forms are measures of catalase activity. In the following experiments, you will determine the degree of catalase action by measuring the thickness of the foam layer.

 ## Assignment 1

Complete item 1 on Laboratory Report 6 that begins on page 75.

Preparation of Tissue Extracts

The following experiments require attention to cleanliness to prevent an extract of one tissue from contaminating other extracts, so be sure you use clean glassware, spatulas, and droppers for each experiment.

You will assess the action of catalase in several types of cells, so it is necessary to rupture the cells to free the enzyme. You will do this by grinding the tissue plus a little water and sand using a mortar and pestle.

Procedure

1. Use a scalpel to cut the tissue to be ground into small cubes about 0.5 cm on a side.
2. Place a pinch of sand and a dropper of tap water in the mortar, and add a cube of the tissue to be ground.
3. Use the pestle to grind the tissue to a pulp.
4. Use 1 drop of the fluid extract in the experiments.
5. *Be sure to wash the mortar, pestle, and dropper after each extraction to avoid contamination that will invalidate your results!*

 ## Assignment 2

Materials

Glass-marking pen
Manganese dioxide (MnO₂) powder
Medicine dropper
Metric ruler
Mortar and pestle
Sand
Scalpel or knife

Spatula
Test tubes
Test-tube racks
Dropping bottles of:
   hydrogen peroxide, 3%
   tap water
Beef liver, 0.5 cm cube

Assess the action of manganese dioxide (MnO_2), an inorganic catalyst, and catalase in liver cells as follows.

1. Number and prepare four clean test tubes as follows:
   Tube 1: a spatula of sand
   Tube 2: 1 dropper (1 ml) of water
   Tube 3: a spatula of MnO₂
   Tube 4: 1 drop of liver extract
2. Add a dropper of water to each tube.
3. Add 1 dropper (1 ml) of H₂O₂ to each tube and record the time.
4. After 1 min, measure the thickness of the foam layer in millimeters (mm).
5. **Complete item 2 on the laboratory report.**
6. Wash all glassware used in the experiment and remove the excess water.

 ## Assignment 3

Materials

Glass-marking pen
Medicine dropper
Metric ruler
Mortar and pestle
Sand
Scalpel or knife
Test tubes
Test-tube racks
Dropping bottles of:
   hydrogen peroxide, 3%
   tap water
Biologicals (0.5-cm cubes)
   apple
   chicken breast
   beef liver
   potato

Compare the action of catalase in different tissues by testing this hypothesis: *The breakdown of H_2O_2 occurs at the same rate in all cells.*

1. Prepare the tubes with 1 drop of tissue extract and 1 dropper (1 ml) of water, as follows:
   Tube 1: apple
   Tube 2: potato
   Tube 3: chicken breast
   Tube 4: beef liver

2. Add 1 dropper (1 ml) of H_2O_2 to each tube and record the time.
3. After 1 min, measure the thickness of the foam layer in millimeters (mm).
4. ***Complete item 3 on the laboratory report.***
5. Wash all glassware used in the experiment and remove the excess water.

Assignment 4

Materials

Beakers, 250 ml
Celsius thermometer
Crushed ice
Glass-marking pen
Hot plate
Medicine dropper
Metric ruler
Mortar and pestle
Sand
Scalpel or knife
Test tubes
Test-tube racks
Dropping bottles of:
   hydrogen peroxide, 3%
   tap water
Beef liver, 0.5 cm cube

Compare the effect of temperature on the action of catalase in liver cells by testing this hypothesis: *Temperature has no effect on catalase activity.*

1. Place 1 drop of liver extract and 1 dropper (1 ml) of water in each of four test tubes numbered 1A, 2A, 3A, and 4A.
2. Place 1 dropper (1 ml) of H_2O_2 in each of four test tubes numbered 1B, 2B, 3B, and 4B.
3. Expose the tubes to designated temperatures for 5 min. Determine and record the temperature of each exposure.

Tubes 1A and 1B: in a beaker of crushed ice
Tubes 2A and 2B: in a test-tube rack at room temperature
Tubes 3A and 3B: in a beaker of water at 70°C
Tubes 4A and 4B: in a boiling water bath

4. After 5 min, pour the peroxide from tubes 1B, 2B, 3B, and 4B into corresponding tubes 1A, 2A, 3A, and 4A. After 1 min, measure the thickness of the foam layer of each tube.
5. ***Complete item 4 on the laboratory report.***
6. Wash all glassware used in the experiment and remove the excess water.

Assignment 5

Materials

Glass-marking pen
Medicine dropper
Metric ruler
Mortar and pestle
Sand
Scalpel or knife
Test tubes
Test-tube racks
Dropping bottles of:
   buffer solutions, pH 2, 7, and 11
   hydrogen peroxide, 3%
   tap water
Beef liver, 0.5 cm cube

1. Determine the effect of pH on the action of catalase in liver cells by designing your own experiment. There are dropping bottles of buffers at pH 2, 7, and 11 for your use.
2. ***Complete item 5 on the laboratory report showing the hypothesis to be tested, the procedures of the experiment, the results, and the conclusion.***
3. ***Complete the laboratory report.***

ENZYMES

Student _____

Lab Instructor _____

1. BACKGROUND

a. What are the chemical subunits of enzymes? _____

b. What determines the specific substrate with which an enzyme can react? _____

_____

c. What determines the three-dimensional shape of an enzyme? _____

_____

d. Explain the basic role of enzymes.

_____

e. For the reaction catalyzed by catalase, indicate the:

substrate _____   products _____

2. INORGANIC AND ORGANIC CATALYSTS

a. Why is the foam layer an indication of the breakdown of H_2O_2? _____

_____

b. Record the thickness (mm) of the foam layer after 1 min.

Tube	Thickness of Foam Layer	Tube	Thickness of Foam Layer
1. Sand		3. MnO_2	
2. Water		4. Liver extract	

c. What is the function of Tubes 1 and 2? _____

What should be done if foam appears in Tubes 1 and 2? _____

_____

_____

d. State a conclusion from your results. _____

_____

_____

_____

_____

_____

_____

_____

_____

_____

3. CATALASE IN ANIMAL AND PLANT TISSUES

_a._ What is the hypothesis to be tested? _____

_____

_b._ Record the thickness (mm) of the foam layer after 1 min.

Tube	Thickness of Foam Layer	Tube	Thickness of Foam Layer
1. Apple		3. Chicken breast	
2. Potato		4. Liver	

_c._ Why was it unnecessary to prepare tubes of sand and water? _____

_____

_____

_d._ Do the results support the hypothesis? _____

_e._ State a conclusion from your results. _____

_____

_____

_____

_f._ Do animal or plant tissues exhibit greater catalase activity? _____

How do you explain this? _____

_____

_____

_____

_____

4. TEMPERATURE AND CATALASE ACTION

_a._ What is the hypothesis to be tested? _____

_____

What is the independent variable? _____

What is the dependent variable? _____

_b._ Record the thickness (mm) of the foam layer after 1 min.

Tube	Temperature	Thickness of Foam Layer
1		
2		
3		
4		

c. Do the results support the hypothesis? _____ State a conclusion from your results. _____

_____

_____

_____

d. Among the temperatures tested, which totally inactivated catalase? _____

_____

5. pH AND CATALASE ACTION

a. State your hypothesis. _____

_____

_____

b. Describe your procedures in a stepwise manner.

_____

_____

_____

_____

_____

_____

_____

_____

_____

_____

_____

_____

_____

_____

_____

c. Record your results.

Tube	pH	Thickness (mm) of Foam Layer
1		
2		
3		
4		
5		

d. Do the results support the hypothesis? _____ State a conclusion from your results. _____

_____

_____

e. Among the pH values tested, which totally inactivated catalase? _____

_____

_____

6. SUMMARY

a. At the molecular level, how is an enzyme inactivated? _____

_____

_____

b. Does catalase seem to be active within a broad or narrow range of:

Temperatures? _____

pH values? _____

c. Of the temperatures and pH values tested, indicate the optimum or the optimum range for catalase activity.

Optimum temperature or temperature range _____

Optimum pH or pH range _____

d. Would you expect the activity of all enzymes to exhibit the same temperature and pH ranges as catalase? _____

_____ Explain. _____

_____

_____

_____

_____

_____

PHOTOSYNTHESIS

OBJECTIVES

After completing the laboratory session, you should be able to:
1. Write the summary equation for photosynthesis and identify the role of each reactant.
2. Identify materials or conditions necessary for photosynthesis.
3. Describe the pathway that each reactant and product takes to reach or leave the photosynthetic cells of a leaf.
4. Separate and identify the chloroplast pigments.
5. Describe the effect of light quantity on the rate of photosynthesis.
6. Define all terms in bold print.

Living organisms require a constant source of energy to operate their metabolic (living) processes. The ultimate source of energy is the sun. Except for the few chemosynthetic bacteria, all organisms depend on the conversion of **light energy** into the **chemical energy** of organic nutrients by **photosynthesis.** The photosynthesizers are plants, plantlike protists, and cyanobacteria. All these organisms contain **chlorophyll,** the green pigment that captures light energy and converts it into the chemical energy of organic molecules. Photosynthetic organisms are **autotrophs** (self-feeders); they are able to synthesize organic nutrients from inorganic materials. **Heterotrophs** (feed on others) must obtain their organic nutrients by feeding on other organisms.

The summary equation for photosynthesis is shown in Figure 7.1. The process is not as simple as the summary equation suggests, however. It actually is a complex series of chemical reactions that may be separated into two parts: (1) a light reaction and (2) a dark reaction.

During the **light reaction,** chlorophyll absorbs light energy and converts it into chemical energy. In the process, water is split, releasing oxygen molecules. The **dark reaction** normally occurs in the presence of light, but it does not *require* light energy—hence its name. In this reaction, chemical energy, which was formed in the light reaction, is used to synthesize glucose by combining (1) carbon dioxide from the atmosphere and (2) hydrogen split from water in the light reaction.

Consult your text for a complete discussion of photosynthesis.

When **glucose,** a six-carbon sugar, is formed in a photosynthesizing cell, much of it is converted into **starch** for temporary storage. Starch is a polysaccharide composed of many glucose units. Using glucose as the base material, plants can synthesize all other organic compounds required for their metabolic and structural needs.

In this exercise, you will experimentally verify the major reactants necessary for photosynthesis. You will use the presence of starch in chlorophyll-containing cells as indirect evidence that photosynthesis has occurred. The presence of reducing sugars—notably, glucose and fructose—is not evidence of photosynthesis because they may result from carbohydrates transported from other parts of the plant.

 Assignment 1

Complete item 1 on Laboratory Report 7 that begins on page 87.

CARBON DIOXIDE AND PHOTOSYNTHESIS

In the experiment to be performed in Assignment 2, you will test the hypothesis: *Carbon dioxide is not necessary for photosynthesis.* You will determine the

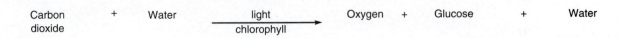

| Carbon dioxide | + | Water | light chlorophyll | Oxygen | + | Glucose | + | Water |

Expressed chemically and showing the reaction of the atoms, the equation appears as follows:

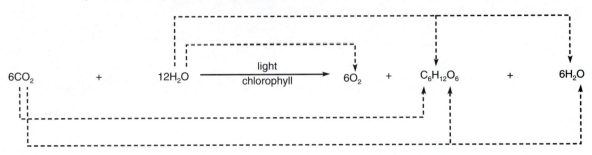

$$6CO_2 \quad + \quad 12H_2O \xrightarrow[\text{chlorophyll}]{\text{light}} 6O_2 \quad + \quad C_6H_{12}O_6 \quad + \quad 6H_2O$$

Figure 7.1 Summary equation of photosynthesis.

occurrence of photosynthesis by the presence or absence of starch in leaves of plants exposed to air with and without carbon dioxide (CO_2). Two healthy plants were placed in darkness for 48 hr and then were placed under separate bell jars. Plant A was exposed to air containing CO_2, and plant B was exposed to air from which CO_2 had been removed by soda lime. Both plants were exposed to light for 24 hr. The setup is shown in Figure 7.2. The presence or absence of CO_2 is the only independent variable.

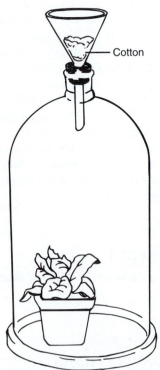

Plant A is exposed to normal air.

Plant B is exposed to air lacking CO_2, because CO_2 has been removed by the soda lime.

Figure 7.2 Setup for the carbon dioxide and photosynthesis experiment.

Assignment 2

Materials

Fluorescent lamps
Hot plates
Beakers, 100 ml, 250 ml
Beaker tongs
Bell jars
Boiling chips
Cotton, nonabsorbent
Dissecting instruments
Petri dishes
Alcohol, isopropyl 70%
Dropping bottles of:
 iodine (I_2 + KI) solution
Soda lime, 8 mesh
Fairy primrose (*Primula malacoides*) in 4-in. pots

1. Remove a leaf from each plant. Cut off the petiole of the leaf from plant A so you can distinguish the leaves.
2. Carefully follow the procedures given here and shown in Figure 7.3 to test each leaf for the presence of starch. A gray to blue-black color indicates the presence of starch. *Caution: Alcohol is flammable; do not get it near an open flame.*
   a. Fill a 250-ml beaker about one-third full with water. Place the beaker on a hot plate, add a few boiling chips, and heat the water to boiling. Add the leaf and boil for 3–5 min to rupture the cells.
   b. Fill a 100-ml beaker about one-third full with alcohol and add a few boiling chips. Use forceps

to transfer the leaf from the boiling water to the beaker of alcohol.
   c. Using beaker tongs, place the beaker of alcohol into the beaker of water on the hot plate to form a water bath. Boil the leaf in alcohol for 3–5 min to extract the chlorophyll. Turn off the hot plate.
   d. Using forceps, transfer the leaf to a petri dish and cover it with iodine solution for 3 min. Rinse the leaf with tap water and examine its coloration.
3. ***Complete item 2 on the laboratory report.***

LIGHT AND PHOTOSYNTHESIS

In the experiment to be performed in Assignment 3, you will test the hypothesis: *Light is not necessary for photosynthesis.* The presence or absence of light is the only independent variable. Healthy green plants were placed in darkness for 48 hr. Leaf shields were placed on the leaves, and the plants were exposed to light and normal air for 24 hr prior to the laboratory session.

 ## Assignment 3

Materials

Fluorescent lamps
Hot plates
Beakers, 100 ml, 250 ml
Boiling chips
Dissecting instruments
Leaf shields

1. Fill 250-ml beaker one-third full with water. Add boiling chips and heat on a hot plate to boiling. Add a leaf and boil for 3–5 min to rupture cells.

2. Fill a 100-ml beaker one-third full with alcohol and add boiling chips. Use forceps to transfer the leaf from the boiling water to the alcohol.

3. Use beaker tongs to place the beaker of alcohol in the beaker of water to form a water bath. Boil the leaf for 3-5 min to remove the chlorophyll.

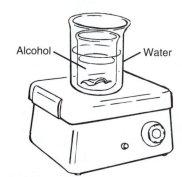

Alcohol — Water

4. Using forceps, transfer the leaf to a petri dish and cover it with iodine solution for 3 min. Rinse the leaf with water and note it's color.

Figure 7.3 Procedure for testing a leaf for the presence of starch.

Petri dishes
Alcohol, isopropyl 70%
Dropping bottles of:
   iodine (I_2 + KI) solution
Fairy primrose (*Primula malacoides*) in 4-in. pots

1. Remove a leaf with a leaf shield from one of the plants. Trace the leaf on a piece of paper, showing the position of the leaf shield.
2. Remove the leaf shield and test the leaf for the presence of starch following the procedures described in Assignment 2 and shown in Figure 7.3. Compare the position of the leaf shield and the distribution of starch.
3. ***Complete item 3 on the laboratory report.***

CHLOROPHYLL AND PHOTOSYNTHESIS

Variegated leaves have unequal distributions of chlorophyll, and occasionally certain leaves of a plant may lack chlorophyll due to a mutation. Such leaves may be used to determine whether chlorophyll is necessary for photosynthesis. In the experiment in Assignment 4, you will test the hypothesis: *Chlorophyll is not necessary for photosynthesis.* On the stock table are *Coleus* plants and normal and albino corn seedlings that were exposed to light for 24 hr prior to the laboratory session.

 Assignment 4

Materials

Fluorescent lamps
Hot plates
Beakers, 100 ml, 250 ml
Boiling chips
Dissecting instruments
Petri dishes
Test tubes
Test-tube holder
Alcohol, isopropyl 70%
Dropping bottles of:
   Benedict's solution
   iodine (I_2 + KI) solution
Coleus plants, green and white, in 4-in. pots
Corn (*Zea*) seedlings, green and albino

1. Remove a variegated leaf from a *Coleus* plant on the stock table. Use a scalpel or scissors to cut the leaf in half along the midrib. Sketch half of the leaf on a piece of paper, showing the distribution of chlorophyll.
2. Test one half of the leaf for the presence of starch as described in Assignment 2 and shown in Figure 7.3, and compare the starch and chlorophyll distributions.

3. Use scissors to separate the green and nongreen portions of the other half of the leaf, and test *each portion* separately for the presence of glucose as described here and shown in Figure 7.4.
   a. Use scissors to cut the portion of the leaf to be tested into small pieces. Place the leaf fragments in a test tube containing 2 droppers (2 ml) of water.
   b. Place the test tube in a 250-ml beaker about one-third filled with water to form a water bath. Place the beaker on a hot plate and boil the water for about 5 min.
   c. Using a test-tube holder, remove the test tube and pour the fluid into another test tube. Add 5 drops of Benedict's solution to the fluid in the second test tube.
   d. Place the second test tube in the beaker of boiling water on the hot plate and boil for 5 min.
   e. Examine the color of the fluid in the test tube. A light green, yellow, or orange coloration indicates the presence of glucose. Compare your results with the distribution of chlorophyll.
4. Observe the corn seedlings, noting that about 25% of them are albinos. Are the albino leaves carrying on photosynthesis?
5. Test green and albino leaves of the corn seedlings for the presence of starch and glucose, as shown in Figures 7.3 and 7.4, respectively.
6. ***Complete item 4 on the laboratory report.***

CHLOROPLAST PIGMENTS

When white light is passed through a prism, the component wavelengths are separated to form a rainbow-like spectrum that we perceive as colors ranging from violet (380 nanometers; nm) through blue, green, yellow, and orange to red (760 nm). Light energy decreases as the wavelengths increase.

When white light strikes a leaf, some wavelengths are absorbed and others are reflected or transmitted. See Figure 7.5. Most leaves appear green because the green wavelengths are reflected and not absorbed by chlorophyll, which is usually the dominant pigment in the leaf. Most leaves also have other light-absorbing pigments in the chloroplasts, notably the yellow **carotenes** and the yellow-orange **xanthophylls.** These pigments absorb different wavelengths of light than chlorophyll and pass the captured energy to chlorophyll.

In Assignment 5, chloroplast pigments will be separated by **paper chromatography,** a technique that takes advantage of slight differences in the relative solubility of the pigments. Your instructor has prepared a concentrated solution of chloroplast pigments by using a blender to homogenize spinach leaves in acetone, filtering the homogenate, and concentrating the filtrate by evaporation.

1. Remove a leaf from a healthy plant exposed to light for 24 hr.

2. With scissors cut the leaf into small pieces.

3. Place leaf fragments in a test tube with 2 droppers (2 ml) of water. Place test tube in a beaker half filled with water and boil 5 min.

4. Pour off the fluid into another test tube.

5. Add 5 drops of Benedict's solution to the fluid.

6. Place test tube in boiling water for 5 min. Yellow or orange color is positive test for glucose.

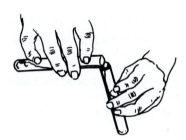

Figure 7.4 Procedure for testing a leaf for the presence of glucose.

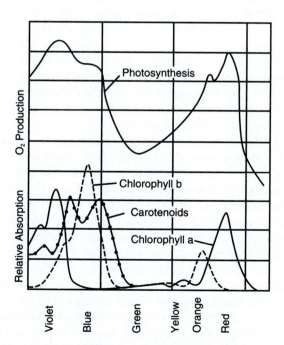

Figure 7.5 Absorption spectrum of chloroplast pigments and action spectrum of photosynthesis.

Assignment 5

Materials

Chromatography jar
Chromatography paper, 1-in. width
Cork stopper for the jar
Paintbrush, fine-tipped
Paper clip
Pliers, pointed-nosed
Scissors
Straight pin
Chloroplast pigment solution
Developing solvent

1. Obtain the jar, cork, straight pin, paper clip, and chromatography paper for the setup shown in Figure 7.6. Handle the paper by the edges because oils from your fingers can adversely affect the results. Cut small notches about 2 cm from one end of the paper, and cut off the corners of this end.

2. Insert a pin into the cork stopper and bend it to form a hook. Attach the paper clip to the paper and

1. Paint a thin line of pigment solution across the paper strip between the notches.

Repeat six times, letting pigment line dry between applications.

2. Pour petroleum ether into a clean jar to a depth of 0.5 cm.

3. Suspend the paper strip from a cork using a paper clip.

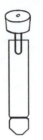

4. Immerse the tip of the paper strip (*but not the pigment line*) in the solvent.

Figure 7.6 Procedure for the paper chromatography setup.

suspend it in the jar *before adding the solvent*. The end of the paper should be *just above the bottom of the jar*. Adjust the paper clip or cut off the paper to achieve this length. Then remove the paper and clip for step 3.

3. Use a small paintbrush to paint a line of chloroplast pigment solution across the paper between the notches. Let it dry. Repeat this step six times to obtain a dark pigment line.

4. Under a fume hood, pour the developing solvent into your jar to a depth of about 1 cm. *Caution: The developing solvent and the pigment solution are volatile and highly flammable. Keep them away from open flames.*

5. Return to your work station, and insert the cork and suspended chromatography paper into the jar. The end of the paper, but not the pigment line, should be in the solvent. *Keeping the jar motionless*, observe the solvent and pigments migrating up the paper.

6. When the solvent has moved to about 2 cm from the top of the paper, remove the paper and place it on a paper towel. Pour the developing solvent into the waste jar *under the fume hood*.

7. Examine your chromatogram. Locate the bright yellow carotene at the top, the blue-green chlorophyll a, and the yellow-green chlorophyll b. The remaining yellow bands are xanthophylls.

8. Calculate and record the R_f (**ratio factor**) **value** for carotene, chlorophyll a, and chlorophyll b. The R_f value is a numerical indicator of the solubility of each pigment in the developing solvent. The higher the R_f value, the more soluble is the solute in the solvent and the farther it is carried by the solvent. The R_f of a given solute varies with the developing solvent that is used.

$$R_f = \frac{\text{distance moved by the pigment (solute)}}{\text{distance moved by the solvent}}$$

For your measurements, measure the distances from the original pigment line to the leading edges of the solvent and pigments.

9. *Complete item 5 on the laboratory report.*

LIGHT QUANTITY AND THE RATE OF PHOTOSYNTHESIS

The quantity of light received by plants in full sun is greater than for plants in the shade. It also varies with the season and as the sun moves across the sky during the day. Does the rate of photosynthesis vary with the changes in light quantity? You will investigate this question in the experiment to be performed in Assignment 6. This experiment will test the hypothesis: *The rate of photosynthesis does not vary with the quantity of light received by a plant.*

As the summary equation of photosynthesis (Figure 7.1) shows, a molecule of oxygen is produced for every molecule of carbon dioxide converted into glucose. Therefore, the rate of oxygen production may be used to determine the rate of photosynthesis. Some of the oxygen produced by photosynthesis is used by plant cells, but the excess oxygen is released into the atmosphere.

In this experiment, you will expose *Elodea*, a water plant, to light from a light source placed at 25, 50, 75, and 100 cm from the plant and measure the rate of oxygen released from the *Elodea* at each distance. The quantity of light received by the *Elodea* decreases as the distance from the light source increases. You will use an experimental setup as shown in Figure 7.7. The experiment is best performed by groups of two or four students.

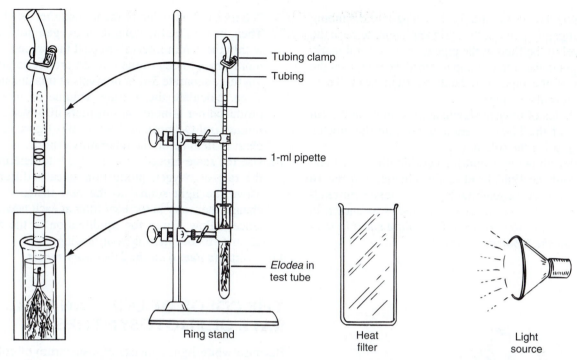

Figure 7.7 Setup for studying the effect of light quantity on the rate of photosynthesis.

Assignment 6

Materials

Heat filter (glass container of water)
Meterstick
Pipette, 1 ml
Plastic tubing, 7-cm lengths
Ring stand
Ring stand test-tube clamps
Scalpel or razor blade
Spot lamp, 300-watt, or fluorescent lamp
Syringe and needle, 5 ml
Test tube
Test-tube rack
Tubing clamp, screw-type
Sodium bicarbonate solution, 2%
Elodea shoots

1. Read the following directions, and then **complete items 6a–6d on the laboratory report.**
2. Fill a test tube about three-fourths full of 2% sodium bicarbonate ($NaHCO_3$) solution, which will provide an adequate concentration of CO_2.
3. Make a diagonal cut through the stem of a leafy shoot of *Elodea* about 10–13 cm from its tip. Use a sharp scalpel, being careful not to crush the stem. Insert the *Elodea* shoot, tip down, into the test tube so the cut end is 1–2 cm below the surface of the solution. If necessary, add more $NaHCO_3$ solution to raise the fluid level to near the top of the test tube. Place the tube in a test-tube holder on a ring stand. If bubbles of oxygen are released too slowly from the cut end of the stem, recut the stem at an angle to obtain a good production of bubbles. (A clean cut and a good production of bubbles is essential to the success of this experiment.)

4. Place a short piece of plastic tubing snugly over the tip (pointed end) of a 1-ml pipette, as shown in Figure 7.7. Place a screw-type tubing clamp on the tubing and tighten it until it is *almost closed*. Place the pipette in a ring-stand clamp with the tip up, and lower the base of the pipette into the test tube so that the cut end of the *Elodea* shoot is inserted into the pipette. Secure the pipette in this position. See Figure 7.7. Study the 0.01-ml graduations on the pipette to ensure that you know how to read them.

5. If using a spot incandescent light source, fill a rectangular glass container with water to serve as a heat filter. The heat filter must be placed between the *Elodea* shoot and the spot lamp, about 10 cm in front of the *Elodea*. A heat filter is not necessary if you are using a fluorescent lamp.

6. You will vary the light quantity by placing the light source at 25, 50, 75, and 100 cm from the tube containing the *Elodea* shoot. Start at the 100-cm distance and arrange your setup as shown in Figure 7.7. Turn off or dim the room lights.

7. Insert the needle of a 5-ml syringe into the tubing and gently pull out the syringe plunger to raise the level of the fluid in the pipette to the 0.9-ml mark. Tighten the screw clamp to close the tubing on the tip of the pipette to hold the fluid level. Then, remove the syringe.

8. Bubbles of oxygen should start forming at the cut end of the *Elodea* stem and rise into the pipette, displacing the solution.

9. After allowing 5 min for equilibration, read and record the fluid level in the pipette, and record the time of the reading. Be sure to take your readings at the bottom of the meniscus as shown in Figure 7.8. ***Record your data in the chart in item 6e on the laboratory report.***

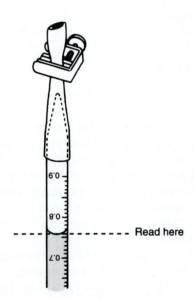

Figure 7.8 Reading the fluid level in the pipette at the meniscus.

Exactly 3 min later, record another reading. Then, determine the volume of oxygen produced within the 3-min interval. Repeat the process two more times to determine oxygen production during three separate 3-min intervals at this distance. Then, calculate the average volume of oxygen produced per minute, in milliliters of oxygen per minute (ml O_2/min), and record this figure in the chart in item 6e on the laboratory report.

10. Use the same procedure as in step 9 to determine the rate of oxygen production at each distance. Move the light source to the new distance and change the water in the heat filter at each new distance. Allow 5 min for equilibration to the new distance before starting your readings.

11. ***Complete item 6 on the laboratory report.***

THE COLOR OF LIGHT AND THE RATE OF PHOTOSYNTHESIS

Because white light consists of a spectrum of colors (violet to red) and because chlorophyll is green, you may wonder how different colors (wavelengths) of light affect the rate of photosynthesis. This can easily be determined with the same setup used in the previous experiment and using colored filters to control the wavelengths of light reaching the *Elodea* shoot.

Assignment 7

1. Design and conduct an experiment to determine the effect of different wavelengths (colors) of light on the rate of photosynthesis. Obtain the colored light filters from your instructor.

2. ***Complete item 7 on the laboratory report showing the hypothesis to be tested, the procedures used, the results, and the conclusion.***

PHOTOSYNTHESIS

Student _____

Lab Instructor _____

1. INTRODUCTION

a. Write the term that matches the phrase.

1. Organisms synthesizing organic nutrients _____

2. Form of energy captured by photosynthesis _____

3. Form of energy formed by photosynthesis _____

4. End product of photosynthesis _____

5. Storage form of carbohydrates in plants _____

6. Source of O_2 formed by photosynthesis _____

7. Source of oxygen atoms in glucose formed by photosynthesis _____

8. Source of hydrogen atoms in glucose formed by photosynthesis _____

9. Source of carbon atoms in glucose formed by photosynthesis _____

b. List the key events in each reaction:

Light reaction _____

_____

_____

Dark reaction _____

_____

2. CARBON DIOXIDE AND PHOTOSYNTHESIS

a. State the hypothesis to be tested. _____

_____

b. Indicate the presence or absence of starch.

Plant A _____   Plant B _____

c. Was the hypothesis supported? _____ State a conclusion from your results. _____

_____

_____

3. LIGHT AND PHOTOSYNTHESIS

a. State the hypothesis to be tested. _____

_____

b. Indicate the presence or absence of starch in the:

exposed leaf tissue _____

shielded leaf tissue _____

c. Was the hypothesis supported? _____ State a conclusion from your results. _____

4. CHLOROPHYLL AND PHOTOSYNTHESIS

a. State the hypothesis to be tested. _____

_____

b. Indicate the presence or absence of starch and sugar in the:

	Starch	*Sugar*
green *Coleus* leaf tissue	_____	_____
nongreen *Coleus* leaf tissue	_____	_____
green corn leaf	_____	_____
nongreen corn leaf	_____	_____

c. What is the source of the sugar in the nongreen plant tissues?

Coleus _____

Corn _____

d. Do your results support the hypothesis? _____ State a conclusion from your results regarding chlorophyll and photosynthesis. _____

_____

5. CHLOROPLAST PIGMENTS

a. Attach your chromatogram and label the original pigment line, top edge of the solvent, and the pigment lines of carotene, chlorophyll a, chlorophyll b, and xanthophyll.

b. Record the R_f values for carotene _____; chlorophyll a _____; chlorophyll b _____

Which chlorophyll was most soluble in the developing solvent? _____

c. Explain why not all leaf pigments are visible in a healthy green leaf.

_____

_____

d. How do you explain the appearance of yellow pigments in many plant leaves in autumn?

_____

_____

e. Examine Figure 7.5 and determine:

1. The three parts of the spectrum responsible for most of photosynthesis.

_____   _____   _____

2. Colors of light absorbed primarily by chlorophyll a.

_____   _____

3. Colors of light absorbed primarily by chlorophyll b.

_____   _____

6. LIGHT QUANTITY AND THE RATE OF PHOTOSYNTHESIS

a. State the hypothesis to be tested. _____

_____

b. What is used to indicate the rate of photosynthesis? _____

_____

Why is this a good indicator of photosynthetic action? _____

_____

c. An assumption in the experiment is that light quantity decreases as distance from the light source increases. Is this a valid assumption? _____ If you wanted to verify it, how would you do it? _____

_____

d. Explain why a heat filter must be used when using a spot lamp.

_____

e. Record your data in the following chart:

LIGHT QUANTITY AND THE RATE OF PHOTOSYNTHESIS

Distance from Light Source (cm)	Readings (ml)		ml O_2/3 min	ml O_2/min
	Start	Stop		
25	____	____	____	
	____	____	____	
	____	____	____	
		Average	____	____
50	____	____	____	
	____	____	____	
	____	____	____	
		Average	____	____
75	____	____	____	
	____	____	____	
	____	____	____	
		Average	____	____
100	____	____	____	
	____	____	____	
	____	____	____	
		Average	____	____

f. Was the hypothesis supported? _____ State a conclusion from your results. _____

_____

_____

g. Explain why your conclusion should be modified to "within the range of values tested."

_____

h. Name two ways that all animals benefit from photosynthesis.

1. _____

2. _____

7. COLOR OF LIGHT AND THE RATE OF PHOTOSYNTHESIS

a. State your hypothesis. _____

_____

_____

b. Describe your procedures in a stepwise manner.

_____

_____

_____

_____

_____

_____

_____

c. Record your results.

COLOR OF LIGHT AND THE RATE OF PHOTOSYNTHESIS

| Color | Readings (ml) | | ml O$_2$/3 min | ml O$_2$/min |
	Start	Stop		
_____	_____	_____	_____	
	_____	_____	_____	
	_____	_____	_____	
		Average	_____	_____
_____	_____	_____	_____	
	_____	_____	_____	
	_____	_____	_____	
		Average	_____	_____
_____	_____	_____	_____	
	_____	_____	_____	
	_____	_____	_____	
		Average	_____	_____

d. Was the hypothesis supported? _____ State a conclusion from your results. _____

_____

_____

_____

_____

_____

_____

_____

CELLULAR RESPIRATION AND FERMENTATION

OBJECTIVES

After completing the laboratory session, you should be able to:
1. Describe the relationship between cellular respiration, ADP, ATP, and cellular work.
2. Compare the end products of cellular respiration and fermentation.
3. Describe the effect of temperature on cellular respiration in frogs and mice.
4. Compare the alcoholic fermentation of glucose, sucrose, and starch.
5. Define all terms in bold print.

All organisms must have a continuous supply of energy from an external source to operate their metabolic functions. The sun is the ultimate energy source, and photosynthesis converts light energy into the **chemical bond energy** of organic nutrients. The three major classes of organic nutrients are **carbohydrates, fats,** and **proteins.** For the energy stored in nutrient molecules to be used for cellular work, their chemical bonds must be broken and the released energy must be captured in **high-energy phosphate bonds (~P).** A high-energy phosphate (~P) combines with **adenosine diphosphate (ADP)** to form **adenosine triphosphate (ATP),** which is the immediate source of energy for cellular work. See Figure 8.1.

There are two processes that extract energy from nutrient molecules to form ATP: cellular respiration and fermentation. **Cellular respiration** is an aerobic (with oxygen) process; **fermentation** is an anaerobic (without oxygen) process. Both involve oxidation (loss of electrons) and reduction (gain of electrons). In contrast to uncontrolled combustion, which is also an oxidation reaction, these enzyme-controlled reactions require little activation energy, occur at a nonlethal temperature, and break nutrient bonds sequentially to release energy in small amounts so much of it can be captured to form ATP.

Cellular respiration begins in the cytosol with *glycolysis,* a process that does not require oxygen and that breaks a glucose molecule into two pyruvate (pyruvic acid) molecules. In the presence of oxygen, pyruvate enters mitochondria, where it is broken down to carbon

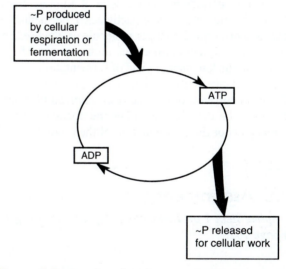

Figure 8.1 Transfer of ~P from respiration to cellular work.

dioxide via the *tricarboxylic acid cycle.* Energy-rich electrons pass along the *electron transport chain,* a series of molecules enabling energy to be captured in ~P to form ATP. The final electron and H^+ acceptor is molecular oxygen yielding the formation of water. Cellular respiration yields a net of 36 ATP molecules for each molecule of glucose respired. See Figure 8.2.

In the absence of oxygen, glucose is degraded to pyruvate by glycolysis, but pyruvate cannot enter mitochondria and the tricarboxylic acid cycle. It remains in

Cellular respiration: 36 ATP, net

Glucose + Oxygen $\longrightarrow$ Carbon dioxide + Water

$C_6H_{12}O_6$ + $6O_2$ $\longrightarrow$ $6CO_2$ + $6H_2O$

Alcoholic fermentation: 2 ATP, net

Glucose $\longrightarrow$ Carbon dioxide + Ethyl alcohol

$C_6H_{12}O_6$ $\longrightarrow$ $2CO_2$ + $2C_2H_5OH$

Lactate fermentation: 2 ATP, net

Glucose $\longrightarrow$ Lactic acid

$C_6H_{12}O_6$ $\longrightarrow$ $2C_3H_6O_3$

Figure 8.2 Summary equations for cellular respiration and fermentation.

the cytosol and serves as the final acceptor of electrons and H^+ removed in glycolysis, a mechanism that enables glycolysis to continue in the absence of oxygen. In *alcoholic fermentation,* which occurs in yeast, pyruvate is converted into ethanol and CO_2. In *lactate fermentation,* which occurs in certain bacteria and archeaens and for brief periods in muscle cells of animals, pyruvate is converted into lactate (lactic acid). Fermentation yields a net of only 2 ATP per molecule of glucose. See Figure 8.2. Only certain bacteria, archaeans, and fungi can live on the low energy yield of fermentation.

In accordance with the summary equations (Figure 8.2), the occurrence of cellular respiration and fermentation may be detected by either the consumption of the reactants or the accumulation of the products.

Assignment 1

Complete item 1 on Laboratory Report 8 that begins on page 101.

CELLULAR RESPIRATION AND CARBON DIOXIDE

In this section, you will determine if cells of living organisms produce carbon dioxide by cellular respiration by testing this hypothesis: *Living organisms do not produce CO_2.* Accumulation of carbon dioxide may be detected by using a dilute bromthymol blue solution, a pH indicator. As carbon dioxide accumulates in a dilute bromthymol solution, it causes a decrease in pH (an increase in acidity) by combining with water to form carbonic acid, which in turn releases hydrogen ions (H^+) as it dissociates. The hydrogen ions react with the bromthymol blue, which causes the solution to turn yellow. See Figure 8.3.

Step 1:　CO_2 + H_2O $\longrightarrow$ H_2CO_3

　　　　Carbon + Water → Carbonic
　　　　dioxide 　　　　　 acid

Step 2:　H_2CO_3 ⇌ H^+ + HCO_3^-

　　　　Carbonic ⇌ Hydrogen + Bicarbonate
　　　　acid 　　　　 ion 　　　　 ion

Figure 8.3 Reaction of carbon dioxide and water.

Assignment 2

Materials

Drinking straws
Glass-marking pen
Glass tubing, 2-cm lengths
Medicine droppers
Rubber-bulb air syringe
Rubber stoppers for test tubes
Test tubes
Test-tube rack
Bromthymol blue, 0.004%, in dropping bottles
Crickets
Pea seeds, germinating

1. Perform Experiment 1 to determine if air exhaled from your lungs contains an accumulation of CO_2. If body cells produce CO_2, it will be carried by blood to the lungs for removal from the body.
   a. Place 3 drops (0.15 ml) of 0.004% bromthymol blue into each of two numbered test tubes.
   b. Exhale your breath through a drinking straw into tube 1 for 3 min.
   c. Use a rubber-bulb syringe to pump atmospheric air into tube 2 for 3 min.
   d. ***Record any color change in the bromthymol blue solutions in the table in item 2a on the laboratory report.***
2. Perform Experiment 2 to determine if germinating seeds and live crickets produce CO_2.
   a. Place 3 drops (0.15 ml) of 0.004% bromthymol blue into each of three numbered test tubes.
   b. Place several short segments of glass tubing into each tube so seeds and crickets will be kept out of the solution.
   c. Place six to ten germinating pea seeds in tube 1, place three crickets in tube 2, and place nothing else in tube 3.
   d. Gently (to prevent a sudden increase in air pressure from injuring the crickets) insert a rubber stopper into each tube. Place the tubes in a test-tube rack for 20 min. Then, observe any color change in the bromthymol blue.
   e. ***Record your results on the laboratory report and complete item 2.***

CELLULAR RESPIRATION AND PHOTOSYNTHESIS

You have learned that living organisms continuously produce carbon dioxide via cellular respiration and that carbon dioxide is used in photosynthesis. Does carbon dioxide accumulate in photosynthesizing cells? Find out by testing this hypothesis: *Carbon dioxide does not accumulate in photosynthesizing cells.*

 ## Assignment 3

Materials

Aluminum foil
Glass-marking pen
Glass tubing, 2-cm lengths
Medicine droppers
Rubber stoppers for test tubes
Test tubes
Test-tube rack
Bromthymol blue, 0.004%, in dropping bottles
Primula or *Viola* plants

1. Place 3 drops (0.15 ml) of 0.004% bromthymol blue into three numbered test tubes.
2. Obtain two leaves from a healthy plant. Roll and insert one leaf into tubes 1 and 2 so the upper surface of the leaf is against the glass. Do not put the leaves into the solutions. Place nothing in tube 3.
3. Insert rubber stoppers into each tube. Wrap tube 1 in aluminum foil to exclude light.
4. Place the tubes in a test-tube rack and expose them to light, but not heat, for 30 min. After 30 min, note the color of the bromthymol blue solution in each tube.
5. **Complete item 3 on the laboratory report.**

CELLULAR RESPIRATION AND HEAT PRODUCTION

Normal body temperature in birds and mammals is produced by heat that is lost as a by-product of cellular respiration. Do simpler organisms produce heat in a similar way? Find out by testing this hypothesis: *Simple organisms do not produce heat as a by-product of cellular respiration.*

A few hours before the start of the laboratory session, your instructor set up three vacuum bottles. Bottle 1 contains germinating pea seeds, bottle 2 contains live crickets, and bottle 3 contains nothing. A one-hole cork stopper with a thermometer was placed in each bottle to measure the temperature inside the bottle.

 ## Assignment 4

Materials

Celsius thermometers
Cotton, nonabsorbent
Glass-marking pen
Vacuum bottles with one-hole stoppers
Crickets
Pea seeds, germinating

1. Read and record the temperatures in the vacuum bottles.
2. **Complete item 4 on the laboratory report.**

RATE OF CELLULAR RESPIRATION

In this section, you will determine the rate of cellular respiration in germinating pea seeds and crickets to see if there is a difference in their respiration rates. You will determine the rate of cellular respiration by measuring the rate of oxygen consumption. The experiment is best done in groups of two to four students. Before starting the experiment, read through the experiment and establish a division of labor.

 ## Assignment 5

Materials

Balance
Celsius thermometer
Cotton, nonabsorbent
Glass-marking pen
Marker fluid in dropping bottles
Pipettes, 1-ml, graduated
Rubber stoppers for test tubes, 1-hole
Soda lime
Test tubes
Test-tube racks
Crickets
Pea seeds, germinating

Test this hypothesis: *The rate of cellular respiration is the same for both germinating pea seeds and crickets.*

1. Obtain two test tubes and number them 1 and 2. *Be certain that the interior of the tubes is completely dry.* Place soda lime in each tube to a depth of about 2 cm. Soda lime absorbs carbon dioxide. Insert a 2-cm pad of lightly packed cotton above the soda lime. Place the tubes in a test-tube rack.
2. Place a second small pad of cotton in the mouth of each tube. These cotton pads will be used later to confine the specimens, and it must be weighed

with the test tubes. Weigh the test tubes to the nearest 0.1 g. Record the weight here.

Tube 1 _____ g   Tube 2 _____ g

3. Obtain 30–40 germinating pea seeds and blot them with a paper towel; *be sure that they are dry.* Remove the cotton pad from the mouth of tube 1 and place the pea seeds in the tube. Then replace the cotton pad to hold them in place.

4. Remove the cotton pad from the mouth of tube 2 and place 3–5 crickets in the tube. Replace the cotton pad to hold them in place.

5. Weigh tubes 1 and 2 again to the nearest 0.1 g.

Tube 1 _____ g   Tube 2 _____ g

6. Calculate the weight of the peas and crickets to the nearest 0.1 g. ***Record their weights on the chart in item 5b on the laboratory report.***

7. Obtain two 1-hole stoppers, that have an inserted 1-ml graduated pipette which is bent 90°. See Figure 8.4. Gently, insert a stopper into each test tube. Place the test tubes in a test-tube rack and orient the pipettes so their graduations are easy to read. *Be sure you know how to read the graduations on the pipettes before proceeding.*

8. Add a drop of colored marker fluid to the tip of each pipette. It will be drawn into each pipette. Its movement will indicate the consumption of oxygen.

9. ***Record the room temperature on the chart in item 5b.***

10. When the front edge of the marker fluid reaches the pipette graduations in *both* tubes 1 and 2, *record the readings from both pipettes in the "0" time column on the chart in item 5b. Record the reading on each pipette at 3-min intervals for 15 minutes.*

11. ***Calculate the oxygen consumption as shown here and complete the chart.***

$$\frac{\text{ml } O_2}{3 \text{ min}} \times \frac{60 \text{ min}}{1 \text{ hr}} = \text{ml } O_2/\text{hr}$$

$$\frac{\text{ml } O_2/\text{hr}}{\text{wt. in grams}} = \text{ml } O_2/\text{hr}/g$$

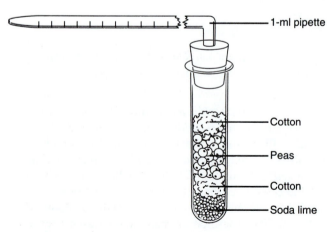

Figure 8.4 Respirometer for pea seeds and crickets.

12. Clean the apparatus and your work station.
13. ***Complete item 5 on the laboratory report.***

TEMPERATURE AND THE RATE OF CELLULAR RESPIRATION

In this experiment, you will use a respirometer similar to the one shown in Figure 8.5 to determine the oxygen consumption of a frog and a mouse at three different temperatures. A frog is a poikilotherm, an animal whose body temperature fluctuates with the ambient temperature. A mouse is a homeotherm, an animal whose body temperature is constant in spite of fluctuations in the ambient temperature. This experiment will allow you to detect any basic differences in the rate of cellular respiration when these animals are subjected to temperatures of 10°C, room temperature, and 40°C.

The manometer (U-tube) of the respirometer connects the experimental and control chambers. A pressure change in one will cause movement of the manometer fluid toward the chamber with the lowest pressure. Because carbon dioxide is absorbed by the soda lime, oxygen consumption can be measured by the movement of fluid toward the experimental chamber. Note that any change in pressure in the control chamber is automatically reflected in the level of the fluid.

 ## Assignment 6

Materials

Balance
Celsius thermometer
Colored water for the manometer
Cotton, nonabsorbent
Respirometer as in Figure 8.5
Ring stand and clamps
Soda lime
Tubing clamps
Water baths
Wire screen
Frogs
White mice

Test this hypothesis: *The rate of oxygen consumption at different ambient temperatures is the same in a frog and a mouse.*

1. Set up a respirometer as shown in Figure 8.5. Place soda lime in the bottom of each chamber to a depth of about 2 cm. Cover the soda lime with about 1 cm of cotton and add a wire screen. Place the respirometer in a water bath at the assigned temperature.

2. Weigh the animal assigned to you to the nearest 0.1 g. (Mice should be picked up by their tails.)

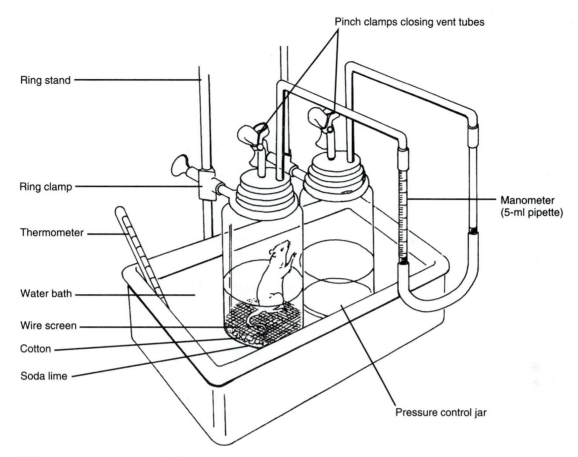

Figure 8.5 Respirometer setup for frog and mouse.

3. Place the animal in the experimental chamber. *With the vent tubes open,* loosely replace the stopper. *Caution: Failure to keep the vent tubes open when inserting the stopper may injure the test animal due to a sudden increase in air pressure.*

4. After 5–10 min for temperature equilibration, *with the vent tubes open,* insert the stopper snugly into the jar. Close the vent tubes, record the time, and take the first reading from the manometer. Exactly 3 min later, take the second reading. *Open the vent tubes, remove the stopper, and place it loosely on the top of the jar.* Subtract the first reading from the second reading to determine the ml O_2 consumed in the first 3-min interval.

5. After 3–5 min, insert the stopper, close the vent tubes, and take the first reading of the second replica. Proceed as before. Repeat to make at least five replicas. ***Record your data in item 6c on the laboratory report.***

6. Calculate the average oxygen consumption (milliliters per 3-min interval) and the ml O_2/hr/g of body weight.

7. Return the animal to its cage, and clean the respirometer and your work station.

8. Exchange data with groups doing the experiment at different temperatures.

9. ***Plot the average respiration rate (ml O_2/hr/g) for each organism studied in item 6d on the laboratory report.***

10. ***Complete item 6 on the laboratory report.***

FERMENTATION IN YEAST

Yeast is a unicellular fungus that can thrive on the low energy yield of fermentation in the absence of oxygen. Yeast ferments glucose to yield ethyl alcohol, carbon dioxide, and 2 ATP net. See Figure 8.2. It is used commercially in the production of alcoholic beverages and bakery products. In bread making, starch is digested to glucose, which is then fermented by yeast. The bubbles of carbon dioxide produced by fermentation causes the dough to rise, and the ethanol is evaporated by the heat of cooking.

In this section, you will determine if different carbohydrates—glucose, sucrose, and starch—affect the fermentation rate of yeast. Glucose is a monosaccharide that is the primary carbohydrate energy source

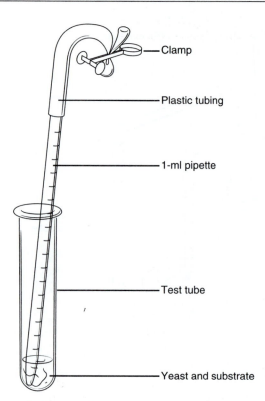

Figure 8.6 Setup for investigating fermentation.

for cells in many organisms. Sucrose (table sugar) is a disaccharide composed of glucose and fructose subunits. Starch is a polysaccharide composed of many glucose subunits, and it is the primary energy storage form in plants. You will use a respirometer similar to the one in Figure 8.6 to measure the rate of carbon dioxide production at room temperature. Read through the experiment and decide on a division of labor among your team members.

 Assignment 7

Materials

Celsius thermometer
Dropping bottles of:
    distilled water
    glucose, 10%
    sucrose, 10%
    starch, 10%
    yeast suspension
Glass-marking pen
Pipettes, 1 ml
Pipette pump
Test tubes
Test-tube rack
Tubing clamps
Tubing, plastic, 12–15 cm in length

Test this hypothesis: *Fermentation of glucose, sucrose, and starch by yeast occurs at the same rate.*

1. Number four clean test tubes 1 through 4 and place them in a test-tube rack.
2. Add to the test tubes the quantity of materials shown in Table 8.1. You may use either "full droppers" or milliliters in dispensing the materials.
3. Shake the tubes from side to side to mix the contents.
4. Obtain four clean 1-ml pipettes that have a short piece of plastic tubing attached. Place a pipette in each of the numbered test tubes. *Be sure you know how to read the graduations on the pipettes.*
5. Attach a pipette pump to the tubing on the pipette in tube 1. Pull the fluid into the pipette to the 0.0 mark by turning the knurled wheel with your thumb. Fold the tubing over and crimp it closed with a pinch clamp to prevent the fluid from running out of the pipette. Remove the pipette pump. After clamping the tube and removing the pipette pump, the fluid level should be between the 0.0 mark and the 0.1 mark. Leave the pipette in the test tube.
6. Use the same procedure to quickly fill the other pipettes.
7. ***Record the time and the first reading from each pipette in the chart under item 7 on the laboratory report.*** Repeat the readings at 5-min intervals for 30 minutes.

    As fermentation occurs, carbon dioxide will form and accumulate at the top of the pipette, displacing the fluid. By reading the fluid level at constant time intervals, you can determine the rate of carbon dioxide production, which indicates the rate of fermentation.
8. Subtract the first reading ("0" time) from the last reading to get the total carbon dioxide produced in 30 min.
9. Record the room temperature here. _____ °C
10. ***Complete the chart and item 7 on the laboratory report.***

INDEPENDENT INQUIRY

The investigations in this exercise should raise some questions in your mind about additional investigations that can be made about cellular respiration and fermentation. In this section, you have the opportunity to design and conduct an experiment to answer a question that has been raised by the preceding experiments. Here is a short list of questions to help you get started.

1. Does temperature affect the respiration rate in peas or crickets?

TABLE 8.1	CONTENTS OF THE FERMENTATION TUBES		
	Droppers or Milliliters		
Tube	Yeast Suspension	Substrate	Distilled Water
1	3	0	3
2	3	3 glucose	0
3	3	3 sucrose	0
4	3	3 starch	0

2. Do temperatures colder or warmer than those tested have a proportionate affect on the respiration rate in frogs or mice?

3. Does yeast ferment other carbohydrates?

4. Does the concentration of glucose affect the rate of fermentation by yeast?

5. Does the type of yeast affect the rate of fermentation?

6. Does temperature affect the rate of glucose fermentation by yeast?

 ## Assignment 8

1. Decide on a question that you want to investigate.

2. State your hypothesis.

3. State a prediction of expected results if your hypothesis is correct.

4. Design your experiment.

5. Conduct your experiment.

6. *Your write-up under item 8 should include:*
   a. *Procedures*
   b. *The results in a chart of your design*
   c. *A statement as to whether the results support your hypothesis*
   d. *A conclusion*
   e. *An explanation of your results*

7. After completing your experiment, clean up your equipment and work station.

CELLULAR RESPIRATION AND FERMENTATION

Student _____

Lab Instructor _____

1. INTRODUCTION

a. Write the summary equation for the aerobic respiration of glucose. Underline the reactants and circle the products.

_____

b. Write the term that matches the phrase.

1. Final electron acceptor in cellular respiration

   _____

2. Source of carbon in carbon dioxide formed

   _____

3. Final electron acceptor in fermentation

   _____

4. Molecules enabling a controlled oxidation

   _____

5. Provides energy for immediate cellular work

   _____

6. Combines with ~P to form ATP

   _____

7. Produces 36 ATP units, net

   _____

8. Produces 2 ATP units, net

   _____

2. CELLULAR RESPIRATION AND CARBON DIOXIDE PRODUCTION

a. Record the results of your experiments in the following table:

CARBON DIOXIDE PRODUCTION IN ANIMALS AND GERMINATING SEEDS

Tube	Contents Plus Bromthymol Blue	Color of Bromthymol Blue After Test Interval	CO_2 Concentration	
			Increase	No Change
Exp. 1: Humans				
1	Exhaled air	_____	_____	_____
2	Atmospheric air	_____	_____	_____
Exp. 2: Germinating Seeds and Crickets				
1	Germinating seeds	_____	_____	_____
2	Live crickets	_____	_____	_____
3	Nothing	_____	_____	_____

b. What is the independent variable in these experiments? _____

_____

c. What is the purpose of tube 2 in Experiment 1 and tube 3 in Experiment 2? _____

_____

d. State a conclusion from Experiment 1. _____

_____

e. State a conclusion from Experiment 2. _____

_____

3. CELLULAR RESPIRATION AND PHOTOSYNTHESIS

a. Record the results of your experiment with photosynthesizing and nonphotosynthesizing leaves in the following table:

CARBON DIOXIDE PRODUCTION AND PHOTOSYNTHESIS

Tube	Contents Plus Bromthymol Blue	Color of Bromthymol Blue After Test Interval	CO_2 Concentration Increase	No Change
1	Nonphotosynthesizing leaves	_____	_____	_____
2	Photosynthesizing leaves	_____	_____	_____
3	Nothing	_____	_____	_____

b. Do the results support your hypothesis? _____

c. State a conclusion from the results. _____

_____

_____

_____

d. Does this experiment prove that cellular respiration does not occur during photosynthesis in green plant cells?

_____ Explain. _____

_____

_____

_____

4. CELLULAR RESPIRATION AND HEAT PRODUCTION

a. Indicate the temperature of each vacuum bottle.

Bottle 1 _____ Bottle 2 _____ Bottle 3 _____

b. Do the results support your hypothesis? _____

c. State a conclusion from the results. _____

_____

5. RATE OF CELLULAR RESPIRATION

a. State the hypothesis to be tested. _____

_____

What is the independent variable? _____

b. Record your results in the following table:

RATE OF OXYGEN CONSUMPTION AT _____ °C

Time	0	3 min	6 min	9 min	12 min	15 min	Average ml O_2/3 min	ml O_2/hr	ml O_2/hr/g
Peas ____ g									
Crickets ____ g									

c. Do the results support your hypothesis? _____

d. State a conclusion from the results. _____

_____

_____

e. Do you think the respiration rates would be the same at a colder temperature? _____ at a warmer temperature?

_____

Explain. _____

_____

_____

_____

6. TEMPERATURE AND THE RATE OF CELLULAR RESPIRATION

a. State the hypothesis to be tested. _____

_____

What is the independent variable? _____

What is the dependent variable? _____

b. Why is it important to open the vent tubes and remove the stopper of the animal chamber between replicas? ____

_____

c. Record the data collected for the rate of respiration in a frog and a mouse at 10°C, room temperature, and 40°C. Add your room temperature to the chart.

OXYGEN CONSUMPTION IN A FROG AND A MOUSE

Organism (weight)	Temp. (°C)	ml O_2/3 min 5 Replicates 1	2	3	4	5	Average ml O_2/3 min	Average ml O_2/hr/g
Frog (____ g)	10							
	*							
	40							
Mouse (____ g)	10							
	*							
	40							

d. Plot the average respiration rate (ml O_2/hr/g) of each organism tested in the graph. Select and record a different symbol for each organism so your graph may be easily read.

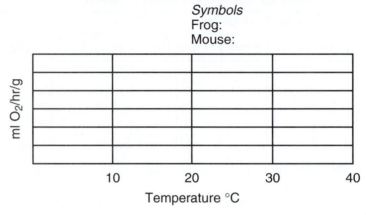

e. Do the results support your hypothesis? _____

f. State a conclusion from the results. _____

_____

_____

g. Explain the different effects of 10°C on respiration rate in the frog and mouse.

_____

_____

_____

7. FERMENTATION IN YEAST

a. When yeast ferments glucose, what is the source of the CO_2 that is produced? _____

What molecule serves as the final electron and H^+ acceptor? _____

What is the source of the carbon atoms in ethyl alcohol? _____

b. What monosaccharide(s) compose sucrose? _____

What monosaccharide(s) compose starch? _____

c. State your hypothesis for this experiment. _____

_____

d. What is the independent variable? _____

e. What is the dependent variable? _____

f. Record your data in the following chart:

RATE OF CO_2 PRODUCTION AT _____ °C

Tube	Pipette Readings (ml)							ml CO_2/30 min	ml CO_2/hr
	0	5 min	10 min	15 min	20 min	25 min	30 min		
1									
2									
3									
4									

g. Do the results support your hypothesis? _____

h. State a conclusion from your results. _____

_____

_____

i. How do you explain the results? _____

_____

_____

_____

_____

8. INDEPENDENT INQUIRY

CELL DIVISION

OBJECTIVES

After completing the laboratory session, you should be able to:
1. Describe the process of binary fission.
2. Name the stages of the cell cycle and describe their characteristics.
3. Name the phases of mitosis and meiosis and describe their characteristics.
4. Identify the phases of mitosis when dividing cells are viewed with a microscope.
5. Compare the processes and end products of mitotic and meiotic cell division.
6. Describe the significance of mitotic and meiotic cell divisions.
7. Define all terms in bold print.

All new cells are formed by the division of preexisting cells. A key part of cell division is the replication of DNA in the **parent cell** and the distribution of DNA to the **daughter cells.** Recall that DNA is the molecule that controls cellular functions, and it contains the hereditary information passed from parent cell to daughter cells. DNA is located in **chromosomes. Genes,** the determiners of inheritance, are relatively short sequences of nucleotides in DNA molecules.

In **prokaryotic cells,** cell division is relatively simple. The process is known as **binary fission.** Binary fission involves (1) replication of the single circular chromosome (a DNA molecule) and attachment of the two chromosomes to separate points on the plasma membrane, (2) separation of the chromosomes by linear growth of the cell, and (3) division of the cell by the inward growth of the plasma membrane and cell wall to form two daughter cells. Figure 9.1 depicts the process of binary fission. The cells are too small for you to observe this process in the laboratory, however.

In **eukaryotic cells,** cell division is more complex. Usually, there are many chromosomes in the cell nucleus, and each eukaryotic chromosome consists of a DNA molecule plus associated protein molecules. Two different types of cell division occur in eukaryotic cells: mitotic cell division and meiotic cell division. Daughter cells formed by **mitotic cell division** contain the same

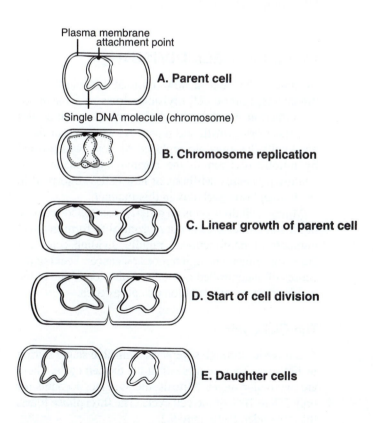

Figure 9.1 Binary fission in a prokaryote cell.

number and composition of chromosomes as the parent cell. In contrast, cells formed by **meiotic cell division** have only one-half the number of chromosomes as the parent cell. Thus, these two types of cell division differ in the way the chromosomes are dispersed to the new cells that are formed. The terms **mitosis** and **meiosis** refer to the orderly process of separating and distributing the replicated chromosomes to the new cells. **Cytokinesis** (division of the cytoplasm) is the process of actually forming the **daughter cells.**

In animals and most plants, body cells contain two sets of chromosomes, and the chromosomes occur in pairs. Both members of a chromosome pair contain similar hereditary information. Therefore, members of a chromosome pair are **homologous chromosomes.** A cell with two sets of chromosomes is said to be **diploid (2n),** whereas a cell with only one set of chromosomes is **haploid (n).** Mitotic division may occur in either diploid or haploid cells, but meiotic division occurs only in diploid cells.

Each organism has a characteristic number of chromosomes in its body cells. For example, fruit flies have 8 chromosomes (4 pairs), onions have 16 (8 pairs), and humans have 46 (23 pairs). Gametes of these organisms are always haploid and contain 4, 8, and 23 chromosomes, respectively. The body cells of most simple organisms are haploid.

MITOTIC CELL DIVISION

In unicellular and a few multicellular eukaryotic organisms, mitotic cell division serves as a means of reproduction. In all multicellular organisms, it serves as a means of growth and repair. As worn-out or damaged cells die, they are replaced by new cells formed by mitotic division in the normal maintenance and healing processes. Millions of new cells are formed in the human body each day in this manner.

Mitotic cell division is an orderly, controlled process, but it sometimes breaks out of control to form massive numbers of nonfunctional, rapidly dividing cells that constitute either a benign tumor or a cancer. Seeking the causes of uncontrolled mitotic cell division is one of the major efforts of current biomedical research.

The Cell Cycle

A cell passes through several recognizable stages during its life span. These stages constitute the **cell cycle.** There are two major stages. **Mitosis,** the M stage, accounts for only 5% to 10% of the cell cycle. The **interphase** forms the remainder. See Figure 9.2.

Interphase has three subdivisions. A growth period, the **G_1 stage,** occurs immediately after mitosis. Cells that will not divide again remain in the G_1 stage and

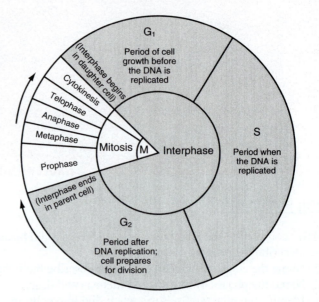

Figure 9.2 Eukaryotic cell cycle.

carry out their normal functions. Cells preparing to divide enter the next stage, the **synthesis (S) stage,** during which chromosome replication occurs. Each replicated chromosome consists of two **sister chromatids** joined at the **centromere.** See Figure 9.3. A second growth stage, the **G_2 stage,** follows and prepares the cell for the next mitotic division. The centrioles replicate in the G_2 stage in animal cells.

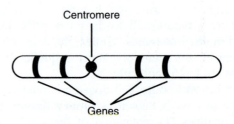

1. An Unreplicated Chromosome

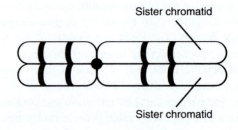

2. A Replicated Chromosome

Figure 9.3 Chromosome structure.

Mitotic Phases in Animal Cells

The process of mitosis is arbitrarily divided into recognizable stages or phases to facilitate understanding, although the process is a continuous one. These phases are **prophase, metaphase, anaphase,** and **telophase.** The characteristics of each phase as observed in animal cells are noted here to aid your study. Interphase is also included for comparative purposes. Compare these descriptions to Figure 9.4.

Interphase

Cells in interphase have a distinct nucleus and, in the G_2 stage, two pairs of **centrioles.** The chromosomes are uncoiled and are visible only as **chromatin granules.**

Prophase

During prophase, (1) the nuclear membrane and nucleolus disappear, (2) the replicated chromosomes coil tightly to appear as rod-shaped structures, (3) each pair of centrioles migrates to opposite ends of the cell, and

(4) **spindle** fibers (microtubules) extend between the pairs of centrioles. Each pair of centrioles and its radiating **astral rays** (microtubules) constitute an **aster** at each end (pole) of the football-shaped spindle. In late prophase, spindle fibers from opposite poles attach to the centromeres of the sister chromatids.

Metaphase

This brief phase is characterized by the chromosomes lining up at the equator of the spindle. It is the lengthening and shortening of the spindle fibers that move the chromosomes to the equator of the spindle.

Anaphase

Anaphase begins with the separation of the centromeres of the sister chromatids. Then, the shortening of the spindle fibers pulls the sister chromatids toward opposite poles. Once the sister chromatids separate and begin moving toward opposite poles of the spindle, they are called **daughter chromosomes.** Thus, a cell in anaphase contains two complete sets of chromosomes.

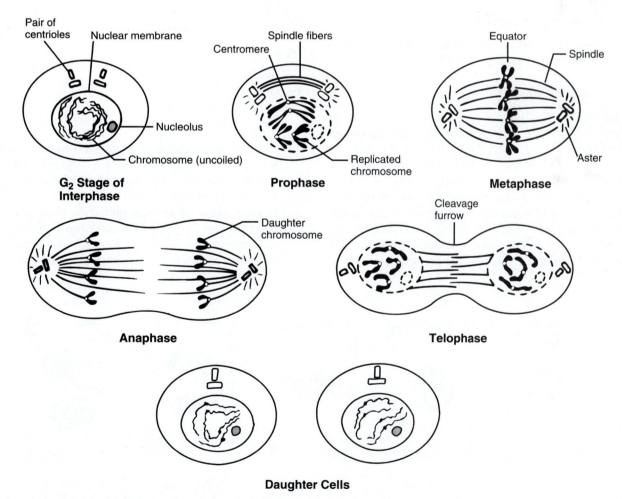

Figure 9.4 Mitotic cell division in an animal cell.

Telophase

In telophase, (1) a new nuclear membrane forms around each set of chromosomes to form two new nuclei, (2) the nucleolus reappears, and (3) the chromosomes start to uncoil. Usually cytokinesis, cytoplasmic division, begins prior to the end of telophase. A **cleavage furrow** forms, which continues to constrict until the parent cell divides to produce two daughter cells.

Mitotic Division in Plant Cells

Mitotic division in plants follows the same basic pattern that occurs in animals, with some notable exceptions. Most plants do not possess centrioles, although a spindle of fibers is present in dividing cells. The rigid cell wall prevents the formation of a cleavage furrow during cytokinesis; instead, a **cell plate** forms to separate the parent cell into two daughter cells, and a new cell wall forms along the cell plate. Cytokinesis usually, but not always, occurs during telophase. See Figure 9.5.

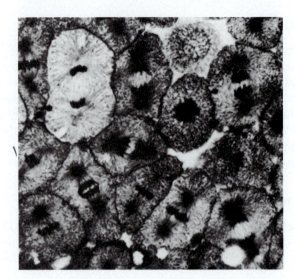

Figure 9.6 Various stages of mitotic division in whitefish blastula cells. (By permission of WARD'S Natural Science Establishment, Inc.)

 Assignment 1

1. Complete items 1, 2a, and 2b on Laboratory Report 9 that begins on page 115.

Microscopic Study

The rapidly dividing cells of whitefish blastula, an early fish embryo, are excellent for studying mitotic division in animals. See Figure 9.6. On the prepared slide that you will use are several thin sections of the blastula, and each contains many cells in various stages of the cell cycle, including mitosis. A prepared slide of onion (*Allium*) root tip is used to study mitotic division in plants. See Figure 9.7. Each slide usually contains three longitudinal sections of root tip. The region of cell division is near the pointed tip.

 Assignment 2

Materials

Compound microscope
Prepared slides of:
  whitefish blastula, x.s.
  onion root tip, l.s.

1. Obtain a prepared slide of whitefish blastula. Locate a section for study with the 4× objective. Then, switch to the 10× objective to find mitotic phases for observation with the 40× objective. *Locate cells in each phase of mitosis, and draw them in the space for item 2c on the laboratory report.* You may have to examine all the sections on your slide, or even other slides, to observe each phase of mitosis.

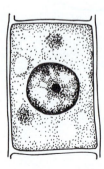 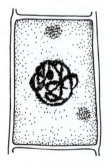   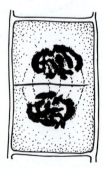

| **A** Interphase | **B** Prophase | **C** Metaphase | **D** Anaphase | **E** Telophase |

Figure 9.5 Mitotic division in a plant cell.

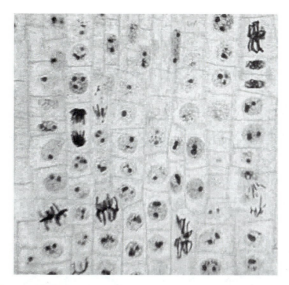

Figure 9.7 Mitotic division in onion root tip cells. (By permission of WARD'S Natural Science Establishment, Inc.)

2. Examine a prepared slide of onion root tip. Locate the cells in mitotic phases near the tip of the section. Observe cells in metaphase, anaphase, and telophase. Note the cell plate. Note how mitotic division in an onion root tip differs from that observed in a whitefish blastula.

3. Return to your slide of whitefish blastula. Using the 10× objective, scan the slide and examine a total of 100 cells. Tabulate the number of cells in interphase and in each phase of mitosis. This will give you an estimate of the percentage of cells in each phase of mitosis or interphase. ***Record your results in item 2g on the laboratory report.***

4. ***Complete items 2d to 2f on the laboratory report.***

MEIOTIC CELL DIVISION

In contrast to mitotic cell division, meiotic cell division consists of *two* successive divisions but only *one* chromosome replication. This results in the formation of four cells that have only half the number of chromosomes of the **diploid** (2n) parent cell. Thus, the daughter cells have a **haploid** (n) number of chromosomes since they each contain only *one member of each chromosome pair*. In addition to reducing the chromosome number in the daughter cells, meiosis also reshuffles the genes, hereditary units formed of small segments of DNA, and this greatly increases the genetic variability among the daughter cells.

In humans and most animals, cells formed by meiotic division become **gametes,** either sperm or eggs. In plants, meiotic cell division results in the formation of meiospores that grow into haploid gametophytes that, in turn, produce gametes by mitotic division. In either case, the basic result of meiosis is the same: haploid cells with increased genetic variation.

Meiotic Phases in Animal Cells

Study Figure 9.8 as you read the following description of meiotic cell division in an animal cell. Chromosome replication occurs in the S stage of interphase prior to the start of meiosis.

Meiosis I

Prophase I exhibits the following characteristics. Each chromosome is composed of two sister chromatids joined together at the centromere. The replicated members of each chromosome pair join together in a side-by-side pairing called **synapsis.** Chromosomes in synapsis are often called **tetrads** because they consist of four chromatids. An exchange of chromosome segments (cross-over) frequently occurs between members of the tetrad and increases the genetic variability of the cells produced by meiotic division. The chromosomes coil tightly to appear as rod-shaped structures, the nuclear membrane and nucleolus disappear, and a spindle forms.

Metaphase I is characterized by the synapsed chromosomes lining up at the equatorial plane, where they attach to spindle fibers by their centromeres.

Anaphase I begins with the separation of the members of each chromosome pair. The centromeres do *not* separate, so each chromosome still consists of two chromatids joined at the centromere. Members of each chromosome pair migrate to opposite poles of the spindle in the replicated state.

Telophase I proceeds to form a nuclear membrane around each set of chromosomes. The chromosomes untwist, and the nucleolus reappears. Cytokinesis separates the parent cell into two daughter cells. Keep in mind that the nucleus of each daughter cell contains only *one member of each chromosome pair* in a replicated state. Thus, each daughter cell is haploid (n).

Meiosis II

Both cells formed by meiosis I divide again in meiosis II, but for discussion purposes we will follow only one of these cells in the second division. In interphase between meiosis I and II, the centrioles replicate but chromosomes do *not* replicate again. Recall that they are already replicated.

Prophase II is characterized by the usual loss of the nuclear membrane and nucleolus, spindle formation, and the appearance of rod-shaped chromosomes.

Metaphase II is characterized by the chromosomes lining up at the equator of the spindle. Each chromosome consists of two sister chromatids joined together at the centromere that is attached to a spindle fiber.

Meiosis I

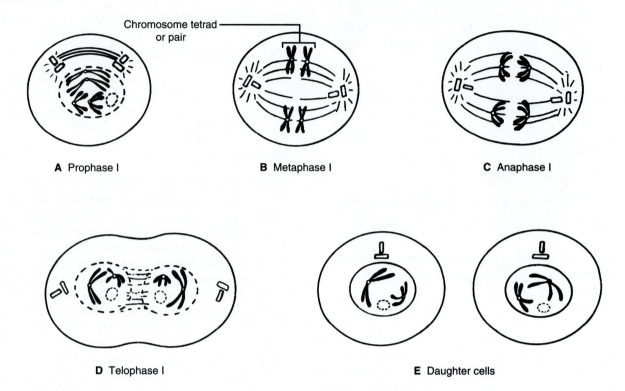

A Prophase I

B Metaphase I

C Anaphase I

D Telophase I

E Daughter cells

Meiosis II

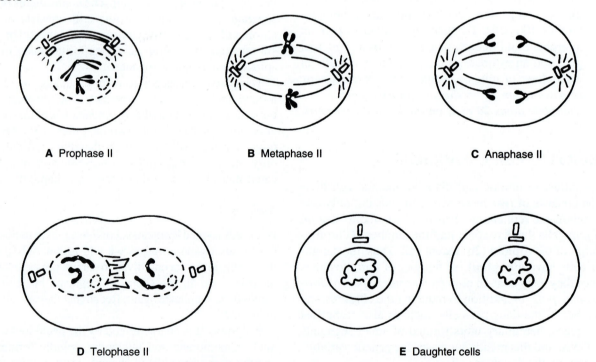

A Prophase II

B Metaphase II

C Anaphase II

D Telophase II

E Daughter cells

Figure 9.8 Meiotic cell division. Each cell produced in meiosis I completes meiosis II, but only one cell is shown here in meiosis II for simplicity.

TABLE 9.1	SIGNIFICANT DIFFERENCES IN MITOTIC AND MEIOTIC CELL DIVISIONS
Mitotic Cell Division	*Meiotic Cell Division*
1. Occurs in both haploid (n) and diploid (2n) cells.	1. Occurs in diploid (2n) cells, but not in haploid (n) cells.
2. Completed when one cell divides to form two cells.	2. Requires two successive cell divisions to produce four cells from the single parent cell.
3. Duplicated chromosomes do not align themselves in homologous pairs during division.	3. Duplicated chromosomes arrange themselves in homologous pairs during the first cell division.
4. The two daughter cells contain (a) the same genetic composition as the parent cell and (b) the same chromosome number as the parent cell.	4. The four daughter cells contain (a) different genetic compositions and (b) one-half the chromosome number of the parent cell.

Anaphase II begins with the separation of the centromeres. The sister chromatids, now called daughter chromosomes, move toward opposite poles of the spindle.

Telophase II proceeds as usual to form the new nuclei, and cytokinesis divides the cell to form two haploid (n) daughter cells.

Because each cell entering meiosis II forms two daughter cells, a total of four haploid (n) cells are produced from the original diploid (2n) parent cell entering meiosis I. Thus, meiotic cell division may be summarized as:

$$1 \text{ cell (2n)} \xrightarrow{\text{MI}} 2 \text{ cells (n)} \xrightarrow{\text{MII}} 4 \text{ cells (n)}$$

 ## Assignment 3

Materials

Colored pipe cleaners, snap beads, or chromosome simulation kits

1. Study Figure 9.8.
2. Using colored pipe cleaners or snap beads to represent chromosomes or a chromosome simulation kit, simulate the replication and distribution of chromosomes in both mitosis and meiosis, where 2n=4.
3. ***Complete item 3 on the laboratory report.***

 ## Assignment 4

1. Study Table 9.1, which summarizes the differences between mitotic and meiotic cell division.
2. ***Complete item 4 on the laboratory report.***

 ## Assignment 5

1. Your instructor has set up several microscopes showing cells in different stages of cell division. Your task is to identify the stage of cell division, whether the cell is an animal or plant cell, and the structure indicated by the pointer.
2. ***Complete item 5 on the laboratory report.***

CELL DIVISION

Student _____

Lab Instructor _____

1. INTRODUCTION

Write the type of cell division that matches each phrase.

1. Occurs in prokaryotic cells _____

2. Occurs in haploid eukaryotic cells _____

3. Occurs in diploid eukaryotic cells _____

4. Forms cells with identical genetic composition _____

5. Forms cells with half the chromosome number of the parent cell _____

2. MITOTIC DIVISION

a. Write the term that matches each phrase.

1. Compose a replicated chromosome _____

2. Division of the cytoplasm _____

3. 5% to 10% of the cell cycle _____

4. 90% to 95% of the cell cycle _____

5. Stage of interphase where chromosomes replicate _____

6. Members of a chromosome pair _____

7. Diploid chromosome number in humans _____

8. Haploid chromosome number in humans _____

b. Write the mitotic stage that matches each phrase.

1. Nuclear membrane and nucleolus disappear. _____

2. Spindle is formed. _____

3. Sister chromatids separate. _____

4. Daughter cells are formed. _____

5. Rod-shaped chromosomes are first visible. _____

6. Chromosomes line up on equatorial plane. _____

7. New nuclei are formed. _____

8. Daughter chromosomes migrate to opposite poles of the cell. _____

c. Draw whitefish blastula cells in the stages noted as they appear on your slide. Label pertinent structures.

Interphase	Prophase	Metaphase

Early Anaphase	Late Anaphase	Telophase

d. Draw onion root tip cells in the following stages as they appear on your slide. Label pertinent structures.

Metaphase	Anaphase	Telophase

e. Describe how mitotic cell division in plants differs from that in animal cells. _____

_____

_____

_____

f. In the onion root tip, do the daughter cells occupy the same column of cells as the parent cell? _____

Is it the same for all cells in the root tip? _____

g. Count 100 whitefish blastula cells, and tabulate the number cells in interphase and in each phase of mitosis.

Interphase	Prophase	Metaphase	Anaphase	Telophase
_____%	_____%	_____%	_____%	_____%

3. MEIOTIC DIVISION

a. Considering meiosis in humans, indicate the:

1. number of chromosomes in the parent cell

   _____

2. number of chromatids in daughter cells formed by first division

   _____

3. number of chromosomes in daughter cells formed by first division

   _____

4. number of chromosomes in daughter cells of second division

   _____

5. number of haploid cells formed by meiotic division of parent cell

   _____

b. Diagram the arrangement of the chromosomes in mitosis and meiosis as they would appear in the phases noted here, where the diploid (2n) chromosome number is 6.

Mitosis

Metaphase   **Anaphase**

Meiosis I

Metaphase I   **Anaphase I**

Meiosis II

Metaphase II   **Anaphase II**

c. Indicate the number of chromosomes in human cells formed by:

mitotic cell division _____   meiotic cell division _____

d. Summarize in your own words the process of meiosis without using the names of the phases.

_____

_____

_____

_____

_____

_____

e. Meiotic division is the process leading to the formation of specific cells in animals and plants. Name these cells.

Animals _____   Plants _____

f. Indicate the number of chromosomes in a human.

Egg cell _____   Sperm _____

g. True/False

The zygote, first cell of the next generation, is formed by the union of egg and sperm. It receives:

a random number of chromosomes from each parent   _____

equal numbers of chromosomes from each parent   _____

one member of each chromosome pair from each parent   _____

both members of each chromosome pair from each parent   _____

4. SUMMARY

Match the type of cell division with the statements.

a. Binary fission   b. Mitotic cell division   c. Meiotic cell division

_____ 1. Produces haploid cells from diploid cells.   _____ 4. Produces gametes in animals.

_____ 2. Cell division in prokaryotes.   _____ 5. Occurs in both haploid and diploid cells.

_____ 3. Enables growth in multicellular organisms.   _____ 6. Produces meiospores in plants.

5. MINIPRACTICUM

Examine the microscope setups and complete the following chart:

Microscope No.	Mitotic Stage	Type of Cell	Structure Indicated by Pointer
1	_____	_____	_____
2	_____	_____	_____
3	_____	_____	_____
4	_____	_____	_____
5	_____	_____	_____
6	_____	_____	_____

Part **III**

Diversity
of Organisms

PROKARYOTES AND PROTISTS

OBJECTIVES

After completing the laboratory session, you should be able to:
1. List in order the taxonomic categories used in classifying organisms.
2. Describe the distinguishing characteristics of the taxonomic groups studied.
3. Describe the patterns of reproduction in prokaryotes and protists.
4. Identify representatives of these groups.
5. Define all terms in bold print.

This is the first of several exercises designed to acquaint you with the diversity of organisms. The exercises focus on distinguishing characteristics, evolutionary relationships, and adaptations of the major organismic groups. But, first, let's consider how organisms are classified.

CLASSIFICATION OF ORGANISMS

Organisms are classified according to their degree of similarity, which, in turn, suggests the degree of their evolutionary relationships. The science of classifying organisms is called **taxonomy.** Organisms are classified using specific **taxonomic categories** that range from the largest category containing the greatest number of species (kinds of organisms) to the smallest category containing only one species.

Taxonomy is a dynamic process because new information tends to alter previous ideas of the evolutionary relationships among organisms. There are eight major taxonomic categories used in classifying organisms. In order of decreasing inclusiveness, they are **Domain, Kingdom, Phylum, Class, Order, Family, Genus,** and **Species.** Each domain consists of a number of related kingdoms. Each kingdom consists of a number of related phyla, each phylum consists of a number of related classes, and so on. The classification of humans is shown in Table 10.1 as an example.

The **scientific name** of an organism consists of both genus and species names. For example, the scientific name for humans is *Homo sapiens*. Note that the first letter of the genus name is capitalized, whereas the first letter of the species name is not. The scientific name is always printed in italics or underlined when handwritten.

The use of a domain as the highest taxonomic category has become generally accepted. Its use has resulted from research on the nucleotide sequence in ribosomal RNA in prokaryotes. This research indicates that prokaryotes are composed of two very different groups of organisms and that they should not be lumped within a single group. Thus, prokaryotes are divided into **Archaea,** which have an unusual biochemistry, and **Bacteria,** a better-known traditional

TABLE 10.1	CLASSIFICATION OF THE HUMAN SPECIES, *HOMO SAPIENS*
Taxonomic Category	Classification of Humans
Domain	Eukarya
Kingdom	Animalia
Phylum	Chordata
Class	Mammalia
Order	Primates
Family	Hominidae
Genus	Homo
Species	sapiens

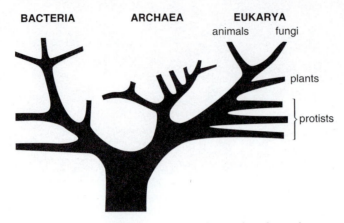

Figure 10.1 Domains Bacteria, Archaea, and Eukarya were the earliest branches in the tree of life. From Audesirk et al., *Life on Earth*, 3rd ed., Fig. 16–4.

group. This places all eukaryotes within the remaining domain, **Eukarya.** The names and numbers of kingdoms within Archaea and Bacteria are currently in a state of flux so we will have to wait to see how this plays out. They may end up with large numbers of kingdoms, but we will not worry about that here. For the time being, the kingdoms in Eukarya are essentially the same as in the prior five-kingdom system of classification.

Figure 10.1 illustrates the three domains of life. The branches within each domain represent kingdoms or potential kingdoms. Most of your study in this book will involve the four kingdoms within the domain Eukarya. A classification of common organisms is listed in Appendix D, page 525.

 Assignment 1

Complete item 1 on Laboratory Report 10 that begins on page 131.

DOMAIN ARCHAEA

Members of the domain Archaea are relatively rare prokaryotes that are distinguished by their unique biochemistry. The composition of the plasma membrane, cell wall, and ribosomal RNA subunits is distinct from both Bacteria and Eukarya. These bacterialike prokaryotes thrive in extremely harsh environments, such as hot springs, deep sea vents, and salt marshes, where few other organisms can survive. The detection of archaeal RNA in seawater from a variety of locations suggests that these organisms may be widespread in the world's oceans. Archaeal RNA nucleotide sequences are more closely related to eukaryan RNA than to bacterial RNA, which suggests an early split into archaeal and bacterial lineages.

DOMAIN BACTERIA

The domain Bacteria consists of the "true bacteria," organisms that have long been recognized as bacteria. There are a multitude of bacterial species. Most are **unicellular,** but some are **colonial** (composed of a group of independently functioning cells). All cells are prokaryotic and lack a nucleus and other membrane-bound organelles. A somewhat rigid cell wall is formed exterior to the plasma membrane. Refer to Figure 3.1 on page 32 for the structure of a bacterial cell.

Bacterial cells exhibit three characteristic shapes: **bacillus** (rodlike shape; plural, bacilli); **coccus** (spherical shape; plural, cocci); and **spirillum** (spiral shape; plural, spirilla). See Figure 10.2 and Plates 1.1–1.3 located between pages 148 and 149. Some species possess **bacterial flagella,** which allow movement. Bacterial flagella are distinctly different from the flagella of eukaryotic cells. Although cell shape and the presence or absence of flagella are used in the identification of bacterial species, a determination of their biochemistry and physiology is required for identification.

Most bacteria are **heterotrophs,** meaning that they must obtain their organic nutrients from the environment. Most heterotrophs are **saprotrophs,** which obtain nutrients by decaying organic matter and dead organisms. Saprotrophs provide a great service by converting organic debris into inorganic chemicals usable by plants. In this way, they clean up the environment and recycle chemicals for repeated use by organisms. A few heterotrophs are **parasites,** which cause disease in other organisms as they obtain organic nutrients from their host's cells and tissues.

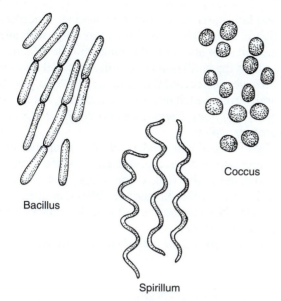

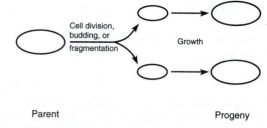

Figure 10.3 Modes of asexual reproduction. Reproduction in bacteria is mainly by binary fission. Some cyanobacteria also reproduce by fragmentation of their colonies.

Figure 10.2 Bacterial cell types.

Some bacteria are **autotrophs,** which can produce their own organic nutrients. **Chemosynthetic bacteria** are able to oxidize inorganic chemicals, such as ammonia or sulfur, and capture the released energy to synthesize organic nutrients. **Photosynthetic bacteria** use light energy to synthesize their organic nutrients. A few photosynthetic bacteria possess *bacteriochlorophyll,* a light-capturing pigment different from chlorophyll that enables them to carry on a nonoxygen-producing form of photosynthesis.

Most photosynthetic bacteria belong to a group known as **cyanobacteria,** which live in freshwater, marine, or moist terrestrial environments. Cyanobacteria lack chloroplasts but possess *chlorophyll a* on thylakoid membranes, which enables them to carry on the same kind of oxygen-producing photosynthesis found in plants. Although cyanobacteria are usually blue-green in color, they also contain other pigments that may cause them to appear yellow, orange, or red. The cells of cyanobacteria differ from other bacterial cells by being larger and either spherical or cylindrical in shape. And, they often have a gelatinous coat external to the cell wall. See Plate 1.4.

Reproduction in Bacteria

Bacteria reproduce by **binary fission,** an asexual form of reproduction. See Figure 9.1 on page 107 and Figure 10.3. This form of cell division enables bacteria to produce massive numbers of individuals within a brief time. When conditions are favorable, a cell may divide to form two cells every 20 min.

Some bacteria are able to form nonreproductive **endospores** within the bacterial cell. An endospore

consists of a bacterial chromosome (DNA) and cytoplasm enveloped by a thick endospore wall. When the bacterial cell dies, the endospore remains viable and is resistant to unfavorable conditions that kill most other bacteria. When conditions become favorable, the endospore grows into a bacterial cell, and reproduction resumes by binary fission. See Plate 1.3.

A sexual process called **conjugation** also occurs in bacteria. Conjugation is the transfer of genetic material (DNA) from one cell to another, and it occurs when adjacent cells are joined by a tiny tube formed by their plasma membranes. The donor DNA combines with, and may replace, homologous segments in the recipient cell's DNA, increasing genetic variability. Subsequently, conjugating bacterial cells separate, and reproduction occurs by binary fission. The transfer of donor DNA and its fusion with DNA of the recipient cell is analogous to the transfer of sperm and fusion of sperm and egg DNAs in higher organisms.

 Assignment 2

Materials

Compound microscope
Antibiotic discs
Medicine droppers
Metric ruler
Microscope slides and cover glasses
Broth cultures of motile bacteria:
   Pseudomonas or *Bacillus cereus*
Cultures of cyanobacteria:
   Gleocapsa
   Oscillatoria
Plates of nutrient agar, sterile
Pour plates of bacteria:
   Escherichia coli B
   Staphylococcus epidermidis

Prepared slides of:
  bacterial types
  Clostridium botulinum with endospores
  Staphylococcus aureus
  Treponema pallidum

1. Examine, without removing the lid, plates of nutrient agar that have been inoculated and incubated for 24 hours. Note the shape, color, and texture of the bacterial colonies. Each colony is composed of thousands to millions of bacteria, which originated from a single cell. Thus, each colony is a *clone* because it is composed of identical cells.

2. Obtain two sterile plates of nutrient agar. Expose one plate by pressing your fingertips or lips against the agar. Replace the lid and use a marking pen to print your initials on the bottom of the plate. Expose the other plate by exposing it to air (with lid off) for 10 min at your work station, outside the building, or in the hallway, restroom, or another location. Replace the lid and print your initials and location of exposure on the bottom of the plate. After exposure, place the plates in an upside-down position in the canister provided. The plates will be incubated for 24 hr, then refrigerated until the next lab session, when you will examine them. Considering all of the plates exposed by your class, which locations had the highest and lowest density of bacteria?

3. Examine a prepared slide of bacterial types at 400×. The slide contains three smears of stained bacteria, and each smear contains bacteria of a specific cell shape. Bacteria are probably smaller than you expect, so focus carefully using reduced illumination. **Complete items 2a–2c on the laboratory report.**

4. Examine the prepared slides of bacteria set up under oil-immersion objectives of demonstration microscopes. Note their small size and shape at 1000×. **Complete item 2d on the laboratory report.**

   a. *Staphylococcus aureus* causes pimples and boils when it gains access to the pores of the skin.
   b. *Treponema pallidum* causes syphilis.
   c. *Clostridium botulinum* produces a potent poison that causes botulism poisoning when it is eaten in contaminated foods. Note the endospore at one end of some cells.

5. Examine the living flagellated bacteria set up in a hanging drop slide under a demonstration microscope. You will not be able to see the flagella, but you can distinguish true motility from Brownian movement.

6. Examine the demonstration agar plates of *Escherichia coli* B and *Staphylococcus epidermidis* on which antibiotic discs were placed prior to incubation. The antibiotic diffuses into the agar and kills susceptible bacteria where its concentration is lethal. The size of the clear no-growth area around each antibiotic disc indicates the effectiveness of the antibiotic against the bacterial species. Measure and record the diameter of the no-growth areas around each disc for each species, but *do not open the plates*. **Complete items 2e–2g on the laboratory report.**

7. Make wet-mount slides of the cyanobacteria *Gleocapsa* and *Oscillatoria* and observe them at 400×. Note the size, color, shape, and arrangement of the cells. Locate the gelatinous sheath, if present. Note the absence of a nucleus and chloroplasts. See Plate 1.4.

8. **Complete item 2 on the laboratory report.**

DOMAIN EUKARYA, KINGDOM PROTISTA

Protists are a heterogeneous group of unicellular or colonial organisms that exhibit animal-like, plantlike, or funguslike characteristics. The major groups probably are not closely related, but are products of evolutionary lines that diverged millions of years ago. Protists and all higher organisms are composed of **eukaryotic cells** that contain a nucleus and other membrane-bound organelles.

Protozoans: Animal-like Protists

These unicellular animal-like protists lack a cell wall and are usually motile. Most are **holotrophs** that engulf food into vacuoles, where it is digested. Some absorb nutrients through the cell membrane, and a few are parasitic. Protozoans occur in most habitats where water is available. Water tends to diffuse into freshwater forms, and some protozoans possess **contractile vacuoles** that repeatedly collect and pump out the excess water to maintain their water balance. Three groups of protozoans are shown in Figure 10.4.

Flagellated Protozoans (Phylum Zoomastigophora)

These primitive protozoans possess one or more **flagella** that provide a means of movement. Food may be engulfed and digested in vacuoles, or nutrients may be absorbed through the cell membrane. Some zooflagellates have established symbiotic relationships with other organisms. *Trypanosoma brucei* is a parasitic form that causes African sleeping sickness. It lives in the blood and nervous system of its vertebrate host and is transmitted by the bite of tsetse flies. *Trichonympha collaris* is a mutualistic symbiont that lives in the gut of termites. It digests the cellulose (wood) to produce simple carbohydrates that can be digested or used by the termite. See Plates 2.1–2.3.

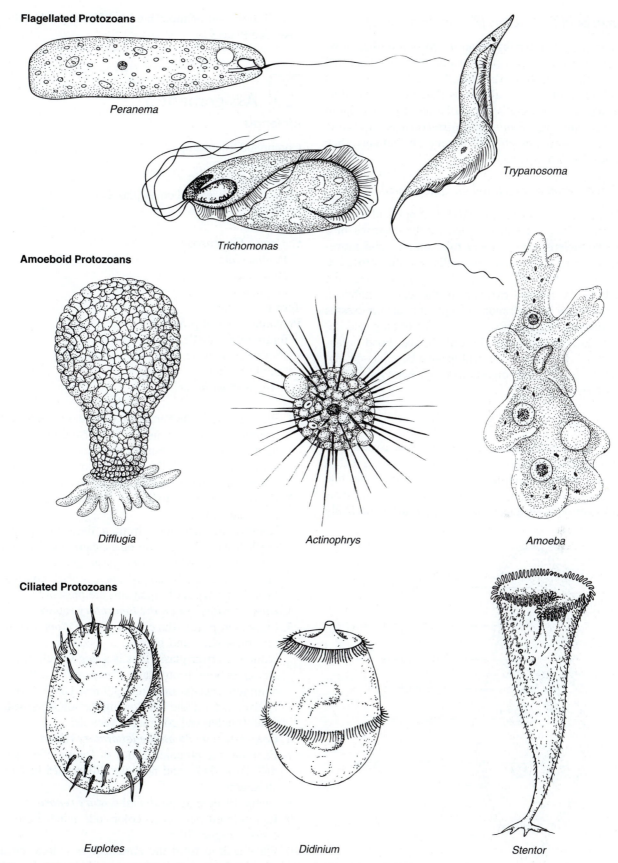

Flagellated Protozoans

Peranema

Trichomonas

Trypanosoma

Amoeboid Protozoans

Difflugia

Actinophrys

Amoeba

Ciliated Protozoans

Euplotes

Didinium

Stentor

Figure 10.4 Examples of flagellated, amoeboid, and ciliated protozoans.

Amoeboid Protozoans (Phylum Sarcodina)

These protists move by means of **pseudopodia,** flowing extensions of the cell. Prey organisms are engulfed and digested in food vacuoles. Some forms secrete a shell for protection. Calcareous shells of foraminiferans and siliceous shells of radiolarians are abundant in ocean sediments. *Entamoeba histolytica* is a parasitic form causing amoebic dysentery in humans. See Plates 2.4, 2.5, 3.1, and 3.2.

Ciliated Protozoans (Phylum Ciliophora)

Ciliates are the most advanced and complex of the protozoans. They are characterized by the presence of a **macronucleus,** one or more **micronuclei,** and movement by means of numerous **cilia,** hairlike processes that cover the cell. A flexible outer covering, the **pellicle,** is located exterior to the cell membrane. *Paramecium* is a common example that also possesses **trichocysts,** tiny dartlike weapons of offense and defense located just under the cell surface. Food organisms are swept down the **oral groove** by cilia and into food vacuoles, where digestion occurs. See Figures 10.5 and 10.6 and Plates 3.3 and 3.4.

Sporozoans (Phylum Sporozoa)

All species lack motility and are internal parasites of animals. *Plasmodium vivax,* a pathogen causing malaria, is a typical example. It is transmitted by the bite of *Anopheles* mosquitoes and infests red blood cells and liver cells of human hosts. Malaria has probably caused the death of more humans than any other disease. See Plate 3.5.

Assignment 3

Materials

Compound microscope
Colored pencils
Medicine droppers
Microscope slides and cover glasses
Protoslo
Toothpicks
Cultures of protozoa:
   Paramecium
   Pelomyxa
   Peranema
Termites, living
Prepared slides of protozoa:
   foraminifera shells
   Plasmodium vivax
   radiolaria shells
   Trypanosoma brucei

1. Place a drop of the *Peranema* culture on a slide and examine the specimens at 100×. Note the use of the flagellum in movement. Use a toothpick to mix some Protoslo with the drop on your slide, add a cover glass, and examine cellular detail at 400×.
2. Examine a prepared slide of *Trypanosoma brucei* in human blood.
3. Remove the abdomen from a live termite and crush it in a drop of water on a clean slide. Discard the abdomen, add a cover glass, and locate the large multiflagellated *Trichonympha collaris*. What other types of organisms are present?
4. ***Complete item 3a on the laboratory report.***
5. Place a drop from the *bottom* of the *Pelomyxa* culture on a slide and observe a specimen at 100× without a cover glass. This is a large, multinucleate amoeboid protozoan. Compare it with the common *Amoeba* in Figure 3.9 that you observed in Exercise 3. Note the flowing motion, food vacuoles, nuclei, and contractile vacuoles.
6. ***Complete item 3b on the laboratory report.***
7. Examine the prepared slides of foraminifera and radiolaria shells and note the differences in their structures.
8. ***Complete item 3c on the laboratory report.***
9. Use colored pencils to color-code labeled structures in Figure 10.5.
10. Place a drop from the *Paramecium* culture on a slide and observe it at 40× or 100× without a cover glass. Note the movement of *Paramecium*. What happens when it bumps into an object?

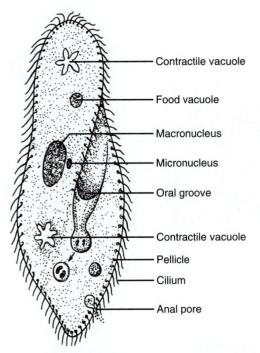

Figure 10.5 *Paramecium caudatum.*

Contractile vacuole

Food vacuole

Macronucleus

Micronucleus

Oral groove

Contractile vacuole

Pellicle

Cilium

Anal pore

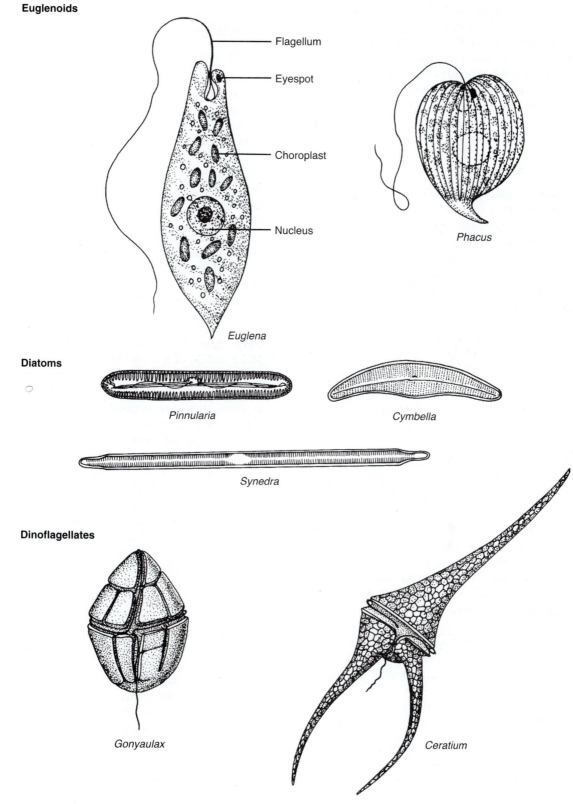

Euglenoids

Flagellum

Eyespot

Choroplast

Nucleus

Euglena

Phacus

Diatoms

Pinnularia

Cymbella

Synedra

Dinoflagellates

Gonyaulax

Ceratium

Figure 10.6 Examples of plantlike protists.

11. Add a drop of Protoslo to your slide, and use a toothpick to mix it with the drop of *Paramecium* culture. Add a cover glass. Locate an entrapped specimen and examine it carefully at 100× and 400×. Locate the structures shown in Figure 10.5.
12. ***Complete item 3 on the laboratory report.***
13. Examine the prepared slide of *Plasmodium vivax* in red blood cells that is set up under a demonstration microscope.

Plantlike Protists

Plantlike protists are collectively known as **algae** (singular, **alga**). In this section, you will study three microscopic and unicellular phyla of algae, which are collectively known as *phytoplankton*. See Figure 10.6. Phytoplankton flourish in marine and freshwater habitats. Nearly 70% of Earth's photosynthesis occurs in marine phytoplankton. The multicellular algae will be considered in Exercise 11.

Euglenoids (Phylum Euglenophyta)

These unicellular protists possess both plantlike and animal-like characteristics. They have **chlorophyll a** and **b** in chloroplasts, a flagellum for movement, and an "eyespot" (stigma) that detects light intensity. A cell wall is absent. See Figure 10.6 and Plates 4.1 and 4.2.

Dinoflagellates (Phylum Dinoflagellata)

Most of these unicellular "fire algae" are marine and photosynthetic with **chlorophyll a** and **c** in their chloroplasts. Some are holotrophic, and a few can live as either holotrophs or photosynthetic autotrophs, depending on the circumstances. Most have a cell wall of cellulose, and all forms have two flagella. One flagellum lies in a groove around the equator of the cell, and the other hangs free from the end of the cell. When nutrients are abundant, certain marine species, such as *Gonyaulax,* reproduce in such enormous numbers that the water turns a reddish color, a condition known as a red tide. Toxins produced by these forms may cause massive fish kills and make shellfish unfit for human consumption. See Plates 4.3 and 4.4.

Diatoms (Phylum Bacillariophyta)

These protists are unicellular and have a cell wall of silica, a natural glass. The cell wall consists of two halves fit together like the top and bottom of a box. Chlorophyll a and c are found in their chloroplasts. When diatoms die, the siliceous walls settle to the bottom. In some areas, they have formed massive deposits of diatomaceous earth that are mined and used commercially in a variety of products such as fine abrasive cleaners, toothpaste, filters, and insulation. Much of the atmospheric oxygen is produced by photosynthesis in marine diatoms. See Plate 4.5.

 ## Assignment 4

Materials

Compound microscope
Colored pencils
Construction paper, black
Medicine droppers
Microscope slides and cover glasses
Petri dish
Protoslo
Toothpicks
Cultures of:
    Euglena
    freshwater dinoflagellates
Diatomaceous earth
Prepared slides of:
    diatoms
    representative dinoflagellates

1. Use colored pencils to color-code the chloroplasts and nucleus of *Euglena* in Figure 10.6.
2. Make a slide of a drop from the *Euglena* culture and examine it at 100× and 400×. Note the size, color, and motility, and the absence of a cell wall. Is more than one method of motility evident? Which end is anterior?
3. Observe the distribution of *Euglena* in the partially shaded petri dish on the demonstration table. *Do not move the dish.* Remove the cover to make your observation, and immediately return the cover to its *exact prior position.*
4. ***Complete items 4a–4c on the laboratory report.***
5. Examine a prepared slide of representative dinoflagellates at 400×. Note their small size, the positions of the two flagella, and the sculptured cell wall.
6. Make a slide of a drop from the mixed dinoflagellate culture and examine it at 400×. Note the spinning movement of the organisms caused by the positions of the flagella.
7. ***Complete item 4d on the laboratory report.***
8. Examine a prepared slide of diatoms. Note the symmetry and pattern of the cell walls.
9. Examine the sample of diatomaceous earth, noting its fine texture. Make a wet-mount slide of a small amount of the dustlike particles, and observe the sample with a microscope. Does your sample contain more than one species?
10. ***Complete item 4e on the laboratory report.***

Slime Molds: Funguslike Protists

The feeding stages of slime molds resemble amoeboid protozoans, but they reproduce asexually by forming sporangia that produce spores. This latter funguslike characteristic explains why they previously were classified as fungi.

Cellular Slime Molds (Phylum Acrasiomycota)

Members of this group live as single-celled amoeboid organisms that engulf bacteria in leaf litter and soil. Poor environmental conditions cause the cells to congregate in what is known as the "swarming stage," which results in the formation of a sporangium containing spores. *Dictyostelium* is an example of this group.

Plasmodial Slime Molds (Phylum Myxomycota)

Feeding stages of plasmodial slime molds consist of strands of protoplasm streaming along in amoeboid fashion and engulfing bacteria on leaf litter, rotting wood, and the like. The large plasmodium is multinucleate. Unfavorable conditions stimulate the plasmodium to migrate and ultimately to form a sporangium with spores. *Physarum* is an example of a plasmodial slime mold. See Plates 5.1 and 5.2.

Assignment 5

Materials

Stereo microscope
Cultures of slime molds:
   Dictyostelium
   Physarum

1. Examine the macroscopic and microscopic demonstrations of cellular and plasmodial slime molds. Note the organization of the nonreproductive stages and the sporangia formed during the reproductive stages.
2. ***Complete item 5 on the laboratory report.***

REPRODUCTION IN PROTISTS

Mitotic cell division is the most common form of asexual reproduction in protists. Some forms also reproduce sexually. Two basic patterns of sexual reproduction occur, depending on the diploid or haploid nature of the organism. Variations of these patterns occur in different species. See Figure 10.7.

1. Haploid (n) protists may become gametes or produce gametes (n) by mitosis. Fusion of the gametes forms a zygote (2n) that divides by meiosis to yield four haploid individuals.
2. Diploid (2n) protists form four haploid (n) gametes by meiosis. Subsequently, the fusion of gametes yields a diploid zygote that divides mitotically to form additional diploid individuals. Note that only the gametes are haploid in this pattern.

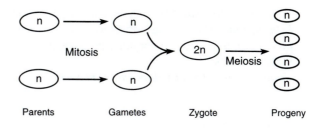

A Haplontic life cycle

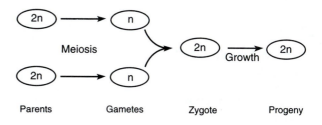

B Diplontic life cycle

Figure 10.7 Two patterns of sexual reproduction occurring in protists. **A.** Haplontic life cycle, where the adults are haploid. **B.** Diplontic life cycle, where the adults are diploid.

Reproduction in *Paramecium*

Paramecium is a diploid protozoan that reproduces asexually by mitotic division and also participates in conjugation occasionally. Although all members of a species appear to be identical, opposite mating types exist, and they join together in conjugation when they encounter each other. Their micronuclei divide by meiosis, and each conjugant transfers a haploid micronucleus to the other, where it combines with a haploid micronucleus in the receiving conjugant to re-form a diploid micronucleus. Then, the conjugants separate and reproduction occurs by mitotic cell division.

Assignment 6

Materials

Prepared slides of:
   Paramecium dividing
   Paramecium conjugating
Cultures of *Paramecium bursaria* mating types

1. Examine a prepared slide of mitotic division in *Paramecium* with your microscope. Is the plane of division random?
2. Examine a prepared slide of *Paramecium* in conjugation. Do the pairs join at a particular site or randomly?

3. Your instructor has mixed mating types of living *P. bursarium.* Prepare and observe a slide of the mixed cultures. Are the conjugating pairs motile? Is the union brief?
4. ***Complete item 6 on the laboratory report.***

APPLICATIONS

In this section, you can use the knowledge you have gained.

 Assignment 7

Materials

Compound microscope
Medicine droppers
Microscope slides and cover glasses
Pond water

1. Make and examine slides of pond water, and observe the monerans and protists that are present. Use reduced illumination for best viewing. ***Draw several observed specimens in item 7a on the laboratory report.*** Make your drawings large enough to show the major features of the organisms.
2. Review what you have learned in the laboratory session, and ***complete item 7b on the laboratory report.***
3. Your instructor has set up several "unknown specimens" for you to identify under numbered demonstration microscopes. ***Match the specimens with the correct responses in item 7c on the laboratory report by placing the numbers of the unknowns in the spaces provided.***

PROKARYOTES
AND PROTISTS

Student _____

Lab Instructor _____

1. INTRODUCTION

Write the term that matches the phrase.

a. The two domains containing prokaryotes.   _____

b. Domain of protists in a three-domain system.   _____

c. Smallest taxonomic category.   _____

d. Compose the scientific name of an organism.   _____

2. DOMAIN BACTERIA

a. Write the term that matches the phrase.

   1. Type of cells composing bacteria   _____

   2. Obtain energy by decomposing organic matter   _____

   3. Obtain energy by oxidizing inorganic matter   _____

   4. Photosynthetic pigment in cyanobacteria   _____

   5. Means of reproduction in bacteria   _____

   6. Sexual process in bacteria   _____

b. Based on sampling by your class, do bacteria seem to be limited or widespread in their distribution? _____

   Explain. _____

c. Draw a few cells of each bacterial shape from your bacterial type slide.

          Bacillus   **Coccus**   **Spirillum**

d. Draw the shape of each of the following and indicate the disease caused by each:

          Clostridium botulinum   ***Staphylococcus aureus***   ***Treponema pallidum***

e. Record the diameter of the no-growth areas.

Antibiotic		Diameter of No-Growth Areas (mm)	
		S. epidermidis	E. coli B
Ampicillin	10 μg	_____	_____
Erythromycin	10 μg	_____	_____
Cephalothin	10 μg	_____	_____
Chloromycetin	10 μg	_____	_____

f. Which antibiotic was most effective against both organisms? _____

g. Which species was more susceptible to the antibiotics? _____

h. What geometric shape best describes the cells of these organisms? _____

  Gleocapsa _____ *Oscillatoria* _____

i. Draw a few cells of each organism at 400× to show their shape and arrangement. Label the gelatinous sheath, if present.

  Gleocapsa   **Oscillatoria**

3. PROTISTA: PROTOZOANS

a. Make a drawing of these zooflagellates from your slides, and label the flagellum and nucleus. Draw a few red blood cells with *T. brucei* to show their relative size.

  Peranema   **Trypanosoma**   **Trichonympha**

b. Use a series of diagrams to show the formation of a pseudopod by *Pelomyxa*.

c. Draw a few "shells" of foraminifera and radiolaria from your slides.

  Foraminifera   **Radiolaria**

d. Is the anterior end of *Paramecium* rounded or pointed? _____

e. Describe the movement of *Paramecium*. _____

_____

f. Do the cilia beat in unison or in small groups? _____

g. What happens when *Paramecium* bumps into an object? _____

_____

h. Describe the function of each:

Oral groove _____

Contractile vacuoles _____

Food vacuoles _____

4. PLANTLIKE PROTISTS

a. List the animal-like and plantlike characteristics of *Euglena*.

Animal-like _____

Plantlike _____

b. Explain the distribution of *Euglena* in the partially shaded dish. _____ _____

_____

_____

c. How does this behavior benefit *Euglena*? _____

_____

d. Are dinoflagellates larger or smaller than *Euglena*? _____

In living dinoflagellates, can you distinguish, at 400×, the:

flagella?_____ cellular contents?_____

What is the color of the living dinoflagellates? _____

What is their form of nutrition? _____

e. Draw the siliceous cell walls of a few diatoms from these sources.

Prepared Slides   **Diatomaceous Earth**

5. PROTISTA: SLIME MOLDS

Make drawings of your observations.

Dictyostelium   **Physarum**

6. REPRODUCTION IN PROTISTS

Paramecium

a. Is the plane of cell division along the longitudinal axis?_____

b. During conjugation, do the conjugants join together at a particular site? _____

_____ Explain. _____

c. After the conjugants separate, what type of cell division produces the progeny from each parent? _____

d. Why is conjugation considered to be a sexual process? _____

_____

Slime Molds

e. Slime molds reproduce by forming spores that grow into the adult organism.

Is this a form of sexual or asexual reproduction? _____ Explain.

_____

Are the spores formed by mitosis or meiosis? _____

7. APPLICATIONS

a. Draw six to eight bacteria and protists observed in slides of pond water. Indicate the magnification used and the division or phylum of each protist, and label the nucleus, chloroplasts, cell wall, cilia, flagella, and vacuoles, if present.

b. Matching

1. Bacteria	2. Cyanobacteria	3. Flagellated protozoans	4. Amoeboid protozoans
5. Ciliates	6. Sporozoans	7. Euglenoids	8. Dinoflagellates
9. Diatoms	10. Slime molds		

_____ Prokaryotic cells   _____ Chloroplasts

_____ Cell wall absent   _____ Silica cell walls

_____ Form sporangia   _____ Chlorophylls a and b

_____ Chlorophyll a only   _____ At least some are parasites

_____ Animal-like protists   _____ Funguslike protists

_____ Most are saprotrophs   _____ Plantlike protists

_____ Nonmotile parasites   _____ Most complex protozoans

c. Unknowns

Write the numbers of the "unknown" specimens in the correct spaces.

_____ Bacteria   _____ Sporozoans

_____ Cyanobacteria   _____ Euglenoids

_____ Flagellated protozoans   _____ Dinoflagellates

_____ Amoeboid protozoans   _____ Diatoms

_____ Ciliated protozoans   _____ Slime molds

GREEN, BROWN, AND RED ALGAE

OBJECTIVES

After completing the laboratory session, you should be able to:
1. Describe the characteristics of the algal groups studied.
2. Identify representatives of the three lines of algae.
3. Identify and state the functions of the parts of the algae studied.
4. Describe the common reproductive patterns in algae.
5. Define all terms in bold print.

Algae are a diverse group of photosynthetic autotrophs that are primarily confined to freshwater or marine habitats. Algae range in size from microscopic unicellular and colonial forms to large multicellular species that may be several meters in length. Some forms grow attached to rocks or other hard objects, some float freely in the water, and a few unicellular or colonial forms possess flagella that enable motility. Algal cells possess chloroplasts containing chlorophyll and cell walls.

Algae reproduce both asexually and sexually, and their sexual cycles range from simple to complex. Green, brown, and red algae compose three evolutionary lines based on the photosynthetic pigments in their chloroplasts. See Table 11.1. Green, brown, and red algae are quite different from each other, and there is a lot of variation within each group.

Biologists consider these algae to be members of the kingdom Protista, primarily because of the similarity of simpler unicellular and colonial algae to other protists.

suggests that green algae are ancestral to higher plants. See Plates 7.1–7.5 and 8.1.

Figure 11.1 shows the microscopic unicellular alga *Chlamydomonas*. It has a cup-shaped chloroplast and stores starch granules in a pyrenoid. An eyespot detects light, and two flagella provide motility, enabling orientation in sufficient light for photosynthesis. Figure 11.2 illustrates *Pandorina,* a colonial alga composed of several cells that resembles *Chlamydomonas. Volvox* is a much larger colonial alga that is composed of hundreds or thousands of *Chlamydomonas*-like cells.

Spirogyra is a filamentous colonial green alga. A filament results when there is a single plane of cell division that causes new cells to be joined end to end. *Spirogyra* is named after the spiral-shaped chloroplast in its cells, as shown in Figure 11.3. Also note in Figure 11.3 the centrally located nucleus and the fluid-filled vacuole that occupies most of the space within a cell.

GREEN ALGAE (PHYLUM CHLOROPHYTA)

Green algae may be unicellular, colonial, or multicellular. Most species occur in freshwater or marine habitats, but a few occur in moist areas on land. The presence of (1) **chlorophylls a and b** in chloroplasts, (2) cellulose cell walls, (3) starch as the nutrient storage form, and (4) whiplash flagella on motile cells

TABLE 11.1	EVOLUTIONARY LINES OF ALGAE
Evolutionary Line	*Characteristic Pigments*
Green	Chlorophylls a and b Carotenoids
Brown	Chlorophylls a and c Fucoxanthin
Red	Chlorophylls a and d Phycobilins

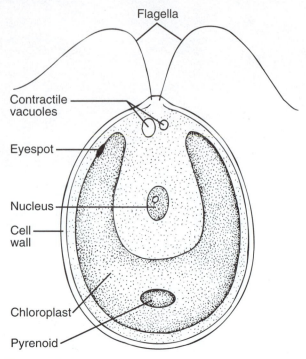

Figure 11.1 *Chlamydomonas.*

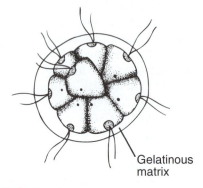

Figure 11.2 *Pandorina.*

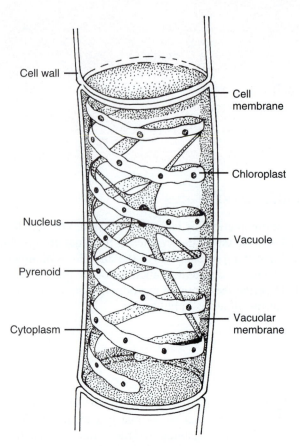

Figure 11.3 *Spirogyra* cell.

of the water. Flagellated cells have one whiplash and one tinsel flagellum. See Plates 8.2, 8.3, and 8.4.

Figure 11.4 illustrates three different species of brown algae, each with a different body form. Note that each species has a well-developed holdfast and that the stipes and blades are distinctly different. Many brown algae are quite large, but the giant kelp *Macrocystis* is the largest. It may be over 100 feet in length.

BROWN ALGAE (PHYLUM PHAEOPHYTA)

These multicellular algae are almost exclusively marine and are often called seaweeds because of their abundance along rocky seacoasts. Their brownish color results from the presence of the pigment **fucoxanthin** in addition to **chlorophylls a and c** in the chloroplasts. The typical structure of a brown alga includes a **holdfast,** which anchors the alga to a rock, a **stipe** (stalk), and **blades,** which are leaflike structures. See Figure 11.4. All parts of the alga carry on photosynthesis, but the blades are the most important photosynthetic organs. Blades of large brown algae often have air-filled bladders associated with them to hold them near the surface

RED ALGAE (PHYLUM RHODOPHYTA)

Most red algae are marine and occur at greater depths than brown algae. They have a delicate body structure, probably because they are not subjected to strong wave action. As with brown algae, they are attached to rocks by holdfasts. See Plate 8.5. The presence of red pigments, **phycobilins,** in addition to **chlorophylls a and d,** allows them to absorb the deeper-penetrating wavelengths of light for photosynthesis. Flagellated cells are absent.

In Figure 11.5, *Polysiphonia* is a deep-water red alga with a soft and delicate body form. *Corallina* is a

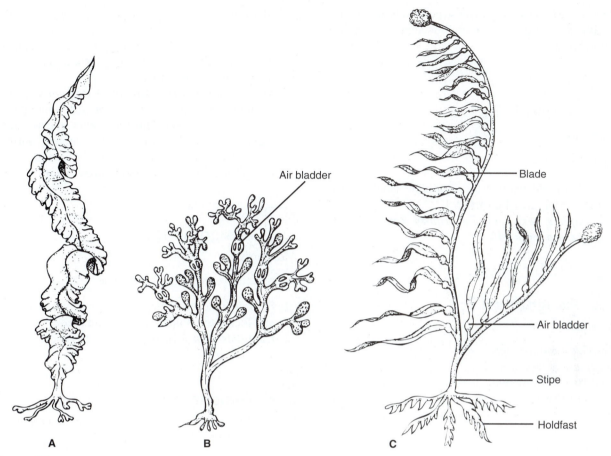

Figure 11.4 Brown algae. The illustrations are not drawn to scale. *Laminaria* (**A**) and *Fucus* (**B**) are much smaller than *Macrocystis* (**C**), a giant kelp.

Polysiphonia

Coraillina

Figure 11.5 Red algae.

robust shallow-water alga whose cell walls are encrusted with calcium deposits. Coralline algae are common on coral reefs.

Assignment 1
Materials

Colored pencils

1. Use colored pencils to color-code labeled structures in Figures 11.1 to 11.4.
2. **Complete item 1 on Laboratory Report 11 that begins on page 143.**

Assignment 2
Materials

Compound microscope
Construction paper, black
Dissecting instruments
Medicine droppers
Microscope slides and cover glasses
Petri dish
Cultures of:
   Chlamydomonas (or *Carteria*)
   Pandorina
   Oedogonium
   Spirogyra
   Volvox
Representative green algae

1. Make a slide of a drop from the *Chlamydomonas* culture and examine it at 400×. Note the color, movement, and size of the organisms. Compare your specimens with Figure 11.1. Note the "eyespot" (stigma), which is sensitive to light intensity, and the two flagella, which enable motility.
2. Observe the distribution of *Chlamydomonas* in a partially shaded petri dish. Remove the cover to make your observation; then, replace it exactly as you found it. How do you explain the distribution? Do you think *Chlamydomonas* prefers bright or dim light? Design and perform an experiment to answer this question. **Complete items 2a–d on the laboratory report.**
3. Make and observe slides of the *Pandorina* and *Volvox* cultures. See Figure 11.2. Note the similarity of the individual cells of these colonial algae to *Chlamydomonas*. Colonial forms are believed to have originated from unicellular organisms that remained attached after cell division. **Complete items 2e and 2f on the laboratory report.**

4. Make and examine a slide of *Spirogyra,* a filamentous green alga. Note the spiral chloroplast, nucleus, and cytoplasm in the cells. Compare your specimen to Figure 11.4. Pyrenoids are sites of starch accumulation.
5. Make and examine a slide of *Oedogonium,* a filamentous green alga, and note the different types of cells in the filament. This represents an early stage leading to the division of labor among various types of cells found in multicellular forms.
6. **Complete item 2 on the laboratory report.**

Assignment 3
Materials

Representative brown and red algae

1. Examine the specimens of brown and red algae and locate the holdfast, stipe, and blade.
2. **Complete item 3 on the laboratory report.**

Assignment 4

Your instructor has set up green, brown, and red algae as unknowns on the demonstration table.

1. Examine and identify the "unknown" algae as green, brown, or red algae.
2. **Complete item 4 on the laboratory report.**

REPRODUCTION IN GREEN ALGAE

Reproduction is quite variable between, and even among, the group of algae. The common reproductive patterns are shown diagrammatically in Figure 11.6. Asexual reproduction includes mitotic cell division in unicellular forms, fragmentation and cell division in colonial forms, and the formation of mitospores in **asexual sporulative reproduction** in colonial and multicellular species. Algae reproduce sexually by **sexual sporulative reproduction,** which exists in two basic forms: zygotic sexual reproduction and sporic sexual reproduction.

The most common pattern of sexual sporulative reproduction is **zygotic sexual reproduction.** In this pattern, the adult alga (n) produces gametes by mitotic cell division. The union of two gametes (n) forms a diploid zygote (2n), the only part of the sexual cycle that is diploid. The zygote undergoes meiotic division to produce meiospores that germinate to form new haploid gametophytes.

Only a few advanced species exhibit **sporic sexual reproduction.** It is characterized by the presence of

Asexual Sporulative Reproduction

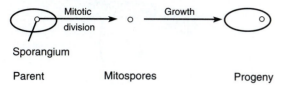

Sporangium

Parent   Mitospores   Progeny

Sexual Sporulative Reproduction
Type 1. Zygotic Sexual Reproduction

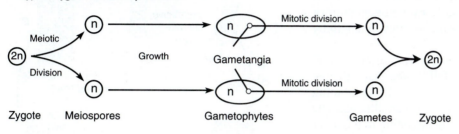

Zygote   Meiospores   Gametophytes   Gametes   Zygote

Type 2. Sporic Sexual Reproduction

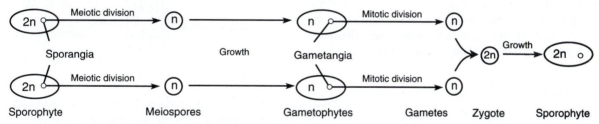

Sporophyte   Meiospores   Gametophytes   Gametes   Zygote   Sporophyte

Figure 11.6 Reproductive patterns in algae. A gametophyte is a gamete-forming adult, and a sporophyte is a meiospore-forming adult. Gametangia and sporangia are specialized structures forming gametes and spores, respectively.

two adults in the life cycle: a **gametophyte** (a gamete-producing plant) and a **sporophyte** (a spore-producing plant). Haploid gametophytes (n) produce gametes (n) by mitosis. Union of sperm and egg forms a zygote (2n) that grows into a diploid sporophyte. The sporophyte (2n) produces meiospores (n) by meiosis. A meiospore germinates to become a haploid gametophyte. Because the gametophyte generation alternates with the sporophyte generation in the life cycle, this form of reproduction is often called **alternation of generations.** See Figures 11.6 and 11.7.

The gametophyte and sporophyte may be similar in appearance in some species but not in others. If they are dissimilar, the larger and longer-lived generation is said to be dominant. In some algae, the gametophyte is dominant, as in *Chara,* a freshwater green alga. In other species, the sporophyte may be dominant, as in *Macrocystis,* a giant marine kelp. In some species, each gametophyte produces both sperm and

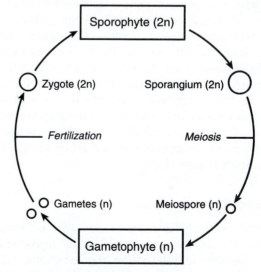

Figure 11.7 Alternation of generations.

eggs, but in others there are separate male and female gametophytes.

We will consider representative reproductive patterns only in green algae because green algae are believed to be the ancestors of land plants. The examples illustrate a trend of increasing complexity in life cycles.

Reproduction in *Spirogyra*

Spirogyra is a haploid, filamentous, colonial, freshwater alga. It reproduces asexually by fragmentation of the filaments, and the new filaments grow in length by mitotic cell division.

Spirogyra reproduces sexually by zygotic sexual reproduction, usually in the fall of the year. Zygote formation is by conjugation because the combining gametes (n) are morphologically similar, although one gamete exhibits a sperm trait by migrating to the other gamete. The zygote (2n) usually forms a thick cell wall to become a zygospore that resists unfavorable winter conditions. In favorable conditions, the zygote nucleus (2n) undergoes meiosis. Three haploid nuclei disintegrate, but the remaining nucleus is functional in the resulting haploid cell, which forms a new filament by mitotic cell division. Study Figure 11.8.

Reproduction in *Oedogonium*

Oedogonium is a haploid, filamentous, colonial, freshwater alga that grows attached to the substrate. Each filament possesses specialized reproductive cells, which is a trend toward a multicellular condition. It reproduces by asexual sporulative reproduction and zygotic sporulative reproduction.

In asexual reproduction, specialized cells called **sporangia** form and release motile **zoospores** (n). A zoospore germinates to form a new haploid filament. In sexual reproduction, specialized cells produce the gametes: sperm and eggs. **Antheridia** produce and release several motile sperm (n), and each **oogonium** produces one egg (n). A sperm is much smaller than a large nonmotile egg. Sperm and egg unite in fertilization to form a zygote (2n), which produces four zoospores by meiotic division. Each zoospore germinates to form a new haploid cell, which attaches to the substrate and produces a new filament by mitotic division. Study Figure 11.9.

Reproduction in *Ulva*

Ulva, a multicellular marine green alga, exhibits an additional advancement: sporic sexual reproduction, which includes an **alternation of generations.** This is a significant advancement, because this reproductive

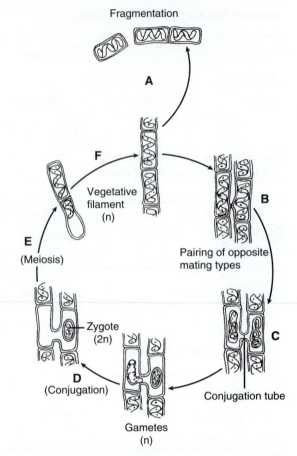

Figure 11.8 Life cycle of *Spirogyra*. **A.** Asexual reproduction is by fragmentation. **B.** In sexual reproduction, filaments of opposite mating types pair. **C.** A conjugation tube forms between paired cells, and the protoplasm in each cell becomes a gamete. **D.** One gamete migrates through the conjugation tube to fuse with the other gamete, forming a zygote (2n). **E.** Meiosis of the zygote nucleus forms a haploid cell. **F.** A new filament (n) is formed by mitotic division.

pattern is characteristic of all plants taxonomically higher than algae.

In *Ulva*, the gametophyte generation consists of separate female and male gametophytes, and the gametophytes have a similar appearance to each other and to the sporophyte. The female and male gametophytes (n) produce motile eggs (n) and sperm (n), respectively, by mitotic division. After fertilization, the zygote (2n) develops into a sporophyte (2n). The sporophyte produces motile zoospores by meiotic division. The zoospores (n) germinate and grow into new female and male gametophytes. Study Figure 11.10.

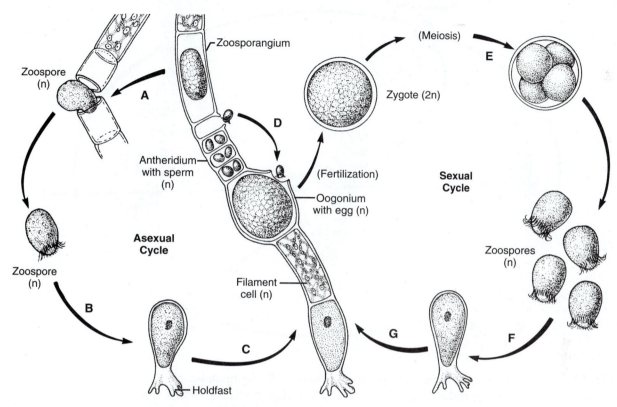

Figure 11.9 Life cycle of *Oedogonium*. **A.** In the asexual cycle, a sporangium forms and releases a motile zoospore (n). **B.** The zoospore germinates to form a haploid cell. **C.** A new filament is formed by mitotic cell division. **D.** In the sexual cycle, a motile sperm (n), released from an antheridium, enters an oogonium and fertilizes the egg (n), forming a zygote (2n). **E.** The zygote undergoes meiotic division, forming four zoospores (n). **F.** A zoopore germinates to form a haploid cell. **G.** A new filament is formed by mitotic division.

Assignment 5

Materials

Colored pencils
Compound microscope
Microscope slides and cover glasses
Culture of *Oedogonium*
Ulva, fresh or preserved
Prepared slides of:
    Spirogyra conjugation
    Oedogonium with antheridia, oogonia, and zoosporangia

1. Study Figures 11.6 to 11.10. Note the characteristics of each life cycle and the evolutionary trends. Color-code diploid stages green.

2. Examine a prepared slide of *Spirogyra* in conjugation. Locate gametes and zygotes. Are the motile gametes on the same filament? ***Complete items 5a–5e on the laboratory report.***

3. Examine prepared slides of *Oedogonium* zoospores, antheridia, and oogonia. Compare your observations to Figure 11.8.

4. Prepare and observe a wet-mount slide of *Oedogonium*. Note the structure of the filament and cells. Are antheridia, oogonia, or sporangia present? ***Complete items 5f–5k on the laboratory report.***

5. Examine the specimens of *Ulva*.

6. ***Complete item 5 on the laboratory report.***

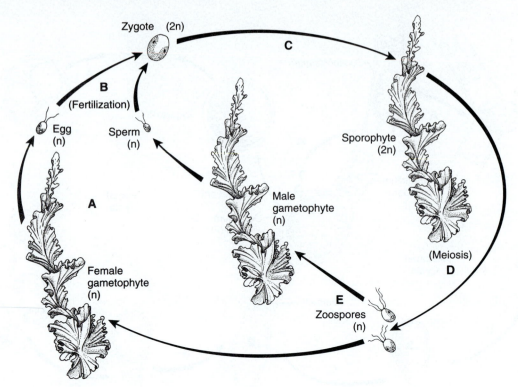

Figure 11.10 *Ulva* reproduces by sporic sexual reproduction. **A.** Separate female and male gametophytes (n) produce motile eggs and sperm, respectively, by mitotic division. **B.** Sperm (n) and egg (n) unite in fertilization, forming a diploid zygote. **C.** The zygote (2n) grows to become a sporophyte (2n). **D.** The sporophyte produces motile zoospores (n) by meiotic division. **E.** A zoospore germinates to form either a male or female gametophyte.

GREEN, BROWN, AND RED ALGAE

Student _____

Lab Instructor _____

1. INTRODUCTION

Write the term that matches the phrase.

1. Type of nutrition in algae _____

2. Material forming the cell wall _____

3. Organelle containing photosynthetic pigments _____

4. Chlorophylls in red algae _____

5. Chlorophylls in green algae _____

6. Chlorophylls in brown algae _____

7. Brown pigment in brown algae _____

8. Red pigment in red algae _____

2. GREEN ALGAE

a. Describe the distribution of *Chlamydomonas* in the partially shaded petri dish. _____
_____

b. Describe the involvement of the flagella and eyespot in this behavior. _____
_____
_____

c. Explain the adaptive advantage of this behavior. _____
_____
_____

d. Do you think *Chlamydomonas* prefers bright or dim light? _____

Perform an experiment to test your hypothesis. Describe your experiment. _____
_____
_____
_____
_____
_____
_____

Describe your results. _____

Do they support your hypothesis? _____ State a conclusion from your results. _____

_____

_____

e. Describe the movement of *Pandorina* and *Volvox*. _____

_____

Which genus is larger? _____

f. What enables this movement? _____ Does

the movement seem coordinated or uncontrolled? _____

Explain your response. _____

_____

g. Draw a few adjacent cells from (1) a *Spirogyra* filament and (2) an *Oedogonium* filament to show the similarities
or differences of the cells in each filament. Label pertinent parts.

Spirogyra	***Oedogonium***

h. What common characteristics of green algae and multicellular land plants suggest that green algae are ancestors
of land plants? _____

_____

_____

3. BROWN AND RED ALGAE

a. Name and give the function of the major parts of a brown alga:

A "rootlike" structure _____

A "stemlike" structure _____

A "leaflike" structure _____

b. What part of a brown alga is most important in photosynthesis? _____

c. What is the function of air-filled sacs in large kelp? _____

_____

d. Feel the brown algae. Is the texture rough and sticky or smooth and slippery? _____

e. Considering their habitat, what advantage does this surface texture and a robust, but flexible, body provide for
brown algae? _____

_____

_____

f. Considering their habitat, explain the delicate body of most red algae as opposed to the robust body of brown algae. _____

_____

g. What is the function of phycobilins in red algae? _____

_____

4. UNKNOWNS

Write the numbers of the "unknown" specimens in the correct spaces.

_____ Green algae   _____ Brown algae   _____ Red algae

5. REPRODUCTION IN GREEN ALGAE

Spirogyra

a. The growth of a filament occurs by _____

and new filaments are formed asexually by _____

b. Are the gametes alike or different in size or shape? _____

c. Describe any pattern in the location of the motile gametes among conjugating filaments. _____

Explain such a pattern, if any. _____

_____

d. How do you distinguish a zygospore from a gamete on your slide? _____

_____

e. What is formed by meiosis in the life cycle? _____

Oedogonium

f. Zoosporangia produce motile haploid _____ by _____

g. From your slides, what are the observable physical differences of sperm and egg? _____

h. What is formed by meiosis in the life cycle? _____

i. Which zoospores have the greatest genetic variability: those formed by mitosis or those formed by meiosis? ____

j. What is the advantage of motile zoospores? _____

k. Draw the following from your slides.

Antheridium with Sperm   **Oogonium with an Egg**

Ulva

l. Describe any distinctive differences in the appearance of the gametophytes and sporophytes. _____

_____

m. In which adult form does meiosis occur? _____

n. What process forms the gametes? _____

o. Are asexual mitospores present in the life cycle? _____

p. What is the advantage of sexual reproduction? _____

_____

q. Does a gametophyte or sporophyte have the greater genetic variability? _____ Explain.

_____

FUNGI

OBJECTIVES

After completing the laboratory session, you should be able to:
1. Describe the distinguishing characteristics of the organismic groups studied.
2. Identify representatives of fungi and lichens.
3. Describe the common reproductive patterns in fungi.
4. Define all terms in bold print.

The **kingdom Fungi** contains a large and diverse group of heterotrophic organisms that occur in freshwater, marine, and terrestrial habitats. Terrestrial fungi reproduce by **spores,** dormant reproductive cells that are dispersed by wind and that germinate to form a new fungus when conditions are favorable. Most fungi are multicellular; only a few are unicellular. Fungi are either **saprotrophs** or **parasites.** Most species are saprotrophs and play a beneficial role in decomposing nonliving organic matter, but some saprotrophs cause serious damage to stored food products. Rusts and mildews are important plant parasites. Ringworm and athlete's foot are common human ailments caused by parasitic fungi.

The vegetative (nonreproductive) body of a multicellular fungus is called a **mycelium,** and it is composed of threadlike filaments, the **hyphae.** Hyphae are formed of cells joined end to end. The cells of hyphae may be separated by cell walls (septate hyphae), or the cell walls may be incomplete or lacking (nonseptate hyphae). The cell walls are formed of chitin. Nutrients are obtained by hyphae secreting digestive enzymes into the surrounding substrate, which is digested extracellularly. The resulting nutrients are then absorbed into the hyphae.

Spores of fungi are formed either from terminal cells of reproductive hyphae or in **sporangia,** enlarged structures at the ends of specialized hyphae. Some fungi form **fruiting bodies** that contain the spore-forming hyphae.

Fungi, which are usually haploid, reproduce both asexually and sexually. Asexual reproduction in unicellular fungi is by budding and cell division, and in multicellular fungi by **asexual sporulative reproduction,** in which reproductive hyphae produce **mitospores** (n) by mitotic division. Mitospores germinate to form new mycelia (n). See Figure 12.1.

Sexual reproduction is more complex and is usually by **zygotic sporulative reproduction.** Within a species, haploid mycelia occur as two mating types, designated (+) and (−). When opposite mating types are in contact, each mating type forms special cells that become **gametes** (n). The gametes fuse, and their nuclei may fuse immediately or at a later time. Fusion of gamete nuclei forms a **zygote** (2n), which produces daughter cells by meiotic division. Depending on the type of fungus, the daughter cells may be **meiospores** (n) or form a sporangium that releases meiospores (n). In either case, the spores germinate to form new haploid mycelia. See Figure 12.2.

 Assignment 1

Complete item 1 on Laboratory Report 12 that begins on page 155.

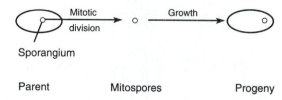

Figure 12.1 Asexual sporulative reproduction is the common asexual reproductive pattern in fungi.

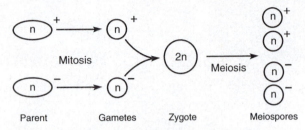

Figure 12.2 Zygotic sporulative reproduction is the sexual reproductive pattern common to fungi.

ZYGOTE FUNGI (PHYLUM ZYGOMYCOTA)

The common black bread mold, *Rhizopus stolonifer,* is an example of this group, which is characterized by the three types of hyphae shown in Figure 12.3. **Stolon hyphae** spread over the surface as the mycelium grows. **Rhizoid hyphae** penetrate the substrate to digest it and to provide anchorage. **Sporangiophores** are upright hyphae that form a **sporangium** at their tips. Asexual spores are formed within the sporangia.

Reproduction in Black Bread Mold

Black bread mold, *Rhizopus stolonifer,* occurs as two mating types designated plus (+) and (–) minus. The life cycle exhibits both asexual and sexual reproduction. Asexual sporulative reproduction occurs in both mating types: Sporangia form and release mitospores that germinate to produce new mycelia.

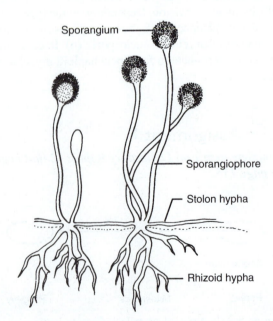

Figure 12.3 *Rhizopus stolonifer,* black bread mold.

Sexual reproduction occurs by zygotic sporulative reproduction only when hyphae of (+) and (–) strains come in contact with each other. When this occurs, adjacent (+) and (–) hyphae form special cells that grow toward each other and become **gametes** (n). Fusion of the gametes and their nuclei (conjugation) produces a **zygote** (2n), which develops a protective cell wall to become a **zygospore.** The zygospore (2n) is resistant to unfavorable environmental conditions. In favorable conditions, the zygospore germinates by meiosis to form a sporangiophore, whose sporangium releases (+) and (–) meiospores (n). Germination of these meiospores yields new haploid (+) and (–) mycelia. See Figure 12.4 and Plates 5.3 and 5.4.

 ## Assignment 2

Materials

Colored pencils
Compound microscope
Dissecting microscope
Dissecting instruments
Medicine droppers
Microscope slides and cover glasses
Water–detergent solution in dropping bottle
Rhizopus cultures
   a. single mating type on bread and Sabouraud's agar
   b. culture of opposite mating types with gametes and zygospores on Sabouraud's agar
   c. *Rhizopus* spores growing on slides coated with Sabouraud's agar
Prepared slides of *Rhizopus* with gametes and zygospores

1. *Complete items 2a–d on the laboratory report.*
2. Study Figure 12.4 and use colored pencils to color-code the labeled structures.
3. Examine the black bread mold growing on agar with a dissecting microscope. Spores were placed in the center of the agar plate to start the colony. Note the pattern of growth.
4. *Complete item 2e on the laboratory report.*
5. Use forceps to remove a small portion of the fungus growing on bread and mount it in a drop of water–detergent solution on a clean slide. Tease away the bread as necessary, and observe at 40× and 100× without a cover glass. Locate the different types of hyphae and sporangia.
6. Examine hyphae growing on an agar-coated microscope slide at 400×. Are they septate? (See Plate 5.6.) Locate a growing tip and observe it for a short time. Is it growing?
7. Examine the *Rhizopus* culture with zygospores set up under a demonstration dissecting microscope.

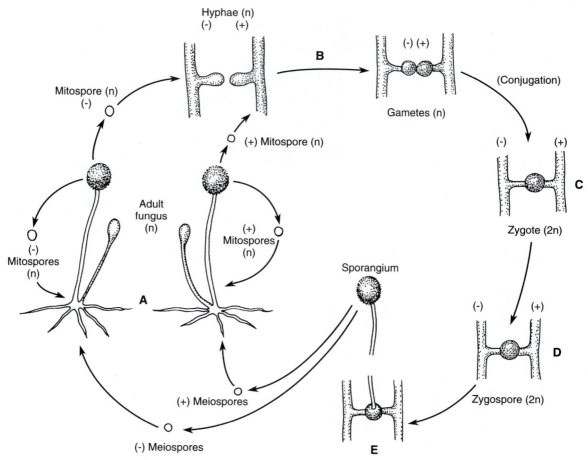

Figure 12.4 Life cycle of black bread mold (*Rhizopus*). **A.** Plus (+) and minus (–) strains (n) reproduce asexually by forming sporangia, which produce mitospores (n) that germinate to form new mycelia. **B.** Sexual reproduction occurs when adjacent (+) and (–) hyphae form (+) and (–) gametes by mitosis. **C.** The gametes (n) fuse to form a zygote (2n). **D.** The zygote forms a protective cell wall to become a zygospore that is resistant to unfavorable conditions. **E.** In favorable conditions, the zygospore germinates by meiosis to form a sporangiophore, whose sporangium releases (+) and (–) meiospores that germinate to form (+) and (–) mycelia, respectively.

Note the color and size of the gametes, zygospores, and sporangia.

8. Examine a prepared slide of gametes and zygospores of *Rhizopus*.
9. ***Complete item 2 on the laboratory report.***

SAC FUNGI (PHYLUM ASCOMYCOTA)

The sac fungi are named for the tiny sacs, called **asci** (singular, **ascus**), in which the meiospores form during sexual reproduction. The sac fungi include serious parasites of plants, such as Dutch elm disease and chestnut blight, and molds that destroy stored food products.

Beneficial forms occur as well. Edible truffles and morels are highly prized food items, and a yeast, *Saccharomyces cerevisiae,* is used in the brewing and wine-making industries to produce ethyl alcohol. Yeast is also used in the baking industry to produce tiny bubbles of carbon dioxide to make the bread rise. In addition, some sac fungi form beneficial associations with plant roots that improve the uptake of water and minerals by plants.

Reproduction in Sac Fungi

Peziza illustrates the pattern of sexual reproduction that is typical of sac fungi. See Figure 12.5. When the haploid hyphae of different mating types are in contact, each hypha forms a specialized reproductive cell that contains numerous nuclei. Fusion of these specialized

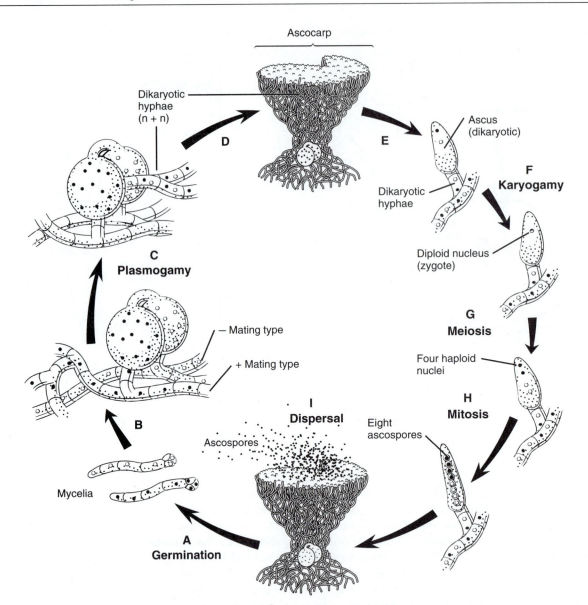

Figure 12.5 Life cycle of *Peziza*. **A.** Germination of ascospores produces hyphae of two mating types. **B.** Hyphae of different mating types form reproductive cells containing many haploid nuclei. **C.** Plasmogamy leads to the formation of dikaryotic hyphae. **D.** Hyphae form a mycelium and an ascocarp. **E.** Dikaryotic hyphae form asci. **F.** Karyogamy forms a diploid nucleus in each ascus. **G.** Meiosis forms four haploid nuclei. **H.** Mitosis forms eight haploid ascospores. **I.** When mature, ascospores are dispersed from the ascocarp.

cells, a process known as **plasmogamy,** leads to the formation of **dikaryotic hyphae** (n + n) by mitotic division. Each cell in dikaryotic hyphae contains two nuclei, one of each mating type. Dikaryotic hyphae, plus the hyphae of both mating types, form the mycelium that develops a cup-shaped **ascocarp,** the fruiting (spore-forming) body. The mycelium penetrates the substrate, often a rotting log, to obtain nutrients. The inner surface of the ascocarp cup is lined with **asci,** enlarged cells that form at the tips of the dikaryotic hyphae. Subsequently, fusion of the haploid nuclei, a process known as **karyogamy,** occurs within the asci. Then, meiosis followed by mitosis forms eight haploid **ascospores** (meiospores) in each ascus—four ascospores of each mating type. Upon maturation, the spores are released and carried by air currents. Each ascospore is capable of germinating to produce a new hypha. See Plate 6.1.

Assignment 3

Materials

Colored pencils
Compound microscope
Dissecting instruments
Medicine droppers
Microscope slides and cover glasses
Peziza cups
Yeast culture in 5% glucose
Representative sac fungi
Prepared slides of *Peziza* cups, l.s.

1. Make a slide of the yeast culture, add a cover glass, and observe at 400×. Locate cells that are forming buds. How do you explain the different sizes of the buds? ***Draw a few cells in item 3b on the laboratory report.***
2. Examine the fruiting body of a cup fungus *Peziza* and compare it with Figure 12.5. Use colored pencils to color-code labeled structures. The sacs (asci) containing the ascospores line the cup.
3. Examine a prepared slide of *Peziza* cup, l.s. Observe the asci, the ascospores, and the two types of hyphae forming the cup. Note the number of nuclei in the hyphal cells. How many nuclei are in the cells of hyphae that form asci? Are the hyphae septate?
4. ***Complete item 3 on the laboratory report.***

CLUB FUNGI (PHYLUM BASIDIOMYCOTA)

The familiar mushrooms, puffballs, and shelf fungi belong to this group. Also included are rusts and smuts, serious plant parasites that cause enormous losses in wheat, corn, and other cereal crops each year. **Basidiospores** are formed after a sexual process by club-shaped **basidia,** terminal cells of reproductive hyphae, which are usually located in a fruiting body, the **basidiocarp.** See Plates 6.2–6.4.

Reproduction in a Mushroom

The life cycle of a mushroom, *Coprinus,* is more complex than that of *Rhizopus* and is representative of the club fungi. See Figure 12.6. The haploid mycelia occur as two mating types, (+) and (–). Whenever opposite mating types come in contact with each other, cells (analogous to gametes) from one mating type grow toward cells of the opposite mating type and fuse together in pairs, but their nuclei remain separate. The cytoplasmic fusion is called **plasmogamy,** and the cell that is formed is a **dikaryotic cell** because it contains two nuclei, one of each mating type.

Mitotic division of a dikaryotic cell yields numerous dikaryotic hyphae, which form the mushroom basidiocarp. A basidiocarp is composed of a cap supported by an upright stalk. The cap bears numerous "gills" on its lower surface, and the gills contain thousands of dikaryotic basidia on their lateral surfaces. The two nuclei in each basidium fuse, a process known as **karyogamy,** to form a diploid zygote nucleus. Then, each basidium produces four haploid basidiospores, two of each mating type, by meiotic division. After release, the spores are distributed by air currents. Germination of a spore produces a haploid mycelium by mitotic division.

Assignment 4

Materials

Colored pencils
Dissecting instruments
Medicine droppers
Microscope slides and cover glasses
Water–detergent solution in dropping bottles
Fresh mushroom basidiocarps
Representative club fungi
Prepared slides of *Coprinus* gills, x.s., with basidia and basidiospores

1. Examine the fruiting body of a mushroom. Locate the parts shown in Figure 12.6. Use colored pencils to color-code labeled structures.
2. Use a scalpel to remove a very thin longitudinal section of mushroom stalk and make a water-mount slide of it. Examine your slide to see if the hyphae are septate.
3. Examine a prepared slide of mushroom gill. Locate the hyphae, basidia, and basidiospores. How many nuclei are in the hyphal cells?
4. Examine other examples of club fungi.
5. ***Complete item 4 on the laboratory report.***

IMPERFECT FUNGI (PHYLUM DEUTEROMYCOTA)

This large group of fungi exhibits considerable diversity, and some species are of importance to humans. It includes fungi that flavor certain cheeses, parasites that cause ringworm and athlete's foot, and the genus *Penicillium,* the source of the first antibiotic.

Reproduction in Imperfect Fungi

The reason members of the Deuteromycota are called "imperfect fungi" is because none have been found to have a sexual reproductive cycle. Reproduction is by asexual means only. As the mycelium develops, erect

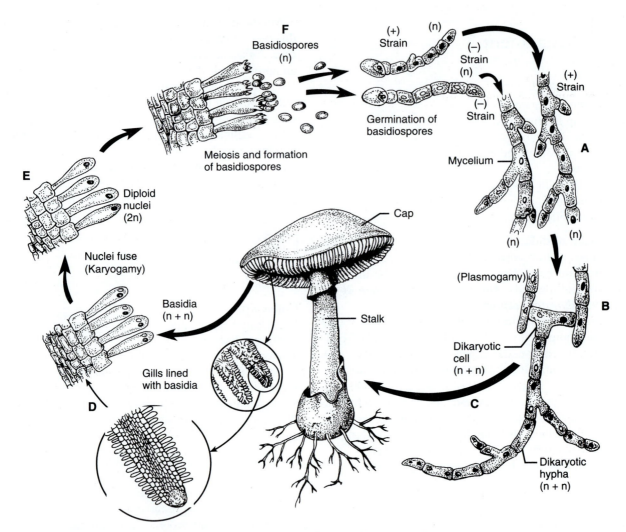

Figure 12.6 Life cycle of a mushroom (*Coprinus*). **A.** Adjacent plus (+) and minus (−) hyphae produce cells that extend toward each other. **B.** The cells fuse (plasmogamy), but the (+) and (−) nuclei remain separate, forming dikaryotic (n + n) hyphae. **C.** Dikaryotic hyphae grow to form a basidiocarp. **D.** The gills of the cap are lined with club-shaped dikaryotic basidia (n + n) **E.** The (+) and (−) nuclei fuse, forming diploid basidia. **F.** Each basidium (2n) produces two (+) basidiospores and two (−) basidiospores by meiotic division. Basidiospores (n) germinate to form (+) and (−) hyphae.

hyphae are formed. These erect hyphae produce mitospores, which are called **conidiospores,** at their tips. A sporangium is not formed. The conidiospores are released and are dispersed by air currents. Germination of a conidiospore produces a new mycelium.

Assignment 5

Materials

Compound microscope
Dissecting microscope
Medicine droppers

Microscope slides and cover glasses
Water–detergent solution in dropping bottles
Penicillium cultures on citrus fruit and Sabauroud's agar

1. Examine a colony of *Penicillium* with a dissecting microscope. Note how the colony grows outward. Locate the older and younger hyphae. What is the color of the younger and older conidiospores?
2. Use the tip of a scalpel to remove a few hyphae with mature conidiospores. Make a water-mount slide using the water–detergent solution. Examine the conidiospores at 100× and 400× to determine the pattern of conidiospore formation.
3. ***Complete item 5 on the laboratory report.***

LICHENS

A lichen is a **symbiotic relationship** of a fungus and an alga. The "organism" resulting from this association has unique characteristics that are different from those of either the alga or fungus composing it. Lichens are usually classified according to the fungus species involved, however. Most of the fungi found in lichens are members of the sac fungi, and most of the algae are green algae.

The fungus of a lichen provides the algal cells with support and protection from high light intensities and desiccation. The alga provides the fungus with organic nutrients produced by photosynthesis. This association allows a lichen to live in habitats that are unfavorable to either fungus or alga if living alone. Most, but not all, biologists consider this symbiotic relationship to be **mutualism,** where both members benefit from the association. Lichens serve as important soil builders by eroding rock surfaces and paving the way for more advanced forms of life.

Lichens have three types of body forms. **Crustose lichens** consist of a thin crust that is tightly attached to rocks. **Foliose lichens** are a bit thicker than crustose lichens but have a flattened body form. They are the most common types of lichens and usually occur on rocks. **Fruticose lichens** have an erect, often branched, body form and may occur on trees and soil as well as on rocks.

 ## Assignment 6

Materials

Representative lichens
Prepared slides of lichen, x.s.

1. Examine the representative lichens and note the type of body form of each one.
2. Examine a prepared slide of lichen, x.s., and locate the fungal hyphae and the algal cells. Is there a pattern to the location of the algal cells?
3. *Complete item 6 on the laboratory report.*

 ## Assignment 7

1. Review your understanding of fungi and the distinguishing characteristics of each group.
2. Your instructor has set up several "unknown specimens" for you to examine and identify as to the fungus group to which they belong.
3. *Complete item 7 on the laboratory report.*

FUNGI

Student _____

Lab Instructor _____

1. INTRODUCTION

 a. Write the term that matches the phrase.

   1. Threadlike filaments composing a fungus   _____

   2. Dormant reproductive cells dispersed by the wind   _____

   3. Nonreproductive body of a fungus   _____

   4. Fleshy reproductive body of a fungus containing spore-forming hyphae   _____

   5. Specialized structures forming spores   _____

 b. Distinguish between saprotrophic and parasitic modes of nutrition.

 Saprotrophic _____

 Parasitic _____

 c. Describe how fungi obtain nutrients from organic substrates. _____

 _____

 _____

2. ZYGOTE FUNGI: *RHIZOPUS*

 a. Contrast the functions of the three types of hyphae in *Rhizopus*.

 Stolon hyphae _____

 _____

 Rhizoid hyphae _____

 _____

 Sporangiophores _____

 _____

 b. What part of the *Rhizopus* life cycle is diploid? _____

 c. How are spores dispersed? _____

 d. What is formed when a spore germinates? _____

 e. Diagram the appearance of a *Rhizopus* colony, showing the location of mature and immature sporangia and the structures that form the outermost portion of the colony. Label the locations of mature sporangia, immature sporangia, and youngest hyphae.

f. From your slides, draw (a) a sporangiophore with a mature sporangium, (b) a few spores, and (c) hyphae, showing relative size and the septate or nonseptate condition of the hyphae. Draw a zygospore and its adjoining hyphae.

Sporangiophore   **Zygospore**

3. SAC FUNGI

a. Draw a few yeast cells with buds at 400×. Label the buds.

Budding is considered to be what type of reproductive pattern? _____

_____

How does budding differ from typical cell division? _____

_____

What is formed by the germination of conidiospores? _____

b. Diagram a small portion of a *Peziza* cup showing the spore-forming hyphae, asci, and ascospores as they appear on your prepared slide. Label your drawings.

c. How do the two types of hyphae in the ascocarp differ? _____

_____

d. After nuclear fusion in an ascus, eight ascospores are formed. Do the asci on your slide contain eight

ascospores? _____ Explain why or why not. _____

_____

4. CLUB FUNGI: MUSHROOMS

a. Draw a few (a) hyphae from a mushroom stalk to show their septate or nonseptate condition and (b) basidia and spores.

Hyphae from Stipe   Basidia and Spores

b. What is the result of plasmogamy? _____

c. What type of hyphal cells form the basidiocarp? _____

   What cells are diploid? _____

d. What process forms basidiospores? _____

e. What is formed by germination of a basidiospore? _____

_____

5. IMPERFECT FUNGI: *PENICILLIUM*

a. Draw an outline of a *Penicillium* colony. Label the region of mature conidiospores and immature conidiospores and the color of each. Draw the appearance of a few hyphae with conidiospores to show how they are formed and the septate or nonseptate condition of the hyphae.

Colony   Hyphae with Conidiospores

b. What process forms conidiospores? _____

c. What is formed by the germination of conidiospores? _____

6. LICHENS

a. Show by diagram the distribution of algal cells and fungal cells observed on the slide of lichen, x.s.

b. In a lichen, describe the role of the:

fungus _____

_____

alga _____

_____

c. Show by diagram the differences between the three types of lichens.

Crustose   **Foliose**   **Fruticose**

7. REVIEW

a. Write the numbers of the "unknown" specimens in the following spaces:

_____ Zygote fungi   _____ Sac fungi   _____ Club fungi

_____ Imperfect fungi   _____ Lichens

b. Write the names of the "unknown" structures in the following spaces:

1. _____   4. _____

2. _____   5. _____

3. _____   6. _____

TERRESTRIAL PLANTS

OBJECTIVES

After completing the laboratory session, you should be able to:
1. Describe the characteristics of the plant groups studied.
2. Identify representatives of moss plants and vascular plants.
3. Describe the life cycles of mosses, ferns, conifers, and flowering plants.
4. Define all terms in bold print.

Terrestrial plants are photosynthetic autotrophs characterized by (1) chlorophylls a and b and carotenoids in their chloroplasts, (2) food stored as starch, (3) cell walls of cellulose, and (4) sporic sexual reproduction involving an alternation of sporophyte and gametophyte generations. See Figures 11.7 and 13.1. They have specialized tissues and organs and lack motility. Some groups of nonseed plants have motile sperm. An increase in complexity is evident from the relatively simple moss plants to the highly specialized flowering plants.

MOSS PLANTS (PHYLUM BRYOPHYTA)

Mosses, liverworts, and hornworts compose the bryophytes. Bryophytes are small plants with relatively simple tissues and organs. Their size is limited because

they lack (1) vascular tissue to transport water, minerals, and nutrients throughout the plant and (2) true roots, stems, and leaves. Bryophytes are usually restricted to moist areas because their flagellated sperm must swim through surface water to reach and fertilize an egg.

Mosses are the most common bryophytes. The basic structure of a moss gametophyte and sporophyte is shown in Figure 13.2. The gametophyte is the **dominant generation,** which means that it is larger and lives longer than the sporophyte. The gametophyte possesses **leaflike** and **stemlike structures** and **rhizoids,** filaments that anchor it to the soil. Gamete-forming organs are located at the tip of the gametophyte. The **sporophyte** is attached to the gametophyte and consists of a **sporangium** supported by a **seta** (stalk). After spores are released, the sporophyte withers and dies.

Study the life cycle of a moss with separate male and female gametophytes in Figure 13.3.

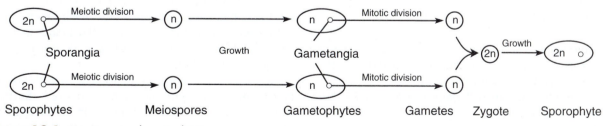

Figure 13.1 Sporic sexual reproduction.

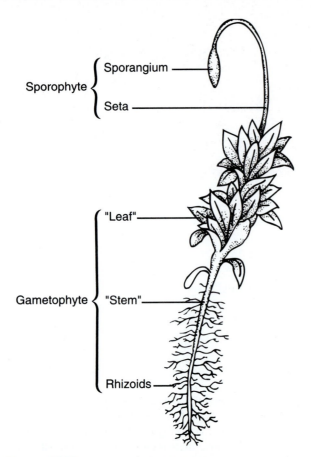

Sporophyte {
Sporangium
Seta
}

Gametophyte {
"Leaf"
"Stem"
Rhizoids
}

Figure 13.2 A moss plant (*Mnium*).

Assignment 1

Materials

Colored pencils
Compound microscope
Stereo microscope
Moss gametophytes with sporophytes
Representative mosses and liverworts
Prepared slides of:
   moss antheridia with sperm
   moss archegonia with egg
   moss capsule with spores
   moss "leaf," x.s.
   moss "stem," x.s.

1. Color-code the zygote and sporophyte stages green in Figures 13.2 and 13.3 to indicate the diploid parts of the life cycle.
2. ***Complete item 1a on Laboratory Report 13 that begins on page 169.***
3. Examine the living moss and liverwort gametophytes. Although their appearance is quite different, they have similar life cycles.

4. Examine a moss gametophyte with sporophyte attached. Use a stereo microscope for your observations, if desirable. Locate the parts labeled in Figure 13.2.
5. ***Make drawings of these structures from your slides in item 1b on the laboratory report.***
   a. Moss "stem," x.s. Note that it is composed of only a few different types of cells. Vascular tissue is absent, but the central column aids water movement.
   b. Moss "leaf," x.s. Only a few different types of cells compose the "leaf."
   c. Antheridium with sperm, l.s.
   d. Archegonium with egg, l.s.
   e. Sporangium with spores, l.s.

VASCULAR PLANTS

Vascular plants are more advanced than moss plants in at least three important ways: (1) **vascular tissue** is present; (2) the sporophyte is the dominant generation; and (3) true **roots, stems,** and **leaves** are found in the sporophyte.

Vascular tissue allows efficient transport of materials within the plant. Without this adaptation, all plants would not be much bigger than mosses and could not exist in drier habitats. There are two vascular tissues: xylem and phloem. **Xylem** transports water and minerals upward from the roots to the shoot system of the plant. **Phloem** carries organic nutrients upward or downward within the plant.

The sporophyte is the dominant generation and is the adult form that is recognizable as a plant. The gametophyte is small and usually microscopic in size.

Ferns (Phylum Pterophyta)

Fern sporophytes typically have an underground stem, a **rhizome,** which is anchored by roots. The large **leaves** are attached to the stem and consist of many leaflets joined to the midrib. On the undersurface of the leaflets of certain leaves are small brown spots, the **sori.** These consist of numerous sporangia that form meiospores. The vascular tissue in ferns is not as well developed as it is in more advanced vascular plants, but it still makes ferns better adapted to terrestrial life than mosses.

Fern gameotophytes are small, flat plants that are anchored to the soil by rhizoids. The gamete-forming organs are located on the lower surface of the gametophytes. Like sporophytes, gametophytes are photosynthetic autotrophs, but they are short-lived. Surface water is required for the sperm to swim to the eggs, as in mosses, so ferns are usually restricted to habitats that are moist for a part of the year.

Study the life cycle of a fern illustrated in Figure 13.4.

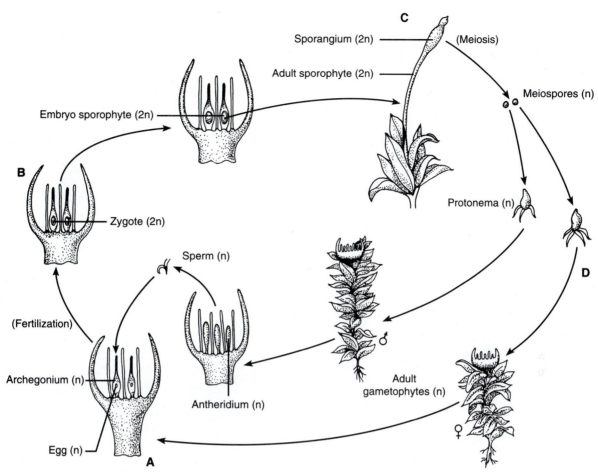

Figure 13.3 Life cycle of a moss. **A.** The gametophytes (n) develop sex organs at their tips: sperm-forming antheridia in the male gametophyte and egg-forming archegonia in the female gametophyte. A flagellated sperm released from an antheridium swims through surface water into an archegonium to fertilize an egg. **B.** The resultant zygote (2n) grows by mitotic division to become a mature sporophyte (2n). **C.** The photosynthetic sporophyte remains attached to the archegonium and obtains some nutrients from the female gametophyte. Meiospores (n) form in the sporangium and are released. **D.** Meiospores germinate to become protonemas, which develop into male or female gametophytes.

 Assignment 2

Materials

Colored pencils
Compound microscope
Fern gametophytes, living
Fern sporophytes with sori, living
Microscope slides and cover glasses
Representative ferns
Prepared slides of:
    fern leaflet with sori, x.s.
    fern gametophyte, w.m., with antheridia and archegonia

1. Study the life cycle of a fern in Figure 13.4. Draw a line across the cycle to separate diploid and haploid components. Color-code the zygote and sporophyte stages green.
2. Examine a fern sporophyte with soil washed from the rhizome and roots. Note the organization of the parts. A leaf is composed of a central midrib and many leaflets. Locate the sori.
3. Remove a leaflet with sori. Place a part of the leaflet on a microscope slide and add a cover glass without water. Examine a sorus at 40×. In some species the sporangia are protected by a thin cover, an indusium. If present, remove the indusium with a dissecting needle and observe

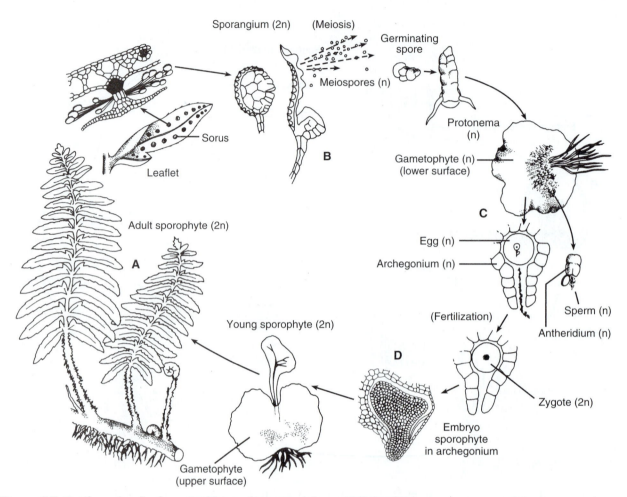

Figure 13.4 Life cycle of a fern. **A.** Mature fern sporophytes (2n) produce sporangia on the undersurface of leaflets on some leaves. The sporangia are grouped together in clusters called sori. **B.** Sporangia produce and release meiospores (n) that germinate to form protonemas, which in turn grow to become gametophytes (n). **C.** Each gametophyte forms both antheridia and archegonia on its undersurface. When surface water is present, flagellated sperm are released from antheridia and swim into archegonia to fertilize the eggs. **D.** After fertilization, a resultant zygote (2n) grows to become a young sporophyte, which becomes an independent adult sporophyte (2n) as the gametophyte dies.

the sporangia. Watch what happens as the sporangia dry out.

4. ***Complete items 2a–2f on the laboratory report.***
5. Examine a prepared slide of fern leaflet and sori, x.s. ***Draw a sorus from your slide in item 2g on the laboratory report.***
6. Examine a living gametophyte, and note its shape, size, and color.
7. Examine a prepared slide of a gametophyte, w.m., at 40×. Locate the archegonia near the notch and the antheridia near the rhizoids. ***Draw an antheridium and archegonium in item 2g on the laboratory report.***

SEED PLANTS

Seed plants are better adapted to terrestrial life than ferns and maintain a life cycle involving alternation of generations. The dominant sporophyte—the generation that you recognize as a plant—consists of a root system and a shoot system. The **root system** usually lies below ground. It provides anchorage, and it absorbs water and minerals. The **shoot system** consists of stems, leaves, and reproductive organs. It produces organic nutrients by photosynthesis.

A seed plant contains well-developed vascular tissue (xylem and phloem) that provides structural support for roots, stems, and leaves as well as the transport of materials. Leaves, the main photosynthetic organs, are supported by **veins** composed of vascular tissue.

Sporophytes of seed plants form two different types of meiospores. **Microspores** are formed in **microsporangia** and mature to form **pollen grains** that, in turn, develop into microscopic **male gametophytes**. **Megaspores** develop within **megasporangia**. Each megasporangium is enveloped by protective sporophytic tissues called **integuments**. The integuments, megasporangium, and megaspore are collectively called an **ovule**, a potential seed. The megaspore develops into a **female gametophyte** containing one or more **egg cells**.

Seed plants are not dependent on water for sperm transport. Instead, pollen grains are usually transferred to the female gametophyte by wind or insects, a process called **pollination**. Then, the pollen grain develops into a male gametophyte with a **pollen tube** that carries **sperm nuclei** to the **egg** in the female gametophyte. Fertilization results in a zygote, the first cell of a new sporophyte generation.

The entire ovule becomes a **seed**, which is composed of the embryo sporophyte, stored nutrients for the future development of the embryo sporophyte, and a seed coat derived from the integuments. The seed coat is resistant to unfavorable environmental conditions. When conditions are favorable, a seed germinates, producing a new sporophyte generation.

Seed plants are divided into two large groups. **Gymnosperms** have seeds borne exposed on the surface of modified leaves. **Angiosperms** produce flowers and have their seeds enclosed within a **fruit**, a ripened (mature) ovary.

Cone-Bearing Plants (Phylum Coniferophyta)

Conifers are the best-known and largest group of gymnosperms. The cones contain the reproductive organs, and two types of cones are formed. **Staminate** (pollen) **cones** are small, with paper-thin scales. **Ovulate** (seed) **cones** are large, with woody scales. Pollen is transferred from staminate cones to ovulate cones by wind. Seeds are borne exposed on the upper surface of the scales of mature ovulate cones. The leaves are either needlelike or scalelike. Conifers may attain considerable size and may live in rather dry habitats because their vascular tissue is well developed, their leaves restrict water loss, and water is not required for sperm transport.

Study the life cycle of the pine illustrated in Figure 13.5.

 Assignment 3

Materials

Colored pencils
Compound microscopes
Stereo microscopes
Cones and leaves of representative conifers
Demonstration microscope setups:
   pine female gametophyte with archegonium and egg
   sectioned pine seed showing embryo sporophyte
Ovulate pine cones with seeds
Staminate pine cones with pollen
Pine seeds
Prepared slides of pine microsporangium with pollen

1. Study the pine life cycle in Figure 13.5. Draw a line across the cycle separating diploid and haploid components. Color-code diploid components green.
2. Examine the representative cones and leaves of conifers. Compare staminate and ovulate cones as to size and weight.
3. ***Complete item 3a on the laboratory report.***
4. Examine a prepared slide of microsporangia with pollen. How is the pollen adapted for wind transport? ***Make a drawing of your observations in item 3b on the laboratory report.***
5. Make a wet-mount slide of a few pollen grains and examine them with your microscope. Do they resemble those on the prepared slide?
6. Examine a prepared slide of a young ovulate cone showing the megasporangium with an archegonium containing an egg cell. ***Make a drawing of your observations in item 3b on the laboratory report.***
7. Examine a mature ovulate cone and locate the two seeds on the upper surface of each scale. How are the seeds dispersed?
8. Obtain a soaked pine seed. Use a scalpel to cut it open longitudinally, and examine it with a stereo microscope. Locate the embryo sporophyte embedded in the tissue of the female gametophyte, which contains nutrients for the embryo sporophyte. What tissues contribute to the formation of a pine seed?
9. ***Complete item 3 on the laboratory report.***

Flowering Plants (Phylum Anthophyta)

Flowering plants are the most advanced plants. Their success in colonizing the land is due to well-developed vascular tissues and **flowers** (reproductive organs) that greatly enhance reproductive success. Most flowers attract insects that bring about pollination, a process leading to the fertilization of egg cells by sperm nuclei. Seeds are enclosed within **fruits** that facilitate dispersal.

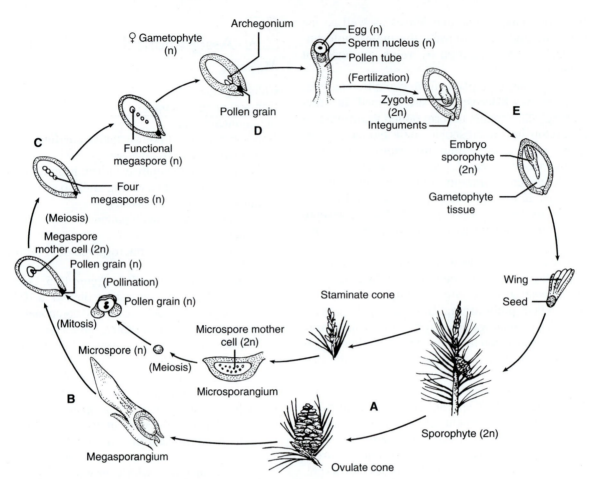

Figure 13.5 Life cycle of a pine. **A.** The sporophyte (2n), a pine tree, produces staminate and ovulate cones, which contain many microsporangia and megasporangia, respectively. **B.** Many microspores are formed by meiosis in microsporangia, and they develop into mature pollen grains (n), the male gametophytes. Pollen grains are released and carried by wind to the megasporangia. **C.** In a megasporangium, four megaspores are formed by meiosis. Three degenerate, and one becomes a functional megaspore (n), which develops into a female gametophyte. **D.** The female gametophyte develops an archegonium containing two eggs. A pollen grain develops a pollen tube, which carries sperm nuclei to the egg cells, resulting in fertilization. **E.** The resulting zygote (2n) grows to become an embryo sporophyte embedded in the female gametophyte containing stored nutrients. The wall of the megasporangium becomes the seed coat of the seed, which is released and dispersed by wind.

The reproductive patterns of flowering plants show some major adaptations over gymnosperms:

1. Reproductive structures are grouped in **flowers** that usually contain both microsporangia and megasporangia.
2. Like gymnosperms, grasses are pollinated by wind, but most flowering plants are pollinated by animal pollinators. Most pollinators are insects, but birds and bats serve as pollinators of certain species. Flower color, flower shape, and nectar attract the pollinators.
3. Portions of the flower develop to form a **fruit** that encloses the seeds and enhances seed dispersal by wind in some plants but by animals in most.

Study the life cycle of a flowering plant illustrated in Figure 13.6.

Flower Structure

The basic structure of a flower is shown in Figure 13.7, but this fundamental organization has many variations.

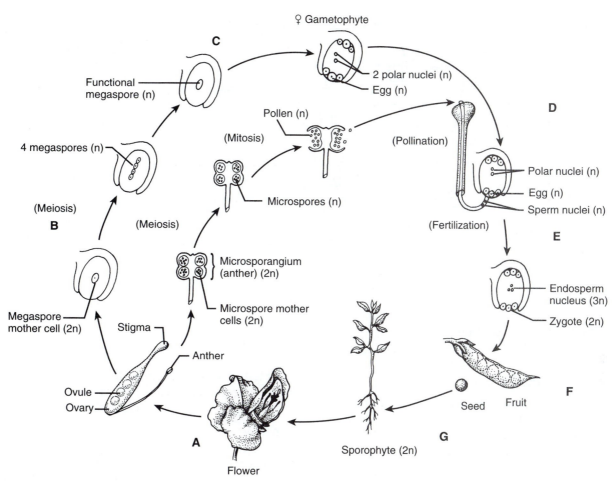

Figure 13.6 Life cycle of a pea plant. **A.** The sporophyte (2n) produces flowers with anthers (microsporangia) and ovules (megasporangia). **B.** Many microspores (n) are produced in anthers by meiosis, and four megaspores (n) are formed by meiosis in ovules. Only one megaspore becomes functional. **C.** Each microspore becomes a pollen grain (male gametophyte). Each functional megaspore becomes a female gametophyte containing one egg and two polar nuclei. **D.** Pollination occurs by wind or insects. Each pollen grain germinates, forming a pollen tube, which carries two sperm nuclei into an ovule. **E.** One sperm nucleus unites with the egg cell in fertilization, and the other sperm nucleus combines with the two polar nuclei to form a triploid endosperm nucleus (3n). **F.** The zygote grows to become an embryo sporophyte, and the endosperm develops to provide stored nutrients for the embryo. The embryo sporophyte, endosperm, and ovule wall become the seed located within the ovary, which ripens to become a fruit. **G.** Fruits and seeds may be distributed by wind or animals. When conditions are favorable, the seed germinates and the embryo sporophyte grows to become an independent sporophyte plant.

The **receptacle** supports the flower on the stem, and the reproductive structures are enclosed in two whorls of modified leaves. The inner whorl consists of **petals,** which are usually colored to attract pollinating insects. The outer whorl consists of **sepals,** which are typically smaller than the petals and are usually green in color.

The **stamens** are the male portions of the flower. Each stamen consists of an **anther** supported by a **filament.** Anthers contain the microsporangia that produce pollen. The **pistil** is the female portion, and it consists of three parts. The basal portion is the **ovary,** which contains **ovules** (megasporangia). The tip of the pistil is the **stigma,** which receives pollen and secretes enzymes promoting pollen germination. The **style** is a slender stalk that joins stigma and ovary. Nectar is secreted near the base of the ovary.

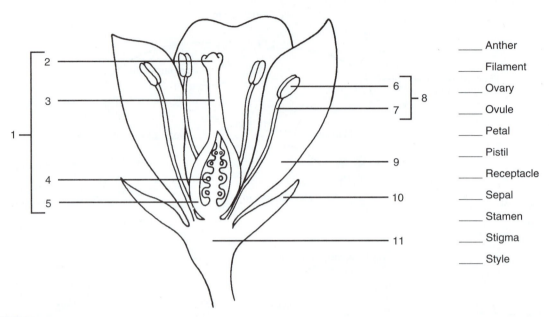

Figure 13.7 Flower structure.

Fruits and Seeds

As the seeds develop in the ovary, the ovary grows and ripens to form a fruit that provides protection for the seeds and facilitates seed dispersal. The three basic types of fruits are shown in Figure 13.8.

1. **Dry dehiscent fruits** split open when sufficiently dry to cast out the seeds, sometimes with considerable force. Pea and bean pods are examples.
2. **Dry indehiscent fruits** do not open, and the ovary wall tightly envelops the seed. Acorns and fruits of corn and other cereals are examples.

3. **Fleshy fruits** remain moist for a considerable period of time and are usually edible and colored. Animals scatter the seeds by feeding on the fruits.

A seed consists of a protective **seed coat** that is derived from the wall of the ovule, stored nutrients, and a dormant **embryo sporophyte.** See Figure 13.9. In monocots and some dicots, the stored nutrients compose the endosperm. In most dicots, the embryonic leaves, **cotyledons,** contain many of the stored nutrients, and the endosperm is reduced. Seeds are able to withstand unfavorable conditions and tend to germinate only when favorable conditions exist.

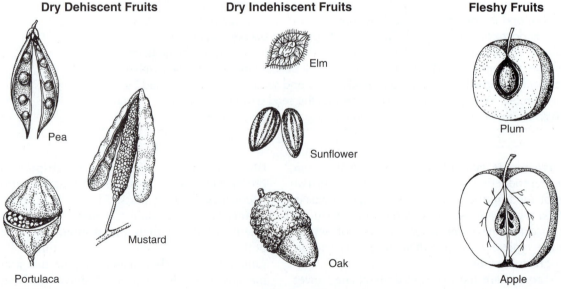

Figure 13.8 Examples of fruits.

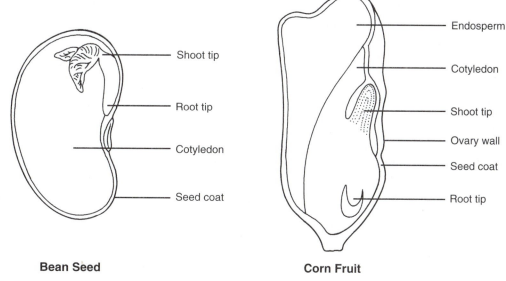

Bean Seed   **Corn Fruit**

Figure 13.9 Structure of bean seed and corn fruit.

Assignment 4

Materials

Colored pencils
Compound microscope
Stereo microscope
Dissecting instruments
Dropping bottles of iodine solution
Microscope slides and cover glasses
Model of a flower
Representative flowering plants
Representative flowers, fruits, and seeds
Lily or *Gladiolus* flowers
Bean seeds, soaked
Corn fruits, soaked
Pea pods, fresh
Prepared slides of lily anthers

1. Study Figure 13.6. Color-code the diploid stages green. Draw a line across the cycle separating diploid and haploid components.
2. Examine the representative flowering plants. Note how their leaves differ from those of conifers.
3. ***Complete item 4a on the laboratory report.***
4. Label Figure 13.7. Color-code the male parts yellow and the female parts green.
5. Compare a lily flower to Figure 13.7 to locate the parts. Observe the anthers with a stereo microscope, and note the pollen. Make a wet-mount slide of pollen and observe it with a compound microscope. Is the pollen shaped like pine pollen? Make a cross-section of the ovary and observe the ovules. How are they attached? ***Make drawings of***

your observations in item 4b on the laboratory report.
6. Examine a prepared slide of lily anther, x.s. Observe the pollen and microsporangia.
7. Examine the representative flowers, and note their variations from the basic flower structure. Make water-mount slides of their pollen and examine them microscopically to compare their structures.
8. Examine the representative fruits, and classify them according to type (Figure 13.8). How do they aid seed dispersal?
9. Color-code the embryo green in Figure 13.9.
10. Split open a soaked bean seed, and locate the parts shown in Figure 13.9. Use a scalpel to cut the soaked corn fruit, and locate the labeled parts. Examine the structures with a stereo microscope.
11. Use a scalpel to scrape gently the exposed surface of the bean cotyledon and the attached embryo sporophyte.
12. Add a drop of iodine solution to the cotyledon and embryo and also to the cut surface of the corn fruit. After 1 min, rinse with water and blot dry. What nutrient is evident? Where is it located?
13. Examine the representative seeds, and note the variations in size and shape.
14. ***Complete item 4 on the laboratory report.***

Assignment 5

Your instructor has set up several "unknown" plants, plant parts, flower parts, fruit types, and seed types to challenge you. ***Identify these unknowns in item 5 on the laboratory report.***

TERRESTRIAL PLANTS

Student _____

Lab Instructor _____

1. MOSS PLANTS

a. Write the term that matches the phrase.

1. Generation producing spores _____

2. Generation producing sperm and eggs _____

3. Dominant generation _____

4. Attaches gametophyte to soil _____

5. Substance required for sperm transport _____

b. Diagram from your slides the following organs. Draw a few cells from each region within the organs to show shape and size. Label sperm, egg, and spores.

Moss "Stem"   **Moss "Leaf"**   **Antheridium**

Archegonium   **Sporangium**

2. FERNS

a. List three distinguishing characteristics of vascular plants.

1. _____

2. _____

3. _____

b. Indicate the function of:

Xylem _____

Phloem _____

c. How are ferns better adapted to terrestrial life than mosses?

_____

_____

d. What restricts ferns to areas of abundant moisture at least during part of a year? _____

_____

e. What happens when sporangia dry out? _____

_____

f. Contrast the function of rhizoids and roots.

Rhizoids _____

Roots _____

g. From your slides, draw the following and label pertinent parts:

Sorus, x.s.   **Antheridium**   **Archegonium**

3. CONE-BEARING PLANTS

a. Write the term that matches each phrase.

1. Dominant generation   _____

2. Cones forming pollen grains   _____

3. Cones with megasporangia   _____

4. Cones with microsporangia   _____

5. Cones forming seeds   _____

6. Develops from functional megaspore   _____

7. Develops from microspores   _____

8. Method of pollination   _____

9. Male gametophyte   _____

10. Method of seed dispersal   _____

b. From your slides and fresh specimens, draw the following structures and label pertinent parts:

Microsporangium with Pollen Grains   **Megasporangium with Archegonium**

Pine Seed (Whole) **Pine Seed (Sectioned Showing Embryo)**

c. How are pollen grains and seeds adapted for dispersal by wind?

Pollen _____

Seeds _____

d. Explain the advantage of pollination in the life cycle. _____

_____

_____

e. Explain the advantage of seeds in the life cycle. _____

_____

_____

4. FLOWERING PLANTS

a. Write the term that matches each phrase.

1. Anthers _____

2. Ovules _____

3. Develops from microspores _____

4. Develops from functional megaspore _____

5. Number of eggs in female gametophyte _____

6. Transfer of pollen from anther to pistil _____

7. Composed of ovule, embryo, and endosperm _____

8. A ripened or mature ovary _____

9. Common pollinating agent in grasses _____

10. Pollinating agent in most flowering plants _____

11. Used to attract insects _____

12. Formed by union of polar and sperm nuclei _____

b. Draw the appearance of these structures of a lily flower as observed with a dissecting microscope. Label pertinent parts.

Anther with Pollen **Ovary, x.s.**

c. List the labels for Figure 13.7.

1. _____   5. _____   9. _____

2. _____   6. _____   10. _____

3. _____   7. _____   11. _____

4. _____   8. _____

d. Why is a corn kernel considered a fruit and not a seed? _____

_____

e. What is the stored nutrient in the bean seed and corn kernel? _____

Where is it concentrated in the:

bean seed? _____   corn kernel? _____

f. Indicate the basic function of a:

Seed _____

Fruit _____

g. List the names and types of the representative fruits.

Name	Type
_____	_____
_____	_____
_____	_____
_____	_____
_____	_____
_____	_____

5. UNKNOWNS

a. Place the numbers of the "unknown" specimens in the correct spaces.

_____ Bryophyte _____ Fern _____ Conifer _____ Flowering plant

b. Write the names of the "unknown" structures in the following spaces:

1. _____   6. _____

2. _____   7. _____

3. _____   8. _____

4. _____   9. _____

5. _____   10. _____

SIMPLE ANIMALS

OBJECTIVES

After completing the laboratory session, you should be able to:
1. Describe the distinguishing characteristics of the taxonomic groups studied and their evolutionary advances.
2. Identify representatives of the groups studied.
3. Identify and state the function of the structural components of the animals studied.
4. Define all terms in bold print.

Animals are members of the **kingdom Animalia** and are classified according to their similarities and evolutionary relationships. All animals are characterized by (1) a **multicellular body** formed of different types of **eukaryotic cells** that lack plastids and cell walls, (2) **heterotrophic nutrition,** and (3) movement by the shortening of **contractile fibers.**

Asexual reproduction in animals is rare. Only a few simple animals reproduce asexually. Animals reproduce sexually by **gametic sexual reproduction.** Diploid adults possess **gonads** that produce haploid gametes by meiotic cell division. **Ovaries** produce eggs, and **testes** produce sperm. Union of sperm and egg at fertilization forms a diploid zygote that grows by mitotic division into an adult. See Figure 14.1.

In this exercise, you will study the distinguishing characteristics of the major groups of "simple animals:" sponges, cnidarians, flatworms, and roundworms. Table 14.1 summarizes their distinguishing characteristics.

MAJOR CRITERIA FOR CLASSIFYING ANIMALS

Biologists use several major criteria and many lesser ones to classify animals. Figure 14.2 shows how certain major criteria are used to establish the presumed evolutionary relationships among the principal animal phyla.

Symmetry

Symmetry refers to the proportion of body parts on each side of a median plane. Animals may exhibit **radial symmetry, bilateral symmetry,** or none. Radially symmetrical animals may be divided into two similar halves by *any* plane passing through the longitudinal axis. Bilaterally symmetrical animals may be divided into two similar halves by only *one* plane, a **midsaggital plane** passing through the longitudinal axis of the body.

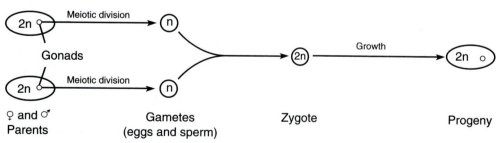

Figure 14.1 Gametic sexual reproduction.

TABLE 14.1	DISTINCTIVE CHARACTERISTICS OF SPONGES, CNIDARIANS, FLATWORMS, AND ROUNDWORMS			
	Sponges	Cnidarians	Flatworms	Roundworms
Symmetry	Asymmetrical or radial symmetry	Radial symmetry	Bilateral symmetry	Bilateral symmetry
Level of organization	Cellular–tissue level	Tissue level	Organ level	Organ system level
Germ layers	None	Ectoderm and endoderm	Ectoderm, mesoderm, and endoderm	Ectoderm, mesoderm, and endoderm
Coelom	N/A	N/A	Acoelomate	Pseudocoelomate
Body plan	N/A	Saclike	Saclike	Tube within a tube

N/A = Not applicable.

Level of Organization

An animal may exhibit a **tissue, organ,** or **organ system** level of body organization.

Embryonic Tissues (Germ Layers)

Animals with an organ level of organization, or higher, possess all three embryonic tissues: **ectoderm, endoderm,** and **mesoderm.** Those with a tissue level of organization possess only ectoderm and endoderm.

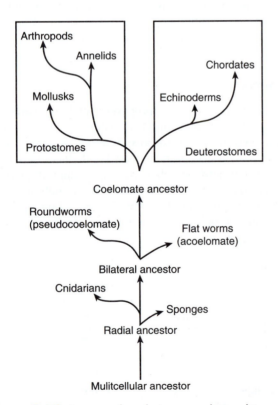

Figure 14.2 Presumed evolutionary relationships of the major animal phyla.

Body Plan

Animals with a tissue level of organization, or higher, exhibit one of two types of body plans. A **saclike body plan** is characterized by a single opening, the mouth, into a gastrovascular cavity. A **tube-within-a-tube body plan** has a mouth at one end of a tubelike digestive tract and an anus at the other.

Body Cavity (Coelom)

Animals with a tube-within-a-tube body plan have a fluid-filled space between the digestive tract and the body wall that is called a **coelom.** If the coelom is completely lined with mesoderm, it is a **true coelom.** If not, it is a **false coelom** or **pseudocoel.** See Figure 14.3.

Fate of the Blastopore

The first opening that is formed in an early embryo is called the **blastopore.** In some coelomate animals (protostomes), the blastopore becomes the mouth; in others (deuterostomes), the blastopore becomes the anus.

Segmentation

Segmentation is the serial repetition of body parts along the longitudinal axis.

SPONGES (PHYLUM PORIFERA)

Sponges are simple, primitive animals that probably evolved from flagellated protozoans. Most forms are marine; only a few live in freshwater. The radially symmetrical or asymmetrical adults live attached to rocks or other objects, but the larvae are ciliated and motile. Figure 14.4 shows the basic structure of a simple sponge. The cells are so loosely arranged into tissues

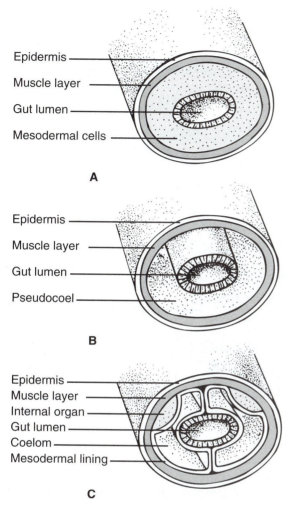

Epidermis
Muscle layer
Gut lumen
Mesodermal cells

A

Epidermis
Muscle layer
Gut lumen
Pseudocoel

B

Epidermis
Muscle layer
Internal organ
Gut lumen
Coelom
Mesodermal lining

C

Figure 14.3 **A.** Acoelomate, **B.** pseudocoelomate, and **C.** eucoelomate conditions in bilaterally symmetrical animals.

that the level of organization is intermediate between distinct cellular and tissue levels, a **cellular–tissue level of organization.**

Sponges are classified according to the type of **spicules,** i.e., skeletal elements, that they possess. Chalk sponges have calcium carbonate spicules, glass sponges have siliceous spicules, and fibrous sponges have a proteinaceous skeleton.

The body wall of a sponge is perforated by numerous pores. The beating flagella of the **choanocytes** (collar cells) maintain a constant flow of water through the **incurrent pores** into the **radial canals,** then into the **spongocoel** (central cavity) and out the **osculum.** Tiny food particles in the water are engulfed by the amoeboid action of the collar cells and digested in food vacuoles. Nutrients are distributed to other cells by diffusion, a process that is aided by the wandering **amoebocytes.**

Most sponges are hermaphroditic, but sperm and eggs are produced at different times, so self-fertilization is unlikely. Some sponges produce asexual reproductive bodies called **gemmules** that are released into the water. They later settle to the bottom and grow into a new sponge. Sponges readily regenerate lost parts, and separated parts can grow into new sponges.

Assignment 1
Materials

Colored pencils
Compound microscope
Stereo microscope
Grantia, preserved
Skeletons of representative sponges
Prepared slides of *Grantia,* l.s.

1. Color-code the collar cells, epidermis, amoebocytes, and spicules in Figure 14.4.
2. Examine whole and longitudinally sectioned *Grantia* with a stereo microscope. Note the incurrent pores, radial canals, spongocoel, and osculum.
3. Examine a prepared slide of *Grantia,* l.s., set up under a demonstration microscope. Observe the collar cells and spicules.
4. Examine the chalk, silica, and protein skeletons of representative sponges.
5. ***Complete item 1 on Laboratory Report 14 that begins on page 185.***

CNIDARIANS (PHYLUM CNIDARIA)

Cnidarians include jellyfish, sea anemones, corals, and hydroids. Most are marine, but a few occur in freshwater. They may live singly or in colonies and are characterized by (1) a saclike body plan with a single opening, the mouth, into a **gastrovascular cavity;** (2) a tissue level of organization, as evidenced by a body wall of two embryonic tissues, the ectoderm (epidermis) and endoderm (gastrodermis); (3) radial symmetry; and (4) **stinging cells** (cnidocytes) that eject dartlike weapons of offense and defense (nematocysts).

Two morphologic types of adults are recognized. See Figure 14.5. **Polyps** have tubular bodies with a mouth and tentacles at one end and usually are attached to the substrate by the other end. **Medusae** are free-swimming jellyfish that have umbrella- or bell-shaped bodies, with a mouth located in the center of the concave side and tentacles hanging from the edge of the bell. Some species exhibit *both* adult forms in an **alteration of generations** (metagenesis). In such a life cycle, medusae, which reproduce sexually by gametes,

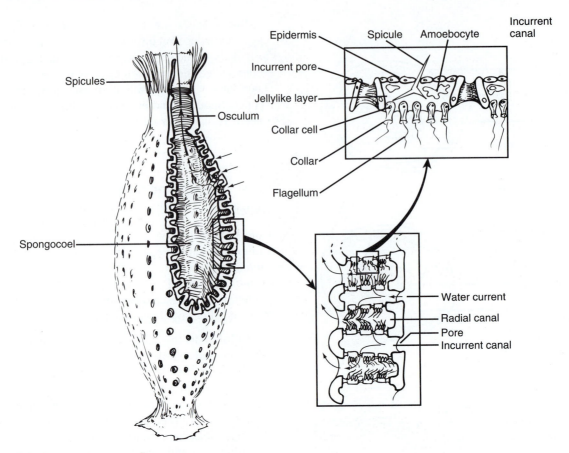

Figure 14.4 *Grantia,* a simple sponge.

alternate with polyps, which reproduce asexually by **budding.** See Figure 14.6.

Figure 14.7 shows the structure of *Hydra,* a common freshwater coelenterate that exhibits the basic characteristics of the phylum. Study it carefully. Locate the cnidarian features described previously, and note the various types of cells in the two layers of the body wall. What is located between the tissue layers?

Cnidarians are capable of simple movements enabled by the contraction of **myofibrils** located in cells of both

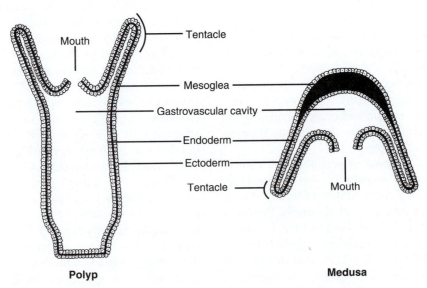

Figure 14.5 Comparison of polyp and medusa.

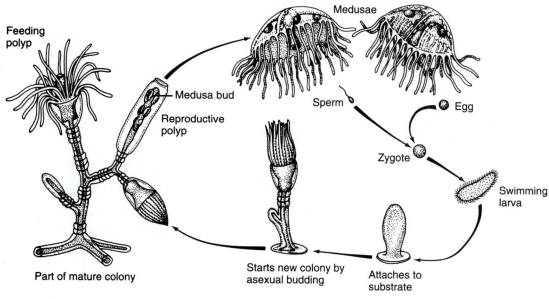

Figure 14.6 Life cycle of *Obelia*.

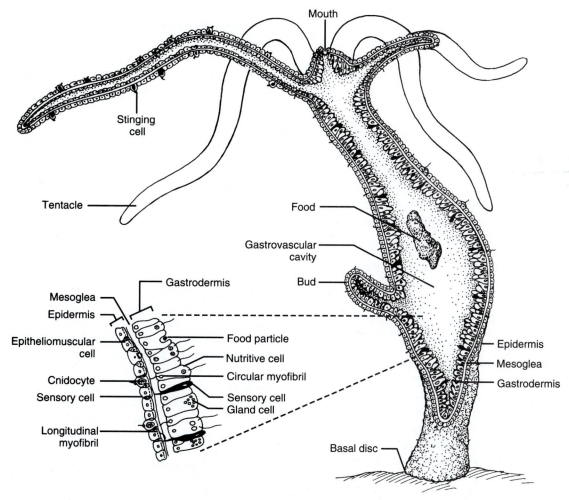

Figure 14.7 Longitudinal section of *Hydra* with detail of body wall.

tissue layers. Contraction of the longitudinally oriented myofibrils in the epidermal cells shortens the body, whereas contraction of the circularly oriented myofibrils in the gastrodermis extends the body. A simple **nerve net** located in the **mesoglea,** a noncellular layer between the two tissue layers, coordinates body functions.

Small prey organisms are paralyzed by poison injected by darts from the stinging cells and are engulfed into the gastrovascular cavity, where digestive enzymes initiate extracellular digestion. Subsequently, minute particles of partially digested food are engulfed by the gastrodermal cells to complete digestion within their food vacuoles. The distribution of nutrients to other cells is by diffusion.

Assignment 2

Materials

Beaker, 1,000 ml
Colored pencils
Compound microscope
Stereo microscope
Medicine droppers
Syracuse dish
Pond water
Daphnia, small, living
Elodea shoots
Hydra, living
Representative cnidarians
Prepared slides of:
   Hydra, l.s.
   Obelia polyps, w.m.
   Obelia medusae, w.m.

1. Color-code the ectoderm blue and the endoderm yellow in Figures 14.5 and 14.7 (detail of body wall only).
2. Examine the representative cnidarians. Learn their names and characteristics. Identify the polyps and medusae. Note the calcium carbonate skeletons of the corals.
3. Examine prepared slides of *Obelia* polyps and medusae. Compare your observations with Figure 14.6.
4. Watch the living *Hydra* in the large beaker or small aquarium capture and engulf *Daphnia.* Look for *Hydra* buds growing out of the side of the adult specimens. Note how a *Hydra* attaches itself by the **basal disc** and usually hangs downward with tentacles extended. What is the feeding strategy?
5. Place a *Hydra* in a Syracuse dish containing water from the culture jar. Then place the dish on the *black surface* of the stage of a stereo microscope for observation. Keep the lamp some distance away so the heat from the bulb does not affect the specimen. Observe how the *Hydra* extends itself when not disturbed and contracts when the dish is tapped. Which movements are more rapid, extensions or retractions? Locate the stinging cells that appear as bumps on the tentacles. Keep the lamp turned off except when observing.
6. Examine a prepared slide of *Hydra,* l.s., and locate the parts shown in Figure 14.7.
7. **Complete item 2 on the laboratory report.**

FLATWORMS (PHYLUM PLATYHELMINTHES)

Flatworms are distinguished by (1) a saclike body plan modified into an elongate, bilaterally symmetrical, and dorsoventrally flattened body; (2) the presence of all three embryonic tissues; (3) an organ level of organization; and (4) the absence of a coelom. The embryonic tissues develop into the tissues and organs of the adult. **Ectoderm** forms the epidermis and nervous tissue, and **endoderm** forms the inner lining of the gastrovascular cavity. **Mesoderm** forms muscles, reproductive organs, excretory organs, and the parenchyma cells that fill the space around the gastrovascular cavity. Animals taxonomically higher than flatworms are composed of all three embryonic tissues. Flatworms may be either free-living or parasitic. All flatworms are **hermaphroditic.** Sperm is exchanged during mating.

Free-Living Flatworms

A common planarian, *Dugesia,* is typical of nonparasitic forms. See Figure 14.8. The mouth and extensible **pharynx** are located at the middle of the ventral surface. Food is passed by the pharynx into the branched gastrovascular cavity, where digestion begins extracellularly and is completed intracellularly, as in cnidarians. Nutrients are distributed to other cells by diffusion.

Figure 14.9 shows the major organs of *Dugesia.* The concentration of neural structures at the anterior end is called **cephalization,** a condition that begins here and reaches its peak in humans. In *Dugesia,* the **auricles** contain touch and chemical receptors, and the **eyespots** are photoreceptors. The **cerebral ganglia** consist of concentrations of nerve cells that coordinate body movements via two widely separated **ventral nerve cords** and transverse connecting nerves. Waste products of metabolism diffuse into the excretory ducts and are flushed from the body through **excretory pores** by the current created by the beating flagella of **flame cells.**

Parasitic Flatworms

Tapeworms are parasitic flatworms that infest humans and other meat-eating animals. Adults live attached to the inner wall of the small intestine by suckers or hooks

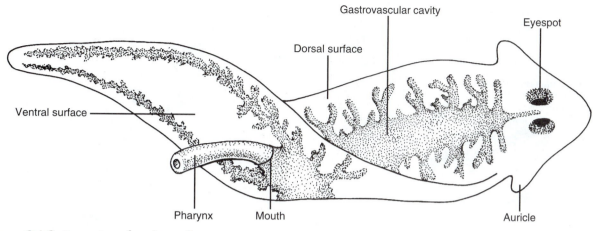

Figure 14.8 *Dugesia,* a free-living flatworm.

of the **scolex** and absorb nutrients through the body wall. The body consists of a series of **proglottids,** "reproductive factories" that are constantly formed by the scolex and broken off at the posterior end when they are full of fertile eggs. A gastrovascular cavity is absent,

and nutrients are absorbed directly through the body wall. Study the life cycle of the beef tapeworm (*Taenia*) shown in Figure 14.10.

Flukes are parasitic flatworms that infest the lungs, liver, and blood of vertebrates. The adult

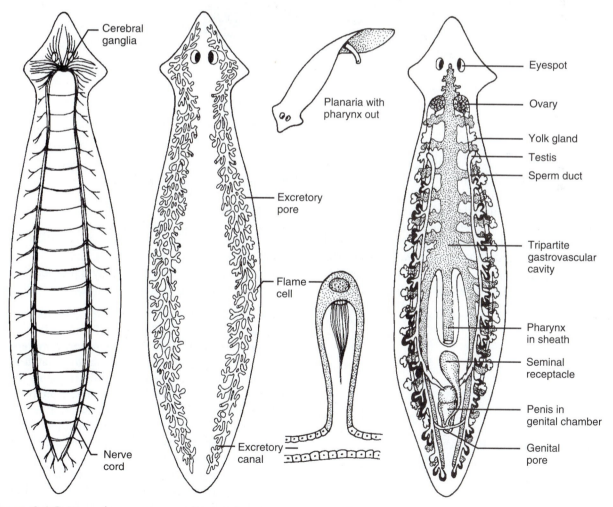

Figure 14.9 Neural, excretory, and reproductive structures in *Dugesia.*

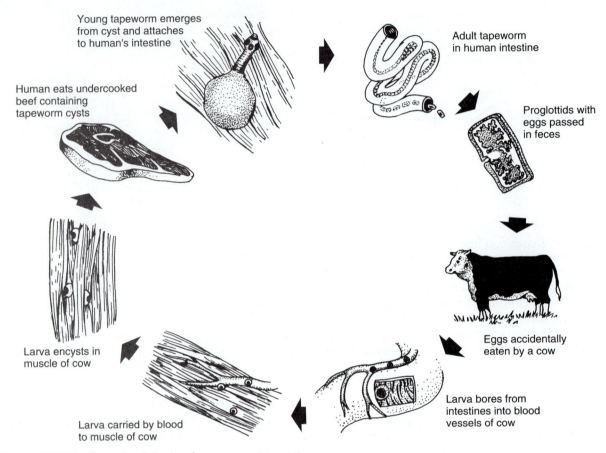

Young tapeworm emerges from cyst and attaches to human's intestine

Human eats undercooked beef containing tapeworm cysts

Adult tapeworm in human intestine

Proglottids with eggs passed in feces

Eggs accidentally eaten by a cow

Larva encysts in muscle of cow

Larva carried by blood to muscle of cow

Larva bores from intestines into blood vessels of cow

Figure 14.10 Life cycle of the beef tapeworm (*Taenia*).

human liver fluke (*Clonorchis*) lives in the bile duct of the human host. It is common in the Orient where the eating of raw fish is practiced. The larval forms are encysted in the muscles of certain fish. The human blood fluke (*Schistosoma*) has become a prominent parasite of humans in the Nile Valley of Egypt since the completion of the Aswan Dam. The dam increased irrigation and allowed the snails serving as the intermediate host of the blood fluke to flourish.

 Assignment 3

Materials

Compound microscope
Stereo microscope
Construction paper, black
Finger bowl
Medicine droppers
Petri dish
Stirring rod
Watch glass

Model of *Dugesia*
Dugesia, living
Tapeworms, preserved
Prepared slides of:
    Clonorchis (human liver fluke), w.m.
    Dugesia, w.m. and x.s.
    Schistosoma (human blood fluke), w.m.
    tapeworm scolex, mature and gravid proglottids
    tapeworm larvae in muscle

1. Examine the model of *Dugesia* and locate the structures shown in Figures 14.8 and 14.9.
2. ***Complete items 3a–3d on the laboratory report.***
3. Place a living planarian, *Dugesia,* in a watch glass with enough water from the culture jar so it can move about. Observe it with a stereo microscope over the white surface of the stage. Locate the external structures, and note the manner of movement. Is there a pattern to the movement? What allows the movement?
4. Examine a prepared slide of *Dugesia,* x.s. Note the gastrovascular cavity, body wall, and cilia on the ventral surface. What is the function of the cilia?

5. Lift the black paper that partially shades the dish of flatworms, and note their distribution. Replace the paper in its exact prior position.

6. On the demonstration table is a dish containing about 25 planarians. Note their position and direction of movement. Swirl the water by stirring it clockwise with a glass rod. As the water continues to swirl, note the direction of movement of the planarians.

7. **Complete items 3e and 3f on the laboratory report.**

8. Examine the preserved tapeworms and estimate their lengths. Locate the proglottids and scolex.

9. Examine a prepared slide of tapeworm scolex and mature proglottids. Note how the scolex attaches to the intestine wall, and locate the fertile eggs in the mature proglottids.

10. Examine a prepared slide of tapeworm larvae encysted in beef muscle. Note the inverted scolex.

11. Examine the demonstration slides of flukes. Note the name and characteristics of each.

12. **Complete items 3g–3j on the laboratory report.**

ROUNDWORMS (PHYLUM NEMATODA)

Roundworms are characterized by (1) a cylindrical, bilaterally symmetrical body that is tapered at each end; (2) a tube-within-a-tube body plan; (3) an organ system level of organization; and (4) a false coelom. Most roundworms are free living, but some are important parasites of animals and plants.

Ascaris is a large roundworm parasite that lives in the small intestine of humans and swine and feeds on partially digested food. Its life cycle is shown in Figure 14.11. Females can be distinguished from

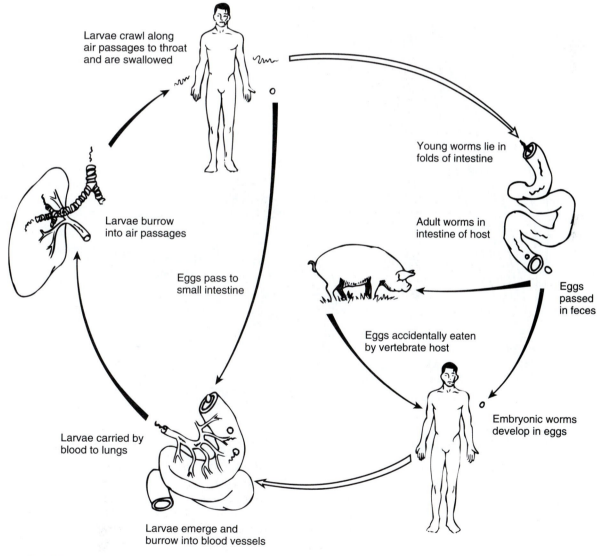

Larvae crawl along air passages to throat and are swallowed

Larvae burrow into air passages

Eggs pass to small intestine

Young worms lie in folds of intestine

Adult worms in intestine of host

Eggs passed in feces

Eggs accidentally eaten by vertebrate host

Embryonic worms develop in eggs

Larvae carried by blood to lungs

Larvae emerge and burrow into blood vessels

Figure 14.11 *Ascaris* life cycle.

males because they are larger and do not have a curved posterior end.

The structure of a female *Ascaris* is shown in dissection and in cross-section in Figure 14.12. The body wall consists of longitudinal muscles, dermis, and a cuticle that is secreted by the epidermis. The **cuticle** protects the worm from the digestive action of enzymes in the host's small intestine. The space between the intestine and the body wall is a **pseudocoelom** (false coelom) or **pseudocoel** because it is *incompletely* lined with mesodermal tissue. The female reproductive system and the ribbonlike intestine occupy most of the

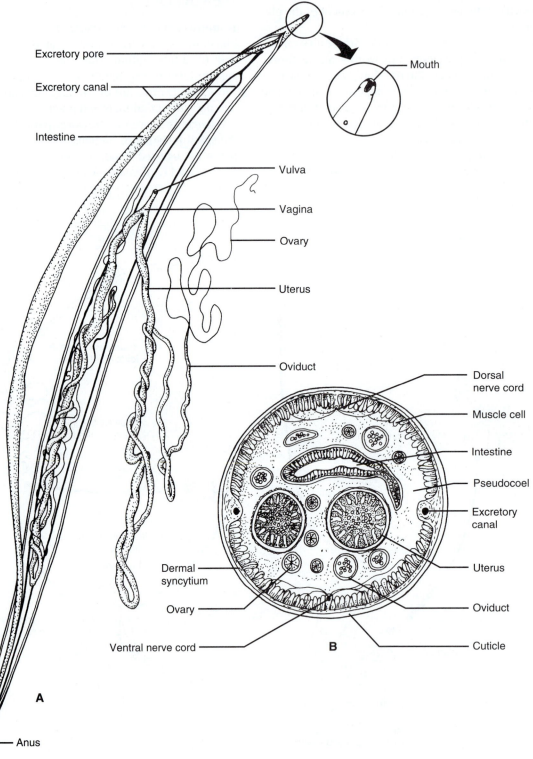

Figure 14.12 Female *Ascaris*. **A.** Dissection. **B.** Cross-section.

pseudocoelom. Note the locations of the nerve cords and the excretory canals.

Assignment 4

Materials

Colored pencils
Compound microscope
Stereo microscope
Dissecting instruments
Dissecting pan, wax bottomed
Dissecting pins
Microscope slides and cover glasses
Protoslo
Toothpicks
Ascaris, preserved
Turbatrix culture
Prepared slides of:
    Ascaris female, x.s
    Ascaris eggs
    Ascaris larvae
    Trichinella spiralis, encysted larvae
    Wuchereria bancrofti, w.m.

1. Perform a dissection of a female *Ascaris* as described here to observe the basic roundworm characteristics.
   a. Pin each end of a preserved female *Ascaris* to the bottom of a wax dissecting pan.
   b. Being careful not to damage the internal organs, slit the body wall from one end to the other with the tips of scissors. Pin out the body wall to expose the internal organs. Add water to the pan to cover the specimen so it does not dry out.
   c. Use a probe to carefully separate the parts of the coiled female reproductive organs. Locate the structures shown in Figure 14.12. The excretory pore is difficult to find. Note the ribbonlike intestine. Organisms with a tube-within-a-tube organization have a **complete digestive tract,** with a mouth at one end and an anus at the other.
2. On Figure 14.12, color-code the major structures.
3. Examine a prepared slide of a female *Ascaris,* x.s., and locate the structures labeled in Figure 14.12.
4. Study the *Ascaris* life cycle in Figure 14.11. Examine the prepared slides of *Ascaris* eggs and larvae.
5. Examine the prepared slides of these roundworms:
   a. *Trichinella spiralis* larvae encysted in muscles of the host. Adults live in the small intestine. Humans are infected by eating encysted larvae in undercooked pork.
   b. *Wuchereria bancrofti,* tiny roundworms that plug lymphatic vessels and cause **elephantiasis.** They are transmitted by the bite of certain mosquitoes.
6. Make and examine a slide of the *Turbatrix* (vinegar eels) culture. Note the motility of these nonparasitic roundworms. Mix a bit of Protoslo with the culture fluid on your slide. Add a cover glass and observe a nonmoving specimen.
7. ***Complete item 4 on the laboratory report.***

Assignment 5

Your instructor has prepared a number of "unknown" specimens for you to identify. ***Record your responses in item 5 on the laboratory report.***

SIMPLE ANIMALS

Student _____

Lab Instructor _____

1. SPONGES

a. Write the term that matches the phrase.

1. Level of organization _____

2. Type of symmetry _____

3. Cells that create the water current _____

4. Cells that engulf and digest food _____

5. Skeletal elements of sponges _____

b. Trace the path of water flowing through *Grantia*.

In → _____ → _____ → _____ → _____ → Out

c. Describe the manner of feeding and the distribution of nutrients. _____

_____

_____

_____

2. CNIDARIANS

a. Write the term that matches the phrase.

1. Type of symmetry _____

2. Type of body plan _____

3. Embryonic tissues present _____

4. Level of organization _____

5. Contraction enables movement _____

6. Coordinates movement _____

7. Special cells used to capture prey _____

8. Method of nutrient dispersal _____

b. Distinguish the following stages by structure and function:

Polyp _____

_____

Medusa _____

_____

c. Show by diagram the feeding process as observed in *Hydra*.

Feeding Position	**Capture of *Daphnia***	**Engulfment**

d. Digestion in cnidarians is a two-step process. Describe it.

1. _____

2. _____

_____

e. Diagram a short segment of *Hydra* tentacle showing the stinging cells.

f. In *Hydra*, is extension or retraction more rapid? _____

3. FLATWORMS

a. Write the term that matches the phrase.

1. Type of symmetry   _____

2. Type of body plan   _____

3. Level of organization   _____

4. Embryonic tissues present   _____

5. Method of nutrient dispersal   _____

6. Enable gliding movement   _____

b. Digestion is a two-step process. Describe it.

1. _____

2. _____

c. What is the advantage of the highly branched gastrovascular cavity? _____

_____

d. What is the advantage of cephalization? _____

_____

e. Describe the distribution of *Dugesia* in the partially covered dish. _____

_____

_____

Interpret this pattern. _____

_____

f. Describe the orientation of *Dugesia* to the direction of water movement after swirling the water in the dish. ____

_____

Interpret this pattern. _____

_____

g. Considering the life cycle of the beef tapeworm, state two ways to prevent infestation in beef-eating humans.

1. _____

2. _____

h. List the names and estimated lengths of the demonstration tapeworms.

_____   _____

i. Draw a scolex, gravid proglottid with eggs, and an encysted larva of a beef tapeworm from your slides. Label pertinent structures.

Scolex   **Gravid Proglottid**   **Encysted Larva**

j. Sketch the flukes observed on the prepared slides. Provide sufficient detail to identify the flukes. Indicate their names and where the adults live in the host.

4. ROUNDWORMS

a. Write the term that matches the phrase.

1. Type of symmetry   _____

2. Type of body plan   _____

3. Level of organization   _____

4. Type of coelom   _____

5. Type of digestive tract   _____

b. Where does the adult *Ascaris* live in the host? _____

What is the function of the cuticle? _____

c. Describe the developmental pathway of *Ascaris* after a human swallows an egg until the adult lies in the small intestine. _____

_____

_____

_____

_____

d. Considering the life cycle of *Ascaris,* list two ways to prevent human infestation.

1. _____

2. _____

e. Draw a few eggs and larvae (in lung tissue) of *Ascaris* from your slides.

Eggs   **Larvae**

f. Draw the outline of one to two specimens of these roundworms from your slides to show relative size. Indicate the magnification used in your observations.

Trichinella spiralis   **Wuchereria bancrofti**   **Turbatrix aceti**

g. For each roundworm parasite observed, indicate (1) where the adult lives in the human host, and (2) how humans are infected.

Ascaris   1. _____

                       2. _____

Trichinella   1. _____

                       2. _____

Wuchereria   1. _____

                       2. _____

_____   1. _____

                       2. _____

5. UNKNOWNS

Place the numbers of the "unknown" specimens in the correct spaces.

_____ Sponge   _____ Tapeworm larva

_____ Cnidarian polyp   _____ Fluke

_____ Cnidarian medusa   _____ Roundworm adult

_____ Tapeworm adult   _____ None of the above

MOLLUSKS, SEGMENTED WORMS, AND ARTHROPODS

OBJECTIVES

After completing the laboratory session, you should be able to:
1. Describe the distinguishing characteristics of the groups and organisms studied.
2. Identify representatives of the animal groups studied.
3. Identify and state the function of the structural components of the animals studied.
4. Define all terms in bold print.

Mollusks, segmented worms, and arthropods compose the **protostomate animals,** in which the blastopore of the early embryo forms the mouth. All these animals exhibit (1) an **organ system level of organization,** (2) all three **embryonic tissues,** (3) **bilateral symmetry,** (4) a **tube-within-a-tube body plan,** and (5) a **true coelom.**

MOLLUSKS (PHYLUM MOLLUSCA)

The molluscan body plan is characterized by (1) a **ventral muscular foot,** (2) a **dorsal visceral mass,** (3) a **mantle** that surrounds the visceral mass and often secretes a **shell,** and (4) a **radula,** a filelike

mouthpart used to scrape off bits of food. This plan has been greatly modified among the different groups of mollusks, however. See Table 15.1. Mollusks are primarily marine organisms, but many occur in freshwater and some are terrestrial.

Clams (Class Bivalvia)

The basic molluscan structure as found in the freshwater clam is shown in Figure 15.1. The body is flattened laterally and enclosed within the two valves (halves) of the shell, which are hinged together at the dorsal surface. Thus, this class of mollusks is commonly called bivalves. The valves may be tightly closed by strong **adductor muscles.** A digging foot is used to bury the clam in mud or sand, often with only the tips

TABLE 15.1	DISTINGUISHING CHARACTERISTICS OF MOLLUSCAN CLASSES
Class	Characteristics
Monoplacophora (*Neopilina*)	Primitive; remnants of segmentation; single shell; dorsoventrally flattened foot; radula; marine
Polyplacophora (chitons)	Shell of eight overlapping plates; dorsoventrally flattened foot; radula; marine
Scaphapoda (tooth shells)	Conical shell open at each end; digging foot; radula; marine
Gastropoda (snails and slugs)	Shell absent or single, often coiled; dorsoventrally flattened foot; radula; marine, freshwater, and terrestrial
Bivalvia (clams, oysters)	Shell of two valves, hinged dorsally; laterally flattened foot; no radula; filter-feeder; marine or freshwater
Cephalopoda (octopi, squids)	Single shell or none; foot modified to form tentacles; image-producing eyes; horny beak and radula; predaceous; water-jet propulsion; marine

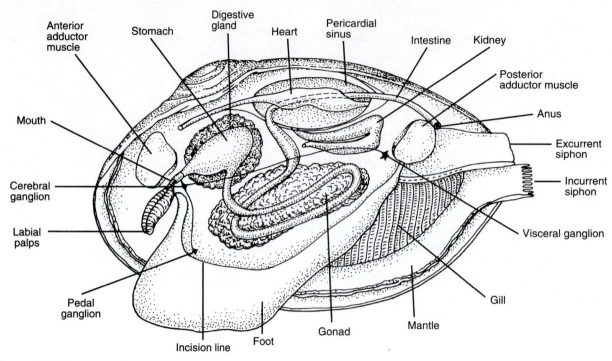

Figure 15.1 Dissected clam (*Anodonta*).

of the siphons protruding. Water is brought into the **mantle cavity** via the **incurrent siphon** and exits via the **excurrent siphon.** Gas exchange occurs between the water in the mantle cavity and blood in the **gills.** Tiny food particles in the water are trapped in mucus coating the gills, and both mucus and food are carried by cilia to the **labial palps** and on into the mouth. Digestion occurs extracellularly in the digestive tract, and nutrients are absorbed into the blood. Blood carries nutrients and oxygen to body cells and carries metabolic wastes and carbon dioxide from the cells to excretory organs. An **open circulatory system** is present. Blood is carried from the heart in vessels that empty the blood into sinuses. The blood moves slowly through the sinuses and back into the heart. Nitrogenous metabolic wastes are removed by a "kidney."

Octopi and Squids (Class Cephalopoda)

Squid and octopi represent the peak of molluscan evolution. In contrast to the filter-feeding clam, cephalopods are predaceous. See Figure 15.2. The foot is modified to form "arms" that have suction cups and are used to capture prey. Specialized arms are called tentacles. Food is eaten with the aid of a **horny beak.** Water enters the mantle cavity via a **siphon** and bathes the gills, enabling gas exchange to occur. Water also can be forcefully ejected through the siphon, which provides a water-jet method of propulsion. Usually squid swim backward, but they also can swim forward,

depending on the direction in which the tip of the siphon is pointing. Ink may be ejected to form a smoke screen when cephalopods are trying to escape from enemies. Octopi lack a shell. In squids, the shell is reduced to a thin, transparent plate called a **pen** that is covered by the mantle and serves to stiffen and support the body. The large **image-forming eyes** superficially resemble vertebrate eyes, but they are formed quite differently.

 ## Assignment 1

Materials

Colored pencils

1. *Complete item 1 on Laboratory Report 15 that begins on page 201.*
2. Color-code the major structures in Figures 15.1 and 15.2.

 ## Assignment 2

Materials

Clam, dissected demonstration
Representative mollusks, living or preserved
Squids, fresh or preserved
Model of dissected clam
Prepared slides of radula

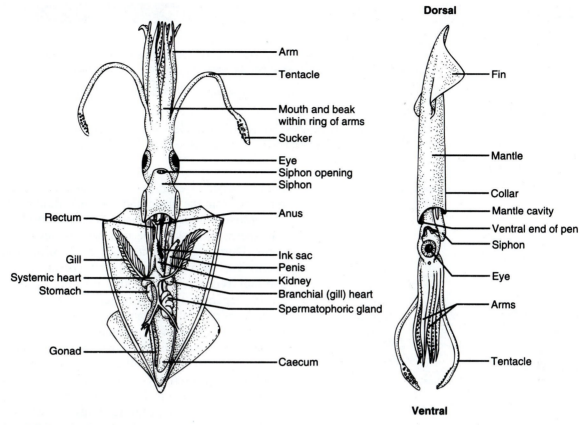

Figure 15.2 Squid (*Loligo*).

1. Examine the representative mollusks. Use Table 15.1 to determine the molluscan class to which each belongs. Note the distinguishing traits of the representatives of each group.
2. Examine a prepared slide of a radula set up under a demonstration microscope. Note the filelike teeth.
3. Use Table 15.1 to determine the class of each "unknown" mollusk.
4. *Complete items 2a–2d on the laboratory report.*
5. Examine the demonstration dissected clam. Compare it with the model of a dissected clam and Figure 15.1. Locate the principal parts.
6. On a preserved or fresh squid, locate the external features shown in Figure 15.2. If time permits, dissect it to observe the internal parts.
7. *Complete item 2 on the laboratory report.*

SEGMENTED WORMS (PHYLUM ANNELIDA)

The major characteristic of these worms is a **segmented body** formed of repeating units called **somites.** Most of the somites are basically similar, including the internal arrangement of muscles, nerves, blood vessels, and excretory ducts. Most annelids are marine, but many occur in freshwater and terrestrial habitats. Earthworms, sandworms, and leeches represent the three major classes of segmented worms. See Table 15.2.

Earthworms

The earthworm exhibits the basic characteristics of segmented worms. Note the external features shown in Figure 15.3. The **prostomium** is a small projection over the mouth. The **clitellum** appears as a smooth band around the worm. It forms a ring of mucus around copulating worms that later becomes the egg case for the fertile eggs. Earthworms are **hermaphroditic** and exchange sperm during copulation.

The internal structure of an earthworm is shown in Figures 15.4 and 15.5. Note the **septa** that separate the segments. Parts of the digestive tract are specialized. The muscular **pharynx** aids in the ingestion of food, which passes through the **esophagus** to the **crop** and then enters the muscular **gizzard.** The gizzard grinds the food into smaller pieces before it enters the intestine, where extracellular digestion occurs. Nutrients are absorbed into the blood and carried to all parts of the body. Annelids, unlike most mollusks, have a **closed circulatory system,** which means that blood is always contained in blood vessels. Contractile

TABLE 15.2	DISTINGUISHING CHARACTERISTICS OF ANNELID CLASSES
Class	Characteristics
Polychaeta (sandworms)	Head with simple eyes and tentacles; segments with lateral extensions (parapodia) and many bristles (setae); sexes usually separate; predaceous; marine
Oligochaeta (earthworms)	No head; segments without extensions and with few, small bristles; hermaphroditic; terrestrial or freshwater
Hirudinea (leeches)	No head; segments with superficial rings and without lateral extensions or bristles; anterior and posterior suckers; hermaphroditic; parasitic, feeding on blood; terrestrial or freshwater

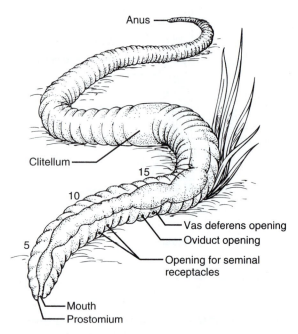

Figure 15.3 Earthworm (*Lumbricus*).

segmental vessels ("hearts"), located around the esophagus, pump the blood throughout the body.

Paired **cerebral ganglia** form the "brain" that lies dorsal to the pharynx. It is connected to the **ventral nerve cord** that runs the length of the body along the ventral midline. A **segmental ganglion** and **lateral nerves** occur in each somite. Metabolic wastes are removed from the blood and coelomic fluid by a pair of **nephridia** in each segment. **Seminal vesicles** contain sperm formed by the two pairs of **testes** within them. During copulation, both earthworms release sperm via **vasa deferentia,** and sperm enters the **seminal receptacles** of both earthworms. After the worms separate, sperm is released from the seminal receptacles to fertilize eggs that are formed by a pair of **ovaries** and released via **oviducts.**

Earthworm structure is shown in cross-section in Figure 15.6. The body wall consists of **circular** and **longitudinal muscle layers, epidermis,** and a **cuticle** formed by the epidermal cells. The dorsal part of the intestine wall, the **typhlosole,** folds inward and increases the surface area of the intestinal lining. Four pairs of **setae** are on each body segment. (What is their function?) Note the blood vessels, ventral nerve cord, and nephridium.

 ## Assignment 3

Materials

Colored pencils
Dissecting instruments
Stereo microscope
Dissecting pan, wax bottomed
Dissecting pins

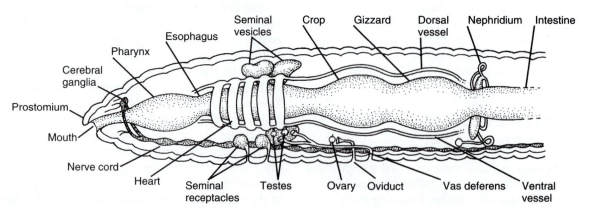

Figure 15.4 Earthworm internal structure, lateral view.

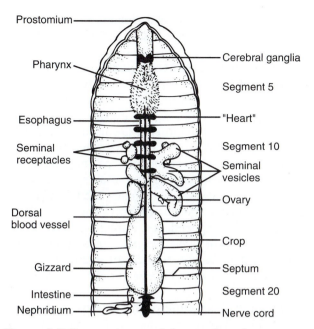

Figure 15.5 Dorsal view of dissected earthworm.

Model of dissected earthworm
Earthworms, preserved
Representative annelids
Prepared slides of earthworm, x.s.

1. Color-code the parts of the digestive tract and the nephridia in Figures 15.4 and 15.5.
2. Examine the representative segmented worms, and note their similarities and differences. Compare the characteristics of the specimens with those given in Table 15.2. Note the adaptations of each specimen.
3. Use Table 15.2 to determine the class of each of the "unknown" annelids.
4. ***Complete items 3a–3c on the laboratory report.***
5. Locate the parts labeled in Figures 15.3, 15.4, and 15.5 on the model of the earthworm.
6. Examine a prepared slide of earthworm, x.s., and locate the structures shown in Figure 15.6.
7. ***Complete items 3d.1 and 3d.2 on the laboratory report.***
8. Perform an earthworm dissection as described here.
   a. Obtain a preserved earthworm and locate the external features, as shown in Figure 15.3.
   b. Pin the earthworm, *dorsal side up,* to the wax bottom of a dissecting pan by placing a pin through the prostomium and another pin through the body posterior to the clitellum. Pin the worm about 5–7 cm from one edge of the pan to facilitate observation of the structures with a stereo microscope. *Note:* The dorsal surface is convex and is usually darker in color.
   c. Use a sharp scalpel to cut *only* through the body wall, along the *dorsal midline* from the clitellum to the prostomium. Be careful to avoid cutting the underlying organs.
   d. Use a dissecting needle to break the septa as you spread out the body wall and pin it to the pan. This will expose the internal organs. Add water to cover the worm to prevent it from drying out.

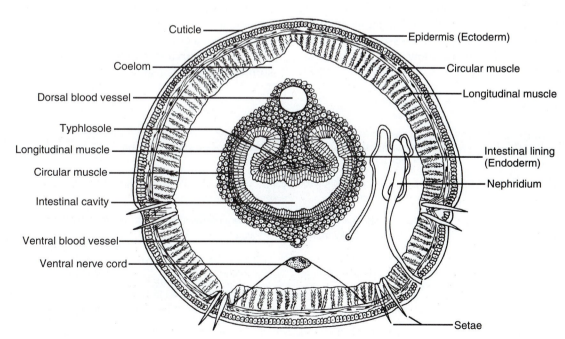

Figure 15.6 Earthworm (*Lumbricus*), x.s.

e. Locate the structures in Figures 15.4 and 15.5. Carefully dissect away surrounding tissue to locate the brain. Use a stereo microscope to observe the smaller structures.

f. Remove the crop, gizzard, and part of the intestine to expose the underlying nerve cord with its segmental ganglia. Cut open the crop and gizzard and compare the thickness of their walls.

9. **Complete item 3d on the laboratory report.**

ARTHROPODS (PHYLUM ARTHROPODA)

Arthropods include more species than any other phylum of animals. They occur in almost every habitat on Earth and usually are considered to be the most successful group of animals because of their vast numbers. Major characteristics of arthropods are (1) a segmented body divided into either **cephalothorax** and **abdomen**, or **head, thorax,** and **abdomen;** (2) **jointed appendages;** (3) an **exoskeleton** of chitin; (4) **simple** or **compound eyes;** and (5) an **open circulatory system.** Table 15.3 lists the characteristics of the major classes.

The Crayfish

The freshwater crayfish is a common crustacean that exhibits the basic characteristics of arthropods. Examine its external structure in Figure 15.7. Most of the exoskeleton is hardened by the accumulation of mineral salts. It provides protection for the soft underlying body parts, and it is flexible at the joints for easy movement of the appendages. The exoskeleton is shed periodically by molting to allow for growth of the crayfish.

The segmented body is divided into a cephalothorax and abdomen. Each body segment has a pair of jointed appendages, and most have been modified for special functions. **Mouthparts** are modified for feeding, and the two pairs of **antennae** contain numerous sensory receptors. The first three pairs of **walking legs** have pincers for grasping, and the first pair of these, the **chelipeds,** are used for offensive and defensive purposes. The **swimmerets** are relatively unmodified appendages on the abdominal segments. In females, they serve as sites of attachment for fertile eggs, and in males, the first two pairs are modified for sperm transport. The **uropods** are the last pair of appendages. Along with the **telson,** the uropods form a "paddle" used in swimming. The **compound eyes** are formed of many individual photosensory units and occur in only certain groups of arthropods.

The internal structure of the crayfish is shown in Figure 15.8. The **heart** consists of a single chamber and lies just under the **carapace** at the dorsal midline in the posterior part of the cephalothorax. **Arteries** carry blood anteriorly, posteriorly, and ventrally into the haemocoel, which spreads the blood so it is near the body tissues for the exchange of materials. Blood returns from the haemocoel into the **pericardial sinus** surrounding the heart and then floats into the heart through tiny, valved openings called **ostia.** Just ventral to the heart lie a pair of elongated **gonads** (testes or ovaries) whose ducts lead to the genital openings. The large digestive glands surround the gonads and stomach and fill much of the cephalothorax.

The digestive system consists of the mouth, esophagus, stomach, digestive glands, intestine, and anus. Food is crushed by **mandibles** and passed into the anterior **cardiac chamber** of the stomach. The stomach contains a **gastric mill** that further grinds

Class	Characteristics
Arachnida (spiders, scorpions)	Cephalothorax and abdomen; four pairs of legs; simple eyes; no antennae; mostly terrestrial
Crustacea (crabs, shrimp)	Cephalothorax and abdomen; compound eyes; two pairs of antennae; five pairs of legs; mostly marine or freshwater
Chilopoda (centipedes)	Elongate, dorsoventrally flattened body; each body segment with a pair of legs; one pair of antennae; simple eyes; terrestrial
Diplopoda (millipedes)	Elongate, dorsally convexed body; segments fused in pairs, giving *appearance* of two pairs of legs per segment; one pair of antennae; simple eyes; terrestrial
Insecta (insects)	Head, thorax, and abdomen; three pairs of legs on thorax; wings, if present, on thorax; simple and compound eyes; one pair of antennae; freshwater or terrestrial

TABLE 15.3   DISTINGUISHING CHARACTERISTICS OF ARTHROPOD CLASSES

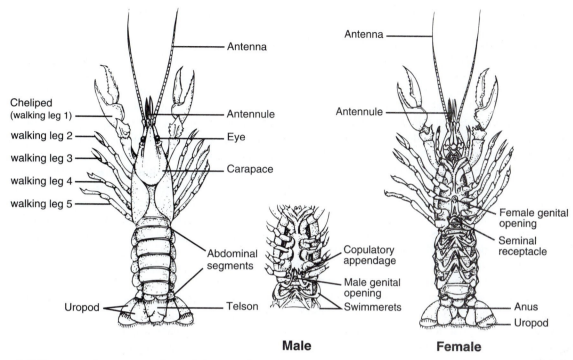

Figure 15.7 External structure of crayfish (*Cambarus*).

the food into particles small enough to pass through a "strainer" formed by bristles into the posterior **pyloric chamber.** The finest particles enter ducts of the digestive glands, where digestion is completed and nutrients are absorbed into the blood. Larger particles pass into the intestine and are removed from the body via the anus.

Metabolic wastes are removed from the blood by the **green glands** (flesh-colored in preserved crayfish)

and excreted via ducts that open at the base of the antennae.

The nervous system includes a pair of **supra-esophageal ganglia** ("brain"), a ventral nerve cord, a series of segmental ganglia, and lateral nerves in each segment. Some of the segmental ganglia have fused to form larger ganglia in the cephalothorax. Crayfish have a variety of sensory receptors, especially on the anterior appendages.

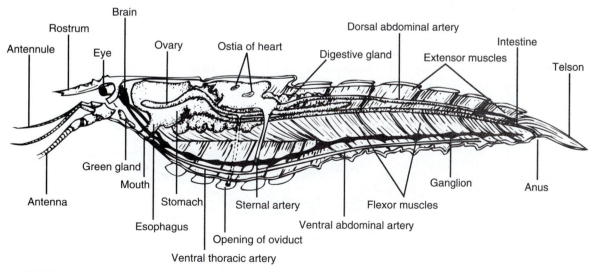

Figure 15.8 Internal structure of female crayfish.

Insects (Class Insecta)

Insects are the largest group of arthropods. They are extremely abundant in terrestrial habitats, and many occur in freshwater. They do not occur in marine habitats, however. Figure 15.9 shows the structure of a grasshopper, a common insect. Note how the body is modified from the more generalized structure of the crayfish. The body is divided into head, thorax, and abdomen. Only three pairs of legs are present, and they are attached to the thorax, which also supports two pairs of **wings.** Abdominal appendages are absent. Both simple and compound eyes are present, but only one pair of antennae occurs. The **tympanum** is a sound receptor. The **spiracles** are openings into the **tracheal system,** a network of tubules that carry air directly to the body tissues.

Internally, the **gastric caeca** are glands that aid in digestion. Metabolic wastes are removed by **malpighian tubules** that empty into the intestine. The dorsal blood vessel contains a series of "hearts" that pump blood anteriorly, where it empties into the haemocoel and then slowly returns to enter the dorsal vessel.

Most insects undergo some form of **metamorphosis,** a hormonally controlled process that changes the body form of an insect at certain stages of the life cycle. Four stages occur in **complete metamorphosis.** The **egg** hatches to yield a wormlike **larva,** which grows through several molts and then turns into a **pupa,** a nonfeeding stage. The **adult** emerges from the pupa. Typically, larvae and adults use different food sources, and unfavorable conditions (e.g., winter) are passed in the egg or pupal stage. Insects with **incomplete metamorphosis** have only three life stages. The egg hatches into a **nymph** (terrestrial forms) or **naiad** (aquatic forms) that passes through several molts, with the sexually mature adult emerging at the last molt. A **nymph** (e.g., grasshopper) resembles a miniature adult, and at each molt, it looks more and more like the winged adult that emerges at the final molt. A **naiad** (e.g., dragonfly) does not resemble the adult. It lives in a different habitat (water) and uses a different food source than the winged adult that emerges at the last molt. Unfavorable conditions are usually passed in the egg or naiad stage.

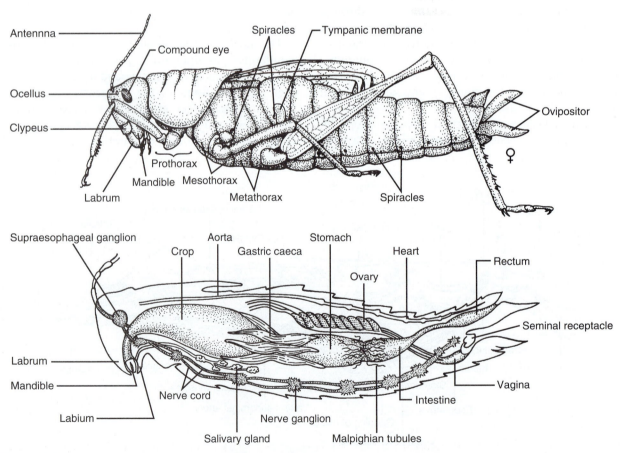

Figure 15.9 Anatomy of a female grasshopper (*Romalea*).

Assignment 4

Materials

Colored pencils
Dissecting instruments
Stereo microscope
Dissecting pans, wax bottomed
Dissecting pins
Examples of insect metamorphosis
Preserved dissection specimens:
   crayfish (*Cambarus*)
   grasshoppers (*Romalea*)
   representative arthropods

1. Color-code the major structures in Figures 15.7–15.9.
2. Examine the representative arthropods. Compare the characteristics of the specimens with those listed in Table 15.3. Note the adaptations of each based on the arthropod body plan.
3. Use Table 15.3 to determine the class of each of the "unknown" arthropods.
4. Examine the insect metamorphosis display. Note the components of each type.
5. *Complete items 4a–4e on the laboratory report.*
6. Perform a dissection of a crayfish as described here.
  a. Obtain a preserved crayfish and locate the external parts, as shown in Figure 15.7. Correlate structure with function. Locate the flexible and rigid parts of the exoskeleton. Note the similarities and differences among the walking legs. Observe an eye and some of the appendages under a stereo microscope. *Draw the appearance of the eye surface in item 4g.4.*
  b. The primary incisions to be made in the dissection of a crayfish are shown in Figure 15.10. Make the first incision on the left side by inserting the tips of scissors under the posterior margin of the carapace and cutting forward to the cervical groove and downward to the ventral edge of the carapace. This removes the left side of the carapace and exposes the gills.
  c. Examine the gills. To what are they attached? Note the thin transparent body wall medial to the gills.
  d. Make the second incision anteriorly, nearly to the eye, and then downward to remove the anterior lateral portion of the carapace. This will expose the heavy mandibular muscle and lateral wall of the stomach.
  e. Make the third incision as shown, and carefully remove the median strip of carapace by gently separating it from the underlying tissues. This will expose the heart and other dorsally located structures. Remove the heart and locate the ostia.

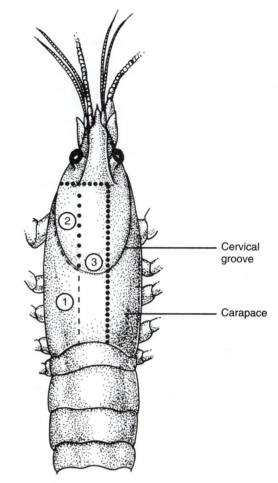

Figure 15.10 Sequential incisions in the crayfish dissection.

  f. Locate the internal structures shown in Figure 15.8. Start with the most dorsal organs, and gently remove them after observation to expose the underlying structures. Proceed in this manner until all the organs have been identified and removed. After removing the stomach, cut it open lengthwise and wash out the contents to expose the gastric mill and the bristles forming the "strainer."
  g. After the internal organs have been removed, you should be able to observe the ganglia along the ventral nerve cord. Try to locate the "brain" dorsal to the esophagus.
  h. Make a transverse cut through the abdomen, and examine the cut surface. Locate the muscles, intestine, and nerve cord.
7. *Complete item 4f on the laboratory report.*
8. Perform a dissection of a grasshopper as described here.
  a. Obtain a preserved grasshopper and locate the external parts shown in Figure 15.9. Note that both legs and wings arise from the thorax. Locate

the spiracles on the abdominal segments. Observe the compound and simple eyes with a compound microscope. ***Draw the appearance of the compound eye in item 4g.4.***

b. To observe the internal structures, make an incision to the right of the midline with your scissors. Start at the posterior end and cut anteriorly. Pin out the body wall to expose the internal organs. When you do this, you will notice the tracheal trunks extending from each spiracle into the body tissues.

c. The dorsal vessel is difficult to locate unless you have made a careful incision. Start with the most dorsal organs and remove them gently after

observation to expose the underlying structures. After the digestive tract has been removed, you will see the large thoracic ganglia and the ventral nerve cord.

9. ***Complete item 4g on the laboratory report.***

 Assignment 5

Your instructor has numbered selected structures in a clam, crayfish, and grasshopper for you to identify as a way to check your understanding. ***Complete item 5 on the laboratory report.***

MOLLUSKS, SEGMENTED WORMS, AND ARTHROPODS

Student _____

Lab Instructor _____

1. PROTOSTOMATES

Write the term that matches the phrase.

1. Embryonic opening forming mouth _____

2. Type of coelom _____

3. Level of organization _____

4. Body plan _____

5. Type of symmetry _____

6. Embryonic tissues present _____

2. MOLLUSKS

a. List the four distinguishing characteristics of mollusks.

1. _____   3. _____

2. _____   4. _____

b. Matching

1. Monoplacophorans   2. Chitons   3. Snails and slugs   4. Tooth shells

5. Clams and mussels   6. Octopi and squids

_____ Image-forming eyes   _____ Conical shell open at each end

_____ Radula present   _____ Shell of eight plates

_____ Remnants of segmentation   _____ Shell of two valves

_____ All predaceous   _____ Filter-feeders

_____ Digging foot   _____ Crawl on flattened foot

_____ Marine only   _____ Foot modified into tentacles

c. Draw a radula from a prepared slide showing the filelike teeth.

d. Place the number of the "unknown" mollusks in the correct spaces.

_____ Monoplacophorans   _____ Snails and slugs

_____ Chitons   _____ Clams, oysters, and mussels

_____ Tooth shells   _____ Octopi and squids

e. Clams

  1. Write the term that matches the phrase.

    a. Covers visceral mass and secretes shell   _____

    b. Muscles closing valves of shell   _____

    c. Brings water into mantle cavity   _____

    d. Used to dig into mud or sand   _____

    e. Organs of gas exchange   _____

  2. Describe the feeding process. _____

_____

_____

_____

f. Squid

List the adaptations for a predaceous, swimming lifestyle. _____

_____

_____

3. ANNELIDS

a. What is the primary distinguishing characteristic of annelids? _____

b. Matching

  1. Oligochaetes   2. Polychaetes   3. Leeches

_____ Head with simple eyes   _____ Blood-sucking parasites

_____ Few setae   _____ Parapodia and many setae

_____ Hermaphroditic   _____ Superficial rings on somites

_____ Suckers for attachment   _____ Marine only

c. Place the numbers of the "unknown" annelids in the correct spaces.

_____ Polychaetes   _____ Oligochaetes   _____ Leeches

d. Earthworms

  1. Write the term that matches the phrase.

    a. Membranes dividing segments   _____

    b. Secretes the egg case   _____

    c. Grinds food into small pieces   _____

    d. Receives sperm during copulation   _____

    e. Removes metabolic wastes   _____

    f. Provide traction for movement   _____

    g. Transports nutrients to body cells   _____

2. Considering that the intestine digests food and absorbs nutrients into the blood, what is the typhlosole and what is its value?

_____

_____

_____

3. As observed in your dissection:

   a. Draw a dorsal view of a short length of the ventral nerve cord, including two segmental ganglia and the attached lateral nerves.

   b. Does the crop or gizzard have a thicker wall? _____

   Relate this to the function of each:

   Crop _____

   Gizzard _____

4. ARTHROPODS

   a. List the major characteristics of arthropods. _____

_____

_____

_____

   b. Matching

   1. Arachnids   2. Crustaceans   3. Centipedes   4. Millipedes   5. Insects

   _____ Three pairs of legs   _____ Compound eyes

   _____ Two pairs of antennae   _____ Four pairs of legs

   _____ Cephalothorax and abdomen   _____ Head, thorax, and abdomen

   _____ Elongate, flattened body   _____ No antennae

   _____ Body segments fused in pairs   _____ Compound and simple eyes

   _____ Elongate, dorsally convex body   _____ Five pairs of legs

   c. List the stages in an insect life cycle with:

   incomplete metamorphosis _____

   complete metamorphosis _____

   d. Considering the complete metamorphosis of a butterfly life cycle as an example, what is the adaptive advantage of:

   1. The different feeding patterns of larvae and adult? _____

_____

_____

   2. The pupal stage? _____

e. Place the numbers of the "unknown" arthropods in the correct spaces.

_____ Arachnids   _____ Centipedes   _____ Insects

_____ Crustaceans   _____ Millipedes

f. Crayfish dissection

1. Indicate the number of longitudinal rows of gills. _____ To what are gills in the outer row attached? _____

   How many gills are on each side? _____

   What protects the delicate gills from injury? _____

2. What are the yellowish organs with a granular texture that fill most of the cephalothorax?

   _____

3. Write the term that matches the phrase.

   a. Form paddle for swimming _____

   b. Body divisions _____

   c. Excretory organs _____

   d. Number of legs _____

   e. Shell-like covering of cephalothorax _____

   f. Number of appendages per body segment _____

   g. Type of eyes _____

4. Draw the following from your dissection specimen. (a) Dorsal view of the brain and the anterior portion of the nerve cord. Label brain, ganglia, nerve cord, and lateral nerves. (b) Cross-section of the abdomen. Label exoskeleton, muscles, intestine, and nerve cord.

   Brain and Nerve Cord   **Abdomen, x.s.**

g. Grasshopper dissection

1. Write the term that matches the phrase.

   a. Body divisions _____

   b. Pairs of legs _____

   c. Excretory organs _____

   d. Respiratory system _____

   e. Type of eyes _____

2. How many body segments compose the thorax? _____

3. What is the function of spiracles? _____

4. Draw several contiguous individual lenses of the compound eye of a crayfish and grasshopper to show their differences in shape.

Crayfish   **Grasshopper**

5. UNKNOWN STRUCTURES

Write the names of the numbered structures in the spaces.

1. _____   6. _____

2. _____   7. _____

3. _____   8. _____

4. _____   9. _____

5. _____   10. _____

ECHINODERMS AND CHORDATES

OBJECTIVES

After completing the laboratory session, you should be able to:
1. Describe the distinguishing characteristics of the animal groups studied and the evolutionary advances of each.
2. Identify representatives of the animal groups studied.
3. Identify and state the function of the structural components of the animals studied.
4. Define all terms in bold print.

Echinoderms and chordates are **deuterostomate animals;** the blastopore of the embryo becomes the anus, and the mouth is formed from the second opening to appear. These groups also are called **enterocoelomate animals** because the coelom is formed from mesodermal pouches that "bud off" the embryonic gut, the enteron. Thus, the embryonic development suggests that echinoderms are more closely related to chordates than to other invertebrates.

ECHINODERMS (PHYLUM ECHINODERMATA)

These "spiny-skinned" animals occur only in marine habitats. Their name is derived from the spines that project from the calcareous plates forming the

endoskeleton just under the thin epidermis. A major distinctive characteristic is the presence of a **water vascular system** that provides a means of locomotion via the associated **tube feet** in most forms.

The **radial symmetry** that is characteristic of adult echinoderms is secondarily acquired because their larvae are bilaterally symmetrical. Adults have no anterior or posterior ends, only **oral** (surface with the mouth) and **aboral** surfaces. A pentamerous (five-part) organization is common in echinoderms. Tiny **dermal gills** project between the spines on the aboral surface. They are involved in gas exchange, as are the tube feet. Microscopic pincerlike structures, **pedicillariae,** protect the gills and keep the aboral surface free of debris. Table 16.1 describes the characteristics of the classes of echinoderms.

TABLE 16.1	CLASSES OF ECHINODERMS
Class	*Characteristics*
Asteroidea (sea stars)	Star-shaped forms with broad-based arms; ambulacral grooves with tube feet that are used for locomotion
Ophiuroidea (brittle stars)	Star-shaped forms with narrow-based arms that are used for locomotion; ambulacral grooves covered with ossicles or absent
Echinoidea (sea urchins)	Globular forms without arms; movable spines and tube feet for locomotion; mouth with five teeth
Holothuroidea (sea cucumbers)	Cucumber-shaped forms without arms or spines; tentacles around mouth
Crinoidea (sea lilies)	Body attached by stalk from aboral side; five arms with ciliated ambulacral grooves and tentacle-like tube feet for food collection; spines absent

Sea Stars (Class Asteroidea)

Sea stars are perhaps the most familiar echinoderms. See Figure 16.1. The radially symmetrical body consists of five **arms** radiating from a **central disc.** The mouth is located in the center of the disc on the **oral surface.** The anus is located on the **aboral surface,** along with the **madreporite,** which filters seawater as it enters the water vascular system.

The water vascular system consists of interconnecting tubes filled with seawater. Water passes from the madreporite through a **stone canal** to the **ring canal** that encircles the mouth. A **radial canal** extends from the ring canal into each arm, where it is concealed by the ambulacral bridge. The oral surface of each arm has an **ambulacral groove** that is bordered with rows of **tube feet,** the locomotor organs. The tube feet are connected to a **radial canal** by short **transverse canals.** Contraction or relaxation of the **ampulla** causes a tube foot to be extended or retracted, respectively, due to changes in the water pressure within the tube foot. A suction cup at the end of the foot enables firm attachment to the substrate.

The central disc is largely filled with the digestive tract. The mouth opens into the **cardiac stomach,** which leads to the **pyloric stomach,** from which a pair of **digestive glands** extends into each arm. A very short intestine leads to the anus. A pair of **gonads** also extends from the central disc into each arm near the oral surface. The nervous and circulatory systems are greatly reduced and will not be considered here.

Sea stars feed on bivalves. Persistent pressure pulling against the valves of the shell weakens the adductor muscles and opens the valves slightly. The sea star then extrudes its stomach into the mantle cavity, so the clam is digested within its own shell.

Assignment 1

Materials

Colored pencils
Dissecting instruments
Dissecting pan
Representative echinoderms

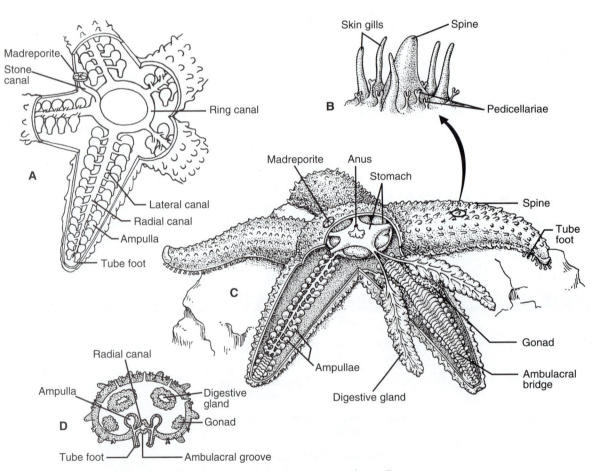

Figure 16.1 A sea star (*Asterias*). **A.** Water vascular system. **B.** Skin gills and pedicillariae. **C.** Dissection. **D.** Arm cross-section.

Sea stars, living in small aquarium
Sea stars, preserved

1. Color-code the water vascular system and digestive organs in Figure 16.1.
2. Examine the representative echinoderms. Note their distinguishing characteristics and modifications of the basic echinoderm body plan. Identify the group to which each belongs. See Table 16.1.
3. Observe the living sea stars in the aquarium. Watch how the tube feet are used in locomotion.
4. ***Complete items 1a–1c on Laboratory Report 16 that begins on page 215.***
5. Perform a dissection of a sea star as described here.
   a. Obtain a preserved sea star and locate the external features shown in Figure 16.1.
   b. Cut off one of the arms and examine the severed end. Locate the digestive glands, gonads, ambulacral groove, radial canal, and tube feet.
   c. Use scissors to remove the aboral surface of the central disc and one of the arms. Locate the stomach, which receives secretions from the five pairs of digestive glands. Note the ampullae of the tube feet, located on each side of the ambulacral bridge.
6. ***Complete item 1 on the laboratory report.***

CHORDATES (PHYLUM CHORDATA)

Chordates possess four distinguishing characteristics that are present in the embryo and may persist in the adult: (1) a **dorsal, tubular nerve cord;** (2) **pharyngeal gill pouches** or **slits;** (3) a **notochord;** and (4) a **postanal tail.** The three major groups of chordates are tunicates, lancelets, and vertebrates.

Tunicates (Subphylum Urochordata)

Figure 16.2 shows the basic structure of larval and adult tunicates (sea squirts). All chordate characteristics are present in the free-swimming larva but not in the immobile adult. Adult tunicates feed by filtering food particles from seawater passing over the gills. The food particles are passed by cilia into the intestine, where digestion occurs. Thus, pharyngeal gills serve in both feeding and gas exchange.

Lancelets (Subphylum Cephalochordata)

The lancelet amphioxus is shown in Figure 16.3. Note that all chordate characteristics are present in the adult. It is a poor swimmer and is usually buried in mud with the anterior end protruding. A true head is lacking. The **cirri** around the **buccal cavity** sweep seawater into the **pharynx.** Water passes from the

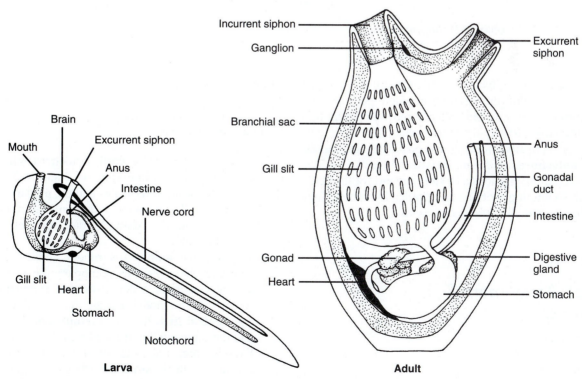

Larva

Adult

Figure 16.2 A tunicate (*Ciona*).

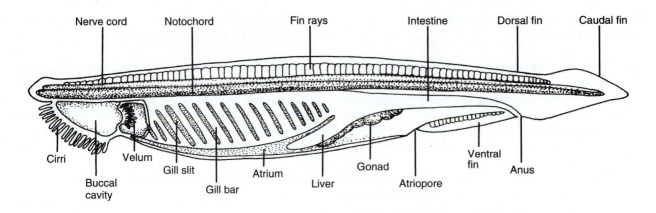

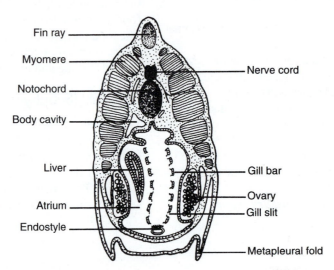

Figure 16.3 Amphioxus (*Branchiostoma*), w.m. and x.s.

pharynx over the **gills** and through the gill slits into the **atrium** and out the **atriopore.** Food particles are trapped in mucus on the gills and moved by cilia into the **intestine,** where digestion occurs. Thus, the gills serve as both filter-feeding and gas-exchange organs. The **notochord,** an evolutionary forerunner of the vertebral column, provides a flexible support for the body. Weak swimming motions are enabled by contractions of the segmentally arranged muscles.

Assignment 2

Materials

Colored pencils
Stereo microscope
Syracuse dishes
Model of amphioxus
Amphioxus adults, preserved

Tunicate adults, preserved
Prepared slides of:
   amphioxus, w.m. and x.s.
   tunicate larva, w.m.

1. Color-code the nerve cord, notochord, gill slits, and digestive tract in Figures 16.2 and 16.3.
2. Examine the adult tunicates and the prepared slide of tunicate larva under a demonstration microscope. Locate the basic chordate features in the larval tunicate.
3. Examine the amphioxus model and locate the chordate characteristics. Compare the preserved amphioxus under a demonstration stereo microscope with the model. Locate the **dorsal, caudal, and ventral fins;** the two **metapleural fin folds;** and the segmentally arranged muscles.
4. Examine prepared slides of amphioxus, w.m. and x.s., and locate the structures labeled in Figure 16.3. Note the positions of the nerve cord, notochord, and pharynx.

TABLE 16.2	CLASSES OF VERTEBRATES
Class	Characteristics
Agnatha (jawless fish)	No jaws; scaleless skin; median fins only; cartilaginous endoskeleton; persistent notochord; two-chambered heart
Chondrichthyes (cartilaginous fish)	Jaws; subterminal mouth; placoid dermal scales; median and paired fins; cartilaginous endoskeleton; two-chambered heart
Osteichthyes (bony fish)	Jaws; terminal mouth; membranous median and paired fins; bony endoskeleton; lightweight dermal scales; two-chambered heart; operculum covers gills; swim bladder
Amphibia (amphibians)	Aquatic larvae with gills; air-breathing adults with lungs; scaleless skin; paired limbs; bony endoskeleton; three-chambered heart
Reptilia (reptiles)	Bony endoskeleton with rib cage; well-developed lungs; epidermal scales; claws; paired limbs in most; leathery-shelled eggs; three-chambered heart
Aves (birds)	Wings; feathers; epidermal scales; claws; endoskeleton of lightweight bones; rib cage; horny beak without teeth; hard-shelled eggs; four-chambered heart
Mammalia (mammals)	Hair; reduced or absent epidermal scales; claws or nails; mammary glands; diaphragm; bony endoskeleton with rib cage; young develop in uterus nourished via placenta; four-chambered heart

5. Study the characteristics of vertebrate classes in Table 16.2.
6. ***Complete item 2 on the laboratory report.***

Vertebrates (Subphylum Vertebrata)

All four chordate characteristics are usually not present in adult vertebrates, but they are present in embryonic stages. An **endoskeleton** of cartilage or bone provides structural support, including a **vertebral column** that partially or completely replaces the notochord.

The anterior part of the dorsal tubular nerve cord is modified to form a **brain** that is encased within a protective **cranium. Skin,** composed of dermal and epidermal layers, protects the body surface. The body is typically divided into **head, trunk,** and **tail,** although a tail does not persist in all adult vertebrates. Most forms have **pectoral** and **pelvic appendages.** Parts of the digestive tract are modified for special functions. Metabolic wastes are removed from the blood by a pair of **kidneys,** and gas exchange occurs by either **pharyngeal gills** or **lungs.** The **closed circulatory system** consists of a ventral heart that is composed of two to four chambers, arteries, capillaries, and veins. Table 16.2 lists the major characteristics of the vertebrate classes.

Jawless Fish (Class Agnatha)

The first vertebrates were small jawless fish, somewhat like amphioxus, but with a true head and a bony endoskeleton. The only living jawless fish, lampreys and hagfish, are degenerate forms that have lost the bony skeleton. They live as scavengers or parasites on other fish. See Figure 16.4. Note the number of gill slits and the absence of paired fins.

Cartilaginous Fish (Class Chondrichthyes)

Cartilaginous fish constitute a side branch from jawed bony fish that appeared about 425 million years ago. Sharks and rays are the best-known members of this group. See Figure 16.4. Paired **pectoral fins** and **pelvic fins** provide stabilization when swimming. The **dermal scales,** characteristic of fish, are of the placoid type, making the skin rough to the touch. The scales provide protection and yet permit freedom of movement. Note the **lateral line system** that runs along each side of the body. It consists of a fluid-filled tube containing sensory receptors for sound vibrations. The number of gill slits is less than in jawless fish because the anterior gill arches form the jaws.

Bony Fish (Class Osteichthyes)

Most modern fish are bony fish. See Figure 16.4. Lightweight dermal scales and fins and a **swim bladder** facilitate swimming. Note the anal fin and the anterior location of the pelvic fins. The number of gills is fewer than in cartilaginous fish, and the gills are covered by a bony **operculum.** Note the lateral line system. Fish and lower animals are **poikilothermic** (body temperature fluctuates with the environmental temperature) and **ectothermic** (source of body heat is external). Fertilization and development are external in most fish.

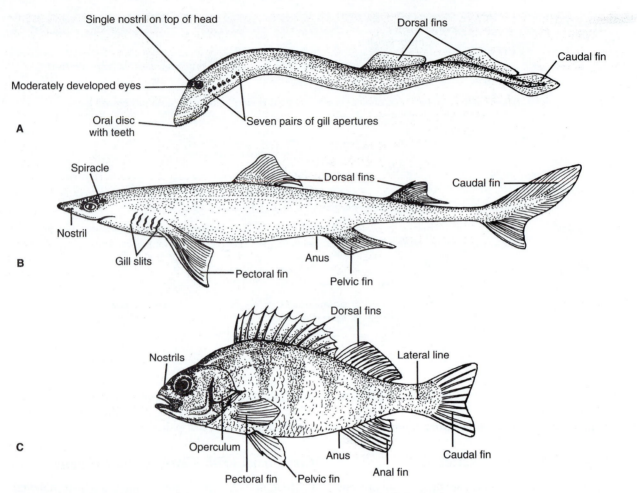

Figure 16.4 Representative fish. **A.** Lamprey (*Petromyzon*). **B.** Dogfish shark (*Squalus*). **C.** Perch (*Perca*).

Assignment 3

Materials

Compound microscope
Stereo microscope
Live fish in an aquarium
Lamprey, dogfish shark, and yellow perch: preserved, whole, and in cross-section
Perch; preserved, skinned on one side
Perch skeleton
Representative fish
Prepared slides of ammocoete (lamprey) larvae, w.m.

1. Observe a prepared slide of an ammocoete (lamprey) larva, and note its similarity to amphioxus. Look for the basic chordate characteristics. Is a notochord present?
2. Examine the whole and sectioned lamprey, shark, and perch. Note the characteristics of each. Look for the presence of a notochord, nerve cord, and cartilaginous or bony skeleton in the cross-sections.

Rub your hand on the skin of each specimen posteriorly and then anteriorly. Do you detect any differences? Examine shark and perch skin with a stereo microscope to see the differences in the dermal scales.
3. Observe the perch skeleton and a skinned perch. Is there any evidence of segmentation? Note how the cranium protects the brain and how the vertebral column protects the spinal cord, but allows flexibility in lateral swimming movements.
4. Observe the live fish in the aquarium. Note how they swim and use the paired fins for stability. Observe their respiratory movements.
5. ***Complete item 3 on the laboratory report.***

Amphibians (Class Amphibia)

Amphibians are believed to have evolved from primitive lobe-finned fish. Fertilization and development are external. The aquatic larvae have functional gills and a tail for swimming. Adults are air breathing and have paired appendages. The moist skin is an auxiliary gas-exchange organ. Some adult amphibians have tails; others do not. Amphibians are poikilothermic and ectothermic.

Reptiles (Class Reptilia)

Reptiles were the first truly terrestrial vertebrates because they can reproduce without returning to water. This is possible because (1) fertilization is internal and (2) a leathery-shelled, evaporation-resistant egg contains nutrients to nourish the embryo until hatching. Pharyngeal gill pouches and a notochord are present only in the embryo, not in adults, of reptiles and higher vertebrates. In most reptiles and in all higher vertebrates, the **ribs** join ventrally to the **sternum** to form a rib cage that protects the heart and lungs.

Epidermal scales first appear in reptiles. Formed by the epidermis of the skin, these scales provide protection and retard evaporative water loss. The feet typically have five toes, and the **claws** are formed from modified epidermal scales. Reptiles have peglike teeth. Like all lower animals, reptiles are poikilothermic and ectothermic.

Birds (Class Aves)

Birds evolved from reptiles, and they still have many reptilian characteristics. Birds possess numerous adaptations for flight, including (1) **feathers,** which are derived from epidermal scales; (2) hollow, lightweight bones; (3) air sacs associated with the lungs; (4) a fused skeleton except for neck, tail, and paired appendages; and (5) pectoral appendages modified into **wings.** The jaws are modified to form a **bony beak** without teeth.

Unlike lower animals, birds are **homeothermic** (maintain a relatively constant body temperature) and **endothermic** (heat produced by cellular respiration maintains the body temperature). This enables birds to be active even when the ambient temperature is low.

Fertilization is internal, and the hard-shelled eggs contain nutrients for the developing embryo. Parents incubate the eggs until hatching and provide considerable care for the young birds until they can care for themselves.

Mammals (Class Mammalia)

Like birds, mammals evolved from reptiles. Key mammalian characteristics are the presence of **hair, mammary glands** (milk-producing glands), and a **diaphragm** that separates the abdominal and thoracic cavities. Mammals are homeothermic and endothermic.

After internal fertilization, most modern mammals nourish the developing embryo *in utero* via a **placenta.** A placenta brings the maternal and embryo blood vessels close together so materials can diffuse from one to the other. **True placental mammals** have a relatively long **gestation period,** so the young mammals are rather well developed at birth. In contrast, **marsupial mammals** have a short gestation period and give birth to underdeveloped young that complete development in an external pouch of the female. Only two species of **egg-laying mammals** have survived to the present: the duck-billed platypus and the spiny anteater. Both occur in Australia, along with most marsupials, and represent a primitive side branch of the main evolutionary line. They are not ancestral to higher mammals.

Assignment 4

Materials

Colored pencils
Compound microscope
Live salamander, lizard, and rat
Representative amphibians, reptiles, birds, and mammals
Study skins of seed-eating, insect-eating, and flesh-eating birds
Prepared slides of:
   frog embryo, neurula stage, x.s.
   chick embryo, 96 hr, w.m.

1. Color-code the three embryonic tissues and their derivatives as follows in Figures 16.5 and 16.6: ectoderm—blue, mesoderm—red, endoderm—yellow.
2. Observe the prepared slide of a frog embryo set up under a demonstration microscope. Locate the notochord and nerve cord, and compare your observations with Figure 16.5.
3. Observe the demonstration slide of a chick embryo, w.m., showing pharyngeal pouches. Compare your observations with Figure 16.6.

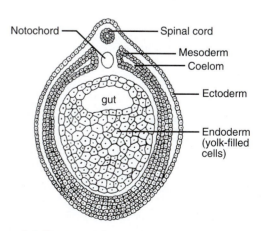

Figure 16.5 Frog embryo, x.s.

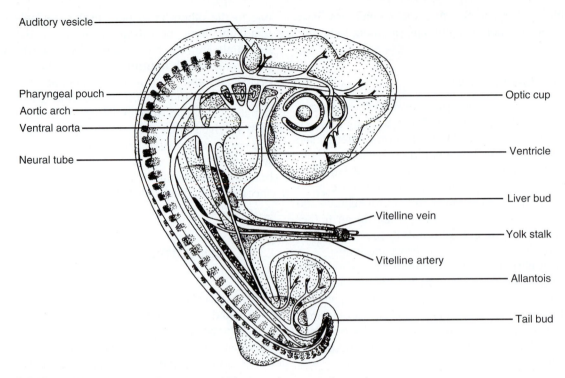

Figure 16.6 Chick embryo (*Gallus*), 96 hr, showing pharyngeal pouches.

4. Examine the representative amphibians, reptiles, birds, and mammals, and note their distinguishing characteristics.
5. Examine a living salamander, lizard, and rat.
   a. Compare the position of the legs in each and determine whether the backbone bends laterally when the animal is walking.
   b. Place the salamander in the small aquarium and watch it swim. Is the movement of the backbone similar when swimming and walking? How does the swimming movement compare with that of fish?
   c. Compare the skin and feet. Look for epidermal scales and claws.
6. Note the adaptations of beaks and feet in the birds. How do these correlate with their different ways of life?
7. ***Complete the laboratory report.***

ECHINODERMS AND CHORDATES

Student _____

Lab Instructor _____

1. ECHINODERMS

a. Write the term that matches the phrase.

1. Fate of the blastopore _____

2. Formed from second embryonic opening _____

3. Type of skeleton _____

4. Major distinctive characteristic _____

5. Structures involved in gas exchange _____

6. Type of symmetry _____

7. Surface on which mouth is located _____

b. Matching

1. Brittle stars   2. Sea cucumbers   3. Sea lilies   4. Sea stars   5. Sea urchins

_____ Star-shaped with broad-based arms   _____ Elongate forms lacking arms and spines

_____ Star-shaped with narrow-based arms   _____ Globose forms with movable spines; mouth with five teeth

_____ Attached by stalk from aboral surface   _____ Ambulacral grooves with tube feet

_____ Arms used for locomotion   _____ Ambulacral grooves without tube feet

c. Write the numbers of the "unknown" echinoderms in the correct spaces.

_____ Brittle stars   _____ Sea cucumbers   _____ Sea lilies

_____ Sea stars   _____ Sea urchins

d. Sea star dissection

1. How many arms are present? _____ How many digestive glands are in each arm? _____

   How many gonads are in each arm? _____

2. What part of the water vascular system extends into each arm? _____

   terminates in a tube foot? _____

3. Estimate the number of tube feet in one arm. _____

2. CHORDATES: GENERAL

a. List the four distinguishing characteristics of chordates. _____

_____

_____

b. In tunicates and cephalochordates, describe the function of:

pharyngeal gills _____

notochord _____

c. Indicate the following characteristics of chordates:

Type of symmetry _____ Fate of the blastopore _____

d. Matching

1. Tunicates   2. Cephalochordates   3. Vertebrates

_____ Filter feeders   _____ Motile larva and nonmotile adult

_____ Head, trunk, and tail   _____ Marine only

_____ Chordate characteristics,   _____ Gills for feeding and gas exchange
            usually in embryo only

                                              _____ Fishlike, headless adult

_____ Endoskeleton   _____ Notochord is only support

3. FISHES

a. Describe the trend from tunicates to bony fish in the:

number of gills _____

function of gills _____

b. Draw the cut surface of a cross-section through the tail of a lamprey, shark, and perch to show the vertebra, nerve cord, and notochord, if present. Label pertinent parts.

Lamprey   **Shark**   **Perch**

c. In what ways is a vertebral column an improvement over a notochord? _____

_____

_____

d. How are bony fish better adapted to a swimming way of life than cartilaginous fish? _____

_____

_____

e. Based on your observations, what evidence of segmentation is found in bony fish? _____

_____

_____

f. Based on your observation of fish in the aquarium, describe the respiratory movements of bony fish. _____

_____

_____

4. AMPHIBIANS TO MAMMALS

a. List two reasons why amphibians are poorly adapted for terrestrial life.

1. _____

2. _____

b. What evidence of a fish ancestry is found in amphibians? _____

_____

c. List three ways that reptiles are adapted to terrestrial life.

1. _____

2. _____

3. _____

d. Explain the presence of pharyngeal pouches in the chick embryo. _____

_____

e. List three adaptations for flight observed in birds.

1. _____

2. _____

3. _____

f. Draw the beaks of seed-eating, insect-eating, and flesh-eating birds to show their adaptations.

Seed-Eating   **Insect-Eating**   **Flesh-Eating**

g. Compare walking movements in a salamander, lizard, and rat by indicating which organism is described by the phrase.

1. Greatest lateral movement of backbone   _____

2. Least lateral movement of backbone   _____

3. Greatest lateral extension of legs   _____

4. Least lateral extension of legs   _____

h. Describe the advantage of homeothermy. _____

_____

Describe any disadvantages. _____

_____

i. Explain the value of hair and feathers in homeothermic animals. _____

_____

j. List two evidences of reptilian ancestry found in the rat.

1. _____   2. _____

5. CHORDATE SUMMARY

a. Matching

1. Jawless fish (modern)   2. Cartilaginous fish   3. Bony fish   4. Amphibians

5. Reptiles   6. Birds   7. Mammals

_____ Diaphragm and hair   _____ Feathers, horny beak

_____ Notochord in adult   _____ Cartilaginous endoskeleton

_____ Leathery-shelled eggs   _____ Intrauterine development

_____ No scales, jaws, or operculum   _____ Lungs, paired limbs, no scales

_____ Epidermal scales   _____ Dermal scales

_____ Operculum, swim bladder   _____ Subterminal mouth, paired fins

_____ Poikilothermic   _____ Homeothermic

_____ Two-chambered heart   _____ Four-chambered heart

_____ Seven pairs of gill slits   _____ Bony endoskeleton

_____ Milk-secreting glands   _____ Claws or nails

_____ Hard-shelled eggs   _____ Lightweight dermal scales

_____ Ribcage   _____ Three-chambered heart

b. Write the numbers of the "unknown" vertebrates in the correct spaces.

_____ Jawless fish   _____ Amphibians   _____ Birds

_____ Cartilaginous fish   _____ Reptiles   _____ Mammals

_____ Bony fish

6. SUMMARY OF THE ANIMAL KINGDOM

Compare the animal groups by placing the correct number(s) in the spaces provided.

1. Sponges   2. Cnidarians   3. Flatworms   4. Roundworms   5. Mollusks

6. Annelids   7. Arthropods   8. Echinoderms   9. Chordates

_____ Bilateral symmetry in adult   _____ Radial symmetry in adult

_____ Saclike body plan   _____ Endoskeleton

_____ Tube-within-a-tube body plan   _____ Organ level

_____ Ectoderm, endoderm, mesoderm   _____ Exoskeleton

_____ Dorsal tubular nerve cord   _____ Ventral nerve cord(s)

_____ Organ system level   _____ Tissue level

_____ Pseudocoelomate   _____ Protostomate

_____ Cellular-tissue level   _____ Pharyngeal gill pouches

_____ Gas-exchange organs   _____ Circulatory system

_____ Excretory organs   _____ Eucoelomate

_____ Notochord   _____ Acoelomate

_____ Water vascular system   _____ Ectoderm and endoderm only

_____ Exoskeleton, jointed appendages   _____ No skeleton, segmentation

Part IV

Animal Biology

DISSECTION OF THE FROG

OBJECTIVES

After completing the laboratory session, you should be able to:
1. Identify the major external features and internal organs of the frog.
2. Describe the relative positions of the internal organs.
3. Define all terms in bold print.

A study of the frog will give you a better understanding of the body organization of vertebrates. The dissection focuses on the major organs of the body cavity. The integumentary, skeletal, muscle, nervous, and endocrine systems are not covered. The dissection will also prepare you for a practical examination on frog or vertebrate structure. Use the scalpel in your dissection *only when absolutely necessary*. Follow the directions carefully and in sequence. The terms of direction in Table 17.1 will help clarify the location of body parts as you perform the dissection. Note that members of each pair have opposite meanings.

Materials

Dissecting instruments and pins
Dissecting pan, wax bottomed
Frogs, live, in terrarium
Frogs, preserved, double-injected
Gloves, protective and disposable

EXTERNAL FEATURES

The frog's body is divided into a **head** and **trunk;** there is no neck. Paired appendages are attached to the trunk. The **posterior appendages** consist of a thigh, lower leg, and foot. The foot has five toes plus a rudimentary medial sixth toe. The elongation of two ankle bones has increased the length of the foot and improved leverage for hopping. The **anterior appendages** consist of an upper leg, lower leg, and foot. The foot has four toes plus a rudimentary medial fifth toe. If your specimen is a male, the medial toes of the front feet will have a swollen pad.

The bulging eyes have immovable upper and lower eyelids. However, a **nictitating membrane,** a third eyelid attached to the inner portion of the lower eyelid, can cover the eye. Observe the nictitating membrane in the live frogs in the terrarium. Locate a large **tympanum** (eardrum) and the paired **external nares** (nostrils).

MOUTH

Use heavy scissors to cut through the bones at the angle of each jaw, so the mouth may be opened, as in Figure 17.1. Locate the structures shown. Note that the tongue is attached at its anterior end. Feel the teeth. Locate the **internal nares** (nostrils). The narrowed

TABLE 17.1	ANATOMICAL TERMS OF DIRECTION IN THE FROG
Anterior	Toward the head
Posterior	Toward the hind end
Cranial	Toward the head
Caudal	Toward the hind end
Dorsal	Toward the back
Ventral	Toward the belly
Superior	Toward the back
Inferior	Toward the belly
Lateral	Toward the sides
Medial	Toward the midline

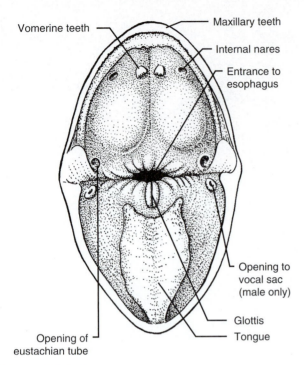

Figure 17.1 The mouth.

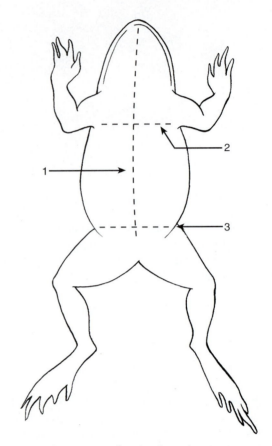

Figure 17.2 Incisions for the frog dissection.

region at the back of the mouth is the **pharynx,** which leads to the **esophagus,** a short tube that carries food into the stomach. Just ventral to the esophageal opening is the **glottis,** a slit through which air passes into the **larynx** and on to the lungs. The **eustachian tubes,** located near the angles of the jaws, communicate with the middle ears, spaces interior to the eardrums. Male frogs have small openings ventral to the eustachian tubes that lead to vocal sacs ventral to the lower jaw. Air-filled vocal sacs increase the volume of vocalizations.

OPENING THE BODY CAVITY

1. Pin the frog on its back to the wax bottom of a dissecting pan by placing pins through its feet.
2. Lift the skin at the ventral midline with forceps and make a small incision with the scissors. Extend the cut anteriorly to the lower jaw and posteriorly to the junction of the thighs, as shown in Figure 17.2. Make incisions 2 and 3 as shown. Then, use a dissecting needle and forceps to separate the skin from the underlying muscles to the lateral margins of the body. Use scissors to cut off the skin flaps to expose the muscles of the ventral body wall. The blue ventral abdominal vein will be visible through the body wall.
3. Being careful to avoid damaging the underlying organs, make incision 1 through the ventral body wall just to the right of the midline to avoid cutting

the **ventral abdominal vein,** which is attached to the body wall at the midline. Free this vein from the body wall with a dissecting needle, and make transverse cuts 2 and 3 through the body wall. Cut off the flaps of body wall.
4. Lift the muscles and bones in the "chest" region with forceps, and use scissors to cut anteriorly through them at the midline. Carefully, separate the tissue from the underlying organs and avoid cutting the major blood vessels. Pin back or remove the severed tissues to expose the heart and other anteriorly located organs.
5. Note how the organs are oriented in the body cavity. The arteries are filled with red latex and the veins with blue latex. The yellow fingerlike structures are fat bodies attached to the anterior portions of the kidneys. You may need to remove most of the fat bodies to get them out of the way. As you locate the internal organs, it will be necessary to move or remove the ventrally located organs to see the more dorsally located organs. However, save those portions attached to blood vessels if you plan to study the circulatory system.

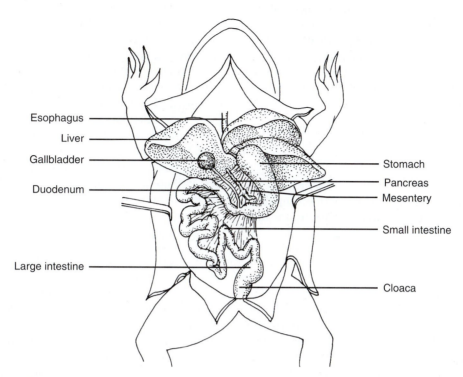

Figure 17.3 Frog digestive system.

DIGESTIVE SYSTEM

Compare your specimen with Figure 17.3. Note the large **liver.** How many lobes does it have? The liver receives nutrient-rich venous blood from the stomach and intestines. It removes excess nutrients from the blood for processing before the blood returns to the heart. Find the **gallbladder,** a greenish sac lying between lobes of the liver. The gallbladder stores bile secreted by the liver and empties it into the small intestine via the **bile duct.** Bile aids fat digestion. Remove most of the liver to expose underlying organs, but keep those portions with blood vessels.

Food passes from the mouth into the very short esophagus, which carries food to the **stomach.** The esophagus is anterior and dorsal to the heart and bronchi. (You can locate the esophagus after studying the circulatory system.) Food enters the stomach through the **cardiac valve,** a sphincter muscle located at the junction of the esophagus and stomach. Locate the stomach, an enlarged portion of the digestive tract that lies along the left side of the body cavity. After partial digestion in the stomach, the food mass passes through the **pyloric valve,** a sphincter muscle, into the **small intestine,** where digestion of food and absorption of nutrients are completed. Locate the small intestine.

Find the **mesentery,** the transparent connective tissue supporting the digestive organs and spleen. It is formed of two layers of the **peritoneum,** a thin membrane that lines the body cavity and covers the internal organs. Locate the narrow, elongate **pancreas** in the mesentery between the stomach and the first portion of the small intestine. It secretes digestive enzymes into the small intestine. The spherical, reddish-brown (blue in injected specimens) **spleen** is also supported by mesentery and lies dorsal and posterior to the stomach. It is a circulatory organ that removes worn-out blood cells from the blood.

The nondigestible food residue passes from the small intestine into the short **large intestine,** which temporarily stores the nondigestible wastes. The large intestine extends into the pelvic region and terminates with the **anus,** which opens into the **cloaca.** You will observe the cloaca when you study the urogenital system.

RESPIRATORY SYSTEM

The reddish-purple lungs are located on each side of the heart and dorsal to the liver. Air flows through the glottis into the very short **larynx** from which two short **bronchi** branch to carry air into the left and right lungs. Inserting a probe through the glottis will help you locate a bronchus.

The lungs in a preserved frog are collapsed and do not show the appearance or structure of a living lung. Amphibian lungs are essentially saclike in structure with ridges on their inner surfaces that increase the lung surface area exposed to air in the lungs. These ridges form a definite pattern, somewhat like the surface of a waffle. Cut open a lung and note its structure. If your instructor has set up a demonstration of an inflated lung under a dissecting microscope, examine it to observe lung structure.

CIRCULATORY SYSTEM

Refer to Figures 17.4, 17.5, and 17.6 as you study the circulatory system. Your dissection will be simplified if you have a good understanding of heart structure and the location and names of *both* arteries and veins *before proceeding.*

The Heart

Examine the structure of the heart in Figure 17.4. The heart is composed of three chambers. The **left atrium** receives oxygenated blood returning from the lungs via **pulmonary veins,** and the **right atrium** receives deoxygenated blood returning from the rest of the body via the **sinus venosus.** Blood enters the sinus venosus from two **anterior venae cavae** and one **posterior vena cava.** Oxygenated and deoxygenated blood are mixed when they enter the single **ventricle,** which pumps blood to both lungs and the rest of the body simultaneously. Blood leaves the ventricle, via the **conus arteriosus,** which quickly divides into left and right branches. Each branch is a **truncus arteriosus** leading to additional arteries.

Locate the heart and remove any overlying tissue to expose it. Locate the **pericardium,** the transparent membrane enveloping the heart and carefully use scissors to remove it. Also remove any excess latex present. Locate the heart chambers and the major vessels shown in the ventral view of the heart in Figure 17.4. You will complete your study of the heart after studying the veins.

Major Arteries

Carefully dissect away the connective tissue to expose the following arteries shown in Figure 17.5. Locate the conus arteriosus and each truncus arteriosus. Each truncus arteriosus divides to form three arterial arches.

1. The anterior arch is the **carotid arch,** which carries blood to the head.
2. The middle arch is the **aortic arch,** which curves around the heart to join posteriorly with the aortic arch from the other side to form the **dorsal aorta.** A **subclavian artery** branches from each aortic arch to supply blood to the shoulder and foreleg.

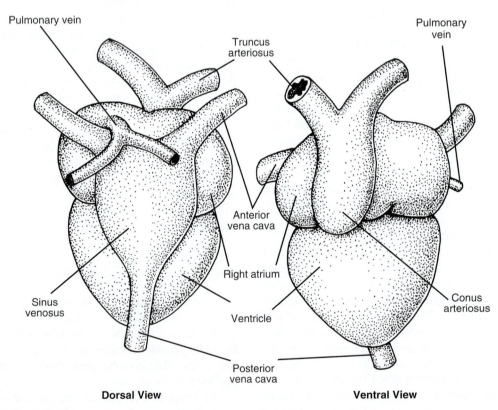

Dorsal View   Ventral View

Figure 17.4 Frog heart.

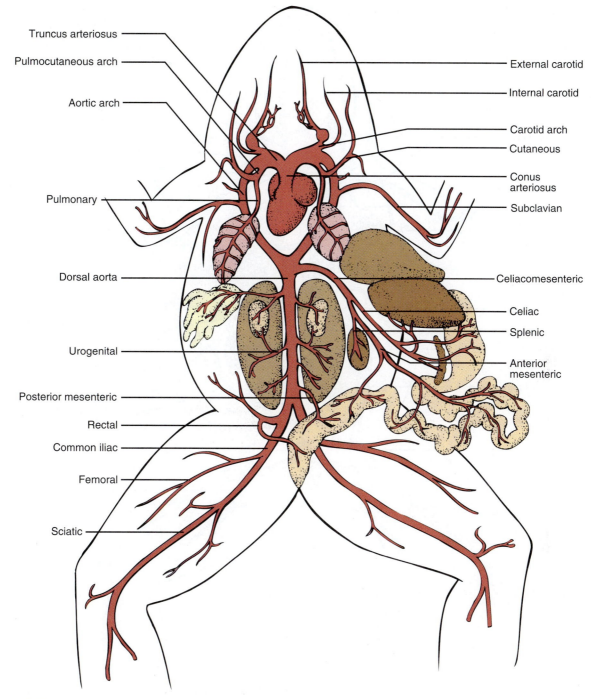

Figure 17.5 Major arteries of the frog, ventral view.

3. The posterior arch is the **pulmocutaneous arch.** It branches to form the **cutaneous artery,** which supplies the skin, and the **pulmonary artery** leading to the lung on that side of the body.

The **celiacomesenteric artery** is the most anterior artery to branch from the dorsal aorta. It is supported by the mesentery and supplies the stomach, small intestine, pancreas, and spleen. Posteriorly, several **urogenital** arteries branch off the aorta to supply the kidneys and gonads. Just posterior to the urogenital arteries, the dorsal aorta gives off a small branch, the **posterior mesenteric artery,** that carries blood to the large intestine. Then, the aorta branches to form the left and right **common iliacs,** which enter the hind legs. Each common iliac forms two major branches. The smaller **femoral artery** serves the thigh, and the larger **sciatic artery** gives off smaller branches and continues to the lower leg.

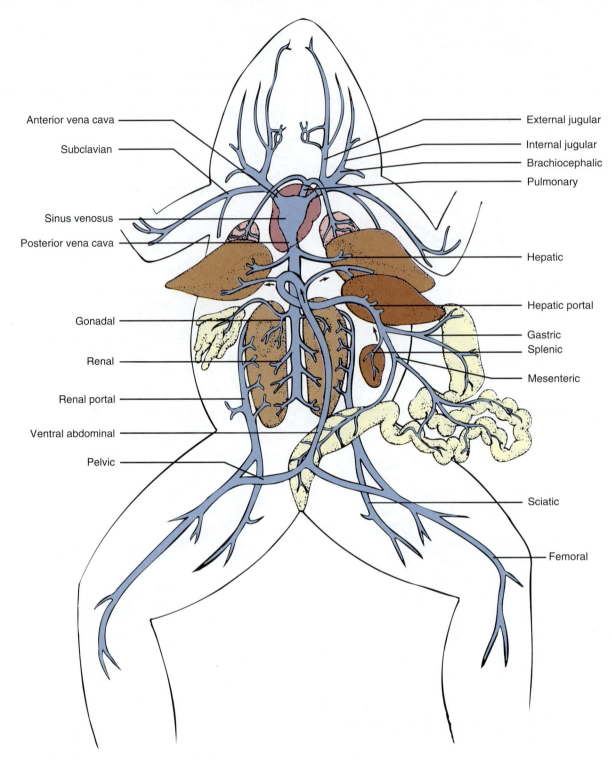

Figure 17.6 Major veins of the frog, ventral view.

Major Veins

Return to the heart. Lift the tip of the heart carefully to see the thin-walled, saclike **sinus venosus,** located on the dorsal surface of the heart. See Figures 17.4 and Figure 17.6. It opens into the right atrium frog receives blood from three large veins, the venae cavae. The **posterior vena cava** drains the body posterior to the heart and enters the sinus venosus posteriorly. The two **anterior venae cavae** enter the sinus venosus laterally, one from each side. Each drains the head, foreleg, body wall, and skin on its side of the body.

Observe the right anterior vena cava. It is formed by the union of three veins arranged clockwise from posterior to anterior.

1. The **subclavian vein** drains the foreleg, body wall, and skin.
2. The **brachiocephalic vein** drains the shoulder and deeper portions of the head.
3. The **external jugular vein** drains the more superficial parts of the head.

Locate the smaller **pulmonary veins** that return oxygenated blood from the lungs to the left atrium.

Locate the posterior vena cava. It receives blood from the liver via two **hepatic veins** located just posterior to the heart and at its posterior origin from the **renal veins** draining the kidneys.

Blood from the digestive tract is carried to the liver by the **hepatic portal vein** for processing and then exits via the hepatic veins. The hepatic portal vein will not be injected with blue latex, so it may be difficult to locate.

The **sciatic** and **femoral veins** drain the hind legs and unite to form the **renal portal veins,** which carry blood to the kidneys. (Renal portal veins are not found in higher vertebrates.) Blood flows through the kidneys and enters the posterior vena cava via the **renal veins.**

The unpaired **ventral abdominal vein** is formed by the union of the **pelvic veins** posteriorly, and it divides anteriorly to enter the liver lobes.

Now, return to the heart and refresh your understanding of the major arteries and veins associated with it. Then, cut these vessels near the heart, leaving short stubs, and remove the heart. Compare it with Figure 17.4. Make a coronal section through the atria and ventricle to observe the interior structures.

UROGENITAL SYSTEM

The urinary and reproductive systems are closely related and are commonly studied as the urogenital system. See Figures 17.7 and 17.8.

Urinary System

Locate the paired **kidneys.** They are the elongate, reddish-brown (blue in injected frogs) organs located on each side of the posterior vena cava against the dorsal body wall. The **fat bodies,** finger-shaped fatty masses attached to the anterior portions of the kidneys, nourish the frog during hibernation. A slender **adrenal body** (difficult to find) lies along the ventral surface of each kidney.

Ureters carry urine formed by the kidneys to the cloaca for removal from the body. Try to find a slender white ureter near the lateral posterior margin of a kidney.

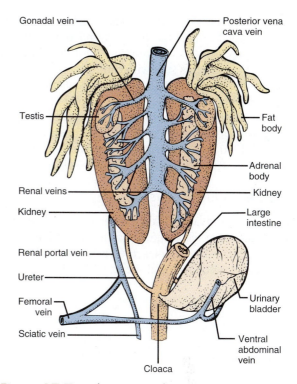

Figure 17.7 Male urogenital system.

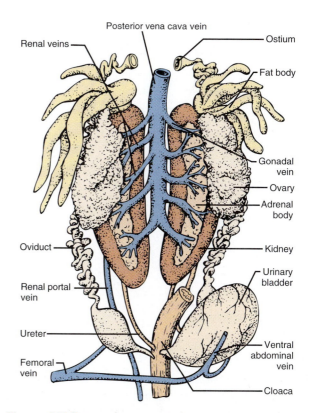

Figure 17.8 Female urogenital system.

To observe the cloaca and urinary bladder, it is necessary to cut through the muscles and bones of the pelvic region at the ventral midline until the cloaca is exposed. Perform this step with care.

Locate the cloaca and find the **urinary bladder,** a ventral pouch attached to the cloaca. Urine in the cloaca backs up into the urinary bladder. The cloaca receives urinary wastes, digestive wastes, and gametes (sperm and eggs). Find the junction of the large intestine, ureters, and oviducts with the dorsal surface of the cloaca.

Male Reproductive System

The yellowish, oval **testes** are located on the anterior ventral surface of the kidneys. Sperm pass into the kidneys via several fine tubules, the **vasa efferentia.** The ureters carry both urine and sperm to the cloaca.

Female Reproductive System

The **ovaries** appear as irregular-shaped sacs located on the anterior ventral surface of the kidneys. Ovaries will vary in size and shape in accordance with the breeding season. Locate the coiled whitish **oviducts** that carry eggs from the ovaries to the cloaca. Trace an oviduct to the cloaca.

CONCLUSION

This completes the dissection. Review as necessary so you can recognize all the organs. When finished, dispose of your specimen as directed by your instructor. Wash and dry your dissecting instruments and pan. There is no laboratory report for this exercise.

DISSECTION OF THE FETAL PIG

OBJECTIVES

After completing the laboratory session, you should be able to:
1. Identify the major internal organs of the fetal pig and indicate to which organ system each belongs.
2. Describe the relative positions of the major organs.
3. Name the body cavities and the organs each one contains.
4. Define all terms in bold print.

In this exercise, you will study the basic body organization of mammals through the dissection of a fetal (unborn) pig. The dissection will focus on the internal organs located in the ventral body cavity. Your dissection will proceed more easily if you understand the common directional terms listed in Table 18.1. These terms are used to describe the relative position of organs within an animal or structures within an organ, assuming that the animal is standing on all four legs. Note that the terms of each pair have opposite meanings and that some terms refer to the same direction. Figure 18.1 shows the relationship of most of the directional terms, as well as the three common planes associated with a bilaterally symmetrical animal. The integumentary, skeletal, muscular, nervous, and endocrine systems will not be studied. Study the organ systems outlined in Table 18.2 so you know their functions and the major organs of each.

GENERAL DISSECTION GUIDELINES

The purpose of the dissection is to expose the various organs for study, and it should be done in a manner that causes minimal damage to the specimen. *Follow the directions carefully and in sequence.* Read the entire description of each incision, and be sure you understand it *before* you attempt it. Work in pairs, alternating the roles of reading the directions and performing the dissection. Use scissors for most of the incisions. *Use a scalpel only when absolutely*

TABLE 18.1	ANATOMICAL TERMS OF DIRECTION IN THE FETAL PIG
Anterior	Toward the front end or head
Posterior	Toward the hind end or tail
Cranial	Toward the head
Caudal	Toward the tail
Dorsal	Toward the back
Ventral	Toward the belly
Superior	Toward the upper surface or back
Inferior	Toward the lower surface or belly
Medial	Toward the midline
Lateral	Toward the sides; away from the midline

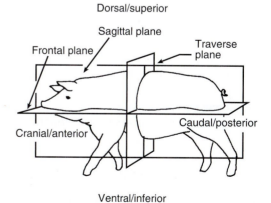

Figure 18.1 Directional terms and body planes.

TABLE 18.2	MAJOR ORGANS AND FUNCTIONS OF THE ORGAN SYSTEMS	
System	Major Organs	Major Functions
Integumentary	Skin, hair, nails	Protection, cooling
Skeletal	Bones	Support, protection
Muscular	Muscles	Movement
Nervous	Brain, spinal cord, sense organs, nerves	Rapid coordination via impulses
Endocrine	Pituitary, thyroid, parathyroid, testes, ovaries, pancreas, adrenal, pineal	Slower coordination via hormones
Digestive	Mouth, pharynx, esophagus, stomach, intestines, liver, pancreas	Digestion of food, absorption of nutrients
Respiratory	Nasal cavity, pharynx, larynx, trachea, bronchi, lungs	Exchange of oxygen and carbon dioxide
Circulatory	Heart, arteries, capillaries, veins, blood	Circulation of blood, transport of materials
Lymphatic	Lymphatic vessels, lymph, spleen, thymus, lymph nodes	Cleansing and return of extracellular fluid to bloodstream
Urinary	Kidneys, ureters, urinary bladder, urethra	Formation and removal of urine
Reproductive Male	Testes, epididymis, vas deferens, seminal vesicles, bulbourethral gland, prostate gland, urethra, penis	Formation and transport of sperm and semen
Female	Ovaries, oviducts, uterus, vagina, vulva	Formation of eggs, sperm reception, intrauterine development of offspring

necessary. Be careful not to damage organs that must be observed later.

Your instructor may have you complete the dissection over several sessions. If so, wrap the specimen in wet paper towels at the end of each session, and seal it in a plastic bag to prevent it from drying out.

Protect your hands from the preservative by applying a lanolin-based hand cream before starting the dissection or by wearing protective disposable gloves. Wash your hands thoroughly at the end of each dissection session and reapply the hand cream.

Materials

Dissecting instruments and pins
Dissecting pan, wax bottomed
Dissecting microscope
Hand cream
String
Gloves, protective and disposable
Fetal pig (*Sus scrofa*), preserved and double-injected

EXTERNAL ANATOMY

Examine the fetal pig and locate the external features shown in Figure 18.2. Determine the sex of your specimen. The **urogenital opening** in the female is immediately ventral to the anus and has a small **genital papilla** marking its location. A male is identified by the scrotal sac ventral to the anus and a **urogenital opening** just posterior to the **umbilical cord.** Two rows of nipples of mammary glands are present on the ventral abdominal surface of both males and females, but the mammary glands later develop only in maturing females. Mammary glands and hair are two distinctive characteristics of mammals.

Make a transverse cut through the umbilical cord and examine the cut end. Locate the two **umbilical arteries** that carry blood from the fetal pig to the placenta and the single **umbilical vein** that returns blood from the placenta to the fetal pig.

POSITIONING THE PIG FOR DISSECTION

Position your specimen in the dissection pan as follows:

1. Tie a piece of heavy string about 20 in. long around both left feet.
2. Place the pig on its back in the pan. Run the strings under the pan, and tie them to the corresponding right feet to spread the legs and expose the ventral surface of the body. This also holds the pig firmly in position.

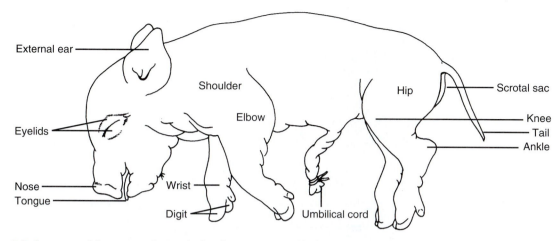

Figure 18.2 External features of a male fetal pig (*Sus scrofa*).

A HEAD AND NECK

This portion of the dissection will focus on structures of the mouth and pharynx. If your instructor wants you to observe the salivary glands, a demonstration dissection will be prepared for you to observe.

Dissection Procedure

To expose the organs of the mouth and pharynx, start by inserting a pair of scissors in the angle of the lips on one side of the head and cut posteriorly through the cheek. Open the mouth as you make your cut and follow the curvature of the tongue to avoid cutting into the roof of the mouth. Continue through the angle of the mandible (lower jaw). Just posterior to the base of the tongue, you will see a small whitish projection, the **epiglottis,** extending toward the roof of the mouth. Hold down the epiglottis and surrounding tissue and continue your incision dorsal to it and on into the opening of the **esophagus.** Now, repeat the procedure on the other side so that the lower jaw can be pulled down to expose the structures, as shown in Figure 18.3.

Observations

Only a few deciduous teeth will have erupted, usually the third pair of incisors and the canines. Other teeth are still being formed and may cause bulging of the gums. Make an incision in one of these bulges to observe the developing tooth.

Observe the tongue. Note that it is attached posteriorly and free anteriorly, as in all mammals. Locate the numerous **papillae** on its surface, especially near the base of the tongue and along its anterior margins. Papillae have numerous microscopic taste buds on their surfaces.

Observe that the roof of the mouth is formed by the anteriorly located **hard palate,** supported by bone and cartilage, and the posteriorly located **soft palate.** Paired nasal cavities lie dorsal to the roof of the mouth. The space posterior to the nasal cavities is called the **nasal pharynx,** and it is contiguous with the **oral pharynx,** the throat, located posterior to the mouth. Locate the opening of the nasal pharynx posterior to the soft palate. The oral pharynx may be difficult to visualize because your incision has cut through it on each side, but it is the region where the glottis, esophageal opening, and opening of the nasal pharynx are located.

Make a midline incision through the soft palate to expose the nasal pharynx. Try to locate the openings of the **eustachian tubes** in the dorsolateral walls of the nasal pharynx. Eustachian tubes allow air to move into or out of the middle ear to equalize the pressure on the eardrums.

When you have completed your observations, close the lower jaw by tying a string around the snout.

GENERAL INTERNAL ANATOMY

In this section, you will open the **ventral cavity** to expose the internal organs. The ventral cavity consists of the **thoracic cavity,** which is located anterior to the diaphragm, and the **abdominopelvic cavity,** which is located posterior to the diaphragm.

Dissection Procedure

Using a scalpel, carefully make an incision, through the skin only, from the base of the throat to the umbilical cord along the ventral midline. This is incision 1 in Figure 18.4. Separate the skin from the body wall along the incision just enough to expose the body wall. Do this by lifting the cut edge of the skin with forceps,

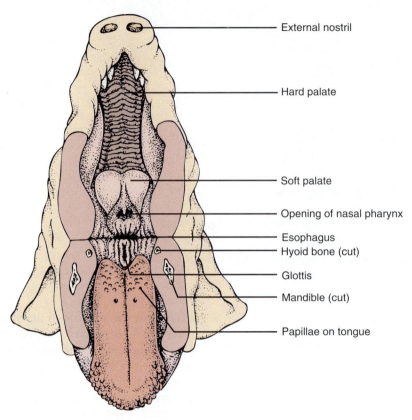

External nostril

Hard palate

Soft palate

Opening of nasal pharynx

Esophagus

Hyoid bone (cut)

Glottis

Mandible (cut)

Papillae on tongue

Figure 18.3 The oral cavity.

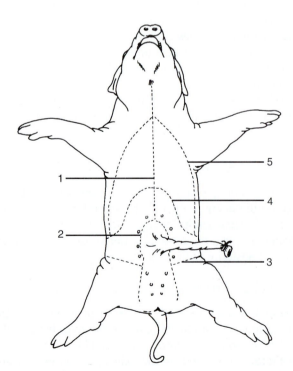

1

2

5

4

3

Figure 18.4 Sequence of the incisions (ventral view).

while separating the loose fibrous connective tissue that attaches the skin to the body wall with the handle of a scalpel or a blunt probe.

While lifting the body wall at the ventral midline with forceps, use scissors to make a small incision through the body wall about halfway between the **sternum** and the **umbilical cord.** (Always cut away from your body when using scissors. This gives you better control. Rotate the dissecting pan as necessary.) Once the incision is large enough, insert the forefinger and middle finger of your left hand and lift the body wall from below with your fingertips as you make the incision. This position gives you good control of the incision and prevents unnecessary damage.

Extend the incision along the ventral midline anteriorly to the tip of the sternum and posteriorly to the umbilical cord. Insert the blunt blade of the scissors under the body wall as you make your cuts. Be careful not to cut any underlying organs.

In this manner, continue cutting through both skin and body wall around the anterior margin of the umbilical cord and posteriorly along each side, as shown for incision 2. Then, make the lateral incisions just in front of the hind legs (incision 3). Make incision 4 from the midline laterally following the lower margin of the ribs as shown. This will allow you to fold out the resulting lateral flaps.

The **umbilical vein,** which runs from the umbilical cord to the liver, must be cut to lay out the umbilical cord and attached structures posteriorly. Tie a string on each end of the umbilical vein so you will recognize the cut ends later. You will have an unobstructed view of the abdominal organs when this is done.

Extend incision 1 anteriorly through the sternum using heavy scissors. Locate the **diaphragm** and cut it free from the ventral body wall. Lift the left half of the ventral thoracic wall as you cut the connective tissue forming the mediastinal septum (Figure 18.5) that extends from the ventral wall to and around the heart. While lifting the left ventral wall, cut through the thoracic wall (ribs and all) at its lateral margin (incision 5), and cut through the muscle tissue at the anterior end of the sternum so the left thoracic ventral wall can be removed. Repeat the process for the right thoracic ventral wall. The heart and lungs are now exposed.

Continue incision 1 anteriorly to the chin, and separate the neck muscles to expose the thymus gland, thyroid gland, larynx, and trachea. Remove muscle tissue as necessary to expose these structures.

Wash out the cavities of the pig in a sink while being careful to keep the organs in place.

THE THORACIC ORGANS

The thoracic cavity is divided into left and right pleural cavities containing the **lungs.** See Figure 18.5. The left lung has three lobes and the right lung has four lobes. The inner thoracic wall is lined by a membrane, the **parietal pleura,** and each lung is covered by a **visceral pleura.** In life, fluid secreted into the potential space between these membranes, the **pleural cavity,** reduces friction as the lungs expand and contract during breathing.

At the midline, the parietal pleurae join with fibrous membranes to form a connective tissue partition between the pleural cavities called the **mediastinum.** The **heart** is located within the mediastinum. The heart is enclosed in a pericardial sac formed of the **parietal pericardium** and an outer fibrous membrane that is attached to the diaphragm. Remove this sac to expose the heart, if you have not already done so. The **visceral pericardium** tightly adheres to the surface of the heart.

In the neck region, locate the **larynx** (voice box), which is composed of cartilage and contains the vocal folds (cords). The **trachea** (windpipe) extends posteriorly from the larynx and divides dorsal to the heart to form the **bronchi,** which enter the lungs. You will see

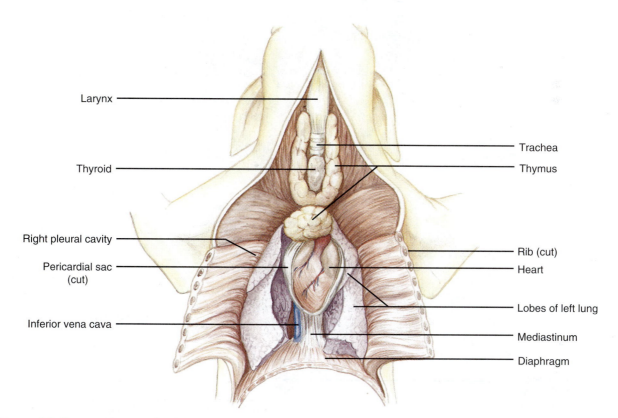

Figure 18.5 Ventral view of the thoracic organs.

these structures more clearly later, after the heart has been removed.

Locate the whitish **thymus gland** that lies near the anterior margin of the heart and extends into the neck on each side of the trachea. The thymus is an endocrine gland, and it is the maturation site for T-lymphocytes, important components of the immune system.

Locate the small, dark-colored **thyroid gland** on the anterior surface of the trachea at the base of the neck. The thyroid is an endocrine gland that controls the rate of metabolism in the body.

Remove the thymus and thyroid glands as necessary to get a better view of the trachea and larynx, but don't cut major blood vessels. Carefully remove the connective tissue supporting the trachea so you can move it to one side to expose the **esophagus** located dorsal to it. The esophagus is a tube that carries food from the

pharynx to the stomach. In the thorax, it descends through the mediastinum dorsal to the trachea and heart and then penetrates the diaphragm to open into the stomach. You will get a better view of it later.

Digestive Organs in the Abdomen

The inner wall of the abdominal cavity is lined by the **parietal peritoneum,** and the internal organs are covered by the **visceral peritoneum.** The internal organs are supported by thin membranes, the **mesenteries,** that consist of two layers of peritoneum between which are located blood vessels and nerves that serve the internal organs.

The digestive organs are the most obvious organs within the abdominal cavity, especially the large liver, which fits under the dome-shaped diaphragm. See Figure 18.6.

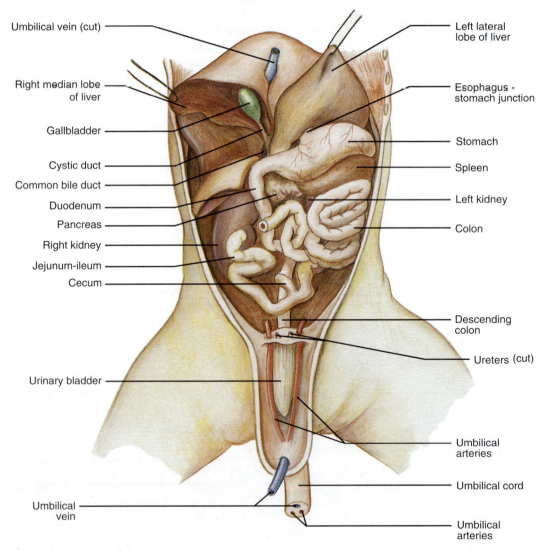

Figure 18.6 Ventral view of the abdominal organs.

Stomach

Lift the liver anteriorly on the left side to expose the **stomach.** The stomach receives food from the esophagus and releases partially digested food into the small intestine. Food is temporarily stored in the stomach, and the digestion of proteins begins there. The saclike stomach is roughly J-shaped. The longer curved margin on the left side is known as the **greater curvature.** The shorter margin between the openings of the esophagus and small intestine on the right side is the **lesser curvature.**

A saclike fold of mesentery, the **greater omentum,** extends from the greater curvature to the dorsal body wall. The elongate, dark organ supported by the greater omentum is the **spleen,** a lymphatic organ that stores and filters the blood in an adult. Locate the **lesser omentum,** a smaller fold of mesentery extending from the lesser curvature of the stomach and small intestine to the liver. The peritoneum also attaches the liver to the diaphragm.

Cut open the stomach along the greater curvature from the esophageal opening to the junction with the small intestine. The greenish material within it and throughout the digestive tract is the **meconium,** which is composed mostly of epithelial cells sloughed off from the lining of the digestive tract, mucus, and bile from the gallbladder. The meconium is passed in the first bowel movement of a newborn pig.

Wash out the stomach and observe the interior of the stomach. Note the numerous folds in the stomach lining that allow the stomach to expand when filled with food. Observe that the stomach wall is thicker just anterior to the junction with the small intestine. This thickening is the **pyloric sphincter.** Usually the sphincter is closed, but it opens to release material from the stomach into the small intestine. Locate a similar, but less obvious, thickening at the esophagus–stomach junction. This is the **cardiac sphincter.** It opens to allow food to enter the stomach, but it usually remains closed to prevent regurgitation.

Small Intestine

The small intestine consists of two parts: the duodenum and the jejunum-ileum. The first part of the small intestine is the short **duodenum.** It extends from the stomach, curves posteriorly, then turns anteriorly toward the stomach. The duodenum ends and the **jejunum-ileum** begins where the small intestine again turns posteriorly. In life, most of the digestion of food and absorption of nutrients occurs within the small intestine.

Lift out the small intestine, and observe how it is supported by mesenteries that contain blood vessels and nerves. Without damaging blood vessels, remove a 1-in. segment of the jejunum-ileum, cut through its wall lengthwise, open it up, and wash it out. The inner lining has enormous numbers of microscopic projections, the **villi,** that give it a velvety appearance. Place a small flattened section under water in a small petri dish and examine the inner lining under a dissecting microscope for a better look. The absorption of nutrients occurs through the villi.

Pancreas and Liver

Locate the pennant-shaped **pancreas** in the mesentery within the curve of the duodenum. It secretes the hormone insulin as well as digestive enzymes that are emptied into the duodenum via a tiny **pancreatic duct.** Try to find this small duct by carefully dissecting away the peritoneum from a little finger of pancreatic tissue that follows along the descending duodenum.

The **liver** is divided into five lobes. It carries out numerous vital metabolic functions and stores nutrients. Nutrients absorbed from the digestive tract are processed by the liver before being released into the general circulation. In addition, the liver produces **bile.** Bile consists of bile pigments, by-products of hemoglobin breakdown, and bile salts. Bile salts help emulsify fats, which facilitates their digestion in the small intestine.

Lift the right median lobe of the liver to locate the **gallbladder** on its undersurface just to the right of where the umbilical vein enters the liver. Careful dissection of the peritoneum will reveal the **hepatic duct,** which carries bile from the liver, and the **cystic duct,** which carries bile to and from the gallbladder. These two ducts merge to form the **common bile duct,** which carries bile into the anterior part of the duodenum. A sphincter muscle at the end of the common bile duct prevents bile from entering the duodenum, except when food enters the duodenum from the stomach. When this sphincter is closed, bile backs up and enters the gallbladder via the cystic duct, where it is stored until needed. Food entering the duodenum triggers a hormonal control mechanism that causes contraction of the gallbladder, forcing bile into the duodenum.

Large Intestine

Follow the small intestine to where it joins with the **large intestine,** or colon. At this juncture, locate the small side pouch, the **cecum,** which is nonfunctional in pigs and humans. The **ileocecal valve** is a sphincter muscle that prevents food material in the colon from reentering the small intestine. Make a longitudinal incision through this region to observe the ileocecal opening and valve.

In life, the large intestine contains substantial quantities of intestinal bacteria that decompose the nondigested food material that enters it. Water is reabsorbed back into the blood, forming the feces that are expelled in defecation.

The pig's large intestine forms a unique, tightly coiled, spiral mass. From the spiral mass, it extends anteriorly to loop over the duodenum before descending posteriorly against the dorsal wall into the pelvic region, where its terminal portion, the **rectum,** is located. The external opening of the rectum is the **anus.** The rectum will be observed in a later portion of the dissection.

CIRCULATORY SYSTEM

The circulatory system consists of the heart, arteries, capillaries, veins, and blood. **Blood** is the carrier of materials that are transported by the circulatory system. It is pumped by the **heart** through **arteries** that carry it away from the heart and into the **capillaries** of body tissues. Blood is collected from the capillaries by **veins** that return the blood to the heart.

In this section, you are to (1) locate the major blood vessels, noting their location and function, and (2) identify the external features of the heart. *In a double-injected fetal pig, the arteries are injected with red latex and the veins are injected with blue latex.* Sometimes a vein may not be injected because venous valves restrict flow of the latex. Such veins will appear as transparent tubes containing embalming fluid or blood. Also, vessels in your specimen may not exactly match the figures.

As you study the heart, major veins, and major arteries, carefully remove tissue as necessary to expose the vessels. This is best done by separating tissues with a blunt probe and by picking away connective tissue from the blood vessels with forceps. Make your observations in accordance with the descriptions and sequence that follow.

Before starting your dissection of the circulatory system, it is important to understand the basic pattern of circulation in an adult mammal, as shown in Figure 18.7, and contrast that with the circulation in a fetal mammal, as shown in Figure 18.8.

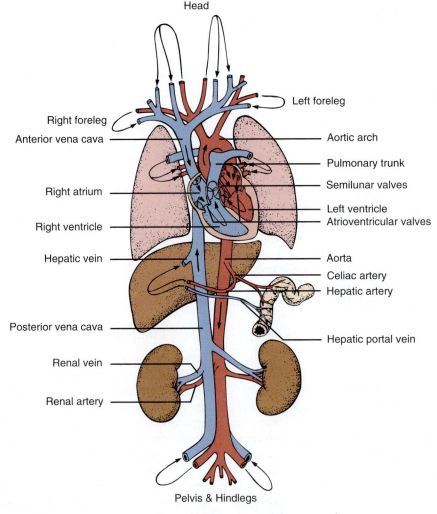

Figure 18.7 Diagrammatic representation of circulation in an adult mammal. Vessels colored blue carry deoxygenated blood. Vessels colored red carry oxygenated blood.

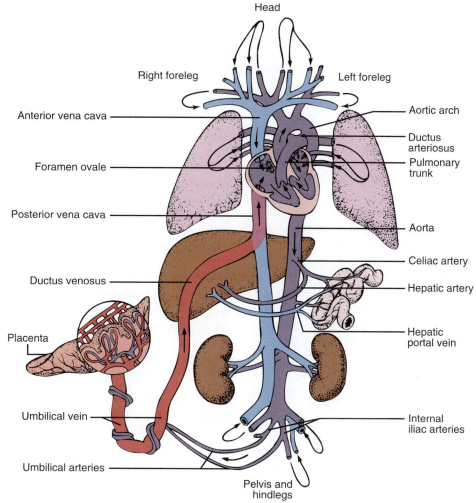

Figure 18.8 Diagrammatic representation of circulation in a fetal mammal. Vessels colored blue carry deoxygenated blood. Vessels colored red carry oxygenated blood. Vessels colored purple carry a mixture of oxygenated and deoxygenated blood.

Adult Circulation

In the adult mammal, deoxygenated blood is returned from the body to the **right atrium** of the heart by two large veins. The **anterior vena cava** returns blood from regions anterior to the heart. The **posterior vena cava** returns blood from areas posterior to the heart. Blood from the digestive tract is carried by the **hepatic portal vein** to the liver, whose metabolic processes modify the nutrient content of the blood before it is carried by the **hepatic vein** into the posterior vena cava.

At the same time when deoxygenated blood enters the right atrium, oxygenated blood from the lungs is carried by the **pulmonary veins** into the **left atrium** of the heart.

When the atria contract, blood from each atrium is forced into its corresponding ventricle. Immediately

thereafter, the ventricles contract. During ventricular contraction, blood pressure in the ventricles closes the **atrioventricular valves,** preventing a backflow of blood into the atria, and opens the **semilunar valves** at the bases of the two large arteries that exit the heart. Deoxygenated blood in the **right ventricle** is pumped through the **pulmonary trunk,** which branches into the **pulmonary arteries,** carrying blood to the lungs. Oxygenated blood in the **left ventricle** is pumped into the **aorta,** which divides into numerous branches to carry blood to all parts of the body except the lungs.

At the end of ventricular contraction, the semilunar valves close, preventing a backflow of blood into the ventricles, and the atrioventricular valves open, allowing blood to flow from the atria into the ventricles in preparation for another heart contraction.

Fetal Circulation

In a fetus, the placenta is the source of oxygen and nutrients, and it removes metabolic wastes from the blood. The lungs, digestive tract, and kidneys are non-functional. Circulatory adaptations to this condition make the circulation of blood in the fetus quite different from that in an adult.

Deoxygenated blood is returned from the body via the anterior and posterior venae cavae, as in the adult. However, blood rich in oxygen and nutrients is carried from the placenta by the **umbilical vein** directly through the liver by a segment called the **ductus venosus** and on into the posterior vena cava. Thus, oxygenated and deoxygenated blood are mixed in the posterior vena cava and returned to the right atrium, where additional mixing occurs as deoxygenated blood returns from organs above the heart.

The fetal heart contains an opening between the right and left atria, called the **foramen ovale** that allows much of the blood entering the right atrium to pass directly into the left atrium, bypassing the right ventricle and the pulmonary circuit. Further, much of the blood pumped into the pulmonary trunk enters the aorta via the **ductus arteriosus**, where it is added to mixed blood pumped from the left ventricle. Very little blood is carried to and from the nonfunctional lungs. These modifications in the fetal heart provide a maximal supply of mixed, oxygen-carrying blood to tissues and organs of the body as quickly as possible.

In the pelvic region, the **umbilical arteries** arise from the internal iliac arteries and carry blood to the placenta, where wastes are removed and oxygen and nutrients are picked up from the maternal blood.

At birth, the lungs are inflated by breathing, and the umbilical blood vessels, ductus venosus, and ductus arteriosus constrict, causing blood pressure changes that close the foramen ovale. These changes produce the adult pattern of circulation. Growth of fibrous connective tissue subsequently seals the foramen ovale and converts the constricted vessels into ligamentous cords.

The Heart and Its Great Vessels

Study the structure of the heart and the great vessels shown in Figures 18.9, 18.10, and 18.11. Be sure you know the relative positions of these vessels before trying to locate them in your specimen.

If you haven't done so, carefully cut away the pericardial sac from the heart and its attachment to the great vessels. Locate the **left** and **right atria**, blood-receiving chambers, and the **left** and **right ventricles**, blood-pumping chambers. Note the **coronary arteries** and **cardiac veins** that run diagonally across the heart at the location of the ventricular septum, a muscular partition separating the two ventricles. Coronary arteries supply blood to the heart muscle, and blockage of these arteries results in a heart attack.

While viewing the ventral surface of the heart, locate the large **anterior vena cava,** which returns blood from the head, neck, and forelegs to the right atrium. By lifting up the apex of the heart you will see the **large posterior vena cava,** which returns blood from regions posterior to the heart to the right atrium. Later, when you remove the heart, you will see where these veins enter the right atrium.

Now, locate the **pulmonary trunk,** which exits the right ventricle, and, just dorsal to it, the **aorta,** which exits the left ventricle. These large arteries appear whitish due to their thick walls. Move the apex of the heart to your left and locate the whitish **ductus arteriosus,** which carries blood from the pulmonary trunk into the aorta, bypassing the nonfunctional fetal lungs. You will see the **pulmonary arteries** and **veins** later when you remove the heart.

Major Vessels of the Head, Neck, and Thorax

As you read the description of each major vessel, locate it first in Figures 18.10 and 18.11, and then locate it in your dissection specimen.

As noted earlier, the large **anterior vena cava** returns blood from the head, neck, and forelegs into the right atrium. It is formed a short distance anterior to the heart by the union of the left and right brachiocephalic veins. The **brachiocephalic veins** are large, but very short. Each one is formed by the union of two jugular veins and a subclavian vein.

There are two jugular veins on each side of the neck. The **external jugular** is more laterally located and drains superficial tissues. The **internal jugular** is more medially located and drains deep tissues, including the brain. External and internal jugular veins may join just before their union with the subclavian vein. A **cephalic vein** drains part of the shoulder and joins with the external jugular near the union of the jugular veins.

The short, paired **subclavian veins** drain the shoulders and forelegs. Each is formed by the merging of a **subscapular vein,** which drains the posterior shoulder muscles, and an **axillary vein,** which receives blood from a **brachial vein** (not shown) of the foreleg.

Inferior to the brachiocephalic veins, the anterior vena cava receives blood from (1) a pair of **internal thoracic veins,** which return blood from the internal thoracic wall; (2) a pair of **external thoracic veins** (not shown), which return blood from the external thoracic wall; and (3) a pair of **costocervical veins,** which return blood from back and neck muscles.

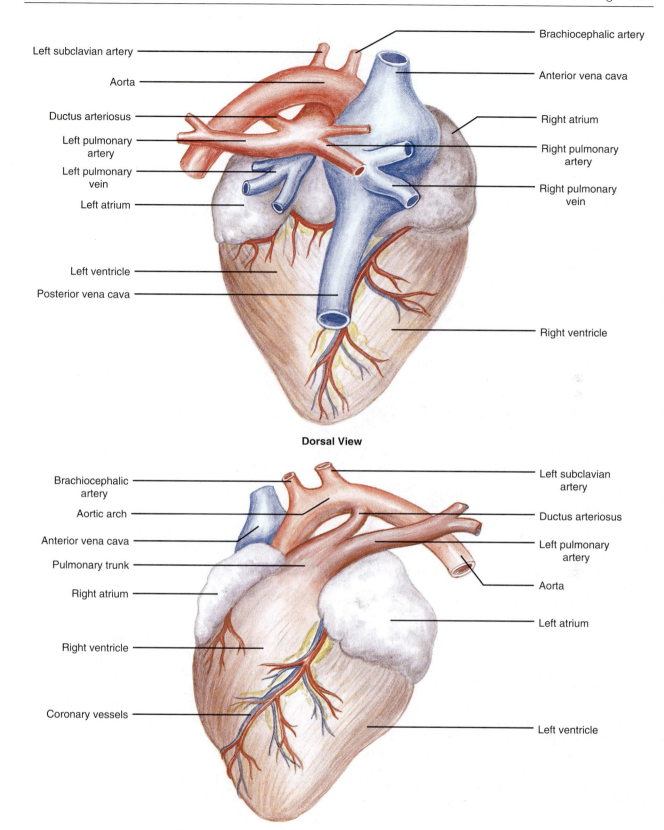

Dorsal View

Left subclavian artery

Aorta

Ductus arteriosus

Left pulmonary artery

Left pulmonary vein

Left atrium

Left ventricle

Posterior vena cava

Brachiocephalic artery

Anterior vena cava

Right atrium

Right pulmonary artery

Right pulmonary vein

Right ventricle

Ventral View

Brachiocephalic artery

Aortic arch

Anterior vena cava

Pulmonary trunk

Right atrium

Right ventricle

Coronary vessels

Left subclavian artery

Ductus arteriosus

Left pulmonary artery

Aorta

Left atrium

Left ventricle

Figure 18.9 Heart of a fetal pig. Veins are colored blue. Arteries are colored red.

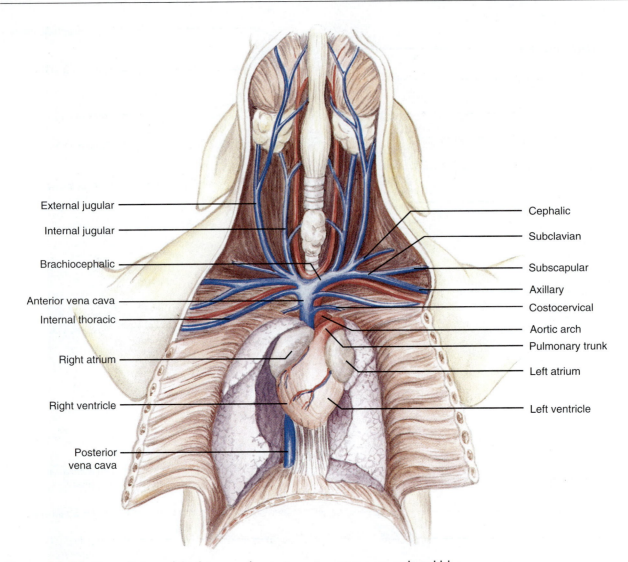

External jugular

Internal jugular

Brachiocephalic

Anterior vena cava

Internal thoracic

Right atrium

Right ventricle

Posterior vena cava

Cephalic

Subclavian

Subscapular

Axillary

Costocervical

Aortic arch

Pulmonary trunk

Left atrium

Left ventricle

Figure 18.10 Ventral view of the heart and anterior veins. Veins are colored blue. Arteries are colored red.

Once you have located these veins, cut the superior vena cava, leaving a stub at the heart, and lift it and the attached veins anteriorly to expose the underlying arteries. Locate the **aorta** and note that it forms the **aortic arch** as it curves posteriorly dorsal to the heart. Two major arteries branch from the aortic arch. The first and larger branch is the **brachiocephalic artery.** The second and smaller branch is the **left subclavian artery.**

Follow the brachiocephalic anteriorly, where it branches to form the **right subclavian artery** and a **carotid trunk** that divides to form a pair of **common carotid arteries.** The common carotid arteries parallel the internal jugular veins to carry blood to the head and neck. At the base of the head, each common carotid artery branches into **internal** and **external carotid arteries.**

Each subclavian artery gives off (1) a **costocervical artery,** which supplies the back and neck; (2) an

external thoracic artery, which supplies the external thoracic wall; (3) a **thyrocervical artery,** which supplies the thyroid gland and neck; and (4) an **internal thoracic artery,** which supplies the internal thoracic wall. Each subclavian becomes an **axillary artery** in the axillary or armpit region.

After you have located the major anterior arteries and veins, carefully cut through the major vessels to remove the heart. Leave stubs of the vessels on the heart and identify them by comparing your preparation with Figure 18.9. Especially locate the pulmonary arteries and veins.

Major Vessels Posterior to the Diaphragm

Follow the aorta and posterior vena cava posteriorly through the diaphragm. Compare your observations with Figures 18.12 and 18.13.

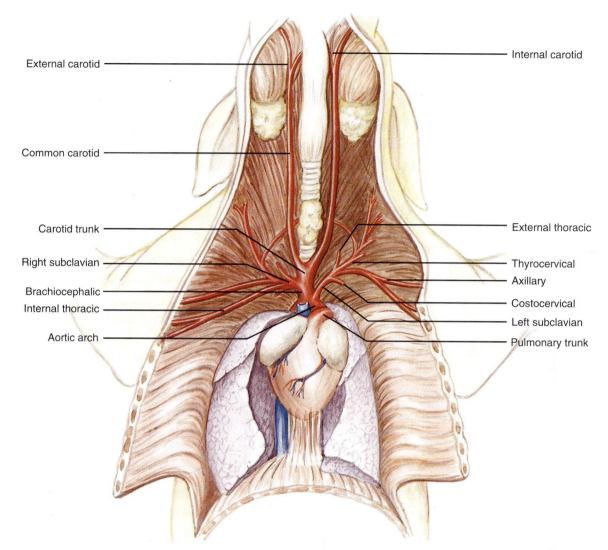

Figure 18.11 Ventral view of the heart and anterior arteries.

Just posterior to the diaphragm, locate the **hepatic vein** from the liver where it enters the posterior vena cava. Immediately posterior to this union, the **ductus venosus** enters the posterior vena cava. Recall that the ductus venosus is an extension of the **umbilical vein** in the fetal pig. Dissect away liver tissue as necessary to expose the unions of the hepatic vein and ductus venosus with the posterior vena cava. Save the liver tissue adjacent to the hepatic vein and ductus venosus, but remove and discard the rest. Cut through the left side of the diaphragm, and push the abdominal organs to your left (specimen's right) to expose the posterior vena cava and aorta.

Lift the stomach anteriorly to expose the first branch from the aorta posterior to the diaphragm. This is the **celiac artery.** It gives off branches to the stomach, spleen, and liver (**hepatic artery**). Posterior to the celiac is a large branch, the **anterior mesenteric artery,** that gives off numerous branches to the intestine. Note

that many small veins from the intestine join to form the **hepatic portal vein,** which runs anteriorly to enter the liver. This vein carries nutrient-rich blood from the intestine to the liver for processing prior to being released into the general circulation.

Now remove most of the intestines, but save a small section associated with the hepatic portal vein. This will expose the lower portion of the aorta and posterior vena cava.

Locate the **renal arteries** and **renal veins** serving the kidneys. Because the kidneys are located dorsal to the parietal peritoneum, you will have to strip away some of this tissue to observe the vessels and kidneys clearly. The **adrenal glands** (endocrine glands), narrow strips of tissue, should be visible on the anterior margin of the kidneys.

Carefully dissect away the connective tissue to expose the posterior portions of the aorta, the posterior vena cava, and the attached vessels. Locate the ureters,

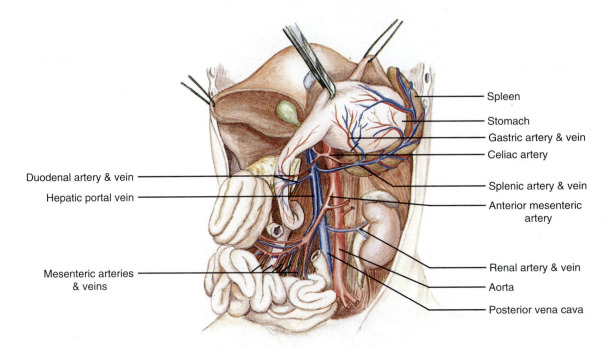

Figure 18.12 Ventral view of the arteries and veins supplying the digestive organs.

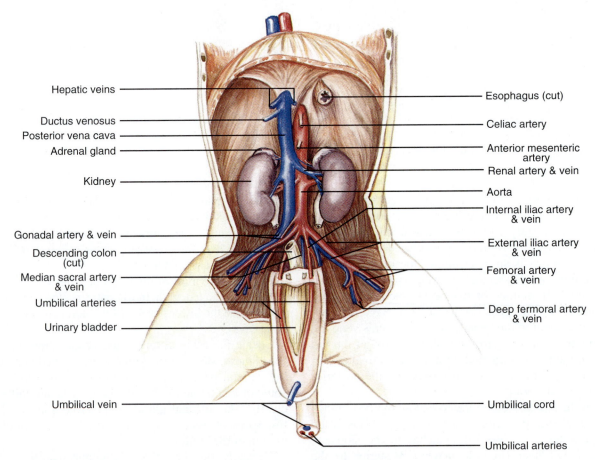

Figure 18.13 Ventral view of arteries and veins in the abdominopelvic cavity.

which carry urine from the kidneys to the urinary bladder. The urinary bladder is located in the reflected tissue containing the umbilical arteries and umbilical cord. Free the ureters from surrounding tissue so they are movable, but don't cut through them.

The small arteries and veins posterior to the renal vessels serve the gonads, ovaries or testes, as the case may be. Just below the gonadal arteries, the **posterior mesenteric artery** arises from the ventral surface of the aorta and supplies most of the large intestine.

Locate the pair of **external iliac arteries** that branch laterally from near the posterior end of the aorta. After giving off small branch arteries to the body wall, each external iliac artery extends into the thigh, where it divides into the medially located **deep femoral artery** and the laterally located **femoral artery.** Each of these arteries is associated with a corresponding vein, the **deep femoral vein** and the **femoral vein,** which drain the leg and merge to form the **external iliac vein.**

Now locate the medial terminal branches of the aorta, the **internal iliac arteries.** The major extensions of these arteries form the **umbilical arteries,** which pass over the urinary bladder to enter the umbilical cord. The other branches of the internal iliac arteries supply the pelvic area and lie alongside the **internal iliac veins,** which drain the pelvic area.

The external and internal iliac veins join to form a **common iliac vein** on each side of the body. The union of the common iliac veins forms the posterior beginning of the posterior vena cava.

You will see between the common iliac veins the **median sacral artery** and **vein,** serving the dorsal wall of the pelvic area.

THE RESPIRATORY SYSTEM

The respiratory system functions to bring air into the lungs, where blood releases its load of carbon dioxide and receives oxygen for transport to body cells. Air enters the lungs via a series of air passageways. Air enters the **nostrils** and flows through the **nasal cavity,** where it is warmed and moistened. It passes into the **nasal pharynx** and on into the oral pharynx, where it enters the **larynx** via a slitlike opening called the **glottis.** You observed these components when you dissected the oral cavity and pharynx. From the larynx, air passes down the **trachea,** which branches to form the **primary bronchi.** The bronchi carry air into the **lungs.**

Return now to the neck and thoracic cavity to observe the respiratory organs. Locate the larynx, trachea, and primary bronchi. Because the heart has been removed, the trachea and primary bronchi can be readily seen by dissecting away some of the surrounding connective tissue and blood vessels. See Figures 18.5 and 18.10.

Make a midventral incision in the larynx, open it, and locate the **vocal folds** on each side. Note how the trachea branches to form the primary bronchi. Each primary bronchus divides within the lung to form smaller and smaller air passages until tiny microscopic **bronchioles** terminate in vast numbers of saclike **alveoli,** the sites of gas exchange. Dissect along a primary bronchus to locate the **secondary bronchi,** which enter the lobes of a lung. Remove a section of trachea and observe the shape of the **cartilaginous rings** that hold it open. Make a section through a lung to observe its spongy nature and the air passages within it.

UROGENITAL SYSTEM

The urinary and reproductive organs compose the urogenital system. They are considered together because the organs are closely interrelated. Refer to Figure 18.14 or 18.15, depending on the sex of your specimen. You are responsible for knowing the urogenital system of each sex, so prepare your dissection carefully and exchange it when finished for a specimen of the opposite sex.

Urinary System

If you haven't removed the intestines, do so now to expose the urinary organs. Leave a stub of the large intestine because you will want to locate the rectum later. Dissect away the parietal peritoneum to expose the **kidneys,** located dorsal to it and against the dorsal body wall. The kidneys are held in place by connective tissue. An **adrenal gland** is located on the anterior surface of each kidney.

Locate the origin of a **ureter** on the medial surface of a kidney near where the renal vessels enter it. As urine is formed by the kidney, it passes into the ureter for transport by peristalsis (wavelike contractions) to the **urinary bladder.** Trace the ureter posteriorly to its dorsally located entrance into the urinary bladder. The urinary bladder lies between the umbilical arteries on the reflected portion of the body wall containing the umbilical cord. Note that the posterior portion of the urinary bladder narrows to form the **urethra,** which enters the pelvic cavity. It will be observed momentarily.

If your instructor wants you to dissect a kidney, make a coronal section to expose its interior. Compare the internal structure with Figure 22.2 in Exercise 22 as you read the accompanying description.

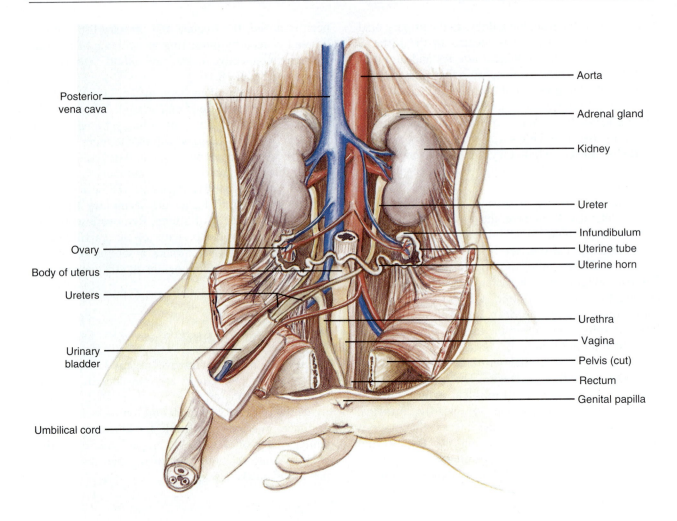

Posterior vena cava

Ovary

Body of uterus

Ureters

Urinary bladder

Umbilical cord

Aorta

Adrenal gland

Kidney

Ureter

Infundibulum

Uterine tube

Uterine horn

Urethra

Vagina

Pelvis (cut)

Rectum

Genital papilla

Figure 18.14 Ventral view of the female urogenital system.

Female Reproductive System

The **uterus** is located dorsal to the urinary bladder. It consists of two **uterine horns** that join posteriorly at the midline to form the **body of the uterus.** Locate the uterine horns and the body of the uterus.

Follow the uterine horns anterolaterally to the small, almost nodulelike **ovaries** just posterior to the kidneys, where they are supported by mesenteries. The uterine horns end at the posterior margin of the ovaries, where they are continuous with the tiny, convoluted **uterine tubes,** which continue around the ovaries to their anterior margins. The anterior end of a uterine tube forms an expanded funnel-shaped opening, the **infundibulum,** that receives the immature ovum when it is released from an ovary. Fertilization occurs in the uterine tubes, and the fetal pigs develop in the uterine horns.

Now, remove or reflect the skin from the ventral surface of the pelvis. Use your scalpel to cut carefully at the midline through the muscles and bones of the pelvic girdle. Spread the legs and open the pelvic cavity to expose the **urethra** and, just dorsal to it, the **vagina,** extending posteriorly from the body of the uterus. You can now see the **rectum,** the terminal portion of the large intestine, located dorsal to the vagina. Note that the vagina and urethra unite to form the **vaginal vestibule** a short distance from the external urogenital opening, which is identified externally by the genital papilla.

Male Reproductive Organs

The **testes** develop within the body cavity just posterior to the kidneys. Later in fetal development, the testes descend through the **inguinal canals** into the **scrotum,** an external pouch. The scrotum provides a temperature for the testes that is slightly less than body temperature. The lower temperature is necessary for the production

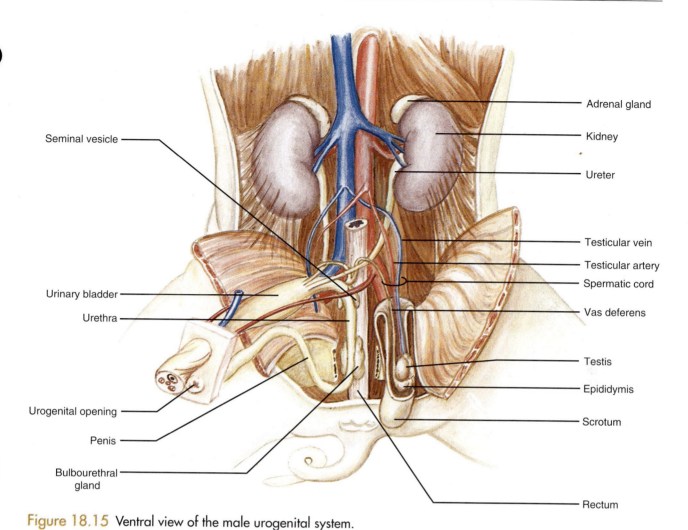

Figure 18.15 Ventral view of the male urogenital system.

of viable sperm. Follow a testicular artery and vein to locate the inguinal canal.

On one side, cut open the scrotum to expose a testis. Locate the **epididymis,** a tortuous mass of tiny tubules that begins on the anterior margin of the testis and extends along the lateral margin to join posteriorly with the **vas deferens,** or sperm duct. Trace the vas deferens anteriorly from the scrotum, through the inguinal canal to where it loops over a ureter to enter the urethra.

Locate the **urogenital opening** just posterior to the umbilical cord in the reflected flap of body wall containing the urinary bladder. The **penis** extends posteriorly from this point. Make an incision alongside the penis and free it from the body wall. Push it to one side, and use your scalpel to make a midline incision through the pelvic muscles and bones. Spread the legs and open the pelvic cavity so you can dissect out the pelvic organs. Locate the **urethra** at the base of the urinary bladder and carefully remove connective tissue

around it, separating it from the **rectum** as it continues posteriorly into the pelvic cavity.

Where the vasa deferentia enter the urethra, locate the small glands on each side of the urethra. These are the **seminal vesicles.** Between the seminal vesicles on the dorsal surface of the urethra is the small **prostate gland.** Follow the urethra posteriorly to locate the pair of **bulbourethral glands** on each side of the urethra where it enters the penis. At ejaculation, these accessory glands secrete the fluids that transport sperm.

CONCLUSION

This completes the dissection. If you have done it thoughtfully, you have gained a good deal of knowledge about the body organization of mammals, including humans. Dispose of your specimen as directed by your instructor. There is no laboratory report for this exercise.

BLOOD AND CIRCULATION

OBJECTIVES

After completing the laboratory session, you should be able to:
1. Distinguish between open and closed circulatory systems.
2. Describe the components and functions of blood and hemolymph.
3. Describe and identify human blood cells.
4. Explain the basis of blood groups and blood typing.
5. Identify the parts of the mammalian heart on charts and in a dissected sheep heart, and describe their functions.
6. Trace the path of blood in humans from the heart to the lungs, intestine, head, kidneys, and back to the heart.
7. Demonstrate the measurement of blood pressure in humans.
8. Define all terms in bold print.

In unicellular animal-like protists and in simple multicellular animals, diffusion alone provides an adequate rate of distribution of materials to meet the metabolic needs of the organism. Larger, more complex animals depend on a **circulatory system** to provide a more rapid transport of materials. Diffusion still plays an important role in the movement of materials between the circulatory fluid and the body cells, however.

Circulatory systems include (1) a circulating fluid (blood or hemolymph); (2) a heart or pulsating vessel, containing one-way valves, that pumps the fluid through the system; and (3) vessels through which the fluid passes. Two types of circulatory systems occur among animal groups: closed systems and open systems.

In **closed circulatory systems,** the circulating fluid is called **blood,** and it is confined within the blood vessels: arteries, capillaries, and veins. Circulation of blood is rapid. Annelids and vertebrates have closed circulatory systems.

In **open circulatory systems,** the circulating fluid is called **hemolymph,** and it is not confined within vessels. It is carried from the pulsating heart through one or more open-ended arteries that empty it into the coelom, which in animals with open circulatory systems is called a **hemocoel.** The hemolymph slowly moves around the internal organs on its way back to the heart. Most mollusks and all arthropods have an open circulatory system.

BLOOD AND HEMOLYMPH

Blood and hemolymph are quite variable among animals. In either case, separated **blood cells** are carried along in a fluid **plasma.** At least some of the cells are capable of amoeboid movement. The function of blood and hemolymph is to transport materials from place to place in the animal's body. These materials include organic nutrients, metabolic wastes, hormones, antibodies, mineral ions, oxygen, and carbon dioxide. Most insects are an exception in that respiratory gases are transported primarily by the tracheal system, not the hemolymph.

Two types of **respiratory pigments** occur that facilitate the transport of oxygen. **Hemocyanin,** a blue-green pigment, is dissolved in the plasma of certain arthropods. **Hemoglobin,** a red pigment, is dissolved in the plasma of annelids and some arthropods, and it is confined within red blood cells in vertebrates.

Assignment 1

Materials

Compound microscope
Prepared slides of:
  amphibian blood
  reptilian blood

1. Examine prepared slides of amphibian and reptilian blood. Observe the red blood cells, which are the most numerous cells on the slide. What is their general shape? Do they have a nucleus? Stain has been used to distinguish the blood cells, so the colors you see are artificial.
2. **Complete items 1a–1c on Laboratory Report 19 that begins on page 261.**

Human Blood

Human blood consists of 45% blood cells and 55% plasma. The formed elements, commonly called blood cells, may be divided into three groups: (1) **erythrocytes,** red blood cells that transport oxygen and carbon dioxide; (2) **leukocytes,** white blood cells that fight infections; and (3) **thrombocytes,** platelets that initiate the clotting process. See Figure 19.1. In 1mm^3 of blood, there are about 5,000,000 erythrocytes, 8,000 leukocytes, and 350,000 thrombocytes, and these constitute only 45% of the volume! Obviously, blood cells are very small and numerous. Millions are destroyed and replaced daily. The types and distinguishing characteristics of white blood cells when stained with Wright's blood stain are presented in Table 19.1.

Assignment 2

Materials

Colored pencils
Compound microscope

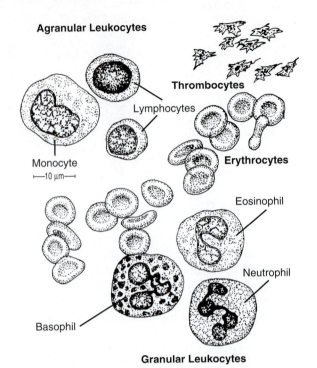

Figure 19.1 Human blood cells.

Prepared slide of human blood
Schilling blood chart

1. Obtain a Schilling blood chart from the stock table. Study the mature blood cells in the bottom row to learn their characteristics and frequency in normal circulating blood. The cells appear as stained with Wright's blood stain. Note that leukocytes are subdivided into **granulocytes** and **agranulocytes,** depending on the presence or absence of cytoplasmic granules. Do the mature red blood cells have a nucleus? This condition is true for all mammals. Note the small size of the platelets, which are nonnucleated fragments of larger parent cells.

TABLE 19.1	LEUKOCYTES	
Name	Characteristics of Stained Cells	Normal Percentage
Granulocytes		
Neutrophils	Nucleus with three to five lobes, very small lavender cytoplasmic granules	60%–70%
Eosinophils	Nucleus with two lobes, red or orange cytoplasmic granules	2%–4%
Basophils	Nucleus with two to three lobes, blue or purple cytoplasmic granules	0.5%–1%
Agranulocytes		
Lymphocytes	Small cell with a large spherical nucleus and small amount of cytoplasm	20%–25%
Monocytes	Large cell with a large irregular or kidney-shaped nucleus	3%–8%

2. Color the blood cells in Figure 19.1 as shown on the Schilling blood chart.
3. Examine a prepared slide of human blood at 400× and locate erythrocytes, neutrophils, lymphocytes, and monocytes. ***Draw a few red blood cells in item 1c.*** Try to find an eosinophil and basophil, but don't spend more than 10 min searching. Compare your observations with the blood chart.
4. ***Complete item 1c and item 2 on the laboratory report.***

Human Blood Groups

Human blood may be classified according to the presence or absence of certain **antigens** (proteins) on the red blood cells. The presence of these antigens is genetically controlled, so an individual's **blood type** is the same from birth to death. See Table 19.2. The antigens most commonly tested for in blood typing tests are A, B, and D (Rh).

Testing for the presence of A and B antigens determines the ABO blood group. The presence or absence of the D (Rh) antigen determines the Rh blood type. The two factors are combined in designating an individual's blood type (e.g., A, Rh$^+$; A, Rh$^-$; AB, Rh$^+$). Table 19.3 indicates the association of these antigens and their corresponding **antibodies** in normal blood.

TABLE 19.2	PERCENTAGE OF ABO BLOOD TYPES	
Blood Type	% U.S. Blacks	% U.S. Whites
A	25	41
B	20	7
AB	4	2
O	51	50

TABLE 19.3	NORMAL ANTIGEN–ANTIBODY ASSOCIATIONS	
Blood Type	Antigen on RBC	Antibody in Plasma
O	None	a, b
A	A	b
B	B	a
AB	AB	None
Rh$^+$	D	None
Rh$^-$	None	d*

*Antibodies are produced by an Rh$^-$ person only after Rh$^+$ red blood cells enter his or her blood.

Not all blood types are compatible with each other. Therefore, knowing the blood types of both donor and recipient in blood transfusions is imperative. The *antigens of the donor* and the *antibodies of the recipient* must be considered in blood transfusions. Transfusion of incompatible blood results in clumping (agglutination) of erythrocytes in the transfused blood, which may plug capillaries and result in death. Clumping of erythrocytes occurs whenever the *antigens of the donor* and the *antibodies of the recipient*, which are designated by the same letter, are brought together. See Table 19.3. The antibodies of the donor are so diluted in the recipient's blood that their effect is inconsequential.

The Rh Factor

About 85% of Caucasians and 99%–100% of Chinese, Japanese, African blacks, and Native Americans possess the D (Rh) antigen and are, therefore, Rh$^+$. If an Rh$^-$ individual receives a single transfusion of Rh$^+$ blood, antibodies are produced against the D (Rh) antigen. However, no clumping of cells occurs, owing to the gradual increase in antibodies and the loss of erythrocytes containing the D (Rh) antigen. If a second transfusion of Rh$^+$ blood is received, clumping will occur, and death may result.

Antibodies against the D (Rh) antigen may be produced by an Rh$^-$ woman carrying an Rh$^+$ fetus, owing to "leakage" of fetal erythrocytes into the maternal blood. The accumulation of antibodies in maternal blood is gradual enough so no complications result during the first pregnancy with an Rh$^+$ fetus. In subsequent pregnancies with Rh$^+$ fetuses, however, the maternal antibodies may diffuse into the fetal blood and destroy the fetal erythrocytes. This pathological condition, *erythroblastosis fetalis*, can be fatal to the fetus. The mother suffers no consequences unless she subsequently receives a transfusion of Rh$^+$ blood.

Erythroblastosis fetalis may be prevented if the Rh$^-$ mother receives an injection of RhoGAM shortly after giving birth to an Rh$^+$ baby. RhoGAM binds with Rh$^+$ antigens and removes them from the blood, preventing the development of antibodies against the Rh$^+$ antigen.

Blood Typing

Testing for the presence of A, B, and D antigens is accomplished by adding **antisera** containing specific antibodies that will combine with these antigens. Because testing for the D antigen requires a temperature of 50°C, you will use a slide warming box and a special blood typing plate to allow simultaneous testing for all three antigens. See Figure 19.2.

Your instructor may choose to have you use either simulated blood or your own blood for blood typing. Follow your instructor's directions carefully.

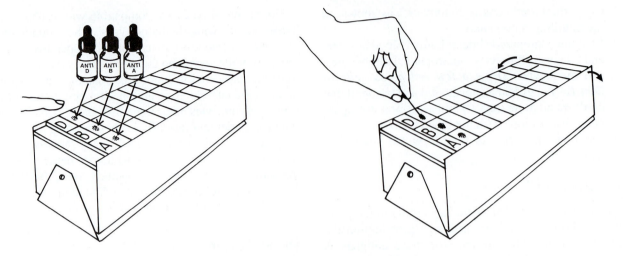

1. Place a drop of blood on the glass slide over each of the three squares.
2. Place a drop of the appropriate antiserum in each blood drop.
3. Quickly stir the antisera and blood, *using a clean toothpick for each drop.*
4. Slowly rock the warming box back and forth for 2 min.

Figure 19.2 Blood typing setup.

Caution: If you use your own blood, care must be taken to prevent self-infection from environmental contaminants and the transmission of blood-borne infections, such as viruses causing hepatitis and AIDS. Typing of your own blood may be done safely if you rigidly follow the safety precautions, typing procedures, and your instructor's directions, because you will be in contact only with your own blood.

Your instructor has set up blood typing stations that are provisioned with the needed supplies.

Safety Precautions

1. Avoid contact with blood of another student.
2. Before piercing your finger, wash your hands with soap and water and disinfect the fingertip with an alcohol wipe (70% alcohol).
3. Do not remove a sterile lancet from its container until you are ready to use it. Do not touch its tip or lay it down.
4. Use a lancet *once* and *immediately* place it in the biohazard sharps container provided. *Never place a used lancet on the tabletop or in a wastebasket.*
5. Place all other materials in contact with blood, such as microscope slides, alcohol wipes, toothpicks, and paper towels, in a biohazard bag *immediately* after use.
6. When finished, wash the tabletop around the typing box with a suitable disinfectant, such as 10% household bleach or 0.5% Amphyl. Wash your hands with soap and rinse with a disinfectant.
7. Perform the typing procedure under the direct supervision of your instructor. Follow your instructor's directions at all times.

 Assignment 3

Materials

Antisera
  anti-A
  anti-B
  anti-D (Rh)
Biohazard bag
Biohazard sharps container
Blood typing box and typing plate
Compound microscope
Cotton, sterile and absorbent
Lancets, sterile and disposable
Microscope slides
Toothpicks, flat
Alcohol wipes
Amphyl solution, 0.5%
Household bleach, 10%

Perform blood typing using the following procedures. Read the procedures completely through before starting.

1. Place the glass typing plate on the warming box and turn on the box. Allow 5 min for temperature build-up.
2. Place a clean microscope slide across the typing plate on the slide warming box.
3. Wash your hands with soap and water and disinfect the fingertip of your choice with an alcohol wipe.
4. Obtain a sterile lancet. Remove it from its container and pierce your fingertip from the side. Place the lancet *immediately* in the biohazard sharps container.

5. Place a drop of blood on the microscope slide over each of the three squares of the typing plate. Blot your fingertip with an alcohol wipe. Place the wipe in the biohazard bag.

6. *Without touching the drop of blood with the dropper* (why?), add a drop of anti-D serum to the blood in the D square, a drop of anti-B serum to the blood in the B square, and a drop of anti-A serum to the blood in the A square. See Figure 19.2.

7. Use a *different* toothpick (why?) to mix the blood–antiserum mixture in each of the three squares. Record the time. Place the toothpicks *immediately* in the biohazard bag.

8. Rock the warming box back and forth for *exactly 2 min*, and then determine the results. If tiny red granules appear in the blood and anti-D serum mixture within 2 min, the blood is Rh$^+$. If not, it is Rh$^-$. The mixture may have to be examined under a dissecting microscope to determine the type. If clumping occurs in either of the other squares used for ABO blood typing, the red granules will be large and obvious. Determine your blood type by referring to Figure 19.3.

9. *Place anything in contact with blood, such as used toothpicks, cotton, and glass slides, in the biohazard bag immediately after use.*

10. **Complete item 3 on the laboratory report.**

Circulation of the Blood

The circulatory system of vertebrates is composed of (1) the **heart,** whose contractions provide the force to circulate the blood; (2) **arteries,** which carry blood away from the heart to all parts of the body; (3) **veins,** which return blood to the heart from the body; and (4) **capillaries,** tiny vessels that connect arteries and veins.

The Heart

The vertebrate heart shows a progressive increase in complexity from fish to birds and mammals. Fish have a two-chambered heart composed of an **atrium,** a receiving chamber, and a **ventricle,** a pumping chamber. A valve between the chambers prevents the backflow of blood. Deoxygenated blood is pumped by the

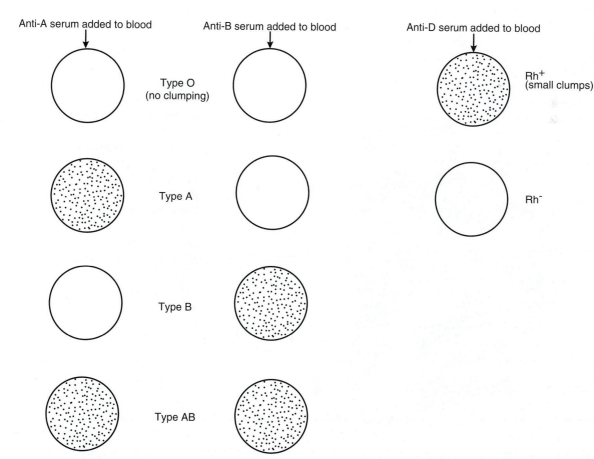

Figure 19.3 **Blood type determination.**

ventricle to the gills, where it is oxygenated before it continues on to the rest of the body.

Amphibians have a three-chambered heart composed of one ventricle and two atria. An interatrial septum separates the atria from each other, so blood cannot flow between them. Blood is pumped to the lungs and body simultaneously by the single ventricle. Deoxygenated blood from the body returns to the right atrium, and oxygenated blood from the lungs enters the left atrium. Blood from each atrium flows into the ventricle, which results in mixing oxygenated and deoxygenated blood in the single ventricle. Thus, the blood nourishing body cells is not highly oxygenated.

Reptiles have a three-chambered heart with an incomplete interventricular septum (a partition of cardiac muscle) that partially divides the ventricle and reduces the mixing of oxygenated and deoxygenated blood. This structure improves the oxygenation of blood supplying the body cells. In birds and mammals, the interventricular septum is complete, so they have four-chambered hearts with two atria and two ventricles that ensure the circulation of highly oxygenated blood to the body cells. Refer to Figures 19.4–19.6 as you study the description of the mammalian heart that follows.

The mammalian heart is a double pump composed of four chambers. The two **atria** receive blood returning to the heart. The two **ventricles** pump blood from the heart. The **right atrium** and **right ventricle** compose one pumping unit, and the **left atrium** and **left ventricle** compose the other. The **interatrial septum** prevents blood from flowing between the two atria, and the **interventricular septum** prevents blood passing between the two ventricles.

The opening between the atrium and ventricle on each side is guarded by an **atrioventricular (AV) valve,** which prevents a backflow of blood from the ventricle into the atrium. The **bicuspid (mitral) valve** separates the left atrium and ventricle. The **tricuspid valve** separates the right atrium and ventricle. The flaps (cusps) of the valves are anchored by tough cords of tissue called **chordae tendineae** to small mounds of muscle, **papillary muscles,** on the inner walls of the ventricles. These cords prevent the valve flaps from being forced into the atria during ventricular contraction and enable the valves to close off the opening. They resemble the lines of a parachute as they restrain the valve flaps.

Semilunar valves are located at the base of both the **pulmonary trunk** and **aorta.** These valves permit blood

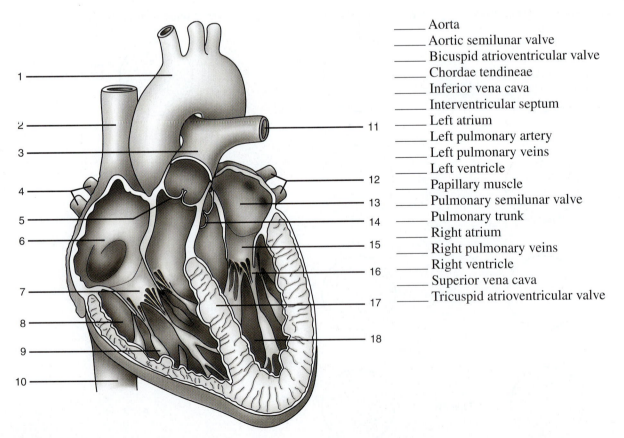

_____ Aorta
_____ Aortic semilunar valve
_____ Bicuspid atrioventricular valve
_____ Chordae tendineae
_____ Inferior vena cava
_____ Interventricular septum
_____ Left atrium
_____ Left pulmonary artery
_____ Left pulmonary veins
_____ Left ventricle
_____ Papillary muscle
_____ Pulmonary semilunar valve
_____ Pulmonary trunk
_____ Right atrium
_____ Right pulmonary veins
_____ Right ventricle
_____ Superior vena cava
_____ Tricuspid atrioventricular valve

Figure 19.4 A coronal section of a sheep heart.

Ventral View

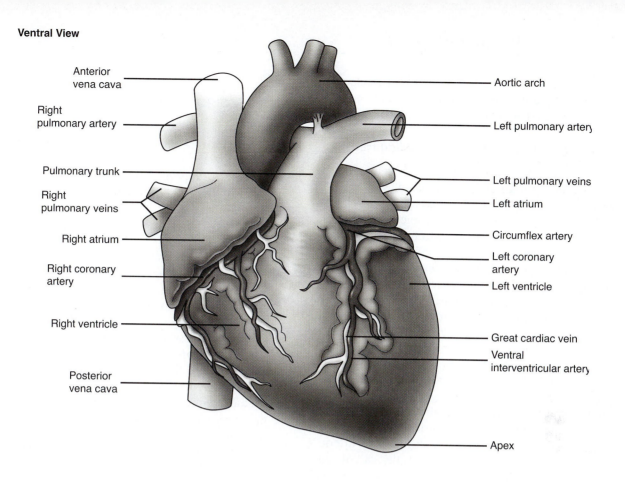

Anterior vena cava

Right pulmonary artery

Pulmonary trunk

Right pulmonary veins

Right atrium

Right coronary artery

Right ventricle

Posterior vena cava

Aortic arch

Left pulmonary artery

Left pulmonary veins

Left atrium

Circumflex artery

Left coronary artery

Left ventricle

Great cardiac vein

Ventral interventricular artery

Apex

Dorsal View

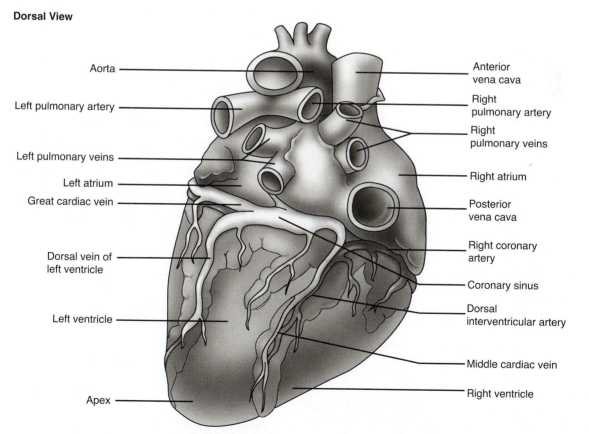

Aorta

Left pulmonary artery

Left pulmonary veins

Left atrium

Great cardiac vein

Dorsal vein of left ventricle

Left ventricle

Apex

Anterior vena cava

Right pulmonary artery

Right pulmonary veins

Right atrium

Posterior vena cava

Right coronary artery

Coronary sinus

Dorsal interventricular artery

Middle cardiac vein

Right ventricle

Figure 19.5 Sheep heart, external structure.

to be pumped into these arteries during ventricular contraction, but prevent a backflow of blood from the arteries into the ventricles during ventricular relaxation.

Flow of Blood Through the Mammalian Heart

The flow of blood through the heart is related to the contraction (systole) and relaxation (diastole) of the atria and ventricles during the heart cycle.

During atrial and ventricular diastole, deoxygenated blood flows into the right atrium from two large veins, the **anterior** (superior) and **posterior** (inferior) **venae cavae,** and oxygenated blood flows into the left atrium from the **left** and **right pulmonary veins.** The contraction of the atria in atrial systole forces blood into the relaxed ventricles until they are filled with blood. In ventricular systole, the ventricles contract and the atria relax. The sudden increase in blood pressure within the contracting ventricles closes the AV valves and pumps blood through the semilunar valves into the arteries. The right ventricle pumps deoxygenated blood into the pulmonary trunk, which leads to the pulmonary arteries and on to the lungs. The left ventricle pumps oxygenated blood into the aorta, which leads to smaller arteries and on to all other parts of the body. At the start of ventricular diastole, the semilunar valves close, and the atrioventricular valves open. The closing of the heart valves produces the characteristic heart sounds.

 ## Assignment 4

Materials

Colored pencils
Heart models of:
  fish
  amphibian
  reptile
  mammal

1. Label Figure 19.4. Color blood vessels and heart chambers carrying oxygenated blood red, and color those carrying deoxygenated blood blue.
2. Trace the flow of blood through the heart in Figure 19.4.
3. Locate the parts of the mammalian heart on a model or chart.
4. Compare the models of vertebrate hearts and note the differences.
5. ***Complete item 4 on the laboratory report.***

The Sheep Heart

A study of the external and internal structure of a sheep heart will extend your understanding of heart structure and function. Refer to Figures 19.4, 19.5, and 19.6 as you proceed.

 ## Assignment 5

Materials

Dissecting instruments and pins
Dissecting pan
Sheep heart, fresh or preserved
Gloves, protective and disposable

Examine the external structure of the sheep heart and perform the sheep heart dissection as described here.

External Structure

1. If the **pericardium,** a membranous sac enclosing the heart, is present, note how it envelops the heart and is attached to the major vessels above the heart. Use scissors to remove it.
2. Position the heart to observe the ventral surface as shown in Figure 19.5. Locate the large, muscular ventricles and the smaller, thin-walled atria. The **interventricular coronary artery** and the **great cardiac vein** lie in a groove (sulcus) that lies over the interventricular septum. Other coronary arteries and cardiac veins lie in the grooves between the atria and ventricles. These vessels are often obscured by fat deposits.
3. Locate the anterior and posterior venae cavae, which carry blood into the right atrium.
4. Locate the pulmonary trunk, which exits the right ventricle and lies next to the left atrium. Insert a probe into it to determine that it comes from the right ventricle.
5. Locate the aorta, which exits the left ventricle. Compare the thickness of the walls of the aorta and venae cavae.
6. Position the heart to observe the dorsal surface. Locate the four small pulmonary veins, which return blood to the left atrium. The pulmonary veins are often embedded in fat on the dorsal surface of the heart. Insert a probe into these veins to determine that they enter the left atrium.
7. Locate the coronary arteries and cardiac veins on the dorsal surface. Cardiac veins drain blood into the coronary sinus, which empties deoxygenated blood into the right atrium.

Dissection Procedures

1. Hold the heart in your left hand with the apex away from you and the dorsal surface up. Insert a blade of the scissors into the anterior vena cava and cut through the right atrium, atrioventricular valve, and to the tip of the right ventricle. If the heart is fresh, you may have to wash out blood clots. Compare the thickness of the atrial and ventricular walls. Observe the atrioventricular valve and the chordae tendineae.
2. Following the procedures in step 1, cut through the left atrium and on to the tip of the left ventricle.

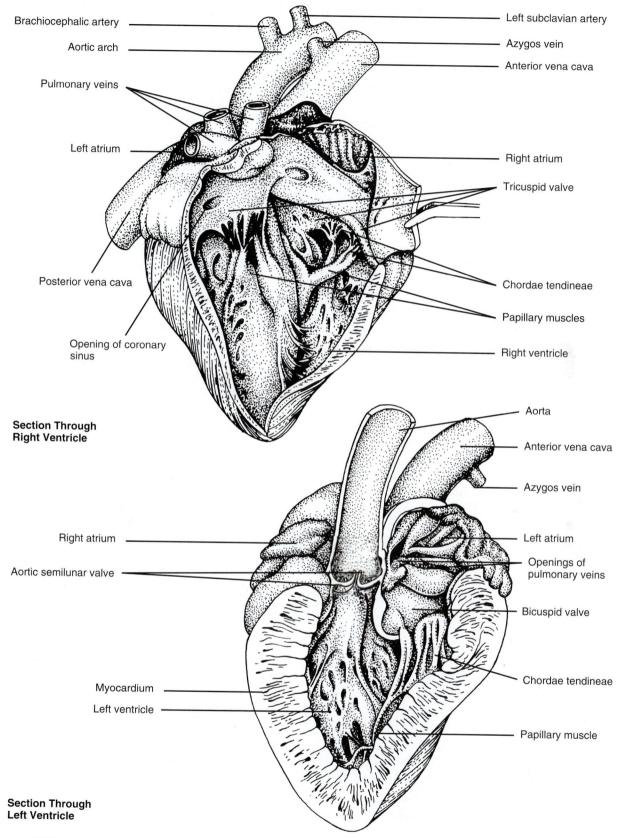

Brachiocephalic artery

Aortic arch

Pulmonary veins

Left atrium

Posterior vena cava

Opening of coronary sinus

Left subclavian artery

Azygos vein

Anterior vena cava

Right atrium

Tricuspid valve

Chordae tendineae

Papillary muscles

Right ventricle

Section Through Right Ventricle

Right atrium

Aortic semilunar valve

Myocardium

Left ventricle

Aorta

Anterior vena cava

Azygos vein

Left atrium

Openings of pulmonary veins

Bicuspid valve

Chordae tendineae

Papillary muscle

Section Through Left Ventricle

Figure 19.6 Sheep heart dissection showing internal structure.

Compare the size and wall thickness of the two ventricles.

3. Insert a probe through the aorta and into the left ventricle. Use scissors to cut along the probe from the left ventricle into the aorta. This cut will expose the aortic semilunar valve. Locate the three membranous pockets that form it. Just above the semilunar valve are the two openings of the coronary arteries where they exit the aorta.

4. Following the procedures in step 3, cut through the pulmonary trunk into the right ventricle. Examine the pulmonary semilunar valve.

5. ***Complete item 5 on the laboratory report.***

Assignment 6

Materials

Compound microscope
Prepared slide of cardiac muscle

1. Examine a prepared slide of **cardiac muscle,** the type of muscle tissue composing the heart. Compare your slide to Figure 19.7 and locate the labeled parts.

2. Note that each cell has a centrally located nucleus and striations and is joined to adjacent cells by **intercalated discs.** Observe how the cells branch and join together to form a meshlike arrangement, which allows cardiac muscle to function as a unit.

3. ***Complete item 6 on the laboratory report.***

The Pattern of Circulation

In vertebrates, blood flows through a system of closed vessels to transport materials to and from body cells. **Arteries** carry blood from the heart, and they divide into smaller and smaller arteries and ultimately lead to **arterioles,** which are nearly microscopic in size. Arterioles lead to the **capillaries,** the smallest blood vessels. The walls of the capillaries are composed of a single layer of squamous epithelial cells. See Figure 19.8. Exchange of materials occurs between blood in capillaries and the body cells. From capillaries, blood flows into tiny veins called **venules** that combine to form larger **veins** that carry blood back to the heart.

Figure 19.9 shows the relationship between an arteriole, capillaries, and a venule. The diameter of the arterioles affects the flow of blood and blood pressure and is controlled by circular smooth muscles that respond to impulses from the autonomic nervous system. The flow of blood into capillaries is controlled by the **precapillary sphincter muscle,** which is governed by impulses from the autonomic nervous system and local chemical stimuli. An increase in the local CO_2 concentration opens the sphincter and increases capillary blood flow. Similarly, a decrease in CO_2 concentration reduces capillary blood flow.

The exchange of materials between the body cells and capillary blood involves **tissue fluid** as an intermediary. Tissue fluid is the thin layer of extracellular fluid that covers all cells and tissues. This is the pattern of the exchange:

Blood in capillaries ⟷ tissue fluid ⟷ body cells

Because the mammalian heart is a double pump, there are two separate pathways of circulation. The **pulmonary circuit** pumps **deoxygenated blood** to the lungs via pulmonary arteries. After the exchange of O_2 and CO_2, **oxygenated blood** returns to the heart via pulmonary veins. The **systemic circuit** pumps oxygenated blood to the body (except the lungs), where O_2 and CO_2 are exchanged between the body cells and capillary blood. Then the deoxygenated blood is returned to the heart. See Figure 19.10.

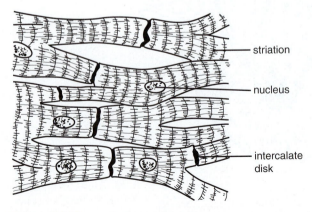

Figure 19.7 Cardiac muscle tissue.

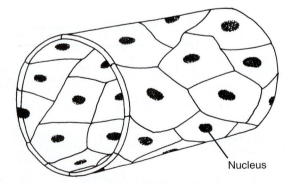

Figure 19.8 A capillary wall is formed of a single layer of epithelial cells.

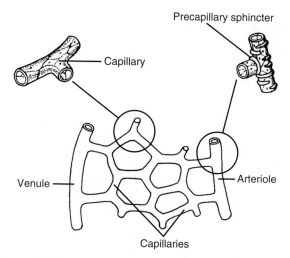

Figure 19.9 Capillary network between an arteriole and a venule.

Assignment 7

Materials

Colored pencils
Compound microscope
Dissecting instruments and pins
Frog board and cloth
Frog, live
Prepared slides of:
   artery and vein, x.s.
   atherosclerotic artery, x.s.

1. ***Complete items 7a and 7b on the laboratory report.***
2. Examine the capillary blood in the webbing of a frog's foot that has been set up under a demonstration microscope. Note the pulsating flow in the feeding arteriole and the smooth flow in the collecting venule. Observe the blood cells moving in single file through a capillary. How would you describe the flow of blood in the capillary? Keep the foot wet with water. Turn off the microscope light when you have finished observing. ***Complete item 7c on the laboratory report.***
3. Examine prepared slides of artery and vein, x.s. Compare the difference in thickness of the walls due to differences in the amount of smooth muscle and connective tissue. Locate the single layer of squamous endothelial cells forming the interior lining. ***Complete item 7d on the laboratory report.***
4. Examine a prepared slide of an atherosclerotic artery, x.s., and note the fatty deposit that partially plugs the vessel. Cholesterol is primarily responsible for this deposit. This type of obstruction in coronary arteries often causes heart attacks and may require coronary bypass surgery.

5. Label Figure 19.10. Add arrows to indicate the direction of blood flow. Color vessels and heart chambers carrying deoxygenated blood blue and color those carrying oxygenated blood red.
6. ***Complete item 7 on the laboratory report.***

BLOOD PRESSURE IN HUMANS

Usually, "blood pressure" refers to the blood pressure within arteries of the systemic circuit. There are two types of blood pressure: systolic and diastolic. **Systolic blood pressure** occurs during ventricular contraction and normally averages 120 ± 10 mm Hg (mercury) when measured in the brachial artery of the upper arm. **Diastolic blood pressure** occurs during ventricular relaxation and normally averages 80 ± 10 mm Hg.

Pulse Pressure

The difference between systolic and diastolic blood pressure produces the **pulse pressure.** The alternating increase and decrease in arterial blood pressure causes a corresponding expansion and contraction of the elastic arterial walls. The pulsating expansion of arterial walls may be detected as the **pulse** by placing the fingers on the skin over a surface artery. See Figure 19.11.

The pulse rate indicates the number of heart contractions per minute. Normal pulse rates usually range between 65 and 80 per minute, but well-conditioned athletes may have rates as low as 40 per minute.

Measurement of Blood Pressure

Blood pressure is most commonly measured in the brachial artery using a **sphygmomanometer** and a **stethoscope.** The sphygmomanometer consists of an inflatable cuff that is wrapped around the upper arm and a pressure gauge to measure the air pressure within the cuff. The stethoscope is used to hear sounds produced by blood rushing through a partially closed brachial artery.

Assignment 8

Materials

Sphygmomanometers
Stethoscopes

1. Determine your pulse rate (beats/min) at rest by placing your fingertips over a radial artery, as shown in Figure 19.11. Count the number of beats for 15 sec and multiply by 4 to get beats/min. ***Record your pulse rate in item 8a on the laboratory report.***

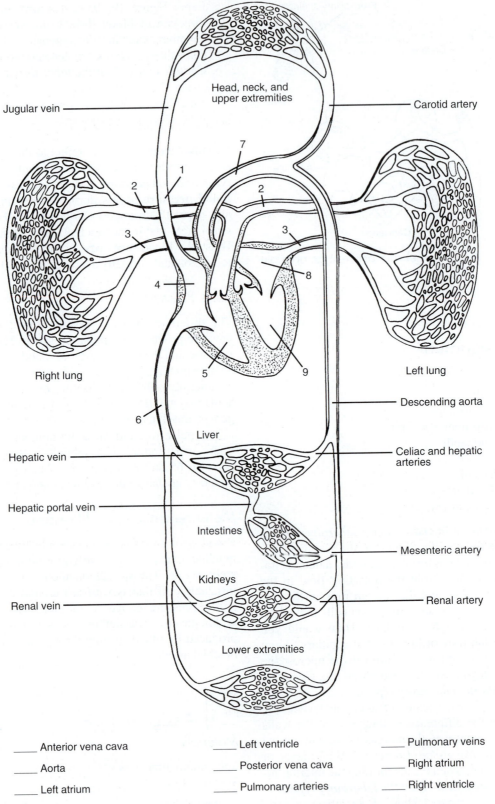

Jugular vein

Head, neck, and upper extremities

Carotid artery

Right lung

Left lung

Descending aorta

Liver

Hepatic vein

Celiac and hepatic arteries

Hepatic portal vein

Intestines

Mesenteric artery

Kidneys

Renal vein

Renal artery

Lower extremities

____ Anterior vena cava

____ Aorta

____ Left atrium

____ Left ventricle

____ Posterior vena cava

____ Pulmonary arteries

____ Pulmonary veins

____ Right atrium

____ Right ventricle

Figure 19.10 Scheme of mammalian circulation.

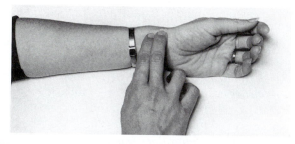

Figure 19.11 Measuring the pulse rate. Place the fingers over the radial artery at the wrist.

2. If you have good health, jog in place for 3–5 min and retake your pulse rate. Determine your pulse rate at 2-min intervals until it returns to the resting rate. The shorter the recovery time, the better is your cardiovascular condition. ***Record your pulse rate in item 8b on the laboratory report.***

3. Working in pairs, determine each other's blood pressure at rest following procedures described below and shown in Figure 19.12. ***Caution: Do not leave the brachial artery compressed for more than 30 sec.*** Read through the following procedures completely before you begin the process.

   a. The arm of the subject should be resting palm up on the table top. Wrap the cuff around the upper arm with the bottom edge of the cuff about an inch above the elbow joint. Secure the cuff with the Velcro® fastener and attach the pressure gauge so the dial can be easily read, as shown in Figure 19.12.

   b. Insert the ear pieces of the stethoscope inward and forward into your ears. Hold the diaphragm of the stethoscope over the brachial artery with your left thumb, as shown in Figure 19.12. With your right hand, gently close the screw valve above the bulb of the sphygmomanometer and squeeze the bulb to inflate the cuff to about 150 mm Hg. This pressure closes the brachial artery.

   c. Open the screw valve slightly to release slowly air from the cuff, decreasing the pressure of the cuff against the artery while listening with the stethoscope for a pulsating sound of blood

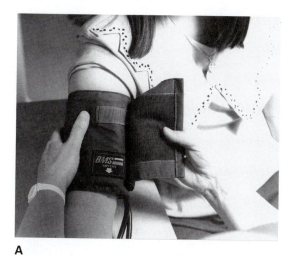

A

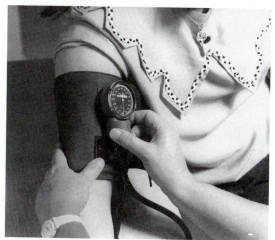

B

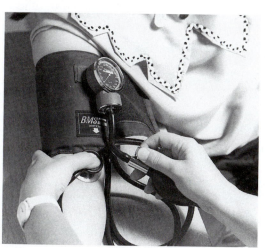

C

Figure 19.12 Measuring blood pressure. **A.** The sphygmomanometer cuff is placed around the upper arm with its lower edge about an inch above the elbow. Attach it snugly with the Velcro® fastener and fold back the leftover flap. **B.** Attach the pressure gauge to the holding strap so it is easily readable. **C.** Close the valve with the set screw and pump air into the cuff to 150 mm Hg. Place the diaphragm of the stethoscope over the brachial artery on the medial half of the anterior elbow joint. Slightly open the valve to slowly release air. When the first pulse sound is heard, read the systolic pressure. Continue releasing air and, when the pulse sound suddenly disappears, read the diastolic pressure.

squirting through the partially closed artery. As soon as you hear the first pulsating sound, read the pressure on the pressure gauge. This is the systolic blood pressure. (The pressure in the cuff equals the systolic pressure forcing blood through the partially closed artery.)

d. Continue to slowly release air from the cuff while listening to the pulsating sound as it deepens and then ceases. As soon as the sound ceases, read the pressure on the pressure gauge. This is the diastolic blood pressure.

e. Release the remaining air from the cuff and remove it from the "patient's" arm.

4. Determine and *record in item 8 on the laboratory report* your blood pressure at rest, after jogging in place for 3–5 min, and at 3-min intervals until it has returned to the resting pressure.

5. *Complete the laboratory report.*

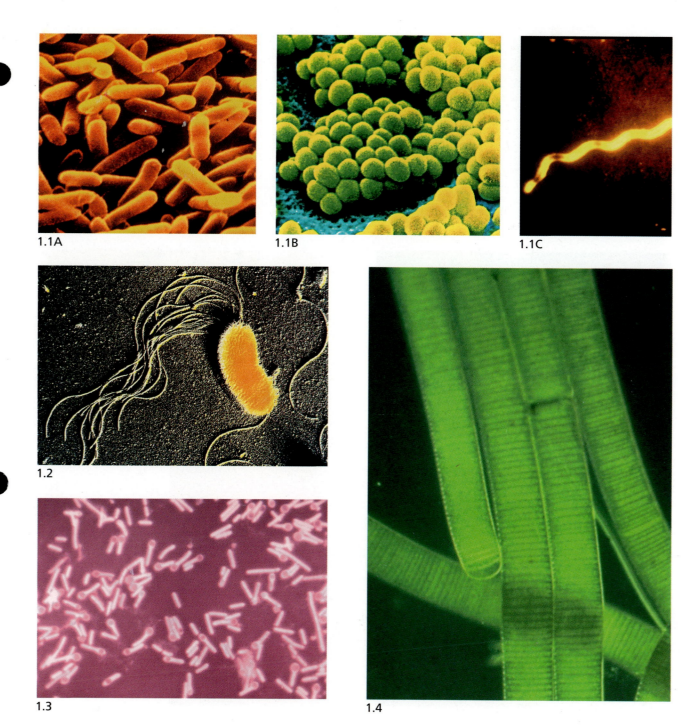

1.1A

1.1B

1.1C

1.2

1.4

1.3

Plate 1.1 The three basic shapes of bacteria.
(A) Rod-shaped (bacillus) cells are shown in this false-color scanning electron micrograph (SEM) of *Pseudomonas aeruginosa. P. aeruginosa* is a chlorine-resistant species that may cause skin infections in swimmers and users of hot tubs. (B) Spherical (coccus) cells are evident in this false-color SEM of *Staphylococcus aureus*, a species causing pimples and abscesses. (C) Spiral-shaped (spirillum) cells are illustrated in this SEM of *Treponema pallidum*, the bacterium causing syphilis.

Plate 1.2 Flagella are clearly visible in this false-color scanning electron micrograph of *Pseudomonas flourescens*, a motile soil bacterium.

Plate 1.3 The rod-shaped bacterium *Clostridium tetani* causes tetanus (lockjaw). Some cells contain an endospore at one end of the cell. An endospore is resistant to unfavorable conditions and grows into a new bacterium when conditions become favorable.

Plate 1.4 Filaments of *Oscillatoria,* a photosynthetic cyanobacterium, are composed of thin, circular cells joined together like stacks of coins.

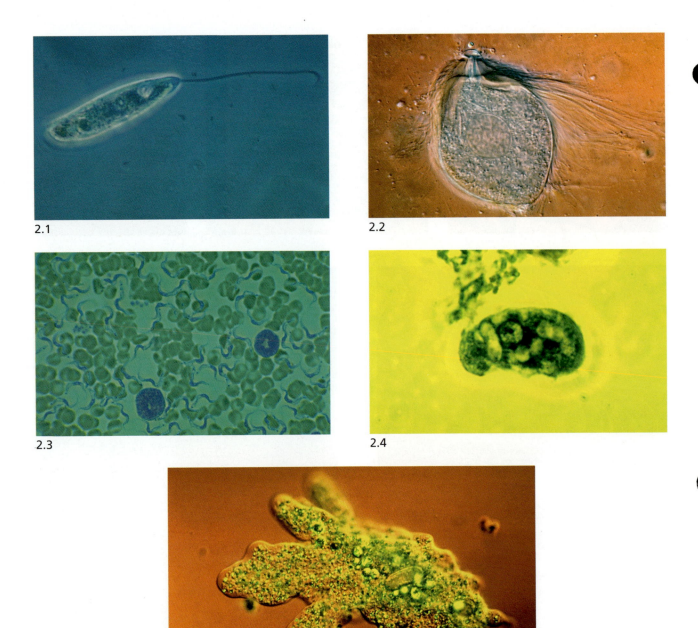

2.1

2.2

2.3

2.4

2.5

Plate 2.1 *Peranema tricophorum*, a freshwater flagellated protozoan, is propelled by a thick anterior flagellum held rigid and straight except for its tip. It engulfs prey through an opening near the base of the flagellum. A thin trailing flagellum is also present but is not visible here.

Plate 2.2 *Trichonympha* is a multiflagellated protozoan that lives in the gut of termites. It digests wood particles, forming nutrients that can be used by the termites.

Plate 2.3 *Trypanosoma brucei* among human red blood cells. It causes African sleeping sickness and is transmitted by the bite of tsetse flies.

Plate 2.4 *Entamoeba histolytica* invades the lining of the intestine, causing amoebic dysentery in humans.

Plate 2.5 *Amoeba proteus* is a common freshwater amoeboid protozoan. Note the pseudopodia. The lighting used here shows the granular nature of the cytoplasm and light-colored food vacuoles.

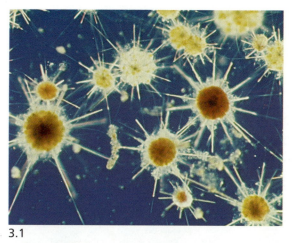

3.1

3.2

Cilia

Contractile
vacuole

Food
vacuoles

3.3

3.4

3.5

Plate 3.1 Radiolarians are marine amoeboid protozoans that secrete a siliceous shell around the central cell body. Slender pseudopodia radiate through tiny holes in the shell and capture tiny food particles that are carried back to the cell body by streaming cytoplasm.

Plate 3.2 A living foraminiferan. Foraminiferans are marine amoeboid protozoans that secrete a calcareous shell. Thin raylike pseudopodia extend through tiny openings in the shell.

Plate 3.3 *Paramecium*, a freshwater ciliate. Beating cilia propel *Paramecium* rapidly through the water. Food

organisms are digested in food vacuoles. Excess water is collected and pumped out by two contractile vacuoles.

Plate 3.4 *Euplotes* is an advanced ciliate. Tufts of cilia unite to form cirri that function almost like legs as the ciliate "walks" over the bottom of a pond. Note the food vacuoles and a contractile vacuole that appears like a cavity in this view.

Plate 3.5 *Plasmodium vivax*, a sporozoan, in a human red blood cell. It causes malaria and is transmitted by the bite of female anopheline mosquitos.

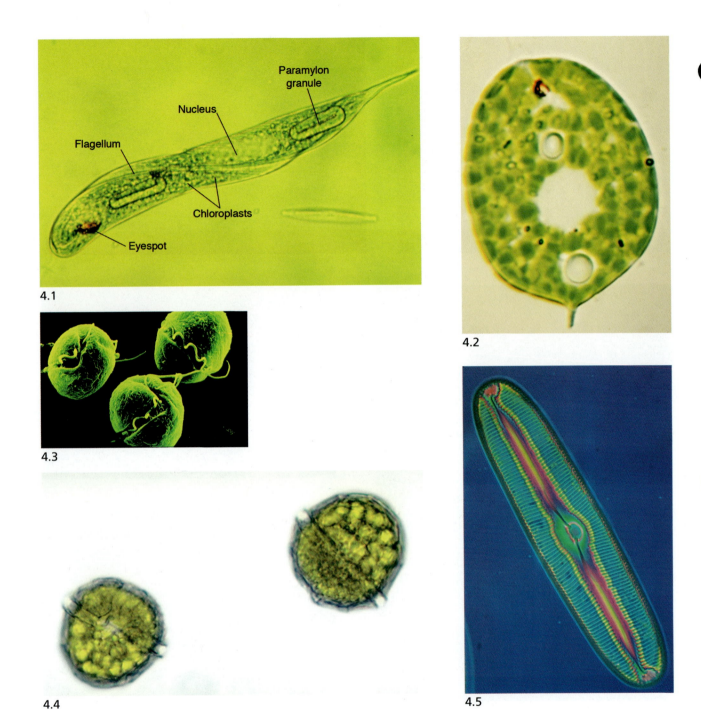

4.1

4.2

4.3

4.4

4.5

Plate 4.1 *Euglena* is a photosynthetic euglenoid. The anteriorly located flagellum provides locomotion. It is recurved along the body in this view. An eyespot, located near the base of the flagellum, detects light intensity. Food reserves are stored as paramylon granules.

Plate 4.2 *Phacus* is a freshwater euglenoid. Note the eye spot and chloroplasts. The nucleus appears as a large clear area in the center of the cell. An anterior flagellum, barely visible in the upper left corner of this photomicrograph, provides locomotion for the cell.

Plate 4.3 Like all dinoflagellates, *Gymnodinium*, a marine form, possesses two flagella. One lies in an equatorial groove in the cell wall, and the other hangs from the end of the cell. The horizontal flagellum is clearly visible in this false-color SEM.

Plate 4.4 *Peridinium* is a freshwater dinoflagellate common in acid-polluted lakes. Here, chloroplasts and equatorial groove are visible, but the flagella are not.

Plate 4.5 The diatom *Pinnularia mobilis*. The siliceous walls of diatoms are formed of two parts that fit together like the top and bottom of a pillbox.

5.1

5.2

5.3

5.4

5.5

5.6

Plate 5.1 The plasmodium of the slime mold *Physarum* lives like a giant, multinucleate amoeba as it creeps along, absorbing nutrients and engulfing food organisms.

Plate 5.2 When either food or moisture becomes inadequate, a *Physarum plasmodium* transforms into spore-forming sporangia supported by short stalks.

Plate 5.3 In *Rhizopus stolonifer*, black bread mold, sporangiophore hyphae support sporangia that form asexual spores. Each sporangiophore is attached to the substrate by rhizoid hyphae.

Plate 5.4 The fusion of (+) and (−) gametes in *Rhizopus stolonifer* results in the formation of a zygospore sandwiched between a pair of gametangia. Later, each zygospore (2n) undergoes meiosis, producing a haploid mycelium.

Plate 5.5 This false-color SEM of the asexual conidiospores of *Penicillium* shows their spherical shape and attachment to a sporangiophore. 5,750×.

Plate 5.6 The hyphae of *Nectria* are composed of separate cells joined end to end. Such hyphae are said to be septate because cross walls separate the cells.

6.1A

6.1B

6.3

6.2

6.4

Plate 6.1 (A) Cup-shaped ascocarps (fruiting bodies) of *Peziza aurantia*, a sac fungus. (B) Saclike asci containing ascospores lines the ascocarps of *P. aurantia*. When the asci rupture, ascospores are released and dispersed by wind.

Plate 6.2 Basidiocarps (fruiting bodies) of shelf fungi, a club fungus, contain spore-forming basidia. The

basidiospores are released through tiny pores on the lower surface of the basidiocarps.

Plate 6.3 Each basidium of a club fungus produces four basidiospores, as shown in this SEM.

Plate 6.4 A discharge of millions of basidiospores from a puffball, a club fungus.

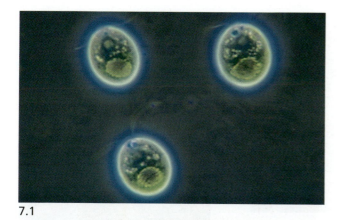

7.1

7.3

7.2

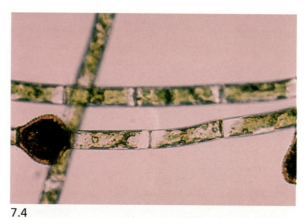

7.4

7.5

Plate 7.1 *Chlamydomonas* is a tiny unicellular green alga that moves by means of two flagella. Only the basal portion of the cup-shaped chloroplast is visible here. An eyespot, appearing as a blue dot, enables detection of light intensity.

Plate 7.2 *Volvox* colonies may consist of up to 50,000 *Chlamydomonas* like cells. *Volvox* may reproduce asexually by forming daughter colonies within the sphere. Some forms also have a few cells specialized for sexual reproduction.

Plate 7.3 The spiral chloroplast is the dominant feature in cells of *Spirogyra*, a filamentous green alga. The enlarged regions on the chloroplast are pyrenoids, sites of starch storage.

Plate 7.4 Cellular specialization for sexual reproduction is evident in *Oedogonium*, a filamentous green alga. The enlarged, dark cells are oogonia that contain an oospore formed after union of sperm and egg cells.

Plate 7.5 Cellular specialization for reproduction also occurs in *Ulothrix*, another filamentous green alga. The cells in the lower horizontal filament are sporangia containing asexual zoospores.

8.1

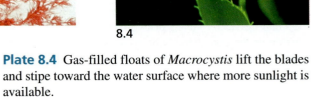

8.2

8.3

8.5

8.4

Plate 8.1 *Ulva* (sea lettuce), a marine green alga, exposed at low tide. A holdfast attaches it to a rock. *Ulva* exhibits an alternation of generations, as do higher plants.

Plate 8.2 Brown marine algae are characterized by a robust body that can withstand strong wave action. Note the holdfast, stipe, and divided blade in *Durvillaea*, a brown alga from the Australian coast.

Plate 8.3 A forest of giant kelp, *Macrocystis*, along the California coast provides an important habitat for marine animals. The long stipes may be over 100 ft. in length.

Plate 8.4 Gas-filled floats of *Macrocystis* lift the blades and stipe toward the water surface where more sunlight is available.

Plate 8.5 The delicate, feathery body structure of red algae sharply contrasts with the robust brown algae. Red algae, like *Callithamnion*, live at ocean depths where wave action is minimal.

BLOOD AND CIRCULATION

Student _____

Lab Instructor _____

1. INTRODUCTION

a. Write the term that matches the phrase.

1. Basic function of circulatory systems   _____

2. Circulatory system with hemolymph   _____

3. Circulatory system with blood   _____

4. Respiratory pigment in annelids   _____

5. Blue-green pigment in certain arthropods   _____

b. Contrast open and closed circulatory systems.

Open system _____

_____

Closed system _____

_____

c. Draw a few red cells from slides of amphibian, reptilian, and mammalian blood.

Amphibian   **Reptilian**   **Mammalian**

2. HUMAN BLOOD

a. Indicate the characteristic(s) that most easily distinguish(es):

Erythrocytes from leukocytes _____

_____

Granulocytes from agranulocytes _____

_____

b. Write the term that matches the phrase.

1. Most abundant blood cell   _____

2. Most abundant white cell   _____

3. Liquid portion of the blood   _____

4. Cells lacking nuclei   _____

5. Cells with lobed nuclei and red
   cytoplasmic granules   _____

6. Cells with a large, kidney-shaped nucleus
   and no cytoplasmic granules   _____

7. Cell fragments that initiate clotting   _____

8. Blood cells that fight infection   _____

9. The smallest white blood cells   _____

10. The rarest white blood cells   _____

3. BLOOD TYPING

a. Indicate your ABO blood group _____   Rh type _____

b. Indicate compatibility (C) and incompatibility (I) of possible blood transfusions shown in the chart.

Blood Type and Antigen of Donor	Blood Type (and Antibodies) of Recipient			
	O (a, b)	A (b)	B (a)	AB (none)
O				
A				
B				
AB				

c. Which ABO blood type may receive blood from the other three types in emergencies? _____

   This is the **universal recipient.**

d. Which ABO blood type may donate blood to the other three types in emergencies? _____

   This is the **universal donor.**

e. Considering *both* ABO and Rh blood types, indicate:

   Universal donor _____   Universal recipient _____

f. Infants suffering from erythroblastosis fetalis may require a massive blood transfusion. What blood type or types

   should be given to a baby with a blood type of A, Rh^+? _____

   Explain your answer. _____

   _____

4. MAMMALIAN HEART

a. How is the mammalian heart more efficient than the frog heart?

   _____

   _____

   _____

b. List the labels for Figure 19.4.

1. _____   10. _____
2. _____   11. _____
3. _____   12. _____
4. _____   13. _____
5. _____   14. _____
6. _____   15. _____
7. _____   16. _____
8. _____   17. _____
9. _____   18. _____

c. Write the term that matches the phrase regarding mammalian heart structure and function.

1. Returns blood to right atrium   _____

2. Returns blood to left atrium   _____

3. Separates ventricles   _____

4. Pumps blood into pulmonary trunk   _____

5. Pumps blood into aorta   _____

6. Prevents backflow of blood into right atrium   _____

7. Prevents backflow of blood into left atrium   _____

8. Prevents backflow of blood into right ventricle _____

9. Prevents backflow of blood into left ventricle _____

10. Contraction phase of heart cycle   _____

11. Relaxation phase of heart cycle   _____

5. SHEEP HEART DISSECTION

a. Explain the difference in size and wall thickness of the ventricles. _____

_____

_____

b. Describe the function of the chordae tendineae. _____

_____

_____

c. Circle the terms that describe the chordate tendineae.

      elastic   nonelastic   thick   thin   pliable   stiff   opaque   transparent

d. Circle the terms that apply to the semilunar valves.

      elastic   nonelastic   thick   thin   pliable   stiff   opaque   transparent

e. Circle the terms that apply to the atrioventricular valves.

      elastic   nonelastic   thick   thin   pliable   stiff   opaque   transparent

6. CARDIAC MUSCLE

Draw a small portion of cardiac muscle from your slide. Label a nucleus and an intercalated disc.

7. PATTERN OF CIRCULATION

a. Write the term that matches the phrase.

1. Vessels carrying blood from the heart _____

2. Vessels carrying blood to the heart _____

3. Smallest blood vessels _____

4. Controls flow of blood into capillary _____

5. Smallest arteries _____

b. By what process does O_2 pass from capillary blood into body cells? _____

_____

c. Describe the flow of blood through a capillary in the frog's foot. _____

_____

d. Draw the following from your observation to show the relative (a) diameter of the vessel lumen and (b) wall thickness.

Normal Vein   **Normal Artery**   **Atherosclerotic Artery**

How do you explain the differences in the thickness of the walls in arteries and veins? _____

_____

Explain the hazards of atherosclerosis. _____

_____

_____

_____

_____

_____

e. Trace the flow of blood from the left ventricle to the intestine and back to the left ventricle.

_____

_____

_____

_____

_____

f. What arteries carry deoxygenated blood? _____

g. What veins carry oxygenated blood? _____

h. List the labels for Figure 19.10.

1. _____   4. _____   7. _____

2. _____   5. _____   8. _____

3. _____   6. _____   9. _____

8. BLOOD PRESSURE IN HUMANS

a. Record your pulse rate in beats per minute. At rest _____

b. Record your pulse rate at these time intervals after exercise.

0 min _____   2 min _____   4 min _____   6 min _____

How long did it take for the pulse rate to return to the resting rate? _____ What can you infer

about your physical condition? _____

c. Record your systolic/diastolic blood pressure in mm Hg.

At rest _____/_____

At these time intervals after exercise.

0 min _____/_____   3 min _____/_____   6 min _____/_____

What can you infer about your physical condition? _____

_____

GAS EXCHANGE

OBJECTIVES

After completing the laboratory session, you should be able to:
1. Identify the necessary characteristics of an effective gas-exchange organ.
2. Describe the role of diffusion in gas exchange.
3. Describe the structure and function of gas-exchange organs of representative aquatic and terrestrial animals.
4. List the components of the human respiratory system and provide their functions.
5. Describe the mechanics of breathing in mammals.
6. Describe the chemical control of breathing in mammals.
7. Determine your tidal volume, vital capacity, and expiratory reserve volume.
8. Define all terms in bold print.

Aerobic cellular respiration requires gaseous oxygen (O_2) and produces carbon dioxide (CO_2). Thus, the continuous exchange of O_2 and CO_2 must occur between the cell and its environment. This exchange always occurs by diffusion across the moist cell membranes.

In unicellular, colonial, and many small multicellular organisms, each cell is either in direct contact with the environment or only a few cells away so the rate of diffusion is sufficient to provide O_2 and to remove CO_2. In larger animals, however, diffusion through the body surface is too slow to provide adequate gas exchange for interior cells.

GAS EXCHANGE IN ANIMALS

The evolution of special organs for gas exchange in animals has taken two basic directions: evaginations (outpocketings), such as **gills,** and invaginations (inpocketings), such as **lungs.** Many variations of these basic patterns exist, but each exhibits certain essential characteristics. The respiratory surface must be (1) moist, (2) protected from injury, (3) of adequate surface area, and (4) thin, for rapid diffusion of gases. Depending on the habitat of the organism, a mechanism exists to pass either water or air over the respiratory surface.

Gases can enter or leave the cells of a gas-exchange organ only through the extracellular fluid that bathes the cells. In aquatic animals, the water of the environment is in contact with the cells and provides this moisture. In terrestrial forms, the ever-present problem of dessication must be overcome while the surface of the organ is kept moist. This problem has been solved in vertebrates by the evolution of internal lungs that maintain a saturated humidity and moist membrane surfaces.

In most cases, the evolution of a circulating internal fluid, either blood or hemolymph, has occurred to transport O_2 and CO_2 between the respiratory organs and the body cells. A close physical relationship exists between the internal fluid and the respiratory surface. The larger the respiratory surface, the faster is the rate of gas exchange between the environment (air or water) and the internal fluid.

Gills

A large number of freshwater and marine animals have gills as gas-exchange organs. For example, molluscs, most segmented worms, many arthropods, echinoderms, fish, and larval amphibians possess gills. The methods used to increase the surface area and to pass water over the gill surface varies from group to group.

Tracheae

Insects possess a unique **tracheal system** for gas exchange that is totally unlike gills or lungs. Tiny openings, the **spiracles,** are present on many segments of an insect's body. The spiracles may be opened and closed. When open, they allow air to enter a series of tiny chitinous tubules called **tracheae** that ramify throughout the insect's body. From the spiracles, tracheae usually lead to longitudinal tracheal trunks that extend the length of the body. Smaller tracheae branch from these trunks leading to ever-smaller tracheae that permeate the body tissues. By this interconnecting system of tubules, air is brought *directly* to the body cells so oxygen diffuses from the tiny tracheae directly into the cells. Carbon dioxide diffuses in the opposite direction, but much of it diffuses through the body surface. Note that this mode of gas exchange is not dependent on hemolymph for the distribution of oxygen and collection of carbon dioxide.

Lungs

Adult amphibians and higher vertebrates possess lungs as gas-exchange organs. The lungs of amphibians are quite simple and little more than thin, air-filled sacs that have an abundant blood supply. In many amphibians, the skin serves as an additional gas-exchange organ. In higher vertebrates, lungs are more complex.

Mammalian lungs are highly developed, with many internal subunits that provide a large surface for rapid gas exchange. The combined respiratory surface is many times greater than the external surface area of the lung.

Assignment 1

Materials

Compound microscope
Stereo microscope
Dissecting instruments and pans
Syracuse dishes
Clams; crayfish; grasshoppers; and fish, living and
    preserved
Prepared slides of human louse, cleared to show tracheae

1. Examine a prepared slide of a human louse that has been cleared to reveal the tracheal system with a stereo microscope and with the 4× objective of your compound microscope. Note the arrangement of the tracheae and the location of the spiracles. Compare your slide with Figure 20.1. Color the tracheal system blue in Figure 20.1.
2. Examine the demonstration dissections of a preserved or freshly killed clam, crayfish, and fish.

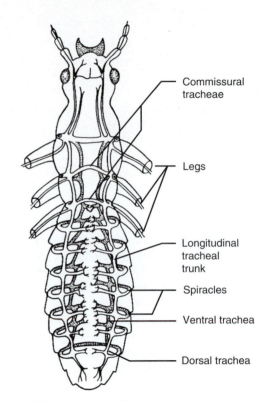

Figure 20.1 Insect tracheal system.

Note the location, structure, and arrangement of the gills. Compare your observations with Figure 20.2. Color the gills red in Figure 20.2.

3. Observe the respiratory movements of the living specimens in the aquarium.
4. Remove a small section of a gill from a clam, crayfish, and fish. Place them in water in a Syracuse dish and examine them with a stereo microscope. Note the gill structure and how a large surface area is provided in each.
5. ***Complete item 1 on Laboratory Report 20 that begins on page 275.***

THE MAMMALIAN RESPIRATORY SYSTEM

Refer to Figure 20.3, which illustrates the basic components of the human respiratory system, as you study this section.

During **inspiration,** air enters the **nasal cavity** through the **nostrils.** The air is filtered and warmed by the moist membranes of the nasal cavity. The surface area of the nasal cavity is increased by the turbinates projecting into the cavity from each side. The nasal cavity is separated from the oral cavity by the **hard** (supported by bone) and **soft palates.**

Air then passes through the **pharynx,** which serves as the passageway for both food and air, and flows

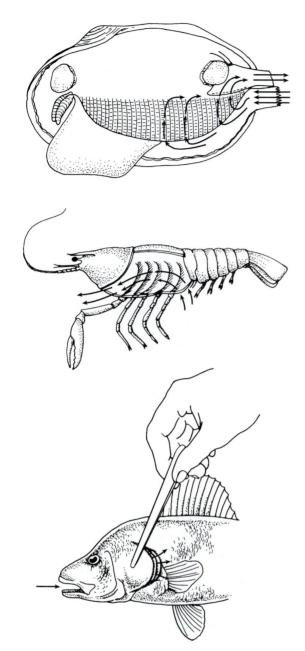

Figure 20.2 Aeration of gills in a clam, crayfish, and fish.

through the **glottis** into the larynx. The **larynx** is a cartilaginous box containing the vocal folds (cords). During swallowing, the larynx is elevated, which causes the **epiglottis** to fold over, closing the glottis and directing food and liquids into the esophagus, the tube leading to the stomach. If food or liquids accidentally enter the larynx, a coughing reflex is triggered to clear out the larynx.

From the larynx, air passes down the **trachea** (windpipe), branches into the **primary bronchi,** and

enters the lungs. The primary bronchi branch into **secondary bronchi,** which branch into smaller and smaller tubules (**bronchioles**) and finally terminate in the **alveoli,** clusters of tiny sacs filled with air. Each human lung has about 300 million alveoli that increase the surface area of the lung to about 75 m^2. Capillary networks surround the alveoli, and gas exchange occurs between the air in the alveoli and the capillary blood. Air exits the respiratory system in reverse order during **expiration.**

Most of the respiratory passages are lined with **pseudostratified ciliated columnar epithelium.** As the name implies, the ciliated, columnar cells forming this tissue appear to be arranged in layers, but actually they are not. See Figure 20.4. The **goblet cells** produce a layer of mucus that coats the surface of the cells. Foreign particles, such as bacteria, viruses, pollens, and dust, are trapped in the mucus, which is swept toward the pharynx, upward from the bronchi and trachea and downward from the nasal cavity, by the beating cilia. The mucus and entrapped particles are then swallowed. In this way, foreign materials are removed from the air passages to prevent infection and to keep the membranes free from contaminants.

Assignment 2

Materials

Beakers, 250 ml
Colored pencils
Compound microscope
Stereo microscope
Human torso model
Dissecting instruments and pans
Frog, pithed
Sheep pluck
Gloves, protective and disposable
Prepared slides of:
   lung tissue, normal and emphysematous
   trachea, x.s.

1. Observe the demonstration dissection of a pithed frog exposing an inflated lung. Examine it under a stereo microscope and note the blood cells flowing through capillaries in the thin wall of the lung. Note how simple the frog lung is compared with the human lung shown in Figure 20.3.
2. Label and color-code the respiratory organs in Figure 20.3. Locate the parts of the respiratory system on a human torso model or chart.
3. Examine the demonstration of the respiratory system of a sheep. Locate the parts that you labeled in Figure 20.3. Observe the **cartilaginous rings** that hold open the trachea. What is their real shape? Cut open the larynx and find the vocal

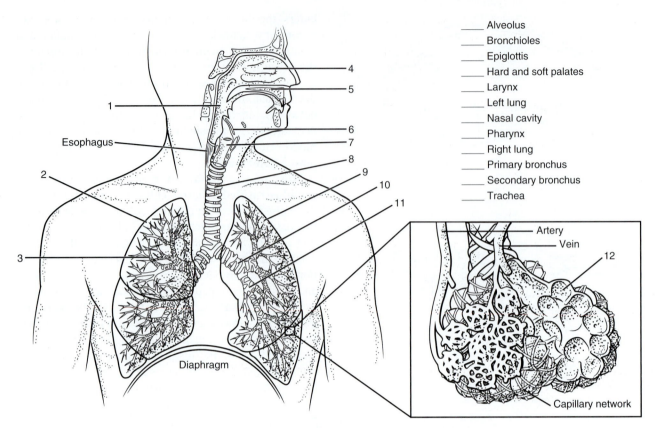

Alveolus
Bronchioles
Epiglottis
Hard and soft palates
Larynx
Left lung
Nasal cavity
Pharynx
Right lung
Primary bronchus
Secondary bronchus
Trachea

Figure 20.3 The human respiratory system. The inset shows clusters of alveoli at the end of terminal bronchioles. Modified from Fig. 26.8 in Krogh, 2nd ed.

folds. Locate the **epiglottis,** a cartilaginous flap that folds over the glottis during swallowing. Feel the lungs to detect their spongy consistency. If the lungs are not preserved, cut off a small piece and place it in a beaker of water. How do you explain what happens? Examine the cut surface of the lung tissue with a stereo microscope. Can you see small air passages?

4. Examine prepared slides of normal and emphysematous lung tissue. Note how the alveoli have broken down to form large spaces in the emphysematous lung tissue. Few people other than smokers develop **emphysema.**

Cilia
Mucus
Nucleus
Goblet cell
Basement membrane
Connective tissue

Figure 20.4 Pseudostratified ciliated columnar epithelium.

5. Examine a prepared slide of trachea, x.s., and observe the ciliated epithelium composing the inner lining. Compare your observations with Figure 20.4.

6. ***Complete item 2 on the laboratory report.***

Mechanics of Breathing

Mammals are unique in that the thoracic and abdominal cavities are separated by a thin sheet of muscle, the **diaphragm,** which is the primary muscle involved in breathing. The **intercostal muscles** between the ribs also are involved, but they play a secondary role.

During inspiration, the diaphragm and external intercostal muscles contract, causing an increase in the volume of the thoracic cavity. When the thoracic cavity expands, the lungs are pulled along, causing the volume within the lungs to increase as well. The increase in lung volume causes a concomitant decrease in air pressure within the lungs. Because the intrapulmonary pressure is less than atmospheric pressure, atmospheric pressure forces air into the lungs.

During expiration, the relaxation of the diaphragm and external intercostal muscles decreases the volume of the thoracic cavity and the lungs. The decreased volume increases the intrapulmonary air pressure.

Because the intrapulmonary pressure is greater than atmospheric pressure, air is forced out of the lungs. A forced expiration requires the contraction of the internal intercostal muscles, which further decreases the volume of the thoracic cavity and lungs.

Assignment 3

Materials

Breathing mechanics model

1. Examine the breathing mechanics model. See Figure 20.5. The balloons represent the lungs, and the glass tubing represents the trachea and primary bronchi. The glass jar corresponds to the thoracic wall, and the rubber sheet simulates the diaphragm. Note that the balloons are in an enclosed space, representing the thoracic cavity.
2. Observe what happens when the rubber sheet is pulled downward and pushed upward. Determine how this works.
3. ***Complete item 3 on the laboratory report.***

Lung Capacity in Humans

Lung capacities vary among males and females, primarily due to variations in the size of the thoracic cavity and the lungs. Capacities also vary with age. Average capacities are shown in Figure 20.6.

In this section, you will use a spirometer to determine your lung volumes. ***Caution:*** *Wipe the spirometer stem with an alcohol wipe or cotton soaked in 0.5% Amphyl before and after using the spirometer, and use a fresh, sterile, disposable mouthpiece. Place all used mouthpieces in the biohazard bag after use.*

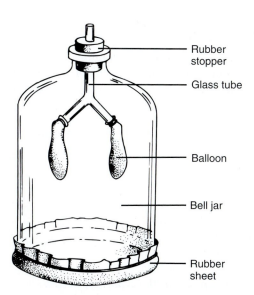

Figure 20.5 Breathing mechanics model.

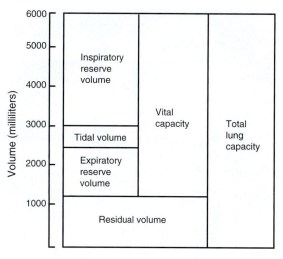

Figure 20.6 Human lung volumes.

Assignment 4

Materials

Biohazard bag
Cotton
Propper spirometer with sterile, disposable mouthpieces
Alcohol wipes
Amphyl solution, 0.5%

Determine your lung volumes as described here, and ***record your results in item 4 on the laboratory report.***

1. The volume of air exchanged during normal quiet breathing is the **tidal volume** (TV). Determine it as described below.
   a. Rotate the dial of the spirometer so the needle is at zero. See Figure 20.7.
   b. Disinfect the stem and add a sterile mouthpiece. Place the mouthpiece in your mouth, keeping the spirometer dial upward.
   c. Inhale through your nose and exhale through the spirometer for five normal, quiet breathing cycles. Record the dial reading and divide by 5 to determine your tidal volume.
2. The amount of air that can be forcefully exhaled after a maximum inhalation is the **vital capacity** (VC). It is often used as an indicator of respiratory function.
   a. Rotate the dial face of the spirometer to place the needle at zero.
   b. Take two deep breaths and exhale completely after each one. Then, take a breath as deeply as possible and exhale through the spirometer. A slow, even expiration is best. Record the reading.

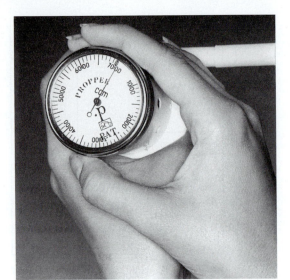

Figure 20.7 Propper spirometer. Slip on a sterile mouthpiece and rotate the dial face to zero before exhaling through the spirometer.

c. Repeat step b two more times, resetting the dial face after each measurement.

d. Record the average of the three measurements as your vital capacity. Compare your vital capacity with those shown in Table 20.1.

3. The volume of air that can be exhaled *after* a normal tidal volume expiration is the **expiratory reserve volume** (ERV).

a. Rotate the dial face of the spirometer so the needle is at 1,000. This is to compensate for the space on the dial between 0 and 1,000.

b. *After* a normal quiet expiration, forcefully exhale as much air as possible through the spirometer. Be sure not to take an extra breath at the start. Subtract 1,000 from the reading to determine your expiratory reserve volume.

c. Based on your preceding determinations, calculate your inspiratory reserve volume: IRV = VC – (TV + ERV).

d. **Complete item 4 on the laboratory report.**

THE CHEMICAL CONTROL OF BREATHING

You are aware that a direct relationship exists between the degree of exercise and the rate of breathing. Do you think that an increase in breathing rate is due to a decrease in oxygen concentration or an increase in carbon dioxide concentration in blood? The following experiments were performed to answer this question.

Study the results carefully so you can form valid conclusions.

TABLE 20.1	EXPECTED VITAL CAPACITIES (ML) FOR ADULT MALES AND FEMALES*					

Males

Height (in.)	Age in Years					
	20	30	40	50	60	70
60	3885	3665	3445	3225	3005	2785
62	4154	3925	3705	3485	3265	3045
64	4410	4190	3970	3750	3530	3310
66	4675	4455	4235	4015	3795	3575
68	4940	4720	4500	4280	4060	3840
70	5206	4986	4766	4546	4326	4106
72	5471	5251	5031	4811	4591	4371
74	5736	5516	5296	5076	4856	4636

Females

Height (in.)	Age in Years					
	20	30	40	50	60	70
58	2989	2809	2629	2449	2269	2089
60	3198	3018	2838	2658	2478	2298
62	3403	3223	3043	2863	2683	2503
64	3612	3432	3252	3072	2892	2710
66	3822	3642	3462	3282	3102	2922
68	4031	3851	3671	3491	3311	3131
70	4270	4090	3910	3730	3550	3370
72	4449	4269	4089	3909	3729	3549

*Data from Propper Mfg. Co., Inc.

Assignment 5

Experiment 1: The concentration of O_2 and CO_2 in arterial blood of a human subject was determined before and after exercise. The results are shown here.

	Results	
	Before Exercise	After Exercise
pO$_2$ (mm Hg)	100	80
pCO$_2$ (mm Hg)	40	45
Breathing rate (cycles/min)	15	45

Complete item 5a on the laboratory report.

Experiment 2: A human subject breathed air from a closed container for 3 min, and the breathing rate was determined as shown in the following charts:

Results

	1 Min	2 Min	3 Min
Breathing rate (cycles/min)	16	21	37

Complete item 5b on the laboratory report.

Experiment 3: The same human subject breathed air from a closed container through a tube containing a chemical filter, which removed CO_2 from the air. The results are shown below.

Results

	1 Min	2 Min	3 Min
Breathing rate (cycles/min)	15	16	15

Complete the laboratory report.

GAS EXCHANGE

Student _____

Lab Instructor _____

1. INTRODUCTION

a. What is the physical process by which gas exchange occurs? _____
_____

b. What is the relationship between the area of the respiratory surface and the rate of gas exchange? _____
_____
_____

c. Describe the function of tracheae in insects. _____
_____
_____

d. From your slide, draw part of a tracheal trunk and its smaller tracheal branches.

e. Diagram a small portion of each type of gill, and show how the surface area is increased in each.

Clam   **Crayfish**   **Fish**

2. MAMMALIAN RESPIRATORY SYSTEM

a. List the labels for Figure 20.3.

1. _____   5. _____   9. _____
2. _____   6. _____   10. _____
3. _____   7. _____   11. _____
4. _____   8. _____   12. _____

b. List in sequence the air passages of the human respiratory system.

Nasal cavity → _____ → _____ → _____ →
_____ → lungs

275

c. Write the term that matches the phrase.

1. Site of gas exchange _____

2. Air passage that enters a lung _____

3. Lining that removes foreign particles _____

4. Filters, humidifies, and warms air _____

5. Tubules carrying air into alveoli _____

d. Draw the microscopic appearance of each:

Normal Lung Tissue   **Emphysematous Lung Tissue**

3. MECHANICS OF BREATHING

a. When the rubber sheet is pulled down, the volume of the bell jar is _____

and the pressure in the bell jar is _____, which causes air to flow

_____ the balloons.

b. When the rubber sheet is pushed up, the volume of the bell jar is _____

and the pressure in the bell jar is _____, which causes air to flow

_____ the balloons.

4. LUNG VOLUMES

Record your:

Tidal volume _____ ml   Vital capacity _____ ml

Expiratory reserve volume _____ ml   Inspiratory reserve volume _____ ml

5. CONTROL OF BREATHING

a. List two hypotheses to explain the increase in breathing rate in experiment 1.

1. _____

2. _____

b. In experiment 2, what changes occurred in the concentration of gases in the container?

O_2 _____ CO_2 _____

How would these changes affect the blood concentrations of:

O_2 _____ CO_2 _____

c. In experiment 3, what changes occurred in the blood concentrations of:

O_2 _____ CO_2 _____

d. State the correct hypothesis based on the results of experiment 3.

_____

_____

DIGESTION

OBJECTIVES

After completing the laboratory session, you should be able to:
1. Locate, identify, and state the functions of the parts of the human digestive system.
2. Identify and state the functions of the four layers of the small intestine.
3. Describe the process of digestion, including the role of enzymes.
4. Describe the effect of temperature and pH on the action of pancreatic amylase.
5. Define all terms in bold print.

Large food molecules of carbohydrates, proteins, and fats must be broken down by **digestion** into smaller, absorbable, nutrient molecules that can be used by cells in cellular respiration. The degradation of complex food molecules into absorbable nutrients occurs by **enzymatic hydrolysis,** the process of splitting large molecules by the addition of water and the catalytic action of enzymes.

Intracellular digestion is commonplace in cells, even photosynthetic cells. Heterotrophs use intracellular digestion or **extracellular digestion** to convert food molecules into absorbable nutrient molecules.

Bacteria and fungi secrete digestive enzymes into the surrounding substrate, and nutrients resulting from the digestion are absorbed. Protozoans and sponges engulf food into vacuoles, where digestion occurs. Cnidarians and flatworms use both extracellular and intracellular digestion. Higher animals have a digestive tract that is used to (1) ingest and digest food extracellularly, (2) absorb nutrients, and (3) remove the nondigestible wastes.

ENZYME ACTION

When reduced to its simplest form, life is a series of chemical reactions. These reactions are complex and are controlled by **enzymes,** organic catalysts that greatly increase the rate of the reactions. Life could not exist without enzymes. Enzymes are usually proteins, and the sequence of their amino acids is determined by the genetic code of DNA molecules.

Most enzymes are specific in the type of chemical reactions that they control. This specificity is determined by the shape of the enzyme, which allows it to fit onto a particular **substrate molecule.** The shape of an enzyme is determined by the kind and sequence of amino acids that compose it. Weak **hydrogen bonds,** which form between amino acids of the chain, are responsible for the three-dimensional shape of the enzyme. Because these bonds are easily broken, the shape of the enzyme may be altered by changes in temperature and pH. Such changes inactivate the enzyme. Most enzymes function best in narrow temperature and pH ranges. Later in the exercise, you will investigate the action of a digestive enzyme to learn more about the nature of enzymes and specifically about the role of enzymes in digestion.

Digestive enzymes are only one type of the thousands of enzymes present in living organisms. The role of digestive enzymes is to speed up the hydrolysis of food molecules. Many different digestive enzymes are required to complete the digestion of food because a particular enzyme acts only on a single type of food molecule. The basic pattern of digestion may be summarized as follows:

$$\text{Large food molecules} \xrightarrow[\text{+}H_2O]{\text{enzymes}} \text{Small nutrient molecules}$$

The interaction of an enzyme and a food molecule is shown in Figure 21.1. The enzyme (E) combines with the **substrate** (food) molecule (S) to form an **enzyme–substrate complex** (ES), where hydrolysis

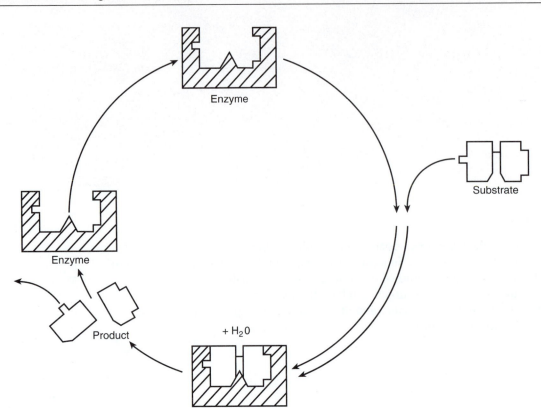

Enzyme

Substrate

Enzyme

Product

+ H$_2$O

Enzyme–substrate complex

Figure 21.1 The mechanism of enzyme action. The enzyme and substrate fit together like a lock and key, forming the enzyme–substrate complex where the reaction occurs. Then the products are released, and the unchanged enzyme is recycled.

occurs. Then, the **product** molecules (P) separate from the enzyme, and the unchanged enzyme is recycled to react with another substrate molecule. Note that the enzyme is not altered in the reaction and may be used repeatedly, so relatively few enzymes can catalyze a great number of reactions. The interaction of enzyme and substrate may be summarized as follows:

$$E + S \longrightarrow ES \rightarrow E + P$$

Assignment 1

Complete item 1 on Laboratory Report 21 that begins on page 285.

DIGESTION IN *PARAMECIUM*

Paramecium exhibits intracellular digestion as found in protozoans. Food particles are collected by the oral groove and are inserted into a food vacuole, where digestion occurs. The vacuole circulates through the cell in a definite pattern, and the nondigestible parts

are expelled through the cell membrane at a site called the anal pore. See Figure 21.2.

In this section, you will observe the digestion of yeast cells by a *Paramecium*. The yeast cells have been stained with congo red, a pH indicator. As digestion proceeds, the yeast cells will change from red to purple and finally to blue. This color change will allow you to follow the progress of digestion.

Assignment 2

Materials

Colored pencils
Compound microscope
Medicine droppers
Microscope slides and cover glasses
Protoslo
Paramecium culture
Yeast cells stained with congo red

1. Make a slide of the *Paramecium* culture containing yeast cells stained with congo red. Add Protoslo to slow down the specimens, and add a cover glass so you can observe them at 100× and 400×. Correlate

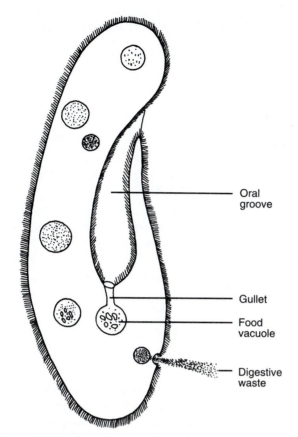

Figure 21.2 Digestion in *Paramecium*.

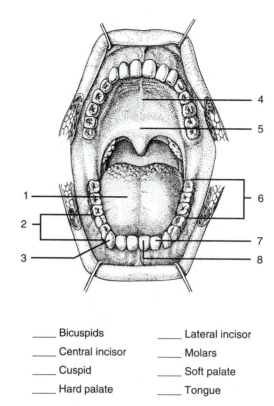

____ Bicuspids   ____ Lateral incisor

____ Central incisor   ____ Molars

____ Cuspid   ____ Soft palate

____ Hard palate   ____ Tongue

Figure 21.3 The oral cavity.

the location of the vacuoles with the color of the yeast cells.

2. Color the vacuoles in Figure 21.2 to show the progress of digestion.

3. ***Complete item 2 on the laboratory report.***

THE HUMAN DIGESTIVE SYSTEM

Refer to Figures 21.3 and 21.4 as you study the human digestive system as an example of a vertebrate digestive system. Table 21.1 lists the major divisions of the digestive tract and their functions. Table 21.2 indicates the end products of digestion for the major food groups.

The Oral Cavity

The mouth contains a number of structures that assist in the digestive process. During chewing, the **teeth** break the food into smaller pieces, while the **tongue** manipulates the food and mixes it with saliva. **Saliva** is produced by three pairs of **salivary glands.** The tongue pushes the food mass back into the **pharynx** to initiate the swallowing reflex, which carries the food into the esophagus.

The roof of the mouth, formed by the anterior **hard palate** and the posterior **soft palate,** divides the oral and nasal cavities. This arrangement allows you to breath while chewing.

There are 32 teeth in a complete set of adult teeth. From front to back in one-half of each jaw and starting at the anterior midline, they are the **central incisor, lateral incisor, cuspid,** first and second **bicuspids** (premolars), and the first, second, and third **molars.** The third molars (wisdom teeth) often become impacted due to the evolutionary shortening of the jaws.

Esophagus to Anus

Food is moved through the digestive tract by **peristalsis,** wavelike contractions of the muscles in its walls. The **esophagus** carries food from the pharynx to the **stomach.** The **cardiac sphincter,** a circular muscle located at the esophagus–stomach junction, opens to allow the passage of food and closes to prevent regurgitation. In the stomach, food is mixed with **gastric juice** and converted to a semiliquid mass called **chyme.** Enzymes in gastric juice begin the digestion of proteins and certain fats. Chyme is then released in small amounts into the small intestine. The **pyloric sphincter** controls the passage of chyme into the small intestine.

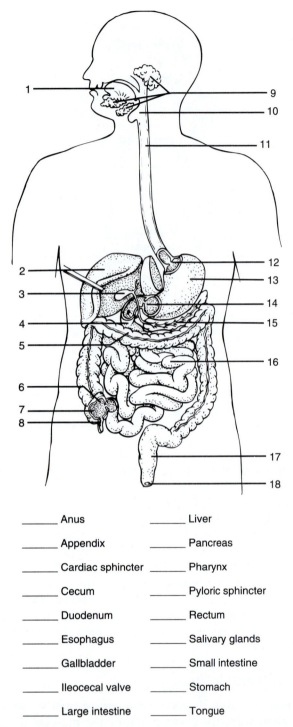

Anus _____   Liver _____

Appendix _____   Pancreas _____

Cardiac sphincter _____   Pharynx _____

Cecum _____   Pyloric sphincter _____

Duodenum _____   Rectum _____

Esophagus _____   Salivary glands _____

Gallbladder _____   Small intestine _____

Ileocecal valve _____   Stomach _____

Large intestine _____   Tongue _____

Figure 21.4 Human digestive system.

TABLE 21.1	MAJOR FUNCTIONS OF THE DIGESTIVE TRACT DIVISIONS
Structure	Major Function
Mouth	Eating, chewing, and swallowing food; digestion of starch begins here
Pharynx	Carries food to esophagus
Esophagus	Carries food to stomach by peristalsis
Stomach	Mixes gastric juice with food to form chyme; protein digestion begins here
Small intestine	Mixes chyme with bile and intestinal and pancreatic juices; digestion and absorption of nutrients completed here
Large intestine	Decomposition of undigested materials by bacteria; reabsorption of water to form feces
Anus	Defecation

temporarily stored in the **gallbladder.** Bile emulsifies fats to facilitate fat digestion. **Pancreatic juice** is produced by the **pancreas,** a pennant-shaped gland located between the stomach and the duodenum. Food entering the duodenum triggers the secretion of bile and pancreatic juice into the duodenum. **Intestinal juice** is secreted by the inner lining (mucosa) of the small intestine. Enzymes in pancreatic juice and intestinal juice act sequentially to complete the digestion of food. Digestion of food and absorption of nutrients into the blood are completed in the small intestine.

The nondigestible material passes into the **large intestine** via the **ileocecal valve,** another sphincter muscle. The **appendix** is a small vestigial appendage attached to the pouchlike **cecum,** the first portion of the large intestine. The large intestine includes ascending, transverse, descending, and sigmoid regions. It ends with the **rectum,** a muscular portion that expels the feces through the **anus.** Decomposition of nondigestible materials by bacteria and the reabsorption of water are the major functions of the large intestine.

TABLE 21.2	END PRODUCTS OF DIGESTION
Food	End Products
Carbohydrates	Monosaccharides
Proteins	Amino acids
Fats	Fatty acids and monoglycerides

In the small intestine, chyme is mixed with bile, pancreatic juice, and intestinal juice. Bile and pancreatic juice enter the **duodenum,** the first portion of the small intestine. The bile and pancreatic ducts join to form a common opening into the duodenum.

Bile is secreted by the **liver,** the large gland in the upper right portion of the abdominal cavity. It is

Histology of the Small Intestine

Examine the structure of the small intestine in cross-section in Figure 21.5. Note the four layers of the intestinal wall.

> **Peritoneum:** The outer protective membrane that lines the coelom and covers the digestive organs.
> **Muscle layers:** Outer longitudinal and inner circular layers whose contractions mix food with digestive secretions and move the food mass by peristalsis.
> **Submucosa:** Connective tissue containing blood vessels and nerves serving the digestive tract.
> **Mucosa:** The inner epithelial lining that secretes intestinal juice and absorbs nutrients.

Locate the **villi,** fingerlike extensions of the mucosa that project into the lumen of the small intestine. Villi increase the surface area of the mucosa, which facilitates the absorption of nutrients. See Figure 21.6.

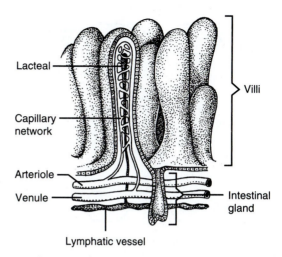

Figure 21.6 A diagrammatic representation of villi in the small intestine.

Assignment 3

Materials

Colored pencils
Compound microscope
Human torso model with removable organs
Prepared slides of small intestine, x.s.

1. Label Figures 21.3 and 21.4 and color-code the digestive organs.
2. Locate the organs of the digestive system on the human torso model. Note their relationships.
3. Examine a prepared slide of small intestine, x.s., and locate the four layers of the intestinal wall.
4. ***Complete item 3 on the laboratory report.***

DIGESTION OF STARCH

In this section, you will assess the effect of temperature and pH on the action of **pancreatic amylase.** This enzyme accelerates the hydrolysis of **starch,** a polysaccharide, to **maltose,** a disaccharide and reducing sugar. See Figure 21.7. The experiments are best done by groups of four students, with each group assigned a different temperature and with all groups sharing their results.

You will dispense solutions from dropping bottles and use a number of test tubes. To avoid contamination, the droppers must not touch other solutions. All glassware must be clean and rinsed with distilled water.

You will dispense "droppers" and "drops" of solutions. As noted earlier, a "dropper" means *one dropper full* of solution (about 1 ml). If you use a pipette to

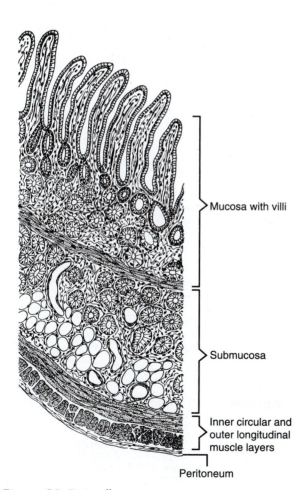

Figure 21.5 Small intestine, x.s.

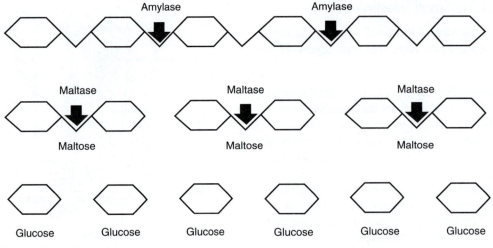

Figure 21.7 Digestion of starch.

dispense the solutions, dispense 1 ml of solution in place of 1 "dropper."

When dispensing solutions by the "drop," hold all droppers at the same angle to dispense equal-size drops. All drops should land squarely in the bottom of the test tube. If they run down the side of the tube, your results may be affected.

Thoroughly mix the solutions placed in a test tube. If a vortex mixer is not available, shake the tube vigorously from side to side to mix the contents.

You will use the iodine test for starch and Benedict's test for reducing sugars to determine if digestion has occurred. See Figure 21.8.

Iodine Test

1. Add 3 drops of iodine solution to 1–2 droppers of solution to be tested.
2. A blue-black color indicates the presence of starch.

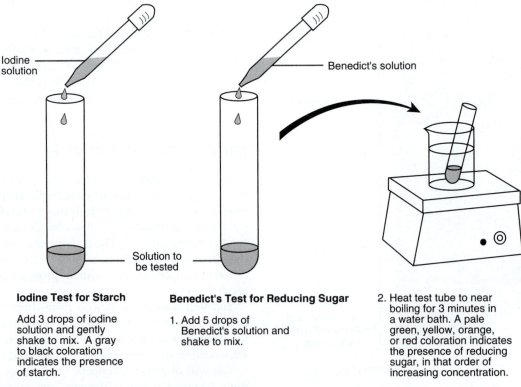

Iodine Test for Starch

Add 3 drops of iodine solution and gently shake to mix. A gray to black coloration indicates the presence of starch.

Benedict's Test for Reducing Sugar

1. Add 5 drops of Benedict's solution and shake to mix.

2. Heat test tube to near boiling for 3 minutes in a water bath. A pale green, yellow, orange, or red coloration indicates the presence of reducing sugar, in that order of increasing concentration.

Figure 21.8 Testing for the presence of starch and sugar.

Benedict's Test

1. Add 5 drops of Benedict's solution to 1 dropper of solution to be tested, and heat the mixture to near boiling in a water bath for 3 min.
2. A light green, yellow, orange, or red color indicates the presence of a reducing sugar, in that order of increasing concentration.

Your instructor has prepared demonstration test tubes showing positive and negative results for starch and reducing sugars. Examine these tubes and use them as standards for interpreting your results.

Tube 1: positive for starch
Tube 2: negative for starch
Tube 3: positive for reducing sugars
Tube 4: negative for reducing sugars

 Assignment 4

Materials

Hot plate
Vortex mixer
Water baths at 5°, 37°, and 70°C
Beaker, 400 ml
Boiling chips
Celsius thermometer
Glass-marking pen
Test tubes
Test-tube holders
Test-tube racks, submersible type
Dropping bottles of:
   Benedict's solution
   buffers, pH 5, 8, 11
   iodine (I_2 + KI) solution
   pancreatic amylase, 0.1%
   soluble starch, 0.1%

1. Label nine test tubes 1–9.
2. Add the pH buffer solutions to the test tubes as follows:
   Tube 1: 2 droppers pH 5
   Tube 2: 2 droppers pH 5
   Tube 3: 3 droppers pH 5
   Tube 4: 2 droppers pH 8
   Tube 5: 2 droppers pH 8
   Tube 6: 3 droppers pH 8
   Tube 7: 2 droppers pH 11
   Tube 8: 2 droppers pH 11
   Tube 9: 3 droppers pH 11
3. Add 5 drops of pancreatic amylase to tubes 1, 3, 4, 6, 7, and 9. Shake the tubes to mix well.

4. Place the test tubes in a test-tube rack and place the rack in the water bath at your assigned temperature for *10 min.*
5. While leaving the test tubes in the water bath, add 1 dropper of starch to tubes 1, 2, 4, 5, 7, and 8. Shake the tubes to mix well. Leave the test tubes in the water bath for 15 min. Table 21.3 shows the contents of each tube for easy reference.
6. After *15 min,* remove the test-tube rack and tubes and return to your work station.
7. Number another set of nine tubes 1B to 9B.
8. Pour half the liquid in tube 1 into tube 1B, pour half the liquid in tube 2 into tube 2B, and so on until you have divided the liquid equally between the paired tubes. You now have nine paired tubes with the members of each pair containing liquid of identical composition.
9. Test all the original nine tubes (1–9) for the presence of starch by adding 3 drops of iodine solution to each tube. ***Record your results in the table in item 4a on the laboratory report.***
10. Test all of the B tubes (1B–9B) for the presence of maltose, a reducing sugar, as follows. Add 5 drops of Benedict's solution to each tube, and place the tubes in a water bath (400-ml beaker half full of water) on your hot plate. Place some boiling chips in the beaker, and heat the tubes to near boiling for 3–4 min. Watch for any color change in the solution. ***Record your results in item 4a on the laboratory report.***
11. Exchange results with other groups using different temperatures. ***Record their results in item 4a on the laboratory report.***
12. ***Complete item 4 on the laboratory report.***

TABLE 21.3 SUMMARY OF TUBE CONTENTS FOR THE STARCH DIGESTION EXPERIMENT

| | Droppers of Solution | | | | |
| | Buffers | | | | Drops of |
Tube	pH 5	pH 8	pH 11	Starch	Amylase
1	2			1	5
2	2			1	
3	3				5
4		2		1	5
5		2		1	
6		3			5
7			2	1	5
8			2	1	
9			3		5

DIGESTION

Student _____

Lab Instructor _____

1. INTRODUCTION

a. Write the term that matches the phrase.

1. Catalyze hydrolysis of foods _____

2. Digestion within cells _____

3. Determines specificity of enzyme _____

4. Enzymatic hydrolysis _____

5. Digestion outside cells _____

b. Summarize the action of digestive enzymes. _____

_____

_____

2. DIGESTION IN *PARAMECIUM*

Show by diagram the location and color of food vacuoles in your specimen.

3. HUMAN DIGESTIVE SYSTEM

a. Write the term that matches the phrase.

1. Produces bile _____

2. Digestive fluid formed by stomach mucosa _____

3. End product of protein digestion _____

4. Where carbohydrate digestion begins _____

5. Site of decay of nondigestible materials _____

6. Absorption of nutrients is completed _____

7. Wavelike contractions of digestive tract _____

8. Separate oral and nasal cavities _____

9. Carries food from pharynx to stomach _____

10. Where digestion of food is completed _____

11. Break food into smaller pieces   _____

12. Pushes food into pharynx in swallowing   _____

13. End product of fat digestion   _____

14. End products of carbohydrate digestion   _____

15. Reabsorbs water from nondigestible material _____

b. List the layers of intestinal wall from outside in.

_____   _____   _____   _____

c. List the labels for Figure 21.3.

1. _____   4. _____   7. _____

2. _____   5. _____   8. _____

3. _____   6. _____

d. List the labels for Figure 21.4.

1. _____   7. _____   13. _____

2. _____   8. _____   14. _____

3. _____   9. _____   15. _____

4. _____   10. _____   16. _____

5. _____   11. _____   17. _____

6. _____   12. _____   18. _____

4. DIGESTION OF STARCH

a. Record your results by placing an "S" for a positive test for starch or an "M" for a positive test for maltose in the appropriate squares.

							Tube Numbers and pH												
			pH 5						pH 8						pH 11			
Temp.	1	1B	2	2B	3	3B	4	4B	5	5B	6	6B	7	7B	8	8B	9	9B
5°C																		
37°C																		
70°C																		

b. List the control tubes expected to give starch tests that are:

positive _____ negative _____

What should be done if these results are not attained? _____

_____

c. Indicate the optimum pH and temperature, among those tested, for amylase activity.

pH _____ Temp. _____° C

Do you think amylase may be active at another pH and temperature? _____

How would you test your hypothesis? _____

_____

d. How do some pH values and temperatures inactivate amylase? _____

_____

e. Would you expect the enzymes of all organisms to work equally well at the optimum temperature determined by

this experiment? _____

Explain. _____

_____

f. Pancreatic amylase is a component of pancreatic juice that is released into the small intestine. What can you

hypothesize about the pH of the digestive fluids in the small intestine? _____

_____

EXCRETION

OBJECTIVES

After completing the laboratory session, you should be able to:
1. Describe the excretory organs of annelids, arthropods, and vertebrates.
2. Identify the components of the mammalian urinary system on a model or chart and state their functions.
3. Identify the parts of a dissected sheep kidney and describe their functions.
4. Describe the structure and function of a nephron in the formation of urine.
5. Perform a urinalysis using a dipstick.
6. Correlate abnormal urine components with disease conditions.
7. Define all terms in bold print.

The metabolic processes of animals produce wastes that are harmful and must be removed. In simple animals, this is accomplished by diffusion alone, but more complex animals use special organs or organ systems. The principal metabolic wastes are **carbon dioxide** (CO_2) formed by cellular respiration, and **nitrogenous wastes,** produced by the breakdown of amino acids composing proteins. Most animals use respiratory organs to expedite the removal of CO_2 and have special excretory organs to get rid of nitrogenous wastes.

In your study of invertebrate animals, you observed three types of excretory organs. Annelids have paired segmental **nephridia** that remove nitrogenous wastes from the coelomic fluid and blood and excrete them through nephridiopores. Crustaceans, such as the crayfish, possess paired **green glands** located at the base of the first antennae. These glands remove nitrogenous wastes from the hemolymph and excrete them through external openings. Similarly, insects have **Malpighian tubules,** which remove nitrogenous wastes from the hemolymph and excrete them into the intestine for passage from the body.

NITROGENOUS WASTES

When amino acids are deaminated, the amine groups tend to form toxic **ammonia.** This substance simply diffuses from small, simple animals, but diffusion is inadequate for its removal in larger, more complex

forms. In aquatic vertebrates, nitrogenous wastes, including ammonia, are flushed from the body with large amounts of water. Terrestrial forms, which must conserve water, convert ammonia into **urea** and **uric acid.** Urea, the primary nitrogenous waste in amphibians and mammals, is excreted in an aqueous solution, but uric acid, the primary nitrogenous waste in insects, reptiles, and birds, is excreted as a semisolid mass with relatively little water.

The process of urinary excretion not only removes the nitrogenous waste products of metabolism but also regulates the concentration of ions, water, and other substances in body fluids. The urinary systems of marine, freshwater, and terrestrial vertebrates are each uniquely adapted to perform these functions in radically different environments.

THE MAMMALIAN URINARY SYSTEM

The mammalian urinary system consists of (1) a pair of **kidneys,** which remove nitrogenous wastes, excess ions, and water from the blood to form **urine;** (2) a pair of **ureters,** which carry urine by peristalsis from the kidney to the urinary bladder; (3) a **urinary bladder,** which serves as a temporary storage container; and (4) a **urethra,** which carries urine from the bladder during urination. See Figure 22.1.

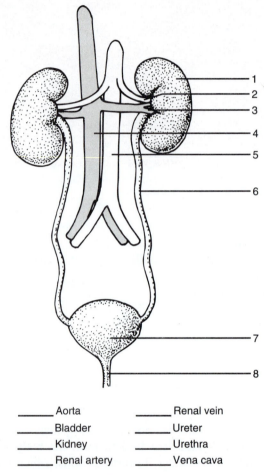

_____ Aorta   _____ Renal vein
_____ Bladder   _____ Ureter
_____ Kidney   _____ Urethra
_____ Renal artery   _____ Vena cava

Figure 22.1 Human urinary system.

The Kidney

The basic structure of a kidney is shown in coronal section in Figure 22.2. The outer portion is the cortex, which contains vast numbers of capillaries and **nephrons.** The inner **medulla** contains the **renal pyramids,** which are composed mainly of **collecting tubules.** The tip (papilla) of each pyramid is inserted into a funnel-like **calyx,** which unites with the **renal pelvis.**

The Nephron

The nephron is the functional unit of the kidney. Each human kidney contains about 1 million nephrons. A nephron consists of (1) a **Bowman's (glomerular) capsule** that surrounds an arteriole capillary tuft, a **glomerulus;** and (2) a tortuous, thin-walled **tubule** that leads to a collecting tubule. The nephron tubule consists of three parts. The **proximal convoluted tubule** leads from the nephron capsule to a U-shaped **loop of Henle** that is contiguous with a **distal convoluted tubule.** Many renal tubules are joined to a single collecting tubule.

Assignment 1

Complete item 1 on Laboratory Report 22 that begins on page 295.

Assignment 2

Materials

Colored pencils
Compound microscope
Dissecting microscope
Model of human kidney
Model of human urinary system
Dissecting instruments and pan
Sheep kidney, fresh or preserved
Sheep kidney, triple injected
Prepared slide of kidney cortex

1. Label and color-code parts of the urinary system in Figure 22.1 and parts of the kidney and nephron in Figure 22.2.
2. Study the models of the urinary system and kidney. Locate the parts shown in Figures 22.1 and 22.2.
3. Examine the demonstration whole kidney. Note its distinctive shape and locate the ureter and renal blood vessels.
4. Obtain a sheep kidney that has been coronally sectioned and locate the parts shown in Figure 22.2A. Correlate the structure of the parts with their functions, which were previously described.
5. Examine the sectioned triple-injected kidney set up under a dissection microscope. Locate the glomeruli and nephron capsules.
6. Examine a prepared slide of kidney cortex. Locate a glomerulus and nephron capsule. Note the thin wall of the tubules. How many cells thick is the tubule wall?
7. *Complete item 2 on the laboratory report.*

Urine Formation

Blood is carried from the aorta to the kidneys by renal arteries and returned from the kidneys by renal veins into the posterior vena cava. As blood passes through the kidneys, the concentration of nitrogenous wastes and other substances in the blood is controlled by (1) the partial removal of diffusible materials in the glomerular blood by **filtration** into Bowman's capsules, (2) **selective reabsorption** of useful substances from the tubules back into the blood, and (3) **secretion** of certain ions from the capillary blood into the tubules.

The renal artery branches into progressively smaller arteries, leading ultimately to the afferent arterioles, which supply blood to the glomeruli. The **afferent**

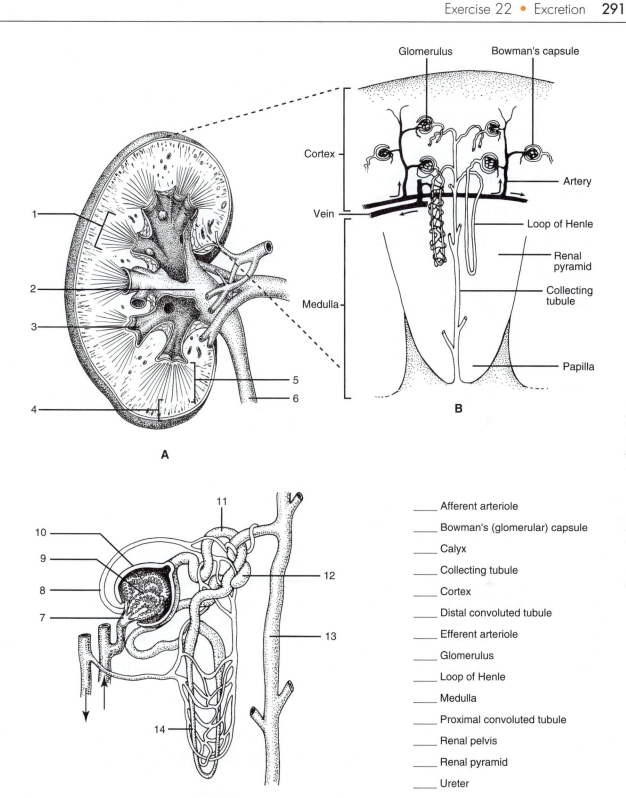

Figure 22.2 The mammalian kidney. **A.** Coronal section. **B.** Detail of cortex and medulla. **C.** Nephron.

arteriole leading to the glomerulus is larger in diameter than the **efferent arteriole** carrying blood from the glomerulus. This causes an increase in blood pressure in the glomerulus and accelerates the movement (filtration) of diffusible materials from the glomerular blood into the capsule. Thus, some of all diffusible materials pass from the glomerulus into Bowman's capsule.

The fluid in the capsule, the **filtrate,** flows through the nephron tubule, which is enveloped by a capillary network. Selective tubule reabsorption of needed materials—especially water, nutrients, and mineral ions—from the filtrate into the capillary blood occurs as the filtrate moves through the tubules. At the same time, certain excess ions (especially H^+ and K^+) may be secreted from the capillary blood into the filtrate. Both **tubule reabsorption** and **tubule secretion** involve active and passive transport mechanisms. The remaining fluid flows from the nephron tubule into a collecting tubule of a renal pyramid, continues into a calyx, and on into the renal pelvis. This fluid, now called **urine,** leaves the kidney via the ureter and is carried to the urinary bladder by peristalsis.

Assignment 3

1. Study Table 22.1. Note the relationships among the daily volumes of blood circulated through human kidneys, filtrate removed, and urine formed. ***Complete items 3a–3d on the laboratory report.***
2. Study Table 22.2. Note how the concentration of selected substances varies in different parts of the nephron and blood vessels. Use Figure 22.2 to orient yourself to the anatomical association of the structures involved. Note that the concentration of urea is constant in columns 1, 2, and 3 of Table 22.2. This indicates that urea is a small

TABLE 22.1	VOLUME OF BLOOD, FILTRATE, AND URINE FORMED IN HUMANS IN A 24-HR PERIOD (LITERS)
Blood flow through the kidneys	1,700
Filtrate removed from blood	180
Urine formed	1.5

molecule that easily diffuses from the glomerulus into Bowman's capsule. How do you explain the change in urea concentration in columns 4 and 5?

3. ***Complete item 3 on the laboratory report.***

URINALYSIS

In humans, the composition of urine is commonly used to assess the general functioning of the body. Diet, exercise, and stress may cause variations in composition and concentration, but significant deviations usually result from malfunctions of the body. See Table 22.3.

Assignment 4

Materials

Biohazard bag
Collecting cup, plastic
Multistix 9 SG reagent strips
Simulated urine samples

In this section, you will use urinalysis dipsticks to analyze several simulated urine samples to determine if they are normal or if their characteristics suggest possible disease. Table 22.3 indicates normal values and some abnormal characteristics. Study it before proceeding.

TABLE 22.2	CONCENTRATION (MG/100 ML) OF DISSOLVED SUBSTANCES IN VARIOUS REGIONS OF THE NEPHRON				
	1	*2*	*3*	*4*	*5*
Materials in Water Solution	*Afferent Arteriole*	*Bowman's Efferent Arteriole*	*Collecting Capsule (Filtrate)*	*Tubule (Urine)**	*Renal Vein*
Urea	30	30	30	2,000	25
Uric acid	4	4	4	50	3.3
Inorganic salts	720	720	720	1,500	719
Protein	7,000	8,000	0	0	7,050
Amino acids	50	50	50	0	48
Glucose	100	100	100	0	98

*Trace amounts are not included.

1. Obtain six Multistix 9 SG reagent strips (dipsticks) and a color chart that indicates how to read the results. Study the color chart carefully to be sure that you understand the time requirements for reading the dipsticks and the location and interpretation of each reagent band. *Note:* The sequence of the reagent bands from the tip to the handle of a dipstick matches the sequence of specific tests listed from top to bottom on the color chart.

2. Obtain six numbered test tubes containing simulated urine samples. Place them in a test-tube rack at your work station.

3. Analyze the urine samples one at a time by dipping a dipstick completely into the sample so all reagent bands are immersed. Remove the dipstick and place it on a paper towel with the reagent bands facing upward. Read your results after 1 min.

4. If you want to test your own urine, obtain a plastic collecting cup and a Multistix 9 SG reagent strip from your instructor and perform the test in the restroom. Dispose of the urine in a toilet, place the collecting cup in the biohazard bag. Wash your hands with soap and water. Read your results. Then, place the dipstick in the biohazard bag.

5. ***Record your results in item 4a on the laboratory report. Compare your results with Table 22.3, and complete item 4 on the laboratory report.***

TABLE 22.3 URINE COMPONENTS EVALUATED BY URINALYSIS

Component	Normal	Abnormal*
Color	Straw to amber	Pink, red-brown, or smoky urine may indicate blood in the urine. The higher the specific gravity, the darker is the color. Nearly colorless urine may result from excessive fluid intake, alcohol ingestion, diabetes insipidis, or chronic nephritis.
Turbidity	Clear to slightly turbid	Excess and persistent turbidity may indicate pus or blood in the urine.
pH	4.8 to 8.0 (average, 6.0)	Acid urine may result from a diet high in protein (meats and cereals) or from a high fever; alkaline urine results from a vegetarian diet or bacterial infections of the urinary tract.
Specific gravity	1.003 to 1.035	Low values result from a deficiency of antidiuretic hormone or kidney damage that impairs water reabsorption. High values result from diabetes mellitus or kidney disease, allowing proteins to enter filtrate.
Blood or hemoglobin	Absent	Presence of intact RBCs may result from lower urinary tract infections, kidney disease allowing RBCs to enter filtrate, lupus, or severe hypertension. Presence of hemoglobin occurs in extensive burns and trauma, hemolytic anemia, malaria, and incompatible transfusions.
Protein	Absent or trace	Proteins are present in kidney diseases that allow proteins to enter the filtrate, and they may be present in fever, trauma, anemia, leukemia, hypertension, and other nonrenal disorders. They may occur due to excessive exercise and high-protein diets.
Glucose	Absent or trace	Presence usually indicates diabetes mellitus.
Ketones	Absent or trace	Presence results from excessive fat metabolism, as in diabetes mellitus and starvation.
Bilirubin	Small amounts	Excess may indicate liver disease (hepatitis or cirrhosis) or blockage of bile ducts.
Nitrite (bacteria)	Absent	Presence indicates a bacterial urinary tract infection.
Leukocytes (pus)	Absent	Presence indicates a urinary tract infection.

* Only a few causes of abnormal values are noted.

EXCRETION

Student _____

Lab Instructor _____

1. NITROGENOUS WASTES

a. Indicate the excretory organs removing nitrogenous wastes in each.

Annelids _____ Crustaceans _____

Insects _____ Mammals _____

b. List three forms of nitrogenous waste excreted by animals.

1. _____ 2. _____ 3. _____

Which compound is most effective in water conservation? _____

Which compound is most toxic? _____

2. MAMMALIAN URINARY SYSTEM

a. List the labels for Figures 22.1 and 22.2.

Figure 22.1	*Figure 22.2*	
1. _____	1. _____	8. _____
2. _____	2. _____	9. _____
3. _____	3. _____	10. _____
4. _____	4. _____	11. _____
5. _____	5. _____	12. _____
6. _____	6. _____	13. _____
7. _____	7. _____	14. _____
8. _____		

b. Describe the function of each.

Kidney _____

Urethra _____

Urinary bladder _____

Ureter _____

c. Diagram the structural relationship between Bowman's capsule, the glomerulus, the afferent arteriole, and the efferent arteriole.

d. List in sequence the parts of a nephron.

1. _____   3. _____

2. _____   4. _____

3. URINE FORMATION

a. Explain the advantage of increased blood pressure in the glomerulus. _____

_____

_____

b. Describe what happens in each.

Glomerular filtration _____

_____

Tubular reabsorption _____

_____

Tubular secretion _____

_____

c. Using the data in Table 22.1, what percentage of the blood passing through the kidneys becomes filtrate? _____

What percentage of filtrate becomes urine?_____

What happens to the rest of the filtrate?_____

d. Why is a low volume of urine production important in terrestrial animals but unimportant in aquatic animals?___

_____

e. Considering Table 22.2, explain how the high concentration of urea in urine (column 4) is attained. _____

_____

f. Does the kidney simply maintain the concentration of urea at a safe level or remove all of it from the blood?____

Explain. _____

g. Comparing filtrate and urine, how many times is urea concentrated in the urine (columns 3 and 4)? _____

How is that accomplished?_____

h. Comparing the concentration of mineral salts in columns 1 and 5, very little is removed from the blood. What happens to most of the salts in the filtrate? _____

i. Explain the protein concentrations in each column.

Column 2 _____

Column 5 _____

j. Does glucose easily diffuse into Bowman's capsule? _____

How do you know this? _____

What happens to the glucose in the filtrate? _____

Explain the slightly lower concentration of glucose in column 5 as compared with column 1. _____

_____

_____

4. URINALYSIS

a. Record the results of your analysis of the simulated urine samples in the following table. Use an X to indicate the presence of a component, and record the color and specific gravity.

Characteristic	Tubes					
	1	2	3	4	5	6
Glucose						
Bilirubin						
Ketone						
Specific gravity						
Blood						
pH						
Protein						
Nitrite						
Color						

b. Indicate for each urine sample the health condition of the "patient" based on your urinalysis. Select the health conditions from the list that follows:

Tube 1 _____

Tube 2 _____

Tube 3 _____

Tube 4 _____

Tube 5 _____

Tube 6 _____

Diabetes insipidis

This disorder is caused by a deficiency of antidiuretic hormone, which results in the inability of kidney tubules to reabsorb water. Symptoms are constant thirst, weight loss, weakness, and production of 4 to 10 liters of urine each day. The urine is very dilute (a low specific gravity) and is nearly colorless.

Diabetes mellitus

This disorder is caused by a deficiency of insulin, which results in a decreased ability of glucose to enter cells. Therefore, glucose accumulates in the blood and fats are used excessively in cellular respiration, producing ketones as a waste product. Symptoms include weakness, fatigue, weight loss, and excessive blood glucose levels. In uncontrolled diabetes, a urine sample contains excess glucose and ketones and has a low pH.

Hepatitis

Hepatitis is usually caused by a viral infection. In type A hepatitis, symptoms may include fatigue, fever, generalized aching, abdominal pain, and jaundice (yellowish tinge to the whites of the eyes due to excessive blood levels of bilirubin). Urine is dark amber in color due to excessive bilirubin.

Glomerulonephritis

In this kidney disease, excessive permeability of Bowman's capsule allows proteins and red blood cells to enter the filtrate. They are present in the urine because they cannot be reabsorbed.

Hemolytic anemia

A number of conditions cause the destruction of red blood cells, thereby releasing hemoglobin into the blood plasma. Such conditions include malaria and incompatible blood transfusions. The urine of such patients contains hemoglobin and may be red-brown or smoky in color.

Normal

See Table 22.3 for the composition of normal urine.

Strenuous exercise and a high-protein diet

Athletes in excellent health may produce urine with an acid pH, a small amount of protein, and an elevated specific gravity due to the presence of the proteins.

Urinary tract infection

Bacterial urinary tract infections may involve the urethra (urethritis), urinary bladder (cystitis), ureters, or kidneys. Production of alkaline urine promotes the growth of infectious bacteria. Urethritis often causes a burning pain during urination. Symptoms may include a low-grade fever and discomfort of the affected region. Urine will test positive for nitrite because bacteria in the urine convert nitrate, a normal component, into nitrite. Severe infections may also cause the urine to contain blood and pus (leukocytes), producing a cloudy urine.

c. Which characteristics would you expect to find in a urine sample if:

there was an acute kidney infection? _____

_____

_____

the glomerular blood pressure was chronically high? _____

_____

_____

d. Explain the relationship between the volume of urine, its specific gravity, and its color. _____

_____

_____

_____

_____

NEURAL CONTROL

OBJECTIVES

After completing the laboratory session, you should be able to:
1. Describe the three functional types of neurons and their basic structure.
2. Explain the mechanism of unidirectional transmission across a synapse.
3. Identify the major parts of the brain and spinal cord, and describe their functions.
4. Describe the components of a reflex arc, and state the function of each.
5. Explain the mode of action of the reflexes studied.
6. Describe the distribution of touch receptors as determined experimentally.
7. Define all terms in bold print.

Neural coordination occurs in all but the simplest animals. For example, a nerve net coordinates movement in cnidarians, widely separated ventral nerve cords occur in flatworms, and a **nervous system** composed of (1) a fused, double ventral nerve cord with segmental ganglia, (2) a "brain," and (3) well-developed sensory receptors occurs in annelids and arthropods.

Vertebrates possess the most highly developed nervous system, and it may be subdivided into two major components. The **central nervous system (CNS)** is composed of a **dorsal tubular nerve cord** and a **brain.** The **peripheral nervous system (PNS)** is composed of **cranial nerves,** which emanate from the brain and extend to the head and certain internal organs, and **spinal nerves,** which extend from the spinal cord to all parts of the body except the head. Of course, well-developed **sensory receptors** are evident.

In this exercise, you will investigate the structure and function of the vertebrate nervous system as found in mammals. The nervous system provides rapid coordination and control of body functions by transmitting **impulses** over neuron processes. The impulses may originate either in the brain or in **receptors,** and they are carried to **effectors** (muscles and glands), where the action occurs.

NEURONS

Neural tissue is composed of two basic types of cells: neurons and neuroglial cells. **Neurons,** or nerve cells, transmit impulses. **Neuroglial cells** provide structural support and prevent contact of neurons, except at certain sites. Although neurons may be specialized in structure and function in various parts of the nervous system, they have many features in common. Figure 23.1 shows the basic structure of neurons, and Table 23.1 identifies the characteristics of the three functional types of neurons.

The **cell body** is an enlarged portion of the neuron that contains the nucleus. Two types of neuron processes or fibers extend from the cell body. **Dendrites** receive impulses from receptors or other neurons and carry impulses *toward* the cell body. **Axons** carry impulses *away from* the cell body. A neuron may have many dendrites, but only one axon is present.

Neuron processes of the peripheral nervous system are enclosed by a covering of **Schwann cells.** In larger processes, the multiple wrappings of Schwann cells form an inner **myelin sheath,** a fatty, insulating material, and the outer layer constitutes the **neurilemma.** The minute spaces between Schwann cells, where the neuron process is exposed, are called **nodes of Ranvier.** Impulses are transmitted more

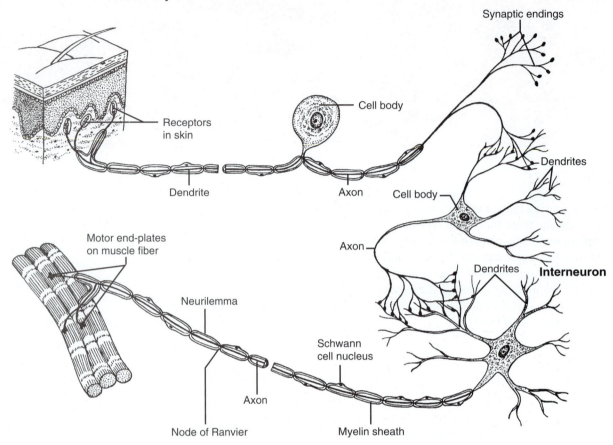

Figure 23.1 The basic structure and relationship of a sensory neuron, an interneuron, and a motor neuron.

rapidly by myelinated fibers than by unmyelinated fibers. Schwann cells provide a pathway for the regeneration of neuron processes and are essential for regrowth. Schwann cells are absent in the CNS, but certain neuroglial cells form the myelin sheath of CNS neurons.

The junction of an axon tip of one neuron and a dendrite or cell body of another neuron is called a **synapse.** See Figure 23.2. Impulses passing along the axon cause the release of a **neurotransmitter** from the axon tip, or

synaptic knob, into the **synaptic cleft,** the minute space between the neurons. The neurotransmitter binds with receptor sites on the postsynaptic membrane, producing either stimulation or inhibition of impulse formation, depending on the type of neurotransmitter involved. Immediately thereafter, an enzyme breaks down or inactivates the neurotransmitter, which prevents continuous stimulation or inhibition of the postsynaptic membrane. Transmission of neural impulses across synapses is always in one direction, axon to

TABLE 23.1	FUNCTIONAL TYPES OF NEURONS	
Neuron Type	Structure	Function
Sensory	Long dendrite, short axon	Carry impulses from receptors to the CNS
Interneuron	Short dendrites, short or long axon	Carry impulses within the CNS
Motor	Short dendrites, long axon	Carry impulses from the CNS to effectors

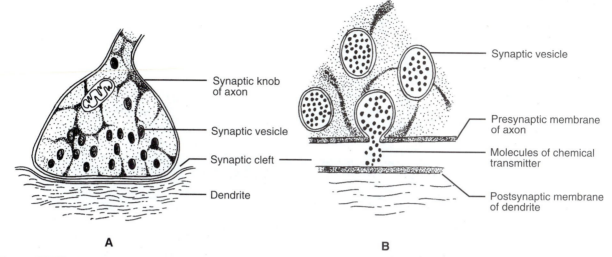

Figure 23.2 The synaptic junction. **A.** Relationship of the synaptic knob of an axon to a dendrite. Synaptic vesicles migrate from the cell body to the synaptic knob. **B.** Synaptic transmission. An impulse moving down the axon causes a synaptic vesicle to release a neurotransmitter that either stimulates or inhibits the formation of an impulse in the dendrite.

dendrite, because only axon tips can release neurotransmitters to activate the adjacent neuron.

Numerous substances are either known or suspected to be neurotransmitters. Among known neurotransmitters, **acetylcholine** and **norepinephrine** are stimulatory neurotransmitters, whereas **glycine** and **gamma-aminobutyric acid (GABA)** are inhibitory neurotransmitters.

 Assignment 1

Materials

Colored pencils
Compound microscope
Prepared slides of:
   giant multipolar neurons
   nerve, x.s.

1. Add arrows to Figure 23.1 to indicate the direction of impulse transmission in motor and sensory neurons. Color-code the three types of neurons.
2. *Complete item 1a on Laboratory Report 23 that begins on page 309.*
3. Examine a slide of giant multipolar neurons. Draw a neuron and label the cell body, nucleus, nucleolus, and neuron processes.
4. Examine a slide of nerve, x.s., and make a drawing of your observations. Note how the axons are arranged in bundles separated by connective tissue. Label an axon, myelin sheath, and connective tissue.
5. *Complete item 1 on the laboratory report.*

THE BRAIN

The **brain** is the control center of the nervous system. It is enclosed within the cranium and covered by the **meninges,** three layers of protective membranes. **Cerebrospinal fluid** within the meninges provides an additional cushion to absorb shocks. Twelve pairs of cranial nerves are attached to the brain. All but one pair innervate structures of the head and neck; vagus nerves innervate the internal organs. Refer to Figures 23.3 to 23.7 as you study the major parts of the brain.

Cerebrum

The **cerebrum** is the largest part of the brain. It consists of right and left **cerebral hemispheres** that are separated by a median **longitudinal fissure.** A mass of neuron fibers, the **corpus callosum,** enables impulses to pass between the two hemispheres. The outer portion of the cerebrum, the **cerebral cortex** or **gray matter,** is composed of neuron cell bodies and unmyelinated neuron processes. Its surface area is increased by numerous **gyri** (ridges) and **sulci** (grooves). The cerebrum initiates voluntary actions and interprets sensations. In humans, it is the seat of will, memory, and intelligence.

Each hemisphere is divided into four lobes by fissures (deep grooves). Table 23.2 indicates the location and major functions of these lobes.

Cerebellum

The **cerebellum** lies just below and posterior to the occipital lobe of the cerebrum. It is divided into left and right hemispheres by a shallow fissure. Muscle

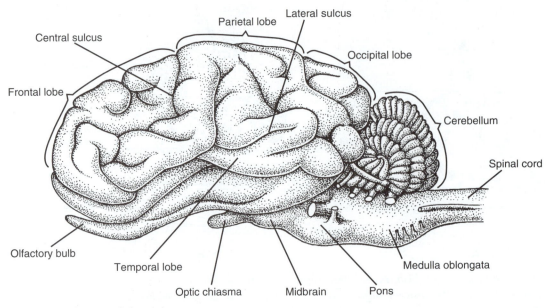

Figure 23.3 Lateral view of sheep brain.

tone and muscular coordination are subconsciously controlled by the cerebellum.

Thalamus and Hypothalamus

The **thalamus,** which is located at the upper end of the brain stem, consists of two lateral globular masses joined by an isthmus of tissue called the *intermediate mass.* It provides an uncritical awareness of sensations such as pain and pleasure and is a relay station between lower brain centers and the cerebrum.

The **hypothalamus** is located just below the thalamus. It plays a major role in homeostasis of the body by controlling things such as appetite, sleep, body temperature, and water balance. A major endocrine gland, the **hypophysis** or **pituitary gland,** is attached to the ventral wall of the hypothalamus by a short stalk, the **infundibulum.** Just anterior to the

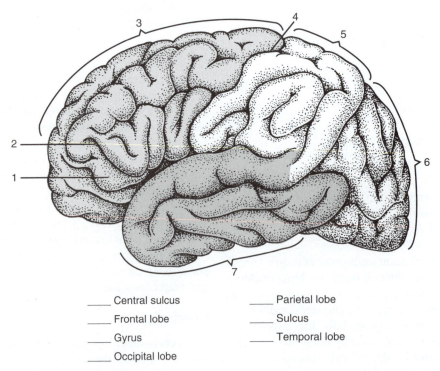

_____ Central sulcus   _____ Parietal lobe

_____ Frontal lobe   _____ Sulcus

_____ Gyrus   _____ Temporal lobe

_____ Occipital lobe

Figure 23.4 Lateral view of human cerebrum.

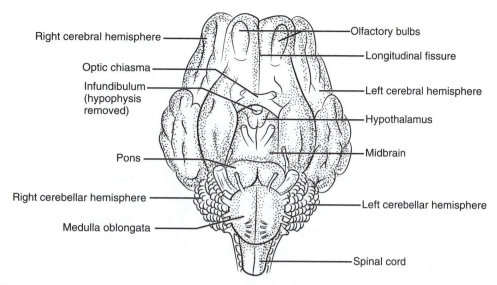

Figure 23.5 Ventral view of sheep brain.

hypophysis lies the **optic chiasma,** where optic nerve fibers cross to the opposite side of the brain.

Brain Stem

The **brain stem** is composed of a number of structures located between the cerebrum and the spinal cord. A midsagittal section is required to locate the various parts. A major function is the linking of higher and lower brain areas, but the individual portions also have some specific functions.

The **midbrain** is a small area between the thalamus and pons that is associated with certain visual reflexes. The **pineal body** is located on the dorsal portion of the midbrain. It secretes melatonin, which helps regulate sleep/wakefulness and reproductive cycles.

The **pons** is a rounded bulge on the ventral side of the brain stem. It is an important connecting pathway for higher and lower brain centers.

The **medulla oblongata** lies between the pons and the spinal cord. It is the lowest part of the brain, and all impulses passing between the brain and spinal cord

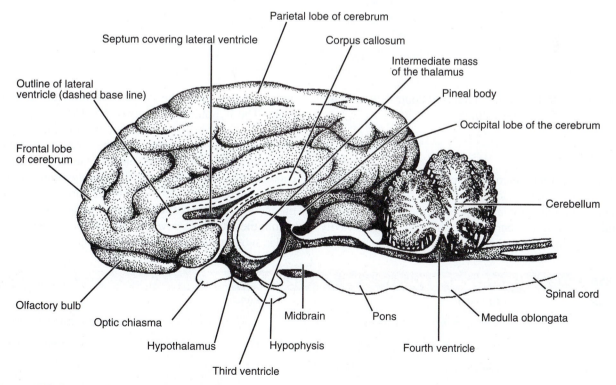

Figure 23.6 Sheep brain, midsagittal section.

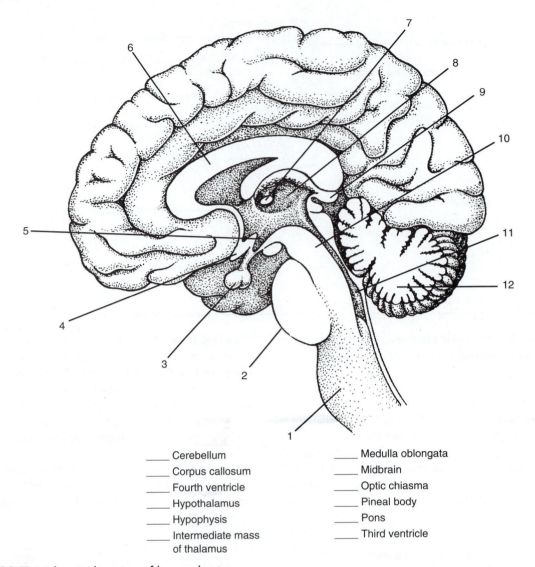

____ Cerebellum
____ Corpus callosum
____ Fourth ventricle
____ Hypothalamus
____ Hypophysis
____ Intermediate mass
        of thalamus

____ Medulla oblongata
____ Midbrain
____ Optic chiasma
____ Pineal body
____ Pons
____ Third ventricle

Figure 23.7 Midsagittal section of human brain.

must pass through it. In addition, it controls heart rate, blood pressure, and breathing.

Ventricles

There are four interconnected **ventricles** (cavities) within the brain. Each hemisphere of the cerebrum contains a relatively large lateral ventricle (ventricles 1 and 2). The third ventricle is a narrow, vertical cavity located at the midline between the lobes of the thalamus and extending from the corpus callosum to the midbrain. The hypothalamus forms the floor of the third ventricle. The fourth ventricle is a narrow channel that lies between the cerebellum and the pons and medulla.

Cerebrospinal fluid is secreted by clumps of special capillaries within each ventricle. It fills the ventricles and flows from the lateral ventricles to the third ventricle and on into the fourth ventricle. Then, it passes into a space within the meninges covering the brain and spinal cord and is subsequently reabsorbed back into the blood.

TABLE 23.2	LOCATION AND FUNCTION OF THE CEREBRAL LOBES	
Lobe	Location	Function
Frontal	Anterior to the central sulcus	Voluntary muscular movements; intellectual processes
Parietal	Between frontal and occipital lobes	Interprets sensations from skin; speech interpretation
Temporal	Inferior to frontal and parietal lobes	Hearing; interprets auditory sensations
Occipital	Posterior part of cerebrum	Vision; interprets visual sensations

Assignment 2
Materials

Colored pencils
Dissecting instruments and pans
Gloves, protective and disposable
Models of sheep and human brains
Sheep brains, preserved

1. Study Figures 23.3, 23.5, and 23.6 until you are familiar with the structures of the sheep brain.
2. Label Figures 23.4 and 23.7, and locate these parts on a model of a human brain. Color-code the cerebral lobes in Figure 23.4 and the parts of the brain in Figure 23.7.
3. Obtain an entire sheep brain for study. Note the meninges, if present, and remove them with scissors. Locate the major parts of the brain shown in Figures 23.3 and 23.5. Observe the ridges and furrows that increase the surface area of the cerebrum. Is there an advantage in this?
4. Obtain half of a sheep brain that has been sectioned along the longitudinal fissure. Locate the structures shown in Figure 23.6. Note that the hypothalamus forms the floor of the third ventricle and that the hypophysis (pituitary gland) projects ventrally from it. The **pineal body** is a remnant of a third eye found in primitive reptiles. Its function in some mammals may be to control seasonal gonadal activity based on photoperiod variation.
5. Examine the demonstration coronal sections of a sheep brain and locate the two **lateral ventricles.** Note that brain tissue appears either white or gray. **White matter** consists of myelinated neuron processes, and **gray matter** consists mostly of neuron cell bodies and nonmyelinated neuron processes. Which type of neural tissue composes the cerebral cortex? The corpus callosum?
6. ***Complete item 2 on the laboratory report.***

THE SPINAL CORD

The **spinal cord** is covered by meninges that are continuous with the meninges covering the brain. It extends from the brain stem down the vertebral canal, where it is protected by both meninges and the surrounding vertebrae. The spinal cord serves as the connecting link between the brain and the **spinal nerves,** which innervate all parts of the body except the head. Unlike the cerebrum, gray matter is located in the interior of the spinal cord and white matter is located around it. Impulses are carried up and down the spinal cord by neuron fibers in the white matter. See Figure 23.8.

In humans, there are 31 pairs of spinal nerves. Each spinal nerve contains both **sensory neuron fibers** that carry impulses from sensory receptors to the spinal cord and **motor neuron fibers** that carry impulses from the spinal cord to an effector (a muscle or gland). Each spinal nerve joins to the spinal cord via a **dorsal root** and **ventral root.** Sensory neuron fibers enter the spinal cord via the dorsal root. Sensory neuron cell bodies are located in the **dorsal root ganglion,** a swelling on the dorsal root. Motor neuron fibers exit via the ventral root, and motor neuron cell bodies are located in the gray matter of the spinal cord.

Assignment 3
Materials

Chart or model of the spinal cord
Compound microscope
Prepared slide of cat spinal cord, x.s.

1. Label Figure 23.8.
2. Examine a prepared slide of cat spinal cord, x.s., at 40×, and observe the gross features. Compare your slide with Figure 23.8. Use 100× to locate the cell bodies of neurons in the gray matter, and examine them at 400×. What do the neuron fibers look like in the white matter?
3. Add arrows to Figure 23.8 to show the path of impulses.
4. ***Complete item 3 on the laboratory report.***

Reflexes

Reflexes are involuntary responses (require no conscious act) to specific stimuli, and they are important mechanisms for maintaining the well-being of the individual. For example, coughing and sneezing are respiratory reflexes controlled by the brain. Spinal reflexes do not involve the brain.

Simple reflexes require no more than three neurons to produce an action to a stimulus, and this is accomplished through a **reflex arc.** See Figure 23.8. A reflex arc consists of (1) a receptor that forms impulses on stimulation, (2) a sensory neuron that carries the impulses to the brain or spinal cord, (3) an interneuron that receives the impulses and transmits them to a motor neuron, (4) a motor neuron that carries impulses to an effector, and (5) an effector that performs the action.

Patellar Reflex

Physicians use reflex tests to assess the condition of the nervous system. The **patellar reflex** is one that is commonly used (Figure 23.9). When the patellar tendon is struck just below the kneecap with a reflex hammer, the reflex action is a slight, instantaneous contraction of the large muscle (quadriceps femoris)

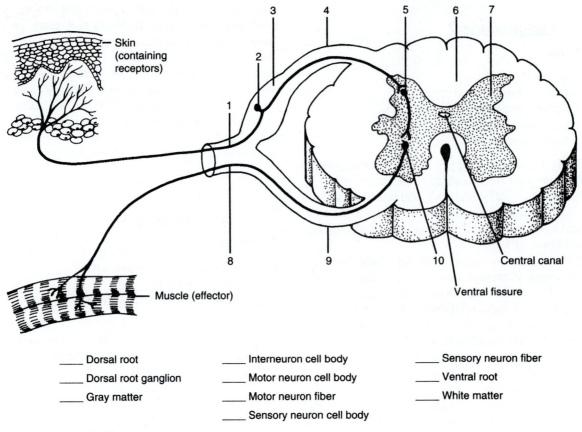

_____ Dorsal root   _____ Interneuron cell body   _____ Sensory neuron fiber

_____ Dorsal root ganglion _____ Motor neuron cell body   _____ Ventral root

_____ Gray matter   _____ Motor neuron fiber   _____ White matter

                           _____ Sensory neuron cell body

Figure 23.8 Spinal reflex arc.

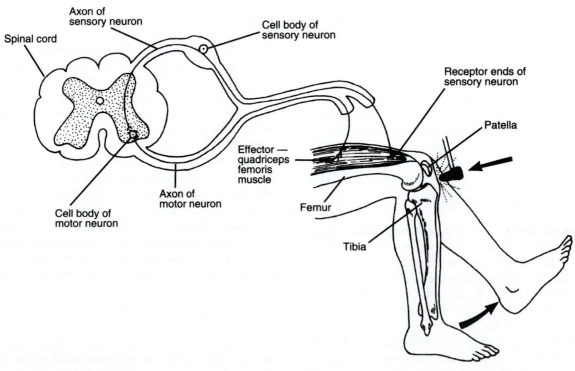

Figure 23.9 The patellar reflex.

on the front of the thigh that extends the lower leg. Striking the patellar tendon causes the muscle to be slightly stretched for an instant. This stretching causes impulses to be formed and carried along a sensory neuron to the spinal cord, where they are passed to a motor neuron that carries the impulses to the muscle, thereby causing a weak, brief contraction.

Photopupil Reflex

This reflex allows a rapid adjustment of the size of the pupil to the existing light intensity, and it is coordinated by the brain. It is most easily observed in persons with light-colored eyes. When a bright light stimulates the retina of the eye, impulses are carried to the brain by sensory neurons. In the brain, the impulses are transmitted to interneurons and on to motor neurons that carry impulses to the muscles of the iris, causing them to contract. Contraction of the iris muscles decreases the size of the pupil and controls the amount of light entering the eye.

 Assignment 4

Materials

Reflex hammer
Laboratory lamp
Penlight

1. Perform the patellar reflex as follows.
   a. The subject should sit on the edge of a table with his or her legs hanging over the edge, but not touching the floor.
   b. Strike the patellar tendon (see Figure 23.9) with the small end of the reflex hammer, and observe the response.
   c. Divert the subject's attention by having the subject interlock the fingers of both hands and pull the hands against each other while you strike the patellar tendon again. Is the response the same as before? If not, how do you explain it?
   d. Test both legs, and record the results.
   e. Add arrows to Figure 23.9 to show the path of impulses.
   f. ***Complete items 4a and 4b on the laboratory report.***
2. Perform the photopupil reflex as follows.
   a. Have the subject sit with eyes closed facing a darkened part of the room for 1–2 min. When the subject opens his or her eyes, note any change in pupil size.
   b. While the subject is looking into the darkened area, shine a desk lamp in his or her eyes (from about 3 ft away). Note any change in pupil size.

   c. To observe the effect of unilateral stimulation, have the subject look into the darkened area. Shine a penlight into the left eye only, noting the effect on the pupil of the left eye. Repeat and note any effect on the right eye. Repeat your observations while stimulating the right eye only.
   d. ***Complete item 4 on the laboratory report.***

REACTION TIME

The nervous system controls and coordinates our reactions to thousands of stimuli each day. Some of these reactions are reflexes, but many are **voluntary reactions,** responses that are consciously initiated. **Reaction time** is the time interval from the instant of stimulation to the instant of a voluntary response. All responses result from the formation of impulses by stimulation and the transmission of these impulses along neurons to effectors that bring about the response.

All sense organs do not form and transmit impulses to the brain at the same speed. The fastest impulses come from proprioceptors in muscles that inform the brain about body positions. They travel at about 390 feet per second, so you always know the position of your arms and legs even if you can't see them. Impulses from touch receptors are the next fastest at about 250 feet per second. Impulses from pain and temperature receptors are the slowest, although intensity of the stimulus alters the speed. Even the speed of visual impulses is affected by light intensity.

In reflexes, impulses flow over predetermined "automated" neural pathways involving very few neurons, and they do not require processing by the cerebral cortex. In contrast, voluntary reactions involve a greater number of neurons and synapses and require processing of impulses by the cerebral cortex. Therefore, reflexes have much shorter response times than voluntary reactions.

The reaction time for a voluntary response is the sum of the times required for

1. A receptor to form impulses in response to a stimulus.
2. Transmission of impulses to an integration center of the cerebral cortex.
3. Processing the impulses in the integration center.
4. Transmission of impulses to effectors.
5. Response of the effectors.

Do you think people differ in their reaction times to the same stimulus? In this section, you will test the hypothesis that there is no difference in the reaction times of different persons in responding to the same stimulus with the same predetermined response.

Measuring Reaction Time

You will measure visual reaction time using a reaction-time ruler and the following procedure:

1. The subject sits on a chair or stool with the experimenter standing facing the subject.
2. The experimenter holds the *release end* of the reaction-time ruler between thumb and forefinger at about eye level or higher. See Figure 23.10.
3. The subject places the thumb and forefinger of his or her dominant hand about an inch apart and on each side of the *thumb line* at the lower end of the ruler. The subject's attention is focused on the ruler at the thumb line.
4. When the subject says he or she is ready, the experimenter, within 10 sec, releases the ruler. The subject seeing the falling ruler catches it between thumb and forefinger as quickly as possible.
5. The reaction time is read in milliseconds at the upper edge of the thumb and recorded.
6. The test is repeated five times, and the average reaction time is calculated. If any reaction time is grossly different, discard it and repeat the test to obtain five results that are fairly consistent.

 Assignment 5

Materials

Reaction-time ruler

1. Have your partner measure your reaction time five times and *record them in item 5a on the laboratory report.*
2. Calculate your average reaction time. Write it on the board for the class tabulation. *Complete items 5b–5f on the laboratory report.*
3. Do you think practice and learning will decrease your reaction time? Repeat the reaction-time test twenty times without recording the reaction time. Then repeat the test five times and record your reaction times.
4. Now put on a pair of dark glasses and dim the lights so you can barely see the reaction-time ruler. Have your partner measure your reaction time as before. *Record the average of five replicates as your results in item 5h.*
5. *Complete item 5 on the laboratory report.*

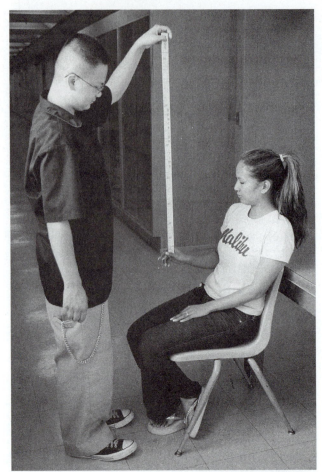

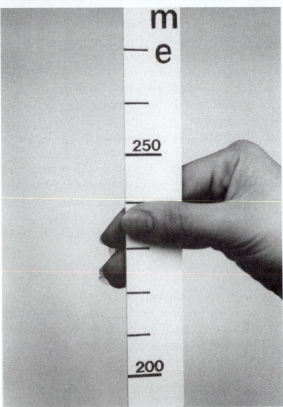

Figure 23.10 Measuring visual reaction time.

NEURAL CONTROL

Student _____

Lab Instructor _____

1. NEURONS

a. Write the term that matches the phrase.

1. Nerve cell _____

2. Neuron carrying impulses to CNS _____

3. Neuron carrying impulses from CNS _____

4. Composed of cranial and spinal nerves _____

5. Composed of brain and spinal cord _____

6. Cells providing support for nerve cells _____

7. Part of neuron containing nucleus _____

8. Process carrying impulses from cell body _____

9. Process carrying impulses toward cell body _____

10. Neuron with all parts located in the CNS _____

11. Junction of axon tip and adjacent neuron _____

12. Secretes neurotransmitter _____

13. Receives neurotransmitter _____

14. Forms myelin sheath in PNS neurons _____

15. Forms myelin sheath in CNS neurons _____

b. Draw a giant multipolar neuron and a nerve, x.s., as observed on the prepared slides. Label pertinent parts.

Neuron Cell Body   **Nerve, x.s.**

2. THE BRAIN

a. List the labels for Figure 23.4.

1. _____   4. _____   6. _____

2. _____   5. _____   7. _____

3. _____

b. List the labels for Figure 23.7.

1. _____   5. _____   9. _____

2. _____   6. _____   10. _____

3. _____   7. _____   11. _____

4. _____   8. _____   12. _____

c. Matching

1. Cerebrum   2. Cerebellum   3. Thalamus   4. Hypothalamus   5. Medulla oblongata   6. Corpus callosum
7. Ventricles

_____ Uncritical awareness of sensations   _____ Muscular coordination

_____ Contains cerebrospinal fluid   _____ Controls water balance

_____ Critical interpretation of sensations   _____ Largest part of mammalian brain

_____ Regulates heart rate   _____ Controls body temperature

_____ Connects cerebral hemispheres   _____ Intelligence, will

_____ Breathing control center

d. Is the relative size of the following larger in humans or sheep?

Cerebrum _____   Cerebellum _____

e. Matching

1. White matter   2. Gray matter

_____ Exterior of cerebrum   _____ Exterior of spinal cord

_____ Myelinated fibers   _____ Nonmyelinated fibers

_____ Neuron cell bodies   _____ Axon tips in spinal cord

3. SPINAL CORD

a. List the labels for Figure 23.8.

1. _____   5. _____   8. _____

2. _____   6. _____   9. _____

3. _____   7. _____   10. _____

4. _____

b. Write the term that matches the phrase.

1. Neuron carrying impulses to effector   _____

2. Neuron receiving impulses from receptor   _____

3. Connects sensory and motor neurons   _____

c. Describe the effect of an injury that cuts the:

dorsal root of a spinal nerve _____

_____

ventral root of a spinal nerve _____

_____

spinal cord _____

_____

d. Diagram from your slide (1) a cat spinal cord, x.s., labeling gray and white matter, ventral fissure, and central canal; and (2) a motor neuron cell body and associated processes, labeling the nucleus, nucleolus, and neurofibrils at the base of the processes.

Cat Spinal Cord, x.s.   **Motor Neuron Cell Body**

4. REFLEXES

a. Rate the strength of your patellar response as strong (+++), moderate (++), weak (+), or none (0):

Left leg: without diversion _____   with diversion _____

Right leg: without diversion _____   with diversion _____

Explain any differences in response. _____

_____

_____

b. If you touch a hot stove, you will reflexively jerk back your hand a split second before you feel the pain. Explain the delayed pain response. _____

_____

_____

_____

c. Describe the effect on the size of the pupil when:

the subject is looking into a darkened area. _____

the subject is looking into a bright light. _____

d. When a penlight is shined into the left eye, what response is seen:

in the left eye? _____

in the right eye? _____

Explain your observations. _____

_____

_____

5. REACTION TIME

a. Record the five replicates of your reaction time in milliseconds (msec).

_____   _____   _____   _____   _____

Calculate your average reaction time.

b. Record the average reaction time (msec) for each class member by gender.

Males   *Females*

1. _____   7. _____   13. _____   1. _____   7. _____   13. _____

2. _____   8. _____   14. _____   2. _____   8. _____   14. _____

3. _____   9. _____   15. _____   3. _____   9. _____   15. _____

4. _____   10. _____   16. _____   4. _____   10. _____   16. _____

5. _____   11. _____   17. _____   5. _____   11. _____   17. _____

6. _____   12. _____   18. _____   6. _____   12. _____   18. _____

c. For each gender and the total class, indicate the range of values and calculate the average of the average reaction times.

   Males: Range _____ msec to _____ msec   Average _____ msec

   Females: Range _____ msec to _____ msec   Average _____ msec

   Class: Range _____ msec to _____ msec   Average _____ msec

d. Do the results support the hypothesis? _____

   Does there seem to be a significant gender difference in reaction times? _____

   Explain. _____

   _____

   _____

e. State a conclusion from the results. _____

   _____

   _____

f. In what types of activities is a faster reaction time advantageous? _____

   _____

   _____

   _____

g. Indicate your reaction times for five replicates after practicing the test twenty times.

   _____   _____   _____   _____   _____

   Calculate your average reaction time. _____

   Did your reaction time decrease with practice? _____

h. Record your average reaction time while wearing dark glasses in dim light. _____ msec

   Was your reaction time in dim light faster or slower than in bright light? _____

   How do you explain this? _____

   _____

SENSORY PERCEPTION IN HUMANS

OBJECTIVES

After completing the laboratory session, you should be able to:
1. Describe the basic structure and function of the eye and ear, and identify the parts on charts and models.
2. Explain the basic characteristics of sensory perception as determined by the tests performed in this exercise.
3. Define all terms in bold print.

Sensations result from the interaction of three components of the nervous system:

1. **Sensory receptors** generate impulses upon stimulation.
2. **Sensory neurons** carry the impulses to the brain or spinal cord, and **interneurons** carry the impulses to the sensory interpretive centers in the brain.
3. The **cerebral cortex** interprets the impulses as sensations.

The type of sensation is determined by the part of the brain receiving the impulses, rather than by the type of receptors being stimulated. For example, the auditory center interprets all impulses it receives as sound stimuli regardless of their origin.

In this exercise, you will study the structure and function of the eye and the ear and perform tests that will demonstrate certain characteristics of sensory perception.

THE EYE

The eyes contain the receptors for light stimuli, and they are well protected by the surrounding skull bones and accessory structures. **Eyelids** protect the anterior surface of the eye. A mucous membrane, the **conjunctiva,** lines the inner surface of the eyelids and covers the anterior surface of the eyes. It contains blood vessels and many pain receptors except where it covers the cornea. The surface of the eyes is kept moist by tears produced by a **lacrimal gland,** which is located above the lateral margin of each eye. Tears are collected into a duct in the medial corner of the

eyes and carried into the nasal cavity. **Eyelashes** grow from the free margins of the eyelids, and they shield the eyes from excessive light. Neuron receptors at the base of the eyelashes are especially sensitive to eyelash movement and trigger reflexive blinking of the eyes. **Eyebrows** consist of coarse, short hairs that overlie the margins of the skull bones above the eyes. They shield the eyes from overhead light and help to prevent perspiration from trickling down into the eyes.

Structure of the Eye

Refer to Figure 24.1 as you study this section.

The eye is a hollow ball, roughly spherical in shape. Its wall is composed of three distinct layers. Most of the outer layer consists of the fibrous **sclera** (label 1), the white of the eye. The extrinsic eye muscles that move the eyeball are attached to the sclera. The sclera provides support and protection for the interior structures of the eye. Anteriorly, the sclera is continuous with the **cornea,** the clear, transparent window through which light rays enter the eye. The greater curvature of the cornea bends light rays as they pass through it.

The middle layer consists of three parts. The **choroid** coat, a black layer, absorbs excess light that passes through the retina and contains the blood vessels that nourish the eye. Anteriorly, the **ciliary body** forms a ring of muscle that controls the shape of the **lens,** which is suspended from the ciliary body by the **suspensory ligaments.** The lens focuses light rays on the retina to enable clear vision. The shape of the lens is changed by contraction or relaxation of muscles in the ciliary body to adjust for near and distance

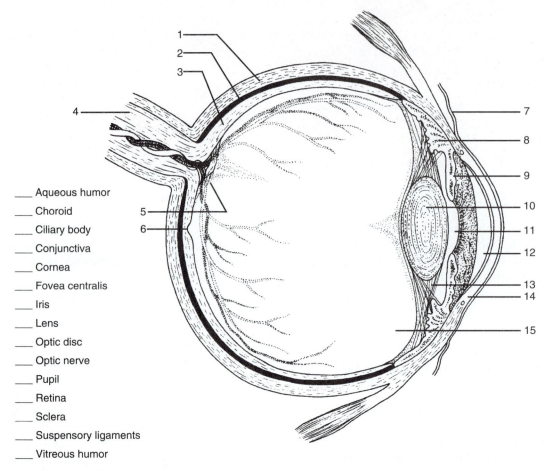

___ Aqueous humor

___ Choroid

___ Ciliary body

___ Conjunctiva

___ Cornea

___ Fovea centralis

___ Iris

___ Lens

___ Optic disc

___ Optic nerve

___ Pupil

___ Retina

___ Sclera

___ Suspensory ligaments

___ Vitreous humor

Figure 24.1 The eye.

vision. The **iris** controls the amount of light entering the eye by controlling the size of the **pupil,** the opening in the center of the iris.

The delicate **retina** composes the inner layer of the eyeball. It contains the **photoreceptors,** special cells that form impulses when stimulated by light. **Rods** are photoreceptor cells involved in black-and-white and dim light vision. **Cones** are photoreceptors for color and bright light vision. Impulses from the photoreceptors pass along neuron fibers in the retina that coalesce at the **optic disc** (label 5) to form the **optic nerve,** which carries the impulses to the brain. The optic disc is known as the blind spot because it has no receptors. A tiny depression just lateral to the optic disc, the **fovea centralis,** contains densely packed cones for sharp, direct vision.

The interior of the eye behind the lens is filled with **vitreous humor,** a transparent, jellylike substance that helps hold the retina in place and gives shape to the eye. A watery fluid, the **aqueous humor,** fills the space between the lens and cornea.

Assignment 1

Materials

Beef eye, preferably fresh
Colored pencils
Dissecting instruments and pan
Eye model
Gloves, protective and disposable

1. Label Figure 24.1, and color-code the sclera, cornea, ciliary body, iris, retina, and lens.
2. Locate the parts of the eye on an eye model.
3. ***Complete items 1a and 1b on Laboratory Report 24 that begins on page 323.***
4. Dissect a beef eye following these procedures:
   a. Examine the external surface of the eye. Locate the optic nerve, and trim away any remnants of the extrinsic eye muscles and conjunctiva. Is the curvature of the cornea greater than that of the rest of the eyeball? What is the shape of the pupil?

b. Use a sharp scalpel to make a small incision in the sclera about 0.5 cm from the edge of the cornea. See Figure 24.2. Holding the eye firmly but gently, insert scissors into this incision and cut through the sclera around the eyeball while holding the cornea upward. The fluid exuded when making this cut is aqueous humor.

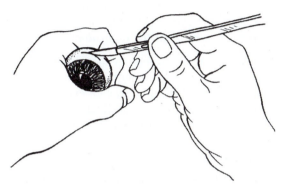

A Holding the eye as shown, use a sharp scalpel to make a small incision through the sclera about 0.5 cm from the edge of the cornea.

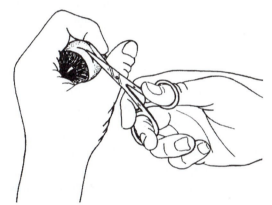

B Insert the pointed blade of a scissors into the incision and continue the incision through the wall of the sclera, completely around the eye.

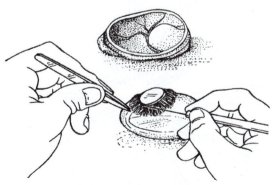

C After pouring the vitreous humor onto the dissecting pan, carefully separate the lens from the ciliary body with a dissecting needle.

Figure 24.2 Three steps in dissecting a beef eye.

c. Now, gently lift off the anterior part of the eye and place it on the dissecting pan with its inner surface upward. In preserved eyes, the lens often remains with the cornea, but in fresh eyes it usually remains with the vitreous humor.

d. Examine the anterior portion. Locate the ciliary body, a thickened black ring, and the iris. There may be a little aqueous humor remaining next to the inner surface of the cornea.

e. Observe the vitreous humor and the attached lens in the posterior part of the eye. Pour the vitreous humor onto the dissecting pan and note its jellylike consistency.

f. Use a dissecting needle to separate the periphery of the lens from the ciliary body on the surface of the vitreous humor. Hold the lens to your eye and look through it across the room. What is unusual about the image? Place it on this printed page. Does it magnify the print?

g. Look at the interior of the posterior portion of the eye to see the thin, beige retina that now is wrinkled since the vitreous humor has been removed. Note that the retina easily separates from the underlying choroid but is attached at the optic disc, the junction of the optic nerve with the retina.

h. Note the iridescent portion of the choroid that aids dim light vision by reflecting light that has passed through the retina back through the retina. This iridescence causes animals' eyes to "shine" at night, reflecting light back to a light source.

i. Dispose of the eye as directed by your instructor.

5. ***Complete item 1 on the laboratory report.***

Function of the Eye

Light waves reflecting from objects are bent as they pass through the cornea. They continue through the pupil and lens, which focuses the light rays on the retina and accommodates for near and distance vision as its shape is changed by muscles in the ciliary body. Photoreceptors in the retina form impulses that are transmitted via the optic nerve to the visual center in the brain, where they are interpreted as visual images.

 Assignment 2

Materials

Metersticks
Color-blindness test plates (Ishihara)
Astigmatism charts
Snellen eye charts

Perform the following visual tests to learn more about visual sensations. Work in pairs. If you wear corrective

Figure 24.3 Blind spot test.

lenses, perform the tests with and without them, and note any differences in your results.

Blind spot: No rods or cones are located at the junction of the retina and optic nerve. This site is known as the optic disc or blind spot. It can be located by using the following procedure:

1. Hold Figure 24.3 about 50 cm (20 in.) in front of your eyes.
2. Cover your left eye and focus with the right eye on the cross. You will be able to see the dot as well.
3. Slowly move the figure toward your eyes while focusing on the cross, until the dot disappears.
4. Have your partner measure and record the distance from your eye to the figure at the point where the dot disappears.
5. Test the left eye in a similar manner, but focus on the dot and watch for the cross to disappear.
6. ***Complete item 2a on the laboratory report.***

Near point: The shortest distance from your eye at which an object is in sharp focus is called the near point. The closer the distance, the more elastic the lens and the greater the eye's ability to accommodate for changes in distance. Elasticity of the lens is greatest in infants, and it gradually decreases with age. See Table 24.1. Accommodation is minimal after 60 yr of age, a condition called presbyopia. How does this relate to the common usage of bifocal lenses by older persons?

1. Hold this page in front of you at arm's length. Close one eye, focus on a word in this sentence, and slowly move the page toward your face until the image is blurred. Then move the page away until the image is sharp. Have your partner measure the distance between your eye and the page.
2. Test the other eye in the same manner.
3. ***Complete item 2b on the laboratory report.***

TABLE 24.1	AGE AND ACCOMMODATION	
	Near Point	
Age (years)	Inches	Centimeters
10	3.0	7.5
20	3.5	8.8
30	4.5	11.3
40	6.8	17.0
50	20.7	51.8
60	33.0	82.5

Astigmatism: This condition results from an unequal curvature of either the cornea or the lens, which prevents all light rays from being focused sharply on the retina.

1. Cover one eye and focus on the circle in the center of Figure 24.4. If the radiating lines appear equally dark and in sharp focus, no astigmatism exists. If astigmatism exists, record the number of the lines that appear lighter in color or blurred.
2. Test the other eye in the same manner.
3. ***Complete item 2c on the laboratory report.***

Acuity: Visual acuity refers to the ability to distinguish objects in accordance with a standardized scale. It may be measured using a Snellen eye chart. If you can read the letters that are designated to be read at 20 ft at a distance of 20 ft, you have 20/20 vision. If the smallest letters that can be read at 20 ft are those designated to be read at 30 ft, you have 20/30 vision.

1. Stand 20 ft from the Snellen eye chart on the wall, while your partner stands next to the chart and points out the lines to read.
2. Cover one eye and read the lines as requested. Record the rating of the smallest letters read correctly.
3. Test the other eye in the same manner.
4. ***Complete item 2d on the laboratory report.***

Color blindness: The sensation of color vision depends on the degree to which impulses are formed by the three types of cones (receptors for red, green, and blue light) in the retina. The most common type of color blindness is red–green color blindness, which is caused by a deficit in cones stimulated by either red or green light. People with such a deficit have difficulty

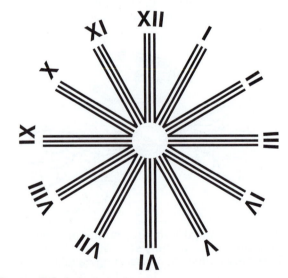

Figure 24.4 Astigmatism test.

distinguishing reds and greens. A totally color-blind person sees everything as shades of gray.

1. If you are not using Ishihara color-blindness test plates, record the normal responses in item 2e on the laboratory report before proceeding.
2. Your partner holds the color-blindness test plates about 30 in. from your eyes in good light and allows you 5 sec to view each plate and give your response. Your partner records your responses in item 2e on your laboratory report.
3. *Complete item 2e on the laboratory report.*

THE EAR

The ear contains not only receptors for sound stimuli but also receptors involved in maintaining equilibrium. For ease of study, the ear may be subdivided into the external ear, middle ear, and internal ear. Refer to Figure 24.5 as you study this section.

External Ear

The **external ear** includes the **auricle** or **pinna,** the flap of cartilage and skin commonly called the "ear," and the **auditory canal** that leads inward through the temporal bone to the **tympanic membrane,** or eardrum.

Middle Ear

The **middle ear,** a small cavity in the temporal bone, is connected to the pharynx by the **auditory tube.** It is filled with air that enters or leaves via the auditory tube, depending on the air pressure at each end of the tube. This enables the air pressure on each side of the tympanic membrane to be equalized. Three small bones, the **ear ossicles,** form a lever system that transmits sound vibrations from the eardrum to the inner ear. In sequence, they are (1) the **malleus** (hammer), which is attached to the ear drum; (2) the **incus** (anvil); and (3) the **stapes** (stirrup).

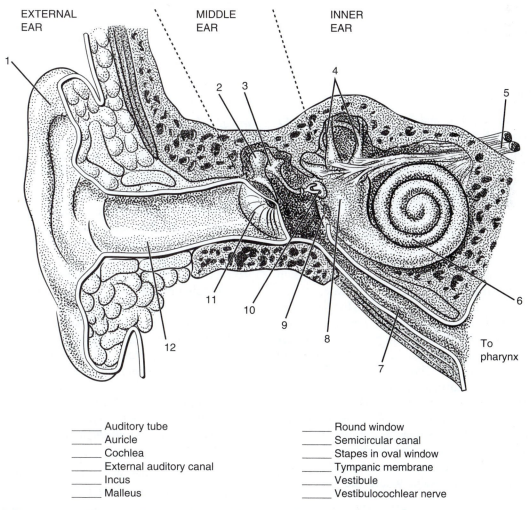

	Auditory tube		Round window
	Auricle		Semicircular canal
	Cochlea		Stapes in oval window
	External auditory canal		Tympanic membrane
	Incus		Vestibule
	Malleus		Vestibulocochlear nerve

Figure 24.5 Anatomy of the ear. The boundaries of the external, middle, and inner ear are roughly indicated by dashed lines. From Martini, A & P, 5e, Fig. 17–22, p. 558.

Inner Ear

The **inner ear** consists of a complex of interconnecting tubes and chambers that are embedded in the temporal bone and that are filled with fluid. It is subdivided into three major parts.

1. The **cochlea** is coiled like a snail shell and contains the receptors for sound stimuli.
2. The **vestibule** is the enlarged portion at the base of the cochlea. The stirrup is inserted into the **oval window** of the vestibule, and the **round window** is covered by a thin membrane. These two windows are involved in the transmission of sound stimuli. In addition, the vestibule contains receptors for static equilibrium that inform the brain of the position of the head.
3. The three **semicircular canals** contain receptors for dynamic equilibrium that inform the brain when the head is turned or when the entire body is rotated.

 ## Assignment 3

Materials

Colored pencils
Model of the ear

1. Label Figure 24.5, and color-code the auditory tube, cochlea, ear ossicles, semicircular canals, and vestibule.
2. Locate the structures shown in Figure 24.5 on the ear model or chart.
3. ***Complete item 3 on the laboratory report.***

Hearing and Balance

The hearing process begins when the auricle channels sound waves into the auditory canal. When sound waves strike the eardrum, it vibrates at the same frequency as the sound waves. This vibration causes a comparable movement of the ear ossicles (hammer, anvil, and stirrup), which, in turn, transfer the movements to the fluid in the vestibule and cochlea of the inner ear. The movement of the fluid is enabled by the thin membrane of the round window, which moves *out and in* in synchrony with the *in and out* movement of the stirrup in the oval window because the fluid cannot be compressed. The movement of the fluid in the cochlea stimulates the sound receptors, which send impulses to the auditory center in the brain, where they are interpreted as sound sensations.

Tone or pitch is determined by which receptors are stimulated and which part of the brain receives the impulses. Loudness is determined by the frequency of impulses formed and transmitted, which, in turn, depends on the intensity of the vibrations.

There are two kinds of hearing loss. **Conduction deafness** results from damage that prevents sound vibrations from reaching the inner ear. It is usually correctable by surgery or hearing aids. **Nerve deafness** is caused by damage to the sound receptors or neurons that transmit impulses to the brain. Nerve deafness usually results from exposure to loud sounds and is not correctable.

As noted earlier, receptors for static and dynamic balance are located in the inner ear, but they are not the only receptors involved in the maintenance of equilibrium. Pressure and touch receptors in the skin, stretch receptors in the muscles, and light receptors in the eyes are also involved. The brain constantly receives impulses from all of these receptors and subconsciously initiates any necessary corrective motor actions.

 ## Assignment 4

Materials

Cotton for ear plugs
Metersticks
Pocket watches, spring wound
Tuning forks

Watch-tick test: This is a simple test to detect hearing loss at a single sound frequency. It requires a quiet area. Work in teams of three students. One student is the subject, another moves the watch, and the third measures and records distances.

1. Have the subject sit in a chair and plug one ear with cotton. The subject is to look straight ahead and indicate by hand signals when the first and last ticks are heard.
2. Start with the watch about 3 ft laterally from the ear being tested and move it *slowly* toward the ear until the subject indicates the first tick is heard. Measure and record the distance from ear to watch.
3. Start with the watch close to the ear and *slowly* move it away from the ear until the last tick is heard. Measure and record the distance from ear to watch.
4. Calculate the average of the two measurements.
5. Test the hearing of the other ear in the same manner.
6. ***Complete item 4a on the laboratory report.***

Rinne test: A tuning fork is used in this test, which distinguishes between nerve and conduction deafness *when some hearing loss exists.* Work in pairs.

1. Have the subject sit in a chair and plug one ear with cotton. The subject is to indicate by hand signals when the sound is heard or not heard.

2. Strike the tuning fork against the heel of your hand to set it in motion. *Never strike it against a hard object.*

3. Hold the tuning fork 6–9 in. away from the ear being tested with the edge of the tuning fork toward the ear as shown in Figure 24.6A.

4. The sound will be heard initially by persons with normal hearing and those with minimal hearing loss. As the sound fades, a point will be reached at which it will no longer be heard. (The subject is to indicate this point by a hand signal.) When this occurs, place the end of the tuning fork against the temporal bone behind the ear. See Figure 24.6B. *When a slight hearing loss exists* and the sound reappears, some conduction deafness is present.

5. Persons with a severe hearing loss will not hear the sound in step 4 or will hear it only briefly. If the sound reappears when the end of the tuning fork is placed against the temporal bone, conduction deafness is evident. If it does not reappear, nerve deafness exists.

6. *Complete item 4b on the laboratory report.*

Static balance: The following test is a simple way to observe the interaction of receptors in the inner ear, muscles, skin, and eyes to maintain static equilibrium. Work in groups of two to four students.

1. Use a meterstick to draw a series of vertical lines on the chalkboard about 5 cm (2 in.) apart. Cover an area about 1 m wide. This will help you detect body movements.

2. Have the subject remove his or her shoes and stand in front of the lined area facing you.

3. With feet together and arms at the side, the subject is to try to stand perfectly still for 30 sec while you watch for any swaying motion. Record the degree of swaying motion as slight, moderate, or great.

4. Repeat the test with the subject's eyes closed. Record the degree of movement. Try it again to see if extending the arms laterally helps the subject to maintain balance. What happens when the subject stands on only one foot with the hands at the side and with the eyes closed?

5. *Complete item 4 on the laboratory report.*

SKIN RECEPTORS

Human skin contains receptors for touch, pain, pressure, hot, and cold stimuli. The following tests will enable you to detect certain characteristics of sensations involving some of these receptors. Work in pairs to perform these experiments.

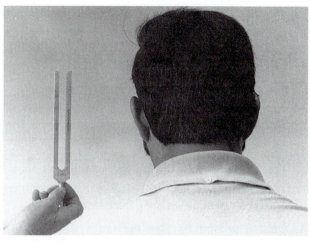

A

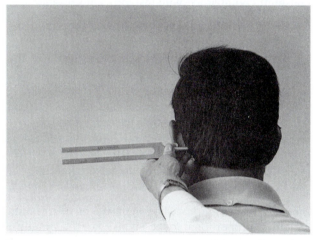

B

Figure 24.6 Rinne test.

Assignment 5

Materials

Beakers, 400 ml, three
Celsius thermometer
Clock or watch with second hand
Coins
Crushed ice
Dividers
Hot plate
Metric ruler

Distribution of touch receptors: For you to perceive two simultaneous stimuli as two touch sensations, the stimuli must be far enough apart to stimulate two touch receptors that are separated by at least one unstimulated touch receptor. This characteristic can be used to determine the density of touch receptors in the skin.

1. With the subject's eyes closed, touch his or her skin with one or two points of the dividers. The

subject reports the sensation as either "one" or "two." Start with the points of the dividers close together and gradually increase the distance between the points until the subject reports a two-point stimulus as a two-touch sensation about 75% of the time. Measure and record the distance between the tips of the dividers as the minimum distance evoking a two-point sensation.

2. Use the procedure in step 1 to determine the minimum distance giving a two-point sensation on the (1) inside of the forearm, (2) back of the neck, (3) palm of the hand, and (4) tip of the index finger.

3. *Complete item 5a on the laboratory report.*

Adaptation to stimuli: Your nervous system has the ability to ignore stimuli or impulses so you are not constantly bombarded with insignificant sensations.

1. Have the subject rest a forearm on the top of the table with the palm of the hand up.

2. With the subject's eyes closed, place a coin on the inner surface of the forearm. The subject is to indicate awareness of the presence of the coin and also the instant the sensation disappears. Record the time between these two events as the adaptation time.

3. Repeat the test using several coins stacked up to make a heavier object that increases the intensity of the stimulus. Determine the adaptation time.

4. *Complete items 5b and 5c on the laboratory report.*

Intensity of sensations: The intensity of a sensation usually is proportional to the intensity of the stimulus. This occurs because the receptors form more impulses when the strength of the stimulus is increased, and the brain interprets the arrival of more impulses as a greater sensation.

1. Fill three 400-ml beakers with ice water, tap water, and warm water (about 50°C), respectively.

2. Place your index finger in the water in each beaker in sequence, and note the sensations. Can you recognize the temperature differences?

3. Use some small beakers to prepare water with slight differences in temperature to determine the smallest difference that is detectable. Record this differential.

4. Now return to the original three beakers. Place your index finger in the ice water, and note the sensation. Then immerse your whole hand, and note the sensation. Repeat this procedure with the warm water.

5. *Complete items 5d–5f on the laboratory report.*

6. Use the three beakers of ice water, tap water, and warm water as in the previous experiment. Be sure the warm water has not cooled.

7. Place one hand in ice water and the other in warm water. Note the sensation. Does it change with time? Explain.

8. After 2 min place both hands in the beaker of tap water. Note the sensation. Does it change with time?

9. *Complete item 5 on the laboratory report.*

TASTE

The taste of food not only increases its appeal but also increases the flow of saliva. In fact, just the thought of delicious food will stimulate salivation. Try it and see. What we usually refer to as taste is a combination of both taste and smell. This is why food "loses its taste" when you have a bad head cold that prevents airborne molecules of food from reaching your odor receptors in the roof of the nasal cavity.

Taste buds, receptors for taste, are located on the sides of very small projections, **papillae,** on the upper surface of the tongue. There are only four basic tastes—sweet, sour, bitter, and salt. Early in the twentieth century, the concept of taste zones on the tongue, as shown in Figure 24.7, was introduced.

As shown in Figure 24.8, the sensory **taste cells** of a taste bud are embedded in epithelial tissue with only their taste hairs protruding through a tiny taste pore.

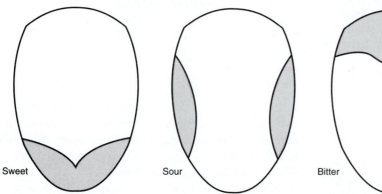
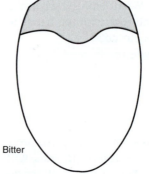
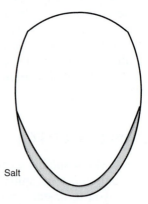

Sweet   Sour   Bitter   Salt

Figure 24.7 Proposed taste zones of the tongue.

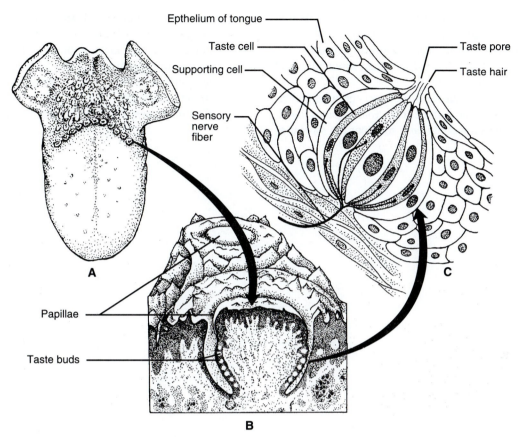

Figure 24.8 **A.** Papillae containing taste buds are located on the upper surface of the tongue. **B.** Taste buds are located along the outer margins of papillae on the tongue. **C.** A taste bud consists of a bulb-shaped cluster of taste cells and supporting cells embedded in epithelial tissue. The hairlike cilia of taste cells protrude slightly through a taste pore.

When the substances capable of activating the taste cells contact the taste hairs, the taste cells are stimulated, forming impulses that are carried by neurons to the brain.

Work in pairs to perform the following experiments, and alternate the roles of subject and experimenter.

 Assignment 6

Materials

Cotton swabs (Q-tips®), 6
Facial tissues, 6
Paper cups, 2
Sucrose granules
Dropping bottles of:
   10% sodium chloride (salt)
   10% sucrose (sweet)
   10% vinegar (sour)

1. In this section, you will perform tests to see if the distribution of taste zones as shown in Figure 24.7 is correct.
  a. Obtain 2 paper cups, several facial tissues, and 6 cotton swabs.
  b. Have the subject blot the upper surface of the tongue with a facial tissue and stick it out.
  c. Place several drops of sucrose solution on a cotton swab. Swab the tip of the tongue. Can the subject detect a sweet taste without withdrawing the tongue into the mouth? Is there any difference in the taste sensation after withdrawing the tongue?
  d. Have the subject rinse his or her mouth with water and blot the tongue and stick it out as before. Now swab an area where taste buds for sweet should be absent. Can the subject detect a sweet taste without withdrawing the tongue? Is there any difference in the taste sensation after withdrawing the tongue?

e. Use the procedure above to verify the distribution of taste receptors for sour and salt.

f. ***Complete item 6a on the laboratory report.***

2. Determine if taste buds can detect molecules that are not in solution.

a. Obtain several facial tissues and about half a teaspoon of sugar granules on a folded paper towel.

b. Have the subject blot and stick out the tongue as before.

c. Sprinkle a number of sugar granules on the "sweet zone" of the tongue. Can the subject detect a sweet taste without withdrawing the tongue into the mouth? After 5 sec, have the subject withdraw the tongue. Is there any difference in the sensation?

d. ***Complete item 6 on the laboratory report.***

SENSORY PERCEPTION IN HUMANS

Student _____

Lab Instructor _____

1. THE EYE

a. List the labels for Figure 24.1.

1. _____   6. _____   11. _____
2. _____   7. _____   12. _____
3. _____   8. _____   13. _____
4. _____   9. _____   14. _____
5. _____   10. _____   15. _____

b. Write the term that matches each meaning.

1. Outer white fibrous layer _____

2. Fills space in front of lens _____

3. Controls amount of light entering eye _____

4. Transparent window in front of eye _____

5. Layer with photoreceptors _____

6. Controls shape of lens _____

7. Opening in center of iris _____

8. Fills space behind lens _____

9. Site of sharp, direct vision _____

10. Receptors for color vision _____

11. Carries impulses from eye to brain _____

12. Receptors for black-and-white vision _____

13. Membrane covering front of eye _____

14. Focuses light on retina _____

15. Junction of retina and optic nerve _____

c. *Circle* the terms that describe the sclera.

   soft   firm   weak   strong   tough   rigid   flexible

d. *Circle* the terms that describe the vitreous humor.

   watery   jellylike   transparent   murky

e. When looking through the lens across the room, what is different about the image? _____

_____

   Does the lens magnify or reduce the print? _____

f. Is the retina thick or thin? _____

   Is the retina firmly attached to the choroid layer? _____

   What holds the retina in place in an intact eye? _____

2. VISUAL TESTS

a. Record the distance from your eye to Figure 24.3 at which the blind spot is detected.

Left eye _____ Right eye _____

Explain any difference in the distances. _____

_____

b. Record the near-point distance for your eyes.

Left eye: without glasses _____ with glasses _____

Right eye: without glasses _____ with glasses _____

Explain any differences in the distances. _____

_____

How do your near-point values compare with your age? See Table 24.1.

_____

c. Is astigmatism present?

Left eye: without glasses _____ with glasses _____

Right eye: without glasses _____ with glasses _____

d. Record the acuity for each eye.

Left eye: without glasses 20/ _____ with glasses 20/ _____

Right eye: without glasses 20/ _____ with glasses 20/ _____

e. Record your responses to the color-blindness test in the table.

RESPONSES TO THE ISHIHARA COLOR-BLINDNESS TEST PLATES

Plate No.	Subject	Responses*		
		Normal	Red–Green Color-Blind	Totally Color-Blind
1	_____	12	12	12
2	_____	8	3	0
3	_____	5	2	0
4	_____	29	70	0
5	_____	74	21	0
6	_____	7	0	0
7	_____	45	0	0
8	_____	2	0	0
9	_____	0	2	0
10	_____	16	0	0
11	_____	Traceable	0	0
			Red Cones Absent	Green Cones Absent	
12	_____	35	5	3	0
13	_____	96	6	9	0
14	_____	Can trace two lines	Purple	Red	0

*O means the subject sees no pattern in the test plate.

Are you color-blind? _____ If so, describe your type of color blindness. _____

_____

3. THE EAR

a. List the labels for Figure 24.5.

1. _____   6. _____   11. _____

2. _____   7. _____   12. _____

3. _____   8. _____

4. _____   9. _____

5. _____   10. _____

b. Write the term that matches each meaning.

1. Contains sound receptors   _____

2. Bone in which inner ear is embedded   _____

3. Contains receptors for static equilibrium   _____

4. Connects middle ear to pharynx   _____

5. Interprets impulses as sound   _____

6. Contain receptors for dynamic equilibrium   _____

7. Conduct vibrations across middle ear   _____

8. Membrane separating outer and middle ear   _____

4. HEARING AND BALANCE

a. Record the average distance for the watch-tick test.

Left ear _____   Right ear _____   Class average _____

b. State your conclusion from the Rinne test.

Left ear _____

Right ear _____

c. Indicate your degree of wavering in the static equilibrium test. Use the terms *slight, moderate,* and *great* to indicate variations.

Standing on both feet with arms at side

Eyes open _____   Eyes closed _____

Standing on both feet with arms extended

Eyes open _____   Eyes closed _____

d. How important are visual clues in maintaining balance?

_____

5. SKIN RECEPTORS

a. Indicate the minimal distances that provided a two-point sensation.

Forearm _____   Back of neck _____

Palm of hand _____   Fingertip _____

What is the value of this distribution? _____

_____

_____

b. Indicate the adaptation time when using:

a single coin _____   several coins _____

What is the value of adaptation to stimuli? _____

_____

_____

c. What is the physiological basis for a boy looking for his hat only to discover that he is wearing it? _____

_____

d. Indicate the smallest temperature differential that you could detect by immersing your finger in water. _____ °C

Indicate the range and average of the class. Range _____ °C　Average _____ °C

e. Was the sensation stronger or weaker when your entire hand was immersed in the water? _____

Explain the sensation. _____

_____

_____

f. Did the sensation change with time when your hands were immersed in:

ice water? _____　warm water? _____

Explain. _____

_____

_____

g. Describe the sensation when both hands were placed in tap water after immersion in ice and warm water, respectively. _____

_____

How do you account for this? _____

_____

_____

h. What is the physiological basis for a girl believing that the temperature of water in a swimming pool is "fine" after testing it with her foot but complaining that it is "freezing cold" after jumping in? _____

_____

6. TASTE

a. For each of the tastes tested, do the locations of your taste receptors agree with those in Figure 24.7?

Sweet _____　Sour _____　Salty _____

Explain any differences. _____

_____

_____

b. Are taste receptors sensitive to dry sugar granules? _____

Explain the role of saliva in taste sensations. _____

_____

c. Indicate the role of the following in sensory perception:

Receptors _____

_____

Brain _____

_____

Nerves _____

_____

CHEMICAL CONTROL IN ANIMALS

OBJECTIVES

After completing the laboratory session, you should be able to:
1. Describe the effect of thyroxine on body weight and metabolic rate as determined in mice.
2. Describe the effect of acetylcholine and epinephrine on the heart rate of a frog.
3. Define all terms in bold print.

Chemical control is widespread among organisms and is responsible for much of the orderly fashion in which living processes are conducted. Much of the chemical control is carried out by chemical messengers called **hormones.**

In animals, chemical control occurs in three distinct ways. **Neurotransmitters** are chemicals released by neuron axons into synaptic junctions that either stimulate or inhibit impulse formation and transmission in *adjacent* neurons. Thus, neural activity depends on these transmitter substances, which are secreted in exceedingly small amounts and exert their effect only at the synapse. Some neurons secrete **neurosecretory hormones** that, like transmitter substances, affect changes in the membrane potential of target cells, but travel greater distances via either diffusion or the bloodstream to affect *nonadjacent* cells. This mode of action occurs in simple animals, such as cnidarians, as well as in higher forms. The so-called **true hormones** occur in mollusks and higher animals and are secreted by **endocrine cells,** which may function independently or may be organized into **endocrine glands.** Hormones secreted by endocrine cells diffuse into body fluids, from which they are picked up by specific *nonadjacent* target cells. These hormones seem to exert their effect by stimulating or inhibiting enzyme activity.

 Assignment 1

Complete item 1 on Laboratory Report 25 that begins on page 331.

THYROXINE AND METABOLIC RATE

Thyroxine is a hormone secreted by the **thyroid gland,** and it is distributed to body cells by the blood. Its primary function is to regulate the rate of metabolism. **Metabolism** is the sum of the chemical and physical reactions of life. **Metabolic rate** may be measured as oxygen (O_2) consumption, carbon dioxide (CO_2) production, or heat (calories) production.

To determine the effect of thyroxine on metabolic rate, you will measure the rate of oxygen consumption in mice (or rats) using a respirometer similar to the one in Figure 25.1. The respirometer consists of two stoppered jars connected by a U-shaped tube (manometer) that is partially filled with fluid. One jar contains the mouse; the other serves as a pressure control. A pressure change in one jar will cause movement of the fluid toward the jar with less pressure. If CO_2 is removed by an absorbent, O_2 consumption can be determined by measuring the movement of fluid in the 5-ml volumetric pipette composing one arm of the U-tube or, if O_2 is available, by injecting, at selected intervals, measured amounts of O_2 into the system with a syringe until the fluid level in the U-tube is returned to the starting point.

Three young male white mice were weighed and placed in separate cages 10 days ago. Food was present at all times, and the amount of food consumed was carefully measured. Mouse A was given drinking water containing 0.02% thyroxine, mouse B was given drinking water containing 0.25% potassium perchlorate ($KCIO_4$), and mouse C was given plain water. Thus, mouse A is experimentally **hyperthyroid,** mouse B is experimentally **hypothyroid,** and mouse C is normal.

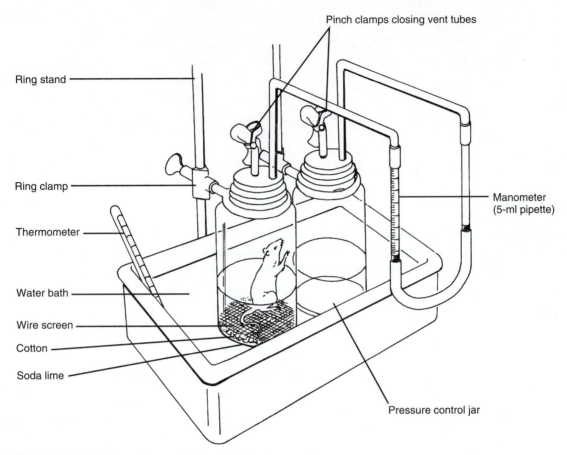

Figure 25.1 Respirometer setup.

Your instructor will provide the following data for each mouse:

1. The weight (in grams) at the start of the 10-day diet.
2. The quantity (in grams) of food consumed.

This experiment is best performed by groups of two to four students. If time is limited, your instructor may assign a different mouse to each student group, with all groups sharing data.

 Assignment 2

Materials

Balance for weighing mice
Respirometer (Figure 25.1)
Water baths
Celsius thermometer
Colored water for manometer
Cotton, absorbent
Wire screen
Potassium perchlorate, 0.25%
Soda lime
Thyroxine, 0.02%

1. Obtain a respirometer or construct one as shown in Figure 25.1. Place soda lime in the bottom of each jar to a depth of about 2 cm, cover it with a layer of absorbent cotton, and add the wire screen.
2. Add colored water to the manometer, if necessary, so that it extends about one-third of the way up each arm.
3. Place the jars in a water bath at room temperature. During the experiment, keep the temperature of the water constant by adding warm or cold water. Why is the maintenance of a constant temperature important?
4. From your instructor, obtain the starting weights and food consumption for each mouse. Weigh each mouse prior to placement in the respirometer. **_Record these data in item 2a on the laboratory report._**
5. Pick up a mouse by its tail and place it in the experimental chamber of the respirometer, and, *with the vent tube open,* loosely replace the stoppers. Allow 5 min for temperature equilibration.
6. *With the vent tubes open,* insert the stoppers. Close the vent tubes with pinch clamps, and record your first reading. Take the second reading after 3 min. Open the vent tubes. **_Record your readings in item 2d on the laboratory report._**

7. Following the procedure in step 6, measure the O_2 consumption in five replicates, each with a duration of 3 min. Open the vent tubes for 3 min between replicates.
8. Discard the lowest and highest readings, and calculate the average O_2 consumption.
9. ***Complete item 2 on the laboratory report.***

CHEMICAL CONTROL AND HEART RATE

The heartbeat in vertebrates is controlled by factors both internal and external to the heart. Internal control is by the rhythmic depolarization of the sinoatrial node (SA node), which causes the rhythmic contractions of the heart. The rate of the contractions is mostly controlled externally by the vagus and accelerator nerves leading to the SA node. Impulses passing down the **vagus nerve** cause the secretion of **acetylcholine** at the junction of the nerve and heart, and impulses in the **accelerator nerve** cause a secretion of **norepinephrine** by the axon tips. These neurotransmitters affect the rate and force of the heart contractions.

The **adrenal glands** secrete the hormone **epinephrine** (adrenalin), a chemical similar to norepinephrine. Epinephrine is carried throughout the body by the blood. It also affects the rate and force of the heartbeat and diverts blood from the digestive tract into the skeletal muscles and nervous system. Epinephrine is the hormone that prepares animals to respond to an emergency or stressful situation. The "butterflies" that you may have experienced before a game or speech are side effects of epinephrine stimulation.

Observe the effect of epinephrine and acetylcholine on the heart rate of a double-pithed frog. A double-pithed frog has had its brain and spinal cord destroyed mechanically so no impulses are sent or received by the central nervous system. The heart continues to beat for a while, however. Not all organs and tissues die at the instant the brain is destroyed or at the same rate. ***Record your data in item 4a on the laboratory report.***

 Assignment 3

Materials

Dissecting instruments and pins
Dissecting pan
Gloves, protective and disposable
Stopwatch

Hypodermic needles, 22 gauge
Hypodermic syringes, 0.5 or 1.0 ml
Acetylcholine, 1:10,000
Epinephrine, 1:10,000
Frog Ringer's solution in dropping bottles
Frog, double-pithed

1. Secure a double-pithed grass frog on its back to the bottom of a dissecting pan by placing pins through the feet.
2. Use scissors to make an incision through the body wall to one side of the midline to avoid the large abdominal vein. Extend this incision anteriorly from the lower abdominal area to the sternum (breastbone). Make lateral cuts just posterior to the sternum and pin back the flaps.
3. Lift the sternum with forceps and carefully cut through its length. Avoid cutting the underlying blood vessels. Pin back the "chest" wall laterally to expose the heart. The liver is also now exposed.
4. Keep the heart and liver wet with frog Ringer's solution throughout the experiment.
5. Record the heart rate (beats per minute). Count the beats for 15 sec and multiply by 4.
6. Fill two hypodermic syringes with 1:10,000 epinephrine and 1:10,000 acetylcholine, respectively, as directed by your instructor.
7. Slowly inject 0.05 ml of 1:10,000 epinephrine into the liver with the hypodermic syringe. To insert the needle easily, "firm up" the liver by holding it between thumb and forefinger. *Exactly 30 sec after the injection, record the heart rate.* The epinephrine is carried to the heart via the hepatic vein and posterior vena cava.
8. Wait 3 min and record the heart rate again.
9. In the same manner as before, slowly inject 0.05 ml of acetylcholine into the liver. Determine the time required for acetylcholine to take effect. Exactly 30 sec after the injection, record the heart rate. Wait 3 min and record the heart rate again.
10. Slowly inject 0.05 ml of 1:10,000 epinephrine into the liver. Determine how long it takes for an effect to be noticed. Again, record the heart rate 30 sec after the injection and 3 min later.
11. Slowly inject 0.25 ml of 1:10,000 epinephrine. How quickly does the effect take place? Record the heart rate after 30 sec and 3 min later.
12. Slowly inject 0.25 ml of 1:10,000 acetylcholine. How quickly does the effect take place? Record the heart rate after 30 sec and 3 min later.
13. ***Complete items 3 and 4 on the laboratory report.***

CHEMICAL CONTROL IN ANIMALS

Student _____

Lab Instructor _____

1. INTRODUCTION

Write the term that matches the phrase.

a. Chemical messengers _____

b. Chemicals released into synaptic junctions _____

c. Chemical messengers formed by endocrine glands _____

d. Chemicals formed by neurons and affecting nonadjacent cells _____

e. Chemicals formed by neurons and affecting adjacent cells _____

f. Aggregations of endocrine cells in mollusks and higher animals _____

2. THYROXINE AND METABOLIC RATE

a. Record the data regarding food consumed and weight gained in the following table.

FOOD CONSUMPTION AND WEIGHT GAINED (GRAMS) IN MICE

Mouse	Starting Weight	Weight Today	Weight Gained	Food Consumed	Weight Gained Food Consumed
A					
B					
C					

b. Which mouse used the food most efficiently? _____

c. Summarize the effect of thyroxine on weight gain in young mice. _____

_____

_____

d. Record the oxygen consumption (milliliters) in the following table.

OXYGEN CONSUMPTION (MILLILITERS) IN MICE

Mouse	Replicates (3 min each)					Average ml O$_2$/min	Average ml O$_2$/g/hr
	1	2	3	4	5		
A							
B							
C							

e. Summarize the effect of thyroxine on the metabolic rate of young mice. _____

_____

_____

_____

3. CHEMICAL CONTROL OF HEART RATE

a. Record the data from the control of heart rate experiment in the following table.

HEART RATE OF THE FROG

Beats per Minute	Condition
_____	Before injection
_____	30 sec after injection of 0.05 ml epinephrine
_____	3 min after injection of 0.05 ml epinephrine
_____	30 sec after injection of 0.05 ml acetylcholine
_____	3 min after injection of 0.05 ml acetylcholine
_____	30 sec after injection of 0.25 ml epinephrine
_____	3 min after injection of 0.25 ml epinephrine
_____	30 sec after injection of 0.25 ml acetylcholine
_____	3 min after injection of 0.25 ml acetylcholine

b. Indicate how long it took for the effect of the chemicals to be recognized after injection.

0.05 ml of epinephrine _____   0.05 ml of acetylcholine _____

Did the effect last 3 min or longer for the injection of:

0.25 ml of epinephrine? _____   0.25 ml of acetylcholine? _____

c. Summarize the effect of the chemicals on the heart rate.

Epinephrine _____

Acetylcholine _____

d. What is the advantage of epinephrine secretion by the adrenal gland during stress? _____

_____

_____

_____

_____

4. REVIEW

Matching

1. Thryoxine   2. Epinephrine   3. Acetylcholine   4. Norepinephrine

_____ Neurotransmitter   _____ Secreted by an endocrine gland

_____ True hormone   _____ Acts on nonadjacent cells

_____ Acts on adjacent cells

THE SKELETAL SYSTEM

OBJECTIVES

After completing the laboratory session, you should be able to:
1. Identify and compare the skeletons of representative animals.
2. Identify the major bones and types of articulations in the human skeleton.
3. Distinguish the sex of human pelvic girdles.
4. Identify the components of a split long bone.
5. Identify and state the function of the components of compact bone when viewed microscopically.
6. Identify when viewed microscopically and state the function of (a) hyaline cartilage, (b) fibrocartilage, and (c) dense fibrous connective tissue.
7. Describe the skeletal adaptations of representative vertebrates for their mode of locomotion.
8. Define all terms in bold print.

Movement in animals requires not only contraction and relaxation of muscles but also a skeleton against which the contraction force can be applied. Three basic types of skeletons occur among animals: hydrostatic skeletons, exoskeletons, and endoskeletons.

A **hydrostatic skeleton** consists of internal body fluids within a limited space. Body fluids resist compression and provide a medium against which muscles can contract. It occurs in many soft-bodied animals, such as cnidarians and annelids. For example, each segment of an earthworm contains a coelomic space filled with fluid. Contraction and relaxation of the muscles in each fluid-filled segment, one after the other, enables the earthworm to move forward.

An **exoskeleton** is a rigid structure formed external to the body and attached to the body surface. Exoskeletons occur in mollusks and arthropods. The exoskeleton of mollusks is usually a hard shell of calcium carbonate that is secreted by the mantle. Mollusk exoskeletons exhibit many variations, but basically the exoskeleton may be a single shell, as in snails; two hinged halves of a shell, as in clams; or eight overlapping plates, as in chitons. The shell enlarges as a mollusk grows, and it is never shed.

The exoskeleton of arthropods is much more complex. It is secreted by the underlying epidermis and closely envelops the entire body like a coat of armor, as previously observed in a crayfish. It is formed of a substance called *chitin*. The exoskeleton is soft and pliable at joints, allowing freedom of movement, but it is hardened by an accumulation of mineral salts over the rest of the body, providing protection, support, and sites of muscle attachment. Because an arthropod's body is tightly encased in its exoskeleton, the exoskeleton must be shed periodically, a process called *molting,* to allow growth. Then, a new and larger exoskeleton is secreted and hardened.

An **endoskeleton** is formed of rigid components within an animal's body. Endoskeletons occur only in echinoderms and vertebrates. The endoskeleton in echinoderms is relatively simple. It consists of calcareous plates and spines located in the body wall just beneath a thin epidermis. The skeleton enlarges as the echinoderm grows.

Among vertebrates, an endoskeleton formed of **cartilage** is found in lampreys, sharks, and rays, but most vertebrates have endoskeletons made of **bone** and associated cartilages. Bones are bound together at joints (articulations) by **ligaments** composed of tough, fibrous connective tissue. The skeleton grows until the vertebrate attains adulthood. A bony skeleton provides (1) protection for vital organs, (2) support for the body, (3) sites for muscle attachment, (4) a storage area for calcium salts, and (5) formation of blood cells.

Assignment 1

Materials

Endoskeletons of echinoderms and vertebrate classes
Exoskeletons of sponges, corals, mollusks, and arthropods
Crayfish, preserved
Dissecting instruments
Dissecting pan

1. Examine the exoskeletons of representative invertebrates. Note the general structure and function provided. In bivalve mollusks, locate the scars at the point of attachment of the adductor muscles used to close the shells. Do molluscan shells grow at the outer edge?
2. Examine the exoskeleton of a preserved crayfish. Note the flexibility at the joints and the hardness elsewhere. Make a transverse cut between the segments of the abdomen. Examine the cut surface and note how tightly the exoskeleton envelops the body. Can you locate the epidermis just inside the exoskeleton? The epidermis secretes the exoskeleton.
3. Cut off a cheliped and use a sharp scalpel or a single-edged razor blade to make lengthwise cuts through the exoskeleton extending across one or two joints. Spread the exoskeleton and observe the way muscles are attached to the interior of the skeleton.
4. Examine the demonstration endoskeletons. Note their general differences and similarities, especially among the vertebrate classes.
5. ***Complete item 1 on Laboratory Report 26 that begins on page 341.***

MACROSCOPIC BONE STRUCTURE

Figure 26.1 depicts the structure of a typical long bone, a human humerus. A long bone is characterized by a shaft of bone, the **diaphysis,** which extends between two enlarged portions forming the ends of the bone, the **epiphyses.** The articular surface of each epiphysis is covered by an **articular cartilage** (hyaline

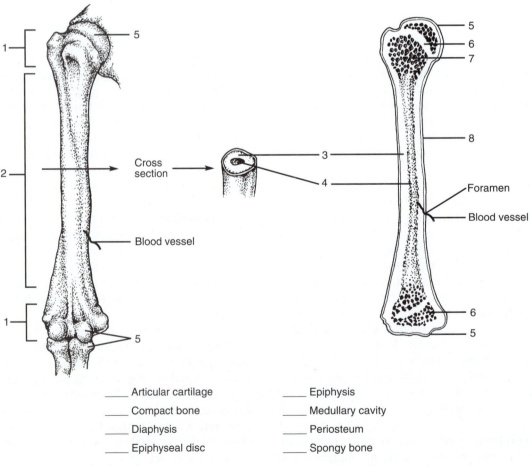

_____ Articular cartilage   _____ Epiphysis
_____ Compact bone   _____ Medullary cavity
_____ Diaphysis   _____ Periosteum
_____ Epiphyseal disc   _____ Spongy bone

Figure 26.1 Structure of a long bone.

cartilage), which reduces friction in the joints and protects the ends of the bone. The rest of the bone is covered by the **periosteum,** a tough, tightly adhering membrane containing tiny blood vessels that penetrate into the bone. Bone deposition by the periosteum contributes to the growth in diameter of the bone. Larger blood vessels and nerves enter the bone through a channel called a **foramen.**

A longitudinal section of the bone reveals that the epiphyses consist of **cancellous** (spongy) **bone** covered by a thin layer of **compact bone,** whereas the diaphysis is formed of heavy, compact bone. **Red marrow** fills the spaces in the spongy bone. It forms blood cells. The **medullary cavity** is lined by a fibrous membrane, the **endosteum,** and filled with the fatty **yellow marrow.** In immature bones, an **epiphyseal disc** of cartilage is located between the diaphysis and each epiphysis; this is the site of linear growth. A mature bone lacks this cartilage because it has been replaced by bone, and only an **epiphyseal line** of fusion remains.

Assignment 2

Materials

Colored pencils
Beef femur or tibia, fresh and split
Human femur, split
Dissecting instruments
Dissecting tray

1. Label Figure 26.1 and color-code the articular cartilages, compact bone, epiphyseal discs, and yellow marrow.
2. Examine a fresh beef bone that has been split. Locate the parts labeled in Figure 26.1. Feel the surface of the articular cartilage. How would you describe it? Is the bone supplied with blood? Locate the stubs of ligaments near the joint. Are they elastic, pliable, and easy to cut?
3. ***Complete item 2 on the laboratory report.***

MICROSCOPIC STUDY

Bones, ligaments, articular cartilages, and intervertebral discs are composed of **connective tissues.** Such tissues are characterized by an abundant intercellular material called **matrix** and relatively few scattered cells.

Bone

As bone tissue forms, **osteocytes** (bone cells) deposit bone matrix (calcium salts) around themselves and become trapped in tiny spaces called **lacunae.** In compact bone, osteocytes in lacunae are arranged in concentric circles around **central (Haversian) canals** containing blood vessels and nerves. Concentric layers of bone matrix, the **lamellae,** are deposited by osteocytes between the rings of lacunae. The central canal and concentric rings of lacunae and lamellae form an **osteon.** Many osteons are fused together by interstitial lamellae in compact bone. Tiny channels between the lacunae, the **canaliculi,** provide passageways for materials to move to and from the bone cells. See Figure 26.2.

Hyaline Cartilage

Like bone cells, **chondrocytes** (cartilage cells) become trapped in lacunae as they deposit the cartilage matrix. Articular cartilages are formed of **hyaline cartilage,** which is characterized by the absence of fibers and a clear, homogeneous matrix. This matrix gives a smooth, glassy appearance to articular cartilages. See Figure 26.3.

Fibrocartilage

The intervertebral discs are formed of **fibrocartilage,** which contains tightly packed collagenous

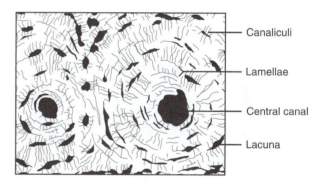

Figure 26.2 Compact bone, x.s.

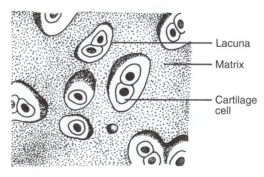

Figure 26.3 Hyaline cartilage.

fibers in the matrix. These nonelastic fibers account for the toughness and resilience of fibrocartilage. See Figure 26.4

Dense Fibrous Connective Tissue

Ligaments, which join bones to bones, and tendons, which attach muscles to bones, are formed of **dense fibrous connective tissue.** It is characterized by rows of **fibroblasts** (fiber-producing cells) between layers of collagenous fibers. The abundant fibers are responsible for the great strength, flexibility, and nonelastic nature of dense fibrous connective tissue. See Figure 26.5

 Assignment 3

Materials

Compound microscope
Prepared slides of:
    compact bone, x.s., ground section

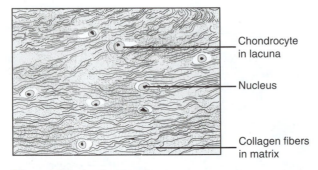

- Chondrocyte in lacuna
- Nucleus
- Collagen fibers in matrix

Figure 26.4 Fibrocartilage.

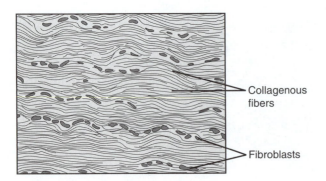

- Collagenous fibers
- Fibroblasts

Figure 26.5 Dense fibrous connective tissue.

hyaline cartilage
fibrocartilage
dense fibrous connective tissue

1. Examine prepared slides of compact bone, hyaline cartilage, fibrocartilage, and dense fibrous connective tissue. Locate the structures shown in Figures 26.2 through 26.5.
2. ***Complete item 3 on the laboratory report.***

THE HUMAN SKELETON

The skeleton consists of two major subdivisions. The **axial skeleton** is composed of the skull, vertebral column, ribs, and sternum. The **appendicular skeleton** consists of the bones of the upper extremities and the pectoral girdle and the lower extremities and the pelvic girdle. See Figure 26.6.

Axial Skeleton

The **skull** is composed of the **cranium** (8 fused bones encasing the brain), 13 fused **facial bones,** and the movable **mandible** (lower jaw).

The **vertebral column** consists of vertebrae separated by **intervertebral discs** composed of fibrocartilage. Vertebrae are subdivided as follows:

> **Cervical vertebrae:** 7 vertebrae of the neck
> **Thoracic vertebrae:** 12 vertebrae of the thorax, to which ribs are attached
> **Lumbar vertebrae:** 5 large vertebrae of the lower back
> **Sacrum:** bone formed of 5 fused vertebrae
> **Coccyx:** tailbone formed of 3–5 fused, rudimentary vertebrae

There are 12 pairs of **ribs.** The first 10 pairs are joined to the **sternum** (breastbone) by costal cartilages to form the thoracic cage. The last 2 pairs are short and unattached anteriorly. They are called *floating ribs.*

Appendicular Skeleton

The **pectoral girdle** supports the upper extremities. It consists of a **clavicle** (collarbone) and **scapula** (shoulder blade) on each side of the body. The clavicle is attached to the sternum on one end and the scapula on the other. The scalpula is supported by muscles that allow mobility for the shoulder.

The **humerus** (upper arm bone) articulates with the scapula at the shoulder and with the **ulna** and **radius** at the elbow. The wrist is composed of eight **carpal bones** that lie between the (1) ulna and radius and (2) **metacarpals,** the bones of the hand. The **phalanges** are the bones of the fingers and thumb.

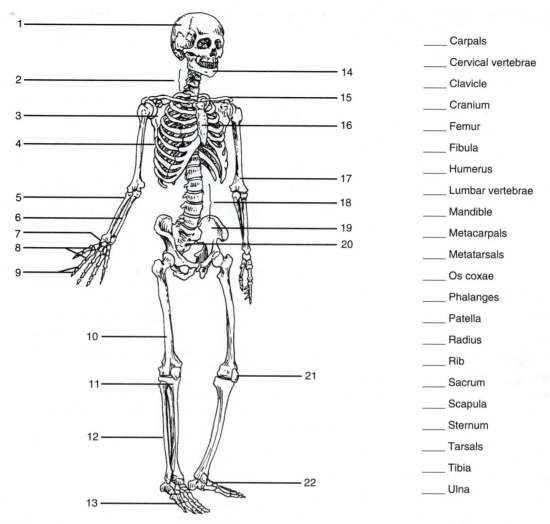

Figure 26.6 Human skeleton.

____ Carpals

____ Cervical vertebrae

____ Clavicle

____ Cranium

____ Femur

____ Fibula

____ Humerus

____ Lumbar vertebrae

____ Mandible

____ Metacarpals

____ Metatarsals

____ Os coxae

____ Phalanges

____ Patella

____ Radius

____ Rib

____ Sacrum

____ Scapula

____ Sternum

____ Tarsals

____ Tibia

____ Ulna

The **pelvic girdle** consists of two **ossa coxae** (singular, os coxae), the hipbones, that join together anteriorly at the pubic symphysis and are fused posteriorly to the sacrum. This provides a sturdy support for the lower extremities. Each os coxae consists of three fused bones: **ilium, ischium,** and **pubis.**

The **femur** (thighbone) articulates with the os coxae at the hip and with the **tibia** (shinbone) at the knee. The **patella** (kneecap) is embedded in the tendon anterior to the knee joint. The smaller bone of the lower leg is the **fibula.** Both tibia and fibula articulate with the **tarsal bones,** which form the ankle and posterior part of the foot. The anterior foot bones are five **metatarsals,** and the toe bones are the **phalanges.**

Articulations

Bones of the skeleton are joined to each other in ways that make the articulations (joints) either movable or immovable. Bones forming movable articulations are bound together at joints by ligaments. Articulations are categorized according to the degree of movement that is possible.

1. **Immovable joints** are rigid, such as those that occur between the skull bones.
2. **Slightly movable joints** allow a little movement, such as those between vertebrae at the pubic symphysis.
3. **Freely movable joints** are the most common and allow the broad range of movement noted in the arms and legs. There are several types:
   a. **Hinge joints** allow movement in one direction only.
   b. **Ball-and-socket joints** allow angular movement in all directions plus rotation.
   c. **Gliding joints** occur where bones slide over each other.
   d. **Pivot joints** allow rotation around only one axis.

TABLE 26.1	COMPARISON OF THE MALE AND FEMALE PELVIS	
Characteristic	Male	Female
General structure	Not tilted forward, narrower and longer, heavier bones	Tilted forward, broader and shorter, lighter bones
Acetabula	Larger and closer together	Smaller and farther apart
Pubic angle	Less than 90°	Greater than 90°
Sacrum	Longer and narrower	Shorter and wider
Coccyx	More curved; less movable	Straighter; more movable
Pelvic brim	Narrower and heart shaped	Wider and oval shaped
Ischial spines	Longer, sharper, closer together, and project more medially	Shorter, blunt, farther apart, and project more posteriorly

Assignment 4

Materials

Colored pencils
Articulated human skeleton

1. Label Figure 26.6. Color-code the labeled bones.
2. Identify the bones of an articulated human skeleton.
3. **Complete items 4a and 4b on the laboratory report.**
4. Locate the various types of joints on the articulated skeleton.
5. **Complete item 4 on the laboratory report.**

SEXUAL DIFFERENCES OF THE PELVIS

The structure of the pelvic girdle is different in males and females, primarily because the female pelvis is adapted for childbirth. Table 26.1 lists some of the major differences between male and female pelvic girdles. Compare these characteristics with Figure 26.7. Because the characteristics of a pelvis may have considerable variations from the ideal, sex determination of a pelvis is based on a combination of characteristics rather than on a single characteristic.

Calculation of a **pelvic ratio** is helpful in determining the sex of a pelvic girdle. Two measurements are required. The ratio is calculated by dividing the first measurement by the second.

1. The distance between the tips of the ischial spines
2. The distance between the inner surface of the pubic symphysis and the upper, inner surface of the sacrum

$$\text{Pelvic ratio} = \frac{\text{first measurement}}{\text{second measurement}}$$

Females usually have a ratio of 1.0 or more; males usually have a ratio of 0.8 or less.

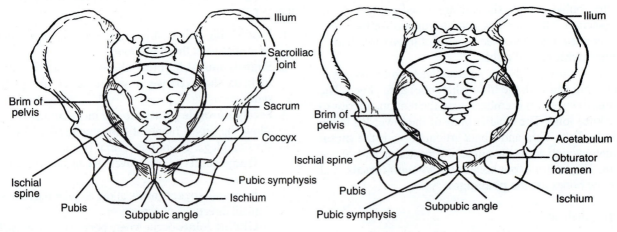

Male Pelvis (anterior view)

Female Pelvis (anterior view)

Figure 26.7 Male and female pelvic girdles.

Assignment 5

Materials

Male and female pelvic girdles
Calipers or dividers
Metric ruler

1. Examine the male and female pelvic girdles provided. Verify the characteristics in Table 26.1 as you compare the male and female pelvic girdles.
2. Use calipers or dividers and metric rulers to make the two pelvic measurements. Then, calculate the pelvic ratio for the male and female pelvic girdles.
3. Determine the pelvic ratio for the pelvis of the articulated skeleton available. Use the characteristics in Table 26.1 and the pelvic ratio to determine the sex of the articulated skeleton.
4. ***Complete item 5 on the laboratory report.***

FETAL SKELETON

The bones of a newborn baby have not completely formed. Skull bones are growing within connective tissue membranes, and the rest of the skeleton is developing within hyaline cartilage. The growth of the skeleton is complete by 25 years of age.

 ## Assignment 6

Materials

Human fetal skeleton

1. Examine a human fetal skeleton. Note the cartilage and bones and the relative size of head and body.
2. ***Complete item 6 on the laboratory report.***

SKELETAL ADAPTATIONS FOR LOCOMOTION

The skeletons of vertebrates are adapted for specific types of locomotion. In this section, you will examine the skeletons provided to correlate major skeletal features with the type of locomotion used by the animals. To assist you in getting started, the skeletons of a frog and a chicken are shown in Figures 26.8 and 26.9, respectively. Correlate the brief discussions that follow with these figures before examining the skeletons.

The hind legs of a frog are adapted to provide extra leverage for jumping. The tibia and fibula have fused to form a single bone, while two "anklebones," the **fibulare** and **tibulare,** have elongated.

The bird skeleton shows a number of adaptations for flight. Bones are hollow and light in weight. The large sternum provides attachment for flight muscles.

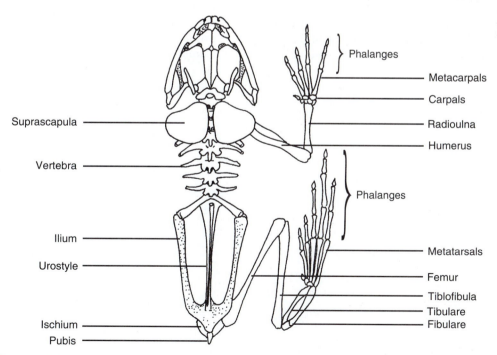

Figure 26.8 Frog skeleton.

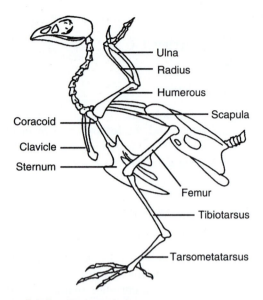

Figure 26.9 Chicken skeleton.

The "wrist" and "hand" bones have been reduced in the wing. The tibia and proximal tarsal bones have fused to form the **tibiotarsus,** and the distal tarsals and metatarsal bones have fused and elongated to form the **tarsometatarsus,** the lowest part of the leg. This extension provides additional leverage for springing into the air and absorbing shocks upon landing. All vertebrae except those in the neck and tail are fused with the ribs and pelvic girdle and provide a rigid framework.

A cat skeleton is provided as an example of a vertebrate adapted for running. Cats are fast runners for short distances. Note how the legs are located under the body rather than extending laterally as in the frog or lizards. Observe the orientation of the bones of the front and hind legs and their attachment to the axial skeleton to provide good leverage and freedom of movement. Note the long tail. What is its function?

The human skeleton shows adaptations for an erect, bipedal mode of locomotion. The major adaptations include (1) a broad pelvis that provides support for the erect body, (2) an almost central inferior attachment of the skull to the vertebral column and the vertebral curvatures that are a result of erect posture, (3) hands that are specialized for handling and grasping objects while the feet have lost this capability in favor of efficient bipedal walking and running, and (4) eye orbits that are forward facing, allowing stereoscopic vision.

 Assignment 7

Materials

Articulated skeletons of:
   cat
   chicken
   frog
   human
   monkey

1. Examine the skeletons of the frog, chicken, and cat, noting the adaptations for their particular modes of locomotion.
2. Examine the articulated human skeleton, and note the major adaptations to erect posture and bipedal locomotion.
3. Compare the human and monkey skeletons, and note the differences in skeletal structure between an erect, bipedal human and a monkey that lives primarily in trees.
4. ***Complete item 7 on the laboratory report.***

THE SKELETAL SYSTEM

Student _____

Lab Instructor _____

1. SKELETAL TYPES

a. Write the term that matches the phrase.

1. Internal skeleton of bone _____

2. Skeleton of body fluids in confined spaces _____

3. Skeleton exterior to the body _____

4. Type of skeleton in mollusks _____

5. Type of skeleton in vertebrates _____

6. Joins bones together at articulations _____

7. Secretes exoskeleton in arthropods _____

8. Hardens arthropod exoskeletons _____

b. Matching.

1. Calcareous exoskeleton   2. Bony endoskeleton   3. Chitinous exoskeleton
4. Cartilaginous endoskeleton   5. Endoskeleton of calcareous plates

_____ Mollusks   _____ Sharks and rays

_____ Echinoderms   _____ Amphibians

_____ Arthropods   _____ Mammals

c. Explain the relationship between molting and growth in arthropods. _____

_____

_____

d. Name three functions that arthropod exoskeletons and vertebrate endoskeletons have in common.

1._____

2._____

3._____

2. MACROSCOPIC BONE STRUCTURE

a. List the labels for Figure 26.1.

1._____   4._____   7._____

2._____   5._____   8._____

3._____   6._____

b. Write the term that matches the phrase.

1. Location of yellow marrow _____

2. Fills spaces in cancellous bone _____

3. Type of bone forming the diaphysis _____

4. Type of bone forming the epiphyses _____

5. Fibrous membrane covering bone _____

6. Forms blood cells _____

7. Cartilage between diaphysis and epiphyses _____

8. Protects articular surface of the bone _____

9. Site of growth in length _____

c. Distinguish between the epiphyseal disc and epiphyseal line.

_____

_____

d. Does the bone seem to have a good supply of blood? _____

e. Does cancellous bone decrease the weight of a bone? _____

Does it significantly decrease the strength of a bone? _____

f. Why is adequate dietary calcium necessary for the development of strong bones? _____

_____

g. What advantage is provided by the articular cartilages? _____

h. Describe the appearance, feel, and function of the articular cartilage.

1. Appearance _____

2. Feel _____

3. Function _____

i. Circle the following terms that describe the ligaments.

pliable   stiff   weak   strong   elastic   nonelastic

3. MICROSCOPIC STUDY

a. Matching.

1. Hyaline cartilage   2. Dense fibrous connective tissue   3. Fibrocartilage   4. Bone

_____ Tendons, ligaments   _____ Connective tissue

_____ Matrix of calcium salts   _____ Many collagenous fibers in matrix

_____ Articular cartilage   _____ Intervertebral discs

b. Draw a portion of these tissues from your slides. Label pertinent parts.

Bone, x.s.   **Hyaline cartilage**

Fibrocartilage   **Dense Fibrous Connective Tissue**

c. Describe the relationship between lacunae and osteocytes. _____

_____

_____

d. What part of bone tissue is nonliving? _____

e. What is the function of canaliculi? _____

4. THE HUMAN SKELETON

a. List the labels for Figure 26.6.

1._____	9._____	16._____
2._____	10._____	17._____
3._____	11._____	18._____
4._____	12._____	19._____
5._____	13._____	20._____
6._____	14._____	21._____
7._____	15._____	22._____
8._____		

b. Matching.

1. Clavicle   2. Humerus   3. Femur   4. Os coxae   5. Scapula   6. Ulna and radius
7. Tibia and fibula   8. Mandible   9. Skull   10. Sacrum   11. Sternum   12. Vertebrae
13. Tarsals   14. Carpals   15. Phalanges

_____ Ankle bones   _____ Form backbone

_____ Bones of lower arm   _____ Bone of upper arm

_____ Finger bones   _____ Bones of lower leg

_____ Shoulder blade   _____ Collarbone

_____ Wrist bones   _____ Breastbone

_____ Thighbone   _____ Lower jawbone

_____ Five fused vertebrae   _____ Hipbones

c. Matching.

1. Immovable   2. Slightly movable   3. Hinge   4. Gliding   5. Pivot   6. Ball-and-socket

_____ Vertebrae 1 and 2   _____ Humerus/ulna

_____ Humerus/scapula   _____ Finger joints

_____ Joints between vertebrae _____ Joints of cranial bones

_____ Femur/os coxae   _____ Wrist and ankle

_____ Femur/tibia   _____ Lower jaw/skull

5. SEXUAL DIFFERENCES OF THE PELVIS

a. Using Table 26.1, record your observations of male and female pelves and of the "unknown" pelvis.

Characteristic	Male	Female	Unknown
General structure			
Acetabula			
Pubic angle			
Sacrum			
Coccyx			
Pelvic brim			
Ischial spines			
Pelvic ratio			

b. What is the gender of the "unknown" pelvis? _____

c. What is the advantage of the adaptations of the female pelvis? _____

_____

6. FETAL SKELETON

a. What portions of the:

long bones are formed of bone? _____

cranial bones are formed of bone? _____

b. What part of the fetal skeleton seems to have the largest circumference? _____

_____

c. What is the advantage of the incomplete ossification of the cranial bones prior to birth? _____

_____

7. SKELETAL ADAPTATIONS

a. In fast-running animals, the legs are:

_____ extended laterally   _____ located under the body

b. Matching: Compare the monkey and human skeletons.

1. Monkey   2. Human

_____ Forward-facing eye orbits   _____ Grasping hands

_____ Heavier, thicker vertebrae   _____ Grasping feet

_____ Broader, heavier pelvis   _____ Narrower, lighter pelvis

_____ More posterior skull attachment   _____ More central skull attachment

_____ Vertebral curvatures   _____ Hind legs shorter than front legs

_____ Many tail vertebrae

MUSCLES AND MOVEMENT

OBJECTIVES

After completing the laboratory session, you should be able to:
1. Compare the structural and functional characteristics of the three types of muscle tissue.
2. Identify and describe the types of levers formed by muscles and bones.
3. Describe and demonstrate the antagonistic action of selected muscles.
4. Determine and describe the action of selected muscles in the hind leg of a frog.
5. Describe the ultrastructure of skeletal muscle fibers.
6. Describe the roles of actin and myosin in muscle contraction.
7. Define all terms in bold print.

Minute **contractile fibrils** are present in many unicellular organisms. They produce a variety of movements, including whiplike beating of flagella and cilia. In plants, such fibrils are found only in the flagella of motile cells, but they have been exploited to a remarkable degree by advanced animals.

In animals, contractile fibrils are incorporated into contractile cells. Sponges, which lack neurons, possess primitive contractile cells that react to direct stimulation by the environment.

In all higher animals, contractile cells are under **neural control.** Other evolutionary trends have been (1) the organization of contractile cells into large groups called *anatomical muscles,* and (2) functional and structural specialization of contractile cells.

VERTEBRATE MUSCLE TISSUES

Muscle tissues of vertebrates can be divided into three distinct groups on the basis of structure, location, and function.

Smooth Muscle Tissue

Smooth muscle tissue occurs in the walls of hollow internal organs such as the digestive tract and blood vessels. Each cell is spindle shaped with a centrally located nucleus. Striations are absent. The cells are not enveloped by a sarcolemma, as in skeletal muscle.

Contractions are slow and untiring. Because smooth muscle is controlled by the autonomic nervous system and is not under conscious control, contractions are said to be *involuntary.* See Figure 27.1.

Skeletal Muscle Tissue

Skeletal muscle tissue composes skeletal muscles, which are distinguished by being attached to bones of the skeleton or to other skeletal muscles. Skeletal muscle tissue consists of numerous muscle fibers bound together by connective tissue. A skeletal muscle fiber may be not a single cell but an aggregation of cells that lack delimiting membranes. The fact that each fiber has several nuclei located on the periphery of the fiber supports this view. A thin membrane, the **sarcolemma,** envelops each fiber but is difficult to see unless it has been separated from the fiber. The

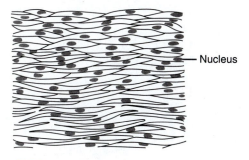

— Nucleus

Figure 27.1 Smooth muscle.

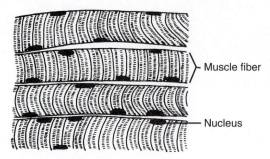

Figure 27.2 Skeletal muscle.

contractile elements are the **myofibrils** that extend the length of the fiber. They exhibit alternating dark and light **striations** (cross-banding), which is the basis for an alternate name for this tissue, *striated muscle.* Skeletal muscle is controlled by the somatic nervous system, which enables its functions to be *voluntary* (under conscious control), although it contracts involuntarily in reflexes. Contractions are rapid, but the muscles are easily fatigued. See Figure 27.2.

Cardiac Muscle Tissue

The muscle of the heart is **cardiac muscle tissue.** It is striated like skeletal muscle, but the fibers form an interwoven network. The boundaries of cells are denoted by *intercalated discs,* and each cell has a single, centrally located nucleus. A sarcolemma envelops each cell. Contractions are rhythmic, untiring, and *involuntary.* Control is by the autonomic nervous system. See Figure 27.3.

 Assignment 1
Materials

Compound microscope
Prepared slides of:
smooth muscle, teased

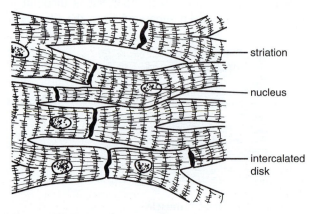

Figure 27.3 Cardiac muscle.

skeletal muscle, teased
cardiac muscle, teased

1. Examine prepared slides of smooth, skeletal, and cardiac muscle tissue, and compare your observations with Figures 27.1, 27.2, and 27.3.
2. ***Complete item 1 on Laboratory Report 27 that begins on page 353.***

SKELETAL MUSCLE FUNCTION

Skeletal muscles and bones of vertebrates are arranged to form a system of levers that work together to produce a variety of movements. Muscles are attached at each end to bones or other muscles by **tendons.** The end that is stationary (does not move during contraction) is the **origin.** The end where the movement occurs is the **insertion.** Further, the muscles are arranged in **antagonistic groups** so contraction of opposing muscles moves a body part in opposite directions. This is necessary because muscles can only contract with force. Examples of antagonistic actions are the following:

> **Flexors and Extensors.** Flexors decrease the angle of bones forming a joint, whereas extensors increase the angle.
> **Abductors and Adductors.** Abductors move a body part away from the midline of the body, whereas adductors move it toward the midline.

When a muscle is stimulated to contract by neural impulses, the antagonist is automatically inhibited from contracting. Consider Figure 27.4. The biceps muscle flexes the forearm, and the triceps extends it. When the biceps contracts, the triceps is relaxed. When the triceps contracts, the biceps is relaxed.

Levers

A lever consists of a rigid rod (a bone) that moves about a fixed point (a joint) called a **fulcrum (F).** Two opposing forces act on a lever. The **resistance (R)** is the weight to be moved; the **contraction force (CF)** is the force applied by a contracting muscle at the point of its insertion. The types of levers differ from each other by the relative positions of the resistance, fulcrum, and contraction force. Figure 27.5 shows the three types of levers. Be sure you know their characteristics before proceeding.

Study Figure 27.4 and note the two types of levers involved in the flexion and extension of the forearm. The fulcrum is always at a joint, and the contraction force is always applied at the site of the muscle insertion. The farther the point of insertion is from the fulcrum, the greater is the mechanical advantage and the greater the resistance that can be moved. The nearer

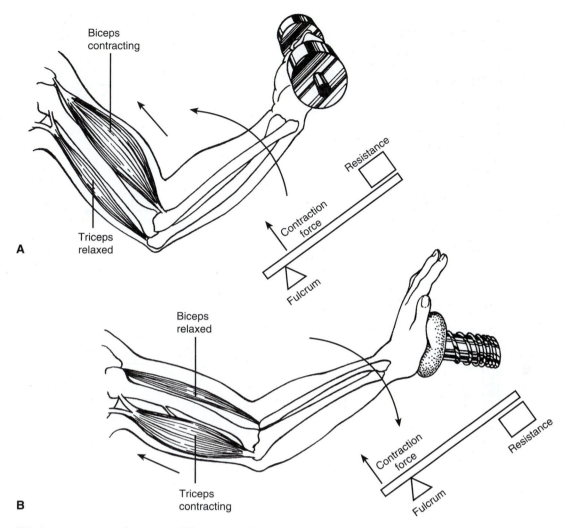

Figure 27.4 Antagonistic functions of biceps and triceps. **A.** Contraction of the biceps flexes the forearm via a third-class lever. **B.** Contraction of the triceps extends the forearm via a first-class lever.

the insertion is to the fulcrum, the greater is the freedom and speed of movement of the body part.

 Assignment 2

1. Determine the types of levers and movements involved in each of the actions shown in Figure 27.6.

Indicate the location of the fulcrum (F), resistance (R), and contraction force (CF) by placing their symbols on the diagrams.

2. *Complete item 2 on the laboratory report.*

Study of Frog Muscles

The hind legs of a frog are suitable subjects for studying the way muscles are arranged to allow movement.

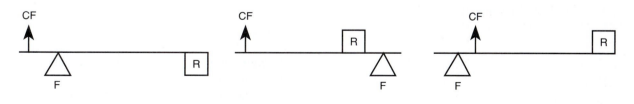

First class   Second class   Third class

Figure 27.5 The three classes of levers.

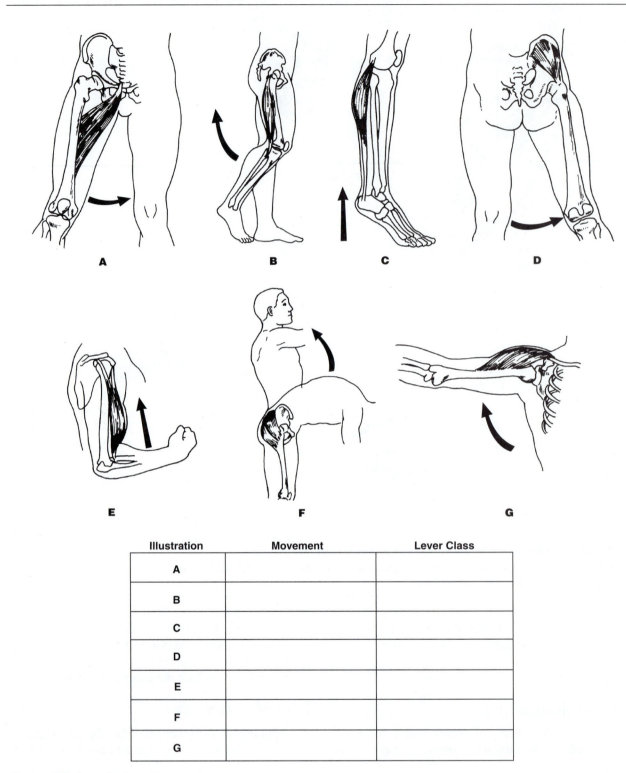

Illustration	Movement	Lever Class
A		
B		
C		
D		
E		
F		
G		

Figure 27.6　Body movements.

 Assignment 3

Materials

Colored pencils
Dissecting instruments and pan
Frog, preserved or freshly killed

1. Using Figures 27.7 and 27.8 as guides, skin a hind leg of a frog.
2. Examine the muscles of the leg. If the frog is freshly killed, note the color and texture of the muscles and connective tissue.
3. Refer to Figure 27.9 to locate the following muscles, and separate them from adjacent muscles

Figure 27.7 Cut the skin around the uppermost part of the thigh.

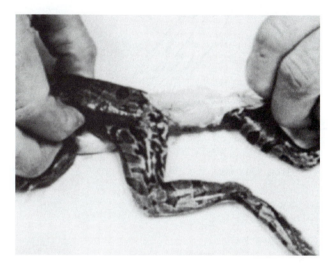

Figure 27.8 Strip the skin off the entire leg.

with a blunt probe. Color-code each muscle in Figure 27.9.

> **Dorsal muscles**
> Triceps femoris
> Semimembranosus
> Gastrocnemius
> Peroneus
> **Ventral muscles**
> Sartorius
> Gracilis major
> Gastrocnemius
> Extensor cruris

4. Note the origins and insertions of the muscles in Table 27.1, and locate their general positions in the frog. Determine and record the action of each muscle by pulling the body of the muscle toward

the origin with forceps while holding the origin in place with your other hand.

5. If time permits, determine the action of other leg muscles in a similar manner.

6. ***Complete item 3 on the laboratory report.***

MYOFIBRIL STRUCTURE AND CONTRACTION

The striations of a skeletal muscle fiber result from the arrangement of actin and myosin **myofilaments** within each myofibril. See Figure 27.10. The thicker **myosin** myofilaments are the main components of the **A band.** A light-colored **H zone** occurs in the center of the A band where only myosin is present. Thinner **actin** myofilaments extend into the A band from the **Z lines.** The Z lines form the boundaries of a **sarcomere,** the contractile unit of a myofibril. A light-colored **I band** lies between the A band and a Z line.

Mechanics of Contraction

Contraction of a muscle fiber results from the interaction of actin and myosin in the presence of calcium and magnesium ions. ATP supplies the required energy.

When a muscle fiber is activated by a neural impulse, the cross-bridges of the myosin myofilaments attach to active sites on the actin myofilaments and bend to exert a power stroke that pulls the actin filaments toward the center of the A band. After the power stroke, the cross-bridges separate from the first active actin sites, attach to the next active sites, and produce another power stroke. This process is repeated until maximal contraction is attained.

Experimental Muscle Contraction

In this section, you will induce muscle fibers to contract using ATP and magnesium and potassium ions. Your instructor has prepared 2-cm segments from a glycerinated rabbit psoas muscle. The segments have been placed in a petri dish of glycerol and teased apart to yield thin strands consisting of very few muscle fibers. The strands should be no more than 0.2 mm thick.

 Assignment 4

Materials

Microscopes, compound and dissecting
Dissecting instruments
Microscope slides and cover glasses
Plastic ruler, clear and flat
Glycerinated rabbit psoas muscle

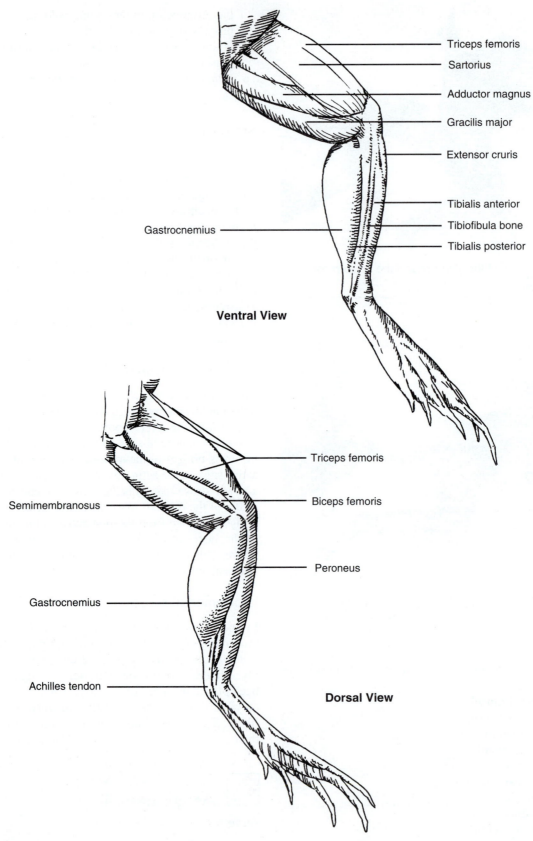

Ventral View

Triceps femoris
Sartorius
Adductor magnus
Gracilis major
Extensor cruris
Tibialis anterior
Tibiofibula bone
Tibialis posterior
Gastrocnemius

Dorsal View

Triceps femoris
Biceps femoris
Peroneus
Semimembranosus
Gastrocnemius
Achilles tendon

Figure 27.9 Superficial muscles of a frog's legs.

TABLE 27.1	SELECTED MUSCLES OF A FROG'S LEG		
Muscle	Origin	Insertion	Action
Triceps femoris	Ilium	Tibiofibula, femur	_____
Semimembranosus	Ischium, pubis	Tibiofibula	_____
Gastrocnemius	Femur	Tarsals	_____
Peroneus	Femur	Tibiofibula	_____
Sartorius	Pubis	Tibiofibula	_____
Gracilis major	Pubis	Tibiofibula	_____
Extensor cruris	Femur	Tibiofibula	_____

Dropping bottles of:
   ATP, 0.25%, in triple-distilled water
   glycerol
   magnesium chloride, 0.001 M
   potassium chloride, 0.05 M

1. Place a few muscle strands in a small drop of glycerol on a microscope slide and add a cover glass.

Observe them with the high-dry or oil-immersion objective. ***Draw the pattern of striations of the relaxed muscle fibers in item 4c of the laboratory report.***

2. Place three to five of the thinnest strands on another slide in just enough glycerol to moisten

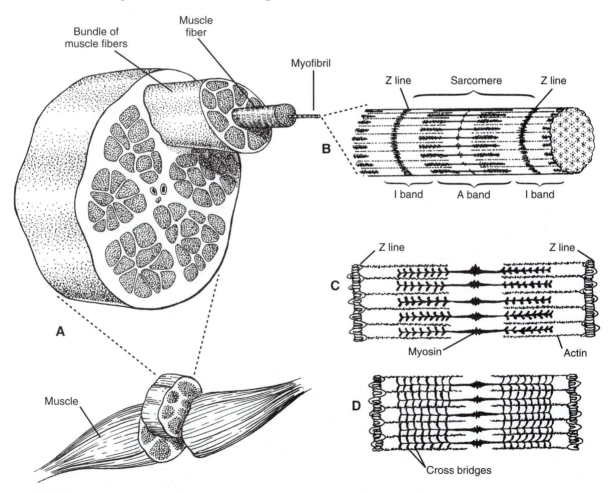

Figure 27.10 Structure of skeletal muscle. **A.** The arrangement of muscle fibers within a muscle. **B.** The ultrastructure of a myofibril, showing the relationship of actin and myosin filaments. Note how the arrangement of actin and myosin differs in **C.** the relaxed state, and **D.** the contracted state.

them. Arrange the strands straight, parallel, and close together.

3. Use a dissecting microscope to measure the length of the relaxed strands by placing the slide on a clear plastic ruler. *Record the length on the laboratory report.* Note the width of the strands.

4. While observing through the dissecting microscope, add 1 drop from each solution: ATP, magnesium chloride, and potassium chloride. Note any changes in length or width of the strands.

5. Remeasure and record the length of the contracted strands.

6. Place a few contracted strands in a small drop of glycerol on another slide and add a cover glass. Observe the strands with the high-power or oil-immersion objective. *Draw the pattern of striations in item 4c on the laboratory report.*

7. *Complete item 4 on the laboratory report.*

MUSCLES AND MOVEMENT

Student _____

Lab Instructor _____

1. VERTEBRATE MUSCLE TISSUES

a. What characteristic of muscle tissue enables movement? _____

b. Matching.

    1. Smooth muscle   2. Skeletal muscle   3. Cardiac muscle

   _____ Attached to bones   _____ Voluntary control

   _____ Heart muscle   _____ Sarcolemma present

   _____ Multinucleate fibers   _____ Striations present

   _____ Involuntary control   _____ In walls of blood vessels

   _____ Rapid contractions, but quickly tiring   _____ Rhythmic, untiring contractions

c. Draw a few teased muscle fibers from your slides. Label pertinent parts and indicate the magnification.

          Smooth   **Skeletal**   **Cardiac**

2. SKELETAL MUSCLE ACTION

a. Indicate the action of the antagonists to muscles that cause:

   flexion _____   adduction _____

b. The stationary end of a muscle is the _____ ; the action end of a muscle is the _____.

c. Considering Figure 27.4, indicate the site of insertion for each.

   Biceps _____   Triceps _____

d. Are most muscle insertions arranged to provide maximum mechanical advantage? _____

   Explain your answer. _____

   _____

   _____

   _____

e. Is the vertebrate body adapted for speed of movement or great strength? _____

Explain your answer. _____

_____

_____

f. Indicate the directional movement and lever class for each illustration in Figure 27.6.

Illustration	*Movement*	*Lever Class*
A	_____	_____
B	_____	_____
C	_____	_____
D	_____	_____
E	_____	_____
F	_____	_____
G	_____	_____

3. FROG MUSCLES

a. Do skeletal muscles extend across a joint, or are they located only between joints?

_____

b. Indicate the actions of these muscles in the frog's leg:

1. Triceps femoris   _____

2. Semimembranosus   _____

3. Gastrocnemius   _____

4. Peroneus   _____

5. Sartorius   _____

6. Gracilis major   _____

7. Extensor cruris   _____

4. STRUCTURE AND CONTRACTION

a. Write the term that matches the phrase.

1. Thick myofilament with cross-bridges   _____

2. Thin myofilaments attached to Z lines   _____

3. Form boundaries of sarcomere   _____

4. Myofilaments composing I band   _____

b. Record the length of the fibers before and after adding ATP, K^+, and Mg^{++}.

Before _____ mm After _____ mm

c. Draw the appearance of striations of the fibers before and after adding ATP, K^+, and Mg^{++}.

Before   **After**

REPRODUCTION IN VERTEBRATES

OBJECTIVES

After completing the laboratory session, you should be able to:
1. Describe the basic reproductive patterns exhibited by vertebrates, and indicate the advantages and disadvantages of each.
2. Describe how reptiles solved the problem of reproduction in a terrestrial environment.
3. Identify stages of gametogenesis on charts and prepared microscope slides, and describe the stepwise process.
4. Identify on charts or models the components of human male and female reproductive systems, and state the function of each.
5. Indicate the mode of action and relative effectiveness of birth control methods.
6. Define all terms in bold print.

All animals exhibit **gametic sexual reproduction,** and, with few exceptions, it is the only method of reproduction in animals. Separate sexes are the general rule, but a few forms are **hermaphroditic,** possessing both male and female sex organs. **Asexual sporulative reproduction** is restricted to the primitive, nonmotile sponges, and **vegetative reproduction** by budding or fragmentation also is associated with simple animals with limited motility: sponges, coelenterates, and flatworms.

REPRODUCTIVE PATTERNS

Vertebrates occur in both aquatic and terrestrial environments, and their reproductive patterns are reflective of the environment that each group has colonized. Thus, some vertebrate groups exhibit internal fertilization, whereas others use external fertilization.

External fertilization occurs in an aquatic environment. It involves the release of both sperm and eggs into the water, where they unite to form the zygote. Most fish and amphibians use external fertilization.

Internal fertilization involves the deposition of sperm in the female reproductive tract, where fertilization occurs. **Copulation** is usually involved in this process. Reptiles, birds, and mammals use internal fertilization.

Types of Gametic Reproduction

The three basic patterns of gametic sexual reproduction in vertebrates are based on (1) the site of embryonic development and (2) the source of nutrients for the embryo. Table 28.1 indicates the most common reproductive patterns for the five classes of vertebrates.

Oviparous reproduction is characterized by external embryonic development following external or internal fertilization. Nutrients for the embryo are contained within the fertilized egg. This pattern occurs in fish, amphibians, reptiles, birds, and egg-laying mammals.

Viviparous reproduction involves both fertilization and development within the female reproductive tract, and the developing embryo receives nutrients from the female parent. This pattern is characteristic of mammals.

Ovoviviparous reproduction occurs when fertilization and development of the embryo take place in the female reproductive tract, but the embryo receives nutrients only from the egg and not from the female parent. This pattern is not common in vertebrates but does occur in some sharks and snakes.

Reproduction and Land Colonization

Among terrestrial vertebrates, amphibians have a limited distribution because they must return to water to reproduce. The frog is a suitable example. Female and

TABLE 28.1   COMMON REPRODUCTIVE PATTERNS OF VERTEBRATES

| | Fertilization | | Reproductive Pattern | | | Reproductive Habitat | | Eggs | | | |
	Ext.	Int.	Ovip.	Ovovip.	Vivip.	Aquatic	Terr.	Size	Number	Protection	Stored Food
Fish	✓	Very few	✓	Very few		✓		Sm.	Many	No shell	Small amt.
Amphibians	✓		✓			✓		Sm.	Many	No shell	Small amt.
Reptiles		✓	✓	Very few			✓	Lg.	Moderate number	Shell and int. mem.	Large amt.
Birds		✓	✓				✓	Lg.	Few	Shell and int. mem.	Large amt.
Placental mammals		✓			✓	Few	✓	Very sm.	Very few	Membranes in mother's uterus (no shell)	None

male simultaneously release eggs and sperm, respectively, into the water, where fertilization occurs. After embryonic development is complete, a larva (tadpole) hatches from the egg. Larvae swim by using the tail in a fishlike manner, and they have functional gills. After suitable growth, the larvae metamorphose into the adult by developing lungs and appendages and by reabsorbing the gills and tail.

Reptiles were the first truly terrestrial vertebrates because they solved the problem of reproduction without returning to water. Their successful adaptations include internal fertilization and the **amniote egg.** The reptilian egg has a leathery shell that prevents excessive water loss and allows an exchange of oxygen and carbon dioxide between the developing embryo and the atmosphere. An adequate supply of stored nutrients (yolk and albumin) allows the development of the embryo to hatching. In addition, special protective membranes enclose the embryo.

The features of the amniote egg are shown in Figure 28.1. Note the four **extra-embryonic membranes.** The **amnion** surrounds the embryo and contains the **amniotic fluid,** which provides the embryo with its own "private pond" in which to develop. Instead of returning to water for embryonic development, reptiles "brought the water to the embryo." The **yolk sac** envelops the yolk and absorbs nutrients for the embryo. The **allantois** is an embryonic urinary bladder, but it also spreads out against the outer membranes and serves as a gas-exchange organ.

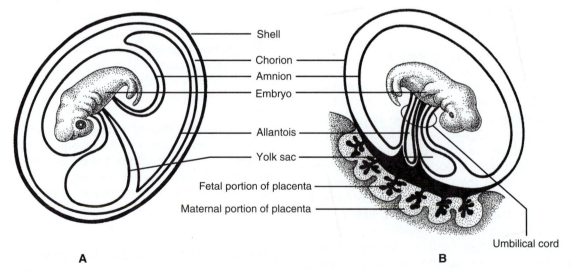

A   B

Figure 28.1 Extra-embryonic membranes of the amniote egg. **A.** Reptile. **B.** Mammal.

The **chorion** is the outermost membrane and encompasses all the others.

The bird egg is very similar to a reptilian egg, but it has a hard calcareous shell. The same basic pattern of extra-embryonic membranes develops as the embryo grows. Birds provide more parental care than reptiles and incubate the eggs.

Except for the primitive platypus and spiny echidna, mammals do not lay eggs. Fertilization is internal, and the embryo develops in the uterus of the female parent. The embryo becomes attached to the uterus by a **placenta** that is formed primarily by the chorion and allantois. The stalk of the allantois is the primary constituent of the **umbilical cord,** although the yolk sac and amnion are also components. The umbilical cord carries blood of the embryo to and from the placenta. Maternal and embryo bloods are separated by thin membranes in the placenta, where an exchange of materials occurs.

$$\text{maternal blood} \xrightleftharpoons[\text{wastes and } CO_2]{\text{nutrients and } O_2} \text{embryo blood}$$

Thus, mammals have capitalized on the extra-embryonic membranes "invented" by reptiles.

Assignment 1

Materials

Compound microscope
Stereo microscope
Fish and amphibian eggs, fresh or preserved
Frog life cycle, preserved or models

Frog larvae, living
Chicken eggs, fresh
Chick embryos showing extra-embryonic membranes
Pregnant cat or pig uterus

1. *Complete items 1a–1d on Laboratory Report 28 that begins on page 363.*
2. Examine the fish, amphibian, and bird eggs. In what ways do the eggs provide protection against environmental hazards and provide for the nutritional needs of the embryo? What is the relationship between parental care and survival of the young?
3. Examine the frog life cycle. Observe living larvae in the aquarium. Note their swimming and feeding behavior.
4. Examine the demonstration of bird and mammalian embryos with extra-embryonic membranes. Note the relationship of the membranes to the embryo. In the mammalian embryos, note the placenta and umbilical cord.
5. *Complete item 1 on the laboratory report.*

HUMAN REPRODUCTIVE SYSTEMS

You will study the human reproductive systems as examples of reproductive systems in mammals. Refer to Figures 28.2 and 28.3.

Male Reproductive System

The male gonads are paired **testes** that are held in the saclike **scrotum.** This arrangement holds the testes outside the body cavity and at a temperature of

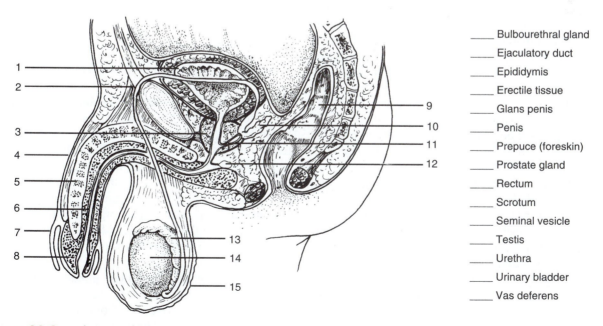

____	Bulbourethral gland
____	Ejaculatory duct
____	Epididymis
____	Erectile tissue
____	Glans penis
____	Penis
____	Prepuce (foreskin)
____	Prostate gland
____	Rectum
____	Scrotum
____	Seminal vesicle
____	Testis
____	Urethra
____	Urinary bladder
____	Vas deferens

Figure 28.2 Male reproductive system.

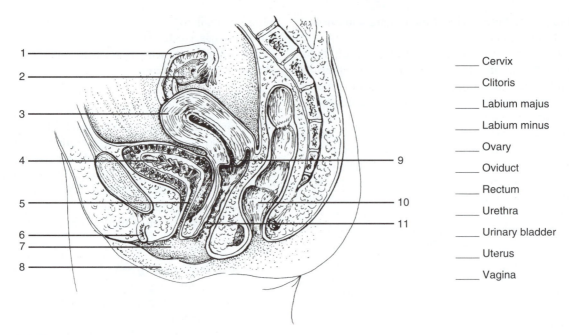

_____	Cervix
_____	Clitoris
_____	Labium majus
_____	Labium minus
_____	Ovary
_____	Oviduct
_____	Rectum
_____	Urethra
_____	Urinary bladder
_____	Uterus
_____	Vagina

Figure 28.3 Female reproductive system.

94–95°F, which is necessary for the production of viable **spermatozoa.** Muscles in the wall of the scrotum relax or contract to position the testes farther from or closer to the body and in this way regulate the temperature of the testes.

A testis contains numerous **seminiferous tubules** that produce the spermatozoa. **Interstitial cells** are located between the tubules and secrete **testosterone,** the male hormone responsible for the sex drive and the development of the sex organs and secondary sexual characteristics. The secretion of the **interstitial cell-stimulating hormone (ICSH)** by the hypophysis (pituitary gland) activates the interstitial cells.

The **penis,** the male copulatory organ, contains three cylinders of spongy **erectile tissue** that fill with blood during sexual excitement to produce an erection. A circular fold of tissue, the **prepuce** (foreskin) covers the **glans penis.** For hygienic reasons, the prepuce of male babies is often removed by a surgical procedure called _circumcision._

Mature, but inactive, sperm are carried down the seminiferous tubules to the **epididymis,** a long, coiled tube on the surface of the testis. Sperm are stored here until they are propelled through the reproductive tract by wavelike contractions during **ejaculation.**

The **bulbourethral glands** open into the urethra below the prostate gland and secrete an alkaline liquid that neutralizes the acidity of the urethra prior to ejaculation. At the male climax, sperm pass from the epididymis into the **vas deferens,** a duct that exits the scrotum and enters the body cavity via the inguinal canal. It continues across the surface of the urinary bladder to join with the **ejaculatory duct** within the **prostate gland** that is located around the urethra just below the bladder. Alkaline secretions from the **seminal vesicles** are mixed with the sperm just before the vase deferentia enter the prostate gland, where sperm-activating prostatic secretions are added. Muscular contractions force **semen,** the mixture of sperm and glandular secretions, out through the urethra.

Female Reproductive System

The external female genitalia consist of (1) two folds of skin surrounding the vaginal and urethral openings, the **labia majora** (outer folds) and **labia minora** (inner folds), and (2) the **clitoris,** a nodule of erectile tissue homologous to the penis in the male. Collectively, these structures are called the **vulva.**

The **vagina** is a collapsible tube extending 4 to 6 in. from the external opening to the uterus. It serves as both the female copulatory organ and the birth canal. The **uterus** is a pear-shaped organ located superior and posterior to the urinary bladder. The **cervix** of the uterus extends a short distance into the upper end of the vagina.

A pair of **ovaries,** the female gonads, are located lateral to the uterus, where they are supported by ligaments. One egg is released from alternate ovaries about every 28 days. The egg is picked up by the expanded end of the **oviduct** and carried toward the uterus by beating cilia of the ciliated epithelium lining the oviduct.

Ovaries secrete two female hormones. **Estrogen** is responsible for the development of the sex organs, the female sex drive, the secondary sex characteristics, and the buildup of the uterine lining. **Progesterone** prepares the uterine lining for the implantation of an early embryo. In turn, ovarian function is controlled by hormones released from the hypophysis (pituitary gland). Consult your text for a discussion of the ovarian and uterine cycles.

 Assignment 2

Materials

Colored pencils
Models of male and female reproductive systems

1. Label and color-code Figures 28.2 and 28.3.
2. Locate the parts of the male and female reproductive systems on the models provided.
3. *Complete item 2 on the laboratory report.*

Gametogenesis

The formation of gametes is called **gametogenesis.** It includes meiotic cell division, which reduces the chromosome number of gametes to one-half that of somatic cells. For example, the diploid (2n) chromosome number in humans is 46, and the haploid (n) gametes contain only 23 chromosomes. If you need to review meiosis, see Exercise 9.

The patterns of gametogenesis described here and illustrated in Figures 28.4 and 28.5 are typical of mammals, although slight variations may occur among individual species.

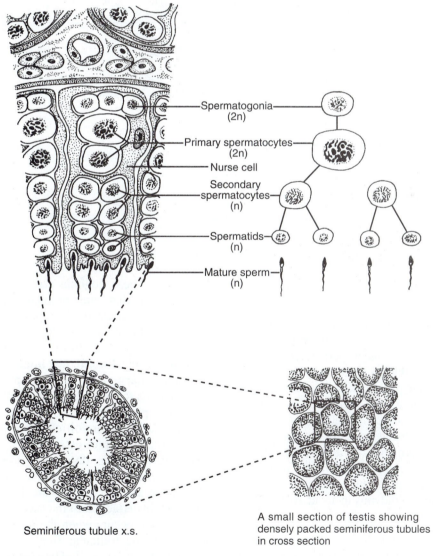

Seminiferous tubule x.s.

A small section of testis showing densely packed seminiferous tubules in cross section

Figure 28.4 Spermatogenesis.

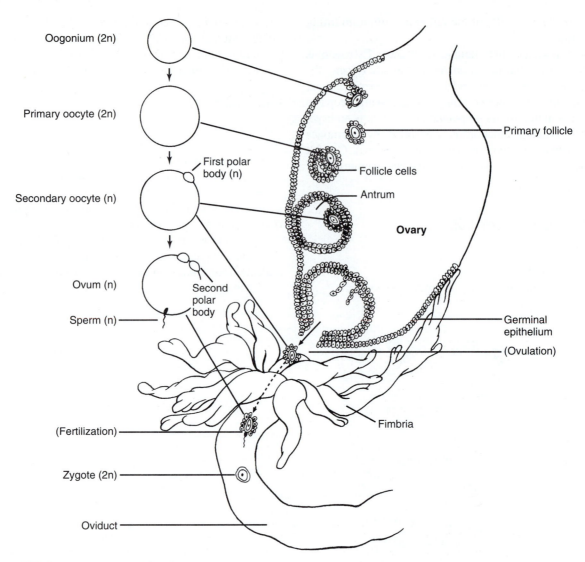

Figure 28.5 Oogenesis and fertilization.

Spermatogenesis

Sperm formation occurs in the seminiferous tubules of the testes. **Spermatogonia** (2n) are the outermost cells of the tubule. They divide mitotically to form a **primary spermatocyte** and a replacement spermatogonium. The primary spermatocyte divides meiotically to yield two **secondary spermatocytes** (n) after meiotic division I and four **spermatids** (n) after meiotic division II. The spermatids attach to "nurse cells" and mature into spermatozoa. Maturation includes the loss of most of the cytoplasm and the formation of a flagellum from a centriole. The sperm head consists mostly of the cell nucleus. Sperm are carried along the seminiferous tubules and reach the epididymis in about 10 days.

Oogenesis

Prior to the birth of a female child, some **oogonia** of the germinal epithelium surrounding each ovary enlarge, become surrounded by follicular cells, and move into the ovary. These oogonia (2n) become the **primary oocytes** (2n) that enter prophase of meiosis I before oogenesis is arrested. The primary oocytes remain inactive at this stage until puberty.

At puberty, the **follicle-stimulating hormone (FSH)** and the **luteinizing hormone (LH)** secreted by the pituitary gland stimulate primary oocytes and follicle cells to further growth. Each month one follicle develops more rapidly than the others to become a mature or Graafian follicle filled with fluid containing a large amount of **estrogen** secreted by the follicular cells. Meiotic division I proceeds to produce (1) a **secondary oocyte** (n) that receives most of the cytoplasm and (2) a much smaller **first polar body** that remains attached to the secondary oocyte.

Ovulation occurs when the mature follicle ruptures and ejects the follicular fluid and secondary oocyte through the ovary wall. The secondary oocyte enters the oviduct and is carried toward the uterus by the ciliated epithelium. Note that while it is common to speak of the "egg" as being released by the ovary in ovulation, a secondary oocyte is actually released. After ovulation, the empty follicle becomes the **corpus luteum,** which produces progesterone to maintain the uterine lining.

No further division occurs unless the secondary oocyte is penetrated by a sperm. If this occurs, the secondary oocyte completes meiotic division II to form the **egg** (n) and another polar body. The first polar body also may complete meiosis II to form an additional polar body. Subsequently, egg and sperm nuclei fuse to form the diploid zygote. The polar bodies disintegrate.

Assignment 3

Materials

Compound microscope
Prepared slides of:
   cat testis, sectioned
   cat ovary, sectioned, with Graafian follicle
   cat ovary, sectioned, with corpus luteum
   human sperm

1. Examine a prepared slide of cat testis. Locate the seminiferous tubules, which produce spermatozoa, and the interstitial cells, which produce testosterone, the male hormone. Compare your observations with Figure 28.4, and locate the cells involved in spermatogenesis.
2. Examine the prepared slide of human sperm. Note their small size. About 350 million sperm are released in an ejaculation.
3. Examine a prepared slide of cat ovary. Compare your slide with Figure 28.5. Locate the germinal epithelium, a primary follicle with a primary oocyte, and a Graafian follicle containing a secondary oocyte.
4. Examine a prepared slide of cat ovary showing a corpus luteum, which secretes progesterone to maintain the uterine lining.
5. ***Complete item 3 on the laboratory report.***

Birth Control

The control of fertility is one of the major concerns of modern society, not only because of the desire to prevent unwanted pregnancies but also because of the need to curb the human population growth rate. Table 28.2 indicates the effectiveness of common birth control methods. Descriptions of some of these methods follow.

Abstinence is not participating in sexual activity. Deciding not to be sexually active until you are ready for the responsibilities it brings is a wise choice.

TABLE 28.2	EFFECTIVENESS OF BIRTH CONTROL METHODS
Method	Pregnancies per 100 Sexually Active Women per Year
Abstinence	0
Norplant	0.05
Vasectomy	0.2
Depo-Provera	0.3
Tubal ligation	0.5
IUD	1.5
Oral contraceptive	5
Condom (high quality)	14
Diaphragm plus spermicide	16
Sponge	17
Rhythm method	24
Spermicide only	25
Withdrawal	26
Condom (poor quality)	30
Douche	60
No birth control method	90

Norplant consists of six tiny rods filled with synthetic progesterone, which are inserted under the skin by a physician. Slow release of the hormone blocks ovulation for up to 5 years. Side effects may include irregular menstrual cycles, headaches, dizziness, and depression.

Vasectomy is a surgical procedure in which a small section of each vas deferens is removed and the cut ends are tied so sperm cannot pass through the vas deferentia.

Depo-Provera is a synthetic progesterone that is administered by injection about four times a year. It is highly effective but may induce side effects of nausea, weight-gain, and headaches.

Tubal ligation is a surgical procedure in which a small section of each oviduct is removed and the cut ends are tied so secondary oocytes cannot pass through the oviducts.

An **intrauterine device (IUD)** is a metal or plastic device placed in the uterus by a physician, and it remains there for long periods of time. An IUD prevents implantation of an early embryo. Some models have been removed from the market due to the possible increased risk of uterine perforation, pelvic inflammatory disease, and sterility.

An **oral contraceptive** (the pill) consists of synthetic estrogens and progesterones that inhibit development of ovarian follicles and ovulation.

A **diaphragm** is a dome-shaped device inserted into the vagina and placed over the cervix prior to sexual intercourse. A **cervical cap** is similar to a diaphragm but smaller. It is slipped over the cervix of the uterus. Both require fitting by a physician and correct positioning by the user; when used with spermicides, they are equally effective.

A **sponge** is a spongelike device that is inserted into the vagina and placed against the cervix prior to sexual intercourse. The sponge contains a spermicide.

A **condom** is a sheath or pouch that collects semen, preventing sperm from entering the vagina during sex. A male condom is a thin sheath of latex worn over the penis. A female condom is a polyurethane pouch with a ring at each end. One ring is inserted deep into the vagina, and the other remains outside. A latex male condom and a female condom are the only birth control devices that provide *some* protection against sexually transmitted diseases, but neither is foolproof.

Spermicides are chemicals that are lethal to sperm. They are marketed as foams and jellies. Spermicides containing nonoxynol-9 are most effective in killing sperm. Spermicides may cause irritation or promote vaginal and urinary infections in some women.

The **rhythm method** involves abstention from sexual intercourse during a woman's fertile period, which extends from a few days before to a few days after ovulation. It requires a woman to take her temperature each morning before arising to detect the 0.2–0.6°F drop in body temperature that occurs just prior to ovulation.

Withdrawal is the removal of the penis from the vagina just prior to ejaculation. Effectiveness is limited because some sperm are often emitted from the penis prior to ejaculation.

A **douche** is the rinsing out of the vagina after sexual intercourse. It is not very effective because sperm can enter the uterus within 1.5 min after being deposited in the vagina.

 Assignment 4

Materials

Demonstration table of birth control devices and spermicides

1. Examine the various birth control devices and spermicides set up as a demonstration. Read the directions for use that accompany each one.
2. ***Complete item 4 on the laboratory report.***

REPRODUCTION IN VERTEBRATES

Student _____

Lab Instructor _____

1. TYPES AND PATTERNS OF REPRODUCTION

a. Write the term that matches each phrase.

1. Type of sexual reproduction in animals

   _____

2. Two types of vegetative reproduction in coelenterates

   _____

3. Animals with asexual sporulative reproduction

   _____

4. Sperm and eggs released into water

   _____

5. Sperm placed in female reproductive tract

   _____

6. Refers to the presence of male and female sex organs in same individual

   _____

b. Define:

Oviparous reproduction _____

_____

_____

Viviparous reproduction _____

_____

_____

Ovoviviparous reproduction _____

_____

_____

c. Matching.

1. Oviparous reproduction   2. Viviparous reproduction   3. Ovoviviparous reproduction

_____ Mammals   _____ Internal fertilization

_____ External fertilization   _____ Embryo nourished by yolk

_____ Embryo nourished by mother _____ Fish and amphibians

_____ Few sharks and snakes   _____ Birds and reptiles

d. Indicate the usual reproductive habitat of organisms that use:

external fertilization _____ internal fertilization _____

e. What reproductive adaptations of amphibians restrict their distribution? _____

_____

f. What reproductive adaptations enabled reptiles to colonize the land? _____

_____

_____

_____

g. Matching.

1. Chorion   2. Allantois   3. Yolk sac   4. Amnion

_____ Envelops embryo   _____ Envelops yolk

_____ Embryonic urinary bladder   _____ Part of umbilical cord

_____ Outermost membrane   _____ Provides "private pond" for embryo

_____ Forms placenta

h. The number of eggs produced by vertebrates decreases progressively from fish to mammals. How do you explain this? _____

_____

_____

_____

i. Describe the function of the placenta and umbilical cord in mammals.

Placenta _____

_____

Umbilical cord _____

_____

2. HUMAN REPRODUCTIVE SYSTEMS

a. List the labels for Figure 28.2.

1. _____   6. _____   11. _____

2. _____   7. _____   12. _____

3. _____   8. _____   13. _____

4. _____   9. _____   14. _____

5. _____   10. _____   15. _____

b. Trace the path of sperm from the seminiferous tubules to the urethra.

_____

_____

c. List the labels for Figure 28.3.

1. _____   5. _____   9. _____

2. _____   6. _____   10. _____

3. _____   7. _____   11. _____

4. _____   8. _____

d. Contrast the function of the urethra in males and females.

Males _____

Females _____

e. Spermatozoa of all animals require a fluid medium in which to swim to the egg. How is this provided in mammals? _____

_____

f. Matching.

1. Testes   2. Ovaries   3. Oviduct   4. Vas deferens   5. Penis   6. Prostate gland
7. Prepuce   8. Vagina   9. Uterus   10. Ejaculatory duct

_____ Secretion that activates sperm   _____ Male copulatory organ

_____ Produces eggs   _____ Carries eggs to uterus

_____ Site of embryo development   _____ Female copulatory organ

_____ Carries sperm to urethra   _____ Removed in circumcision

_____ Produces spermatozoa   _____ Lined with ciliated cells

g. Is there an advantage in the sperm swimming against the current in the oviduct to reach the egg? _____

Explain your answer. _____

_____

h. Indicate the site of production and the action of each:

Testosterone _____

Estrogen _____

Progesterone _____

3. GAMETOGENESIS

a. From your slides, draw a small portion of (1) a seminiferous tubule showing the cells involved in spermatogenesis and (2) the appearance of human sperm. Label pertinent parts.

Seminiferous Tubule   **Human Sperm**

b. Matching.

1. Diploid   2. Haploid   3. Formed by first meiotic division
4. Formed by second meiotic division

_____ Spermatogonium   _____ Oogonium

_____ Spermatid   _____ Secondary oocyte

_____ First polar body   _____ Spermatozoa

_____ Secondary spermatocyte   _____ Primary oocyte

c. From your slides, draw (1) the appearance of the ovary section at 40×, (2) a portion of the ovary showing a Graafian follicle with its secondary oocyte, and (3) a portion of an ovary showing a corpus luteum. Label pertinent parts.

Ovary Section	Graafian Follicle	Corpus Luteum

4. BIRTH CONTROL

a. Contraceptives are birth control methods that prevent the union of sperm and a secondary oocyte. Which of the methods in Table 28.2 are *not* contraceptives? _____

_____

b. Which contraceptives provide a barrier to the entrance of sperm into the uterus? _____

_____

c. Which contraceptives are chemicals that kill sperm? _____

_____

d. Which contraceptives use hormones to prevent ovulation? _____

_____

e. Which birth control method prevents the implantation of an early embryo? _____

_____

f. Explain why the rhythm method is not very effective. _____

_____

_____

g. Does using a condom guarantee that you will not be infected with HIV or pathogens of other STDs? _____

Explain. _____

_____

_____

_____

h. Induced abortion is the removal of a developing embryo or fetus from the uterus prior to term. Is it better to prevent unwanted pregnancies by birth control or to rely on induced abortion? Explain. _____

_____

_____

_____

_____

_____

FERTILIZATION AND DEVELOPMENT

OBJECTIVES

After completing the laboratory session, you should be able to:

1. Describe activation and cleavage, and identify activated eggs and cleavage stages when observed with the microscope.
2. Describe and identify a blastula and a gastrula.
3. Describe the formation of the germ layers and identify the germ layers in a late gastrula.
4. Indicate the adult tissues and organs formed from the germ layers.
5. Distinguish between the embryonic and fetal stages of development in humans.
6. Define all terms in bold print.

The union of gametes and early embryological development are difficult to study in many animals, especially chordates. Echinoderms are good subjects for such a study, however, because the gametes are easy to procure, and minimal care is needed for the adults and embryos, and because embryological development in echinoderms is similar to that in chordates. Here is a brief synopsis of early development in both chordates and echinoderms to prepare you for the study of these processes.

The penetration of an egg by a sperm is called **activation.** The subsequent fusion of egg and sperm nuclei is **fertilization,** and this process forms the diploid **zygote.** A series of mitotic divisions called **cleavage** begins and produces progressively smaller cells. As cleavage progresses, a solid ball of cells, the **morula,** is formed, and continued cleavage produces the **blastula,** a hollow ball of cells. This concludes the cleavage process. The blastula is not much larger than the zygote.

Mitotic divisions continue and transform the blastula into a **gastrula** by a process called gastrulation. This stage of early development is completed by the formation of the three **embryonic tissues** or **germ layers,** from which all other tissues develop.

ACTIVATION AND EARLY DEVELOPMENT IN THE SEA URCHIN

In this section, you will study activation and early development in the sea urchin, a common marine echinoderm that lives in the intertidal zone.

Activation

During the reproductive season, male and female sea urchins release their **gametes,** sperm and eggs, into the seawater so union of sperm and egg is essentially random. Both sperm and eggs release chemicals called **gamones,** which attract sperm to eggs and promote attachment of a sperm to an egg membrane. Similar substances are released by human gametes. The first sperm to attach near the **animal pole** of the egg causes activation. The egg extends a **fertilization cone** (really an activation cone) through the egg membranes to engulf the sperm head and draw it into the egg. The sperm tail remains outside the egg. A rapid release of substances from cytoplasmic vesicles into the space between the **inner** and **outer egg membranes** immediately follows and results in the rapid inflow of fluid into this space. The accumulation of fluid pushes the outer membrane farther outward and prevents penetration of another sperm. The outer membrane is now called the **fertilization membrane** (really an activation membrane). Study Figure 29.1. Subsequently, the egg and sperm nuclei fuse to form the diploid nucleus of the zygote in the process of **fertilization.**

Early Embryology

After fertilization, the **zygote** begins a series of mitotic divisions that produce successively smaller cells. These divisions are called **cleavage,** and the first division occurs about 45–60 min after activation. Succeeding divisions occur at approximately 30-min intervals. The cells formed by the cleavage divisions are called **blastomeres.** See Figure 29.2.

The first cleavage division passes through the **animal** and **vegetal poles** of the zygote to yield cells of equal size. The second division also passes through both poles to form 4 cells of equal size.

The third division is perpendicular to the polar axis and forms 8 cells. The cells of the **animal hemisphere** are slightly smaller than those of the **vegetal hemisphere** due to slightly more **yolk** in the vegetal hemisphere. The fourth division forms 16 cells; however, it forms 4 large cells and 4 tiny cells in the vegetal hemisphere. The tiny cells are called **micromeres.**

The **blastula** is formed about 9 hr after activation and is no longer enveloped by the fertilization membrane. The inner cavity, the **blastocoele,** is filled with fluid. Cilia develop on the outer surfaces of the cells, and their beating produces a rotational motion of the blastula.

Continued division of the cells results in the inward growth (invagination) of cells at the vegetal pole led by the micromeres. The **early gastrula** consists of two cell layers. The inner cell layer, the **endoderm,** forms the **embryonic gut** (archenteron), which opens to the exterior via the **blastopore.** The outer cell layer is the **ectoderm.** In the **late gastrula,** pouches bud off the endoderm to form the **mesoderm.** All three germ layers (embryonic tissues) are now present, and all later-appearing adult tissues and organs are derived from them. See Table 29.1. In both echinoderms and chordates, the blastopore becomes the anus, and a mouth forms later from another opening at the other end of the embryonic gut. In roundworms, mollusks, annelids, and arthropods, the blastopore becomes the mouth and an anal opening forms later.

Assignment 1

Complete item 1 on Laboratory Report 29 that begins on page 375.

Procurement of Gametes

Your instructor has obtained gametes from male and female sea urchins prior to the laboratory session. Several sea urchins may be needed to find a male and female because sex cannot be easily determined by external examination. The gametes have been obtained in the following manner.

Materials

Beakers, 50 ml and 100 ml
Dropping bottles
Finger bowls
Hypodermic needles, 22 gauge

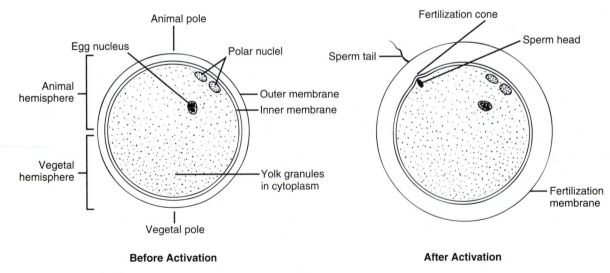

Before Activation   **After Activation**

Figure 29.1 Sea urchin egg.

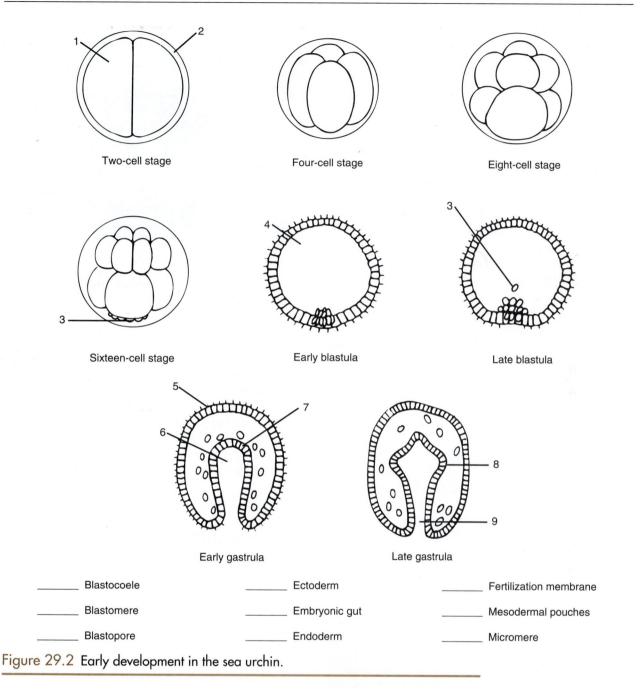

Figure 29.2 Early development in the sea urchin.

TABLE 29.1	EXAMPLES OF TISSUES AND ORGANS FORMED FROM THE GERM LAYERS IN HUMANS	
Endoderm	*Mesoderm*	*Ectoderm*
Linings of the	Skeleton	Epidermis, including hair and nails
Urinary bladder	Muscles	Inner ear
Digestive tract	Kidneys	Lens, retina, and cornea of the eye
Respiratory tract	Gonads	Brain, spinal cord, nerves, and adrenal medulla
Liver	Blood, heart, and blood vessels	
Pancreas	Reproductive organs	
Thyroid, parathyroids, and thymus	Dermis of the skin	

Hypodermic syringes, 5 ml
Medicine droppers
Syracuse dishes
Potassium chloride solution, 0.5 M
Seawater at 20°C
Sea urchins

Procedure

1. Inject 1 ml of the 0.5 M potassium chloride solution into each of three or four sea urchins. Insert the hypodermic needle through the membranous region around the mouth, as shown in Figure 29.3.
2. Place the urchins on paper towels with the oral (mouth) side down. Watch for the release of the gamete secretions from the aboral surface. The sperm secretion is white, and the egg secretion is pale buff in color.
3. As soon as a female starts shedding, place her on a beaker full of cold seawater, aboral side down so that the aboral surface is in the water. Release of all the eggs will take several minutes. See Figure 29.4.
4. After the eggs have been released, discard the female sea urchin. Swirl the eggs and water to wash the eggs, and allow them to settle to the bottom of the beaker. Pour off the water and add fresh seawater. Repeat this washing procedure twice. It will facilitate the activation process.
5. After the final washing, swirl the water to disperse the eggs. Then pour about 25 ml of seawater and eggs into each of five or six finger bowls and keep them at 20°C until used. The eggs will remain viable for 2–3 days at 20°C if the bowls are stacked to reduce evaporation.
6. Allow several minutes for the sperm secretion to accumulate on the aboral surface of a male. Then

Figure 29.4 Shedding female on beaker of seawater.

remove the secretion with a medicine dropper and place it in a Syracuse dish. See Figure 29.5. Undiluted sperm secretion in a covered dish will be viable for 2–3 days at 20°C. Prepare a sperm solution, just before use, by placing 2 drops of sperm secretion in 25 ml of seawater and dispense in a dropping bottle.

7. Keep gametes at 20°C until used.

 ## Assignment 2

Materials

Colored pencils
Compound microscope
Depression slides and cover glasses
Glass-marking pen
Medicine droppers
Toothpicks
Ward's culture gum
Developing embryos at 3, 6, 12, 24, 48, and 96 hr
Unfertilized eggs in finger bowl of seawater at 20°C
Sperm solution in dropping bottle at 20°C
Prepared slides of sea urchin blastula, gastrula, and larval stages

1. **Complete item 2a on the laboratory report.**
2. Place 1 drop of the egg and seawater mixture (8–12 eggs) in a depression slide, and observe without a cover glass. Use the 10× objective. Compare the eggs with Figure 29.1. **Draw two to three eggs in the space for item 2b on the laboratory report.**
3. While the slide is on the microscope stage, add 1 drop of the sperm mixture at the edge of the depression, record the time, and quickly observe

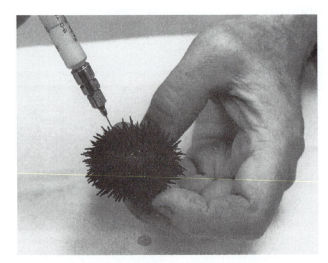

Figure 29.3 Injection of sea urchin.

Figure 29.5 Removing sperm with a dropper.

with the 10× objective. Note how the motile sperm cluster around the eggs. Why? Observe the rapid formation of the fertilization membrane. When most of the eggs have been activated, record the time. *Draw two to three activated eggs in the space for item 2b on the laboratory report.*

4. Use a toothpick to place a small amount of Ward's culture gum around the depression on the slide, and add a cover glass. This prevents evaporation of water but allows passage of O_2 and CO_2. Write your initials on the slide with a glass-marking pen.
5. Place your slide, egg mixture, and sperm mixture in the refrigerator at 20°C.
6. *Complete item 2c on the laboratory report.*
7. Label Figure 29.2. Color the cells as follows to distinguish the embryonic tissue:
   ectoderm—blue
   mesoderm—red
   endoderm—yellow
8. Examine your slide of activated eggs at 15–20-min intervals, and try to observe the division of the zygote. Keep the slide at 20°C when not observing it.
9. Prepare and observe, in age sequence, slides of different stages of sea urchin development. Make only one slide at a time. Label it by age and with your initials. Keep it at 20°C when not observing it. Examine these slides at 15–20-min intervals to observe a cell division. Compare your observations with Figure 29.2.
10. Compare the living blastula, gastrula, and larval stages with the prepared slides.
11. When finished, clean the slides, cover glasses, microscope stage, and objectives to remove all traces of seawater.
12. *Complete item 2 on the laboratory report.*

CHORDATE DEVELOPMENT

The early development of amphioxus, a primitive chordate, is similar to that observed in the sea urchin. Homologous stages are easily recognized because the eggs of both organisms contain relatively little yolk. The homologous stages are more difficult to recognize in eggs with more yolk, especially in bird eggs. This problem exists because cell divisions are slower in regions where yolk is abundant, and the cells become much larger.

 Assignment 3

Materials

Colored pencils
Compound microscope
Prepared slides of:
   amphioxus embryonic development
   frog embryo, x.s., neural tube stage

1. Study Figure 29.6. Note the similarity in development of amphioxus to that of the sea urchin up to the gastrula. Note that (a) the mesoderm is formed from pouches that bud off the endoderm, (b) the mesoderm destined to be the notochord is located between the mesodermal pouches, and (c) the neural plate is formed by the dorsal part of the ectoderm and folds up to form the neural tube.
2. Color the embryonic tissues in Figure 29.6, parts K through O.
   ectoderm—blue
   neural tube—green
   mesoderm—red
   endoderm—yellow
3. Examine a prepared slide of amphioxus development and locate stages like those in Figure 29.6.
4. Examine a prepared slide of frog embryo, x.s., in the neural tube stage and compare it with a similar stage in amphioxus development shown in Figure 29.6. Locate the ectoderm, neural tube, mesoderm, coelom, endoderm, and embryonic gut. Note the large, yolk-filled endodermal cells forming the ventral part of the gut wall.
5. *Complete item 3 on the laboratory report.*

HUMAN DEVELOPMENT

A human embryo (blastocyst) is implanted in the uterus about 5 days after fertilization, and germ layers and extra-embryonic membranes are evident by the 14th day. See Figure 29.7A. Chorionic villi (projections) attach the embryo firmly to the uterine lining, and some of them will become part of the embryonic

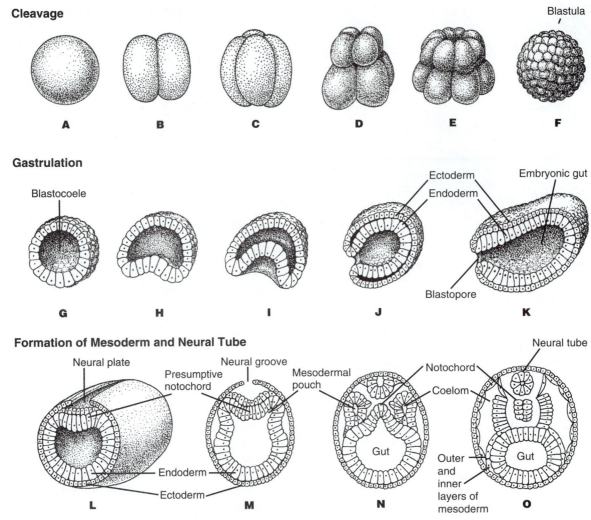

Cleavage

A   B   C   D   E   F

Blastula

Gastrulation

Blastocoele

Ectoderm
Endoderm
Embryonic gut
Blastopore

G   H   I   J   K

Formation of Mesoderm and Neural Tube

Neural plate
Presumptive notochord
Neural groove
Mesodermal pouch
Notochord
Neural tube
Coelom
Endoderm
Ectoderm
Gut
Outer and inner layers of mesoderm
Gut

L   M   N   O

Figure 29.6 Early development in amphioxus.

portion of the placenta. The first 8 weeks of development are known as the **embryonic period,** and the remainder of pregnancy is known as the **fetal period.**

At 8 weeks, the **fetus** is recognizably human, with all organ systems in rudimentary form. It is about 3 cm ($1\frac{1}{4}''$) in length and 1 g in weight. Some recognizable characteristics include the following: (1) The limbs are recognizable as arms and legs, and the fingers and toes are formed; (2) Bone formation begins, and internal organs continue to form; (3) The head is nearly as large as the body. All major brain regions are present, and the eyes are far apart, with the eyelids fused; (4) The cardiovascular system is functional.

Subsequent development results in a baby being born about 280 days after fertilization (conception). When born prematurely, a fetus has about a 15% chance of survival at 24 weeks but nearly a 100% chance at 30 weeks.

 Assignment 4

Materials

Colored pencils
Demonstration of pregnant cat or pig uterus
Model of a pregnant human female torso
Models of human developmental stages
Preserved human fetuses of various ages

1. Study Figure 28.1 to be sure that you understand the location of the extra-embryonic membranes.
2. Study Figure 29.7A, noting the formation of the extra-embryonic membranes and germ layers. Color-code the germ layers.
    ectoderm—blue
    mesoderm—red
    endoderm—yellow

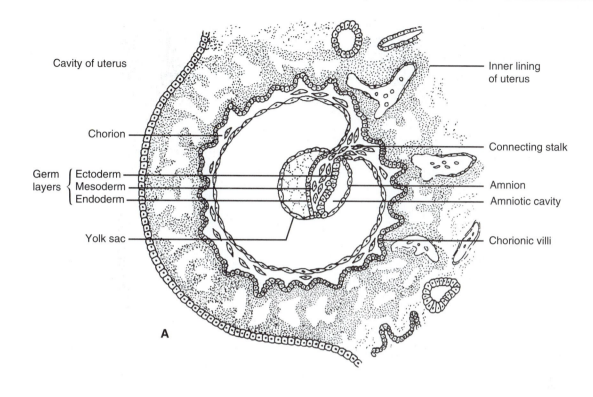

Cavity of uterus

Chorion

Germ layers { Ectoderm Mesoderm Endoderm }

Yolk sac

A

Inner lining of uterus

Connecting stalk

Amnion

Amniotic cavity

Chorionic villi

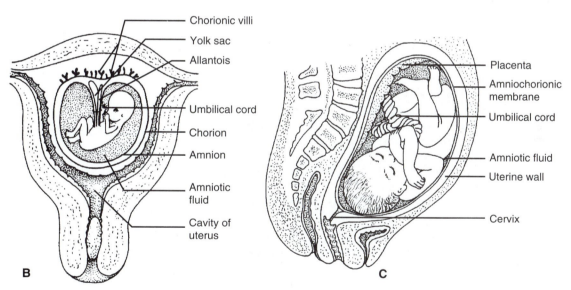

Chorionic villi

Yolk sac

Allantois

Umbilical cord

Chorion

Amnion

Amniotic fluid

Cavity of uterus

B

Placenta

Amniochorionic membrane

Umbilical cord

Amniotic fluid

Uterine wall

Cervix

C

Figure 29.7 Selected stages in human development. **A.** A human embryo, about 14 days old, implanted in the uterus. Note the germ layers and extra-embryonic membranes. **B.** Uterus with a fetus about 10 weeks old. Note the extra-embryonic membranes and the umbilical cord. **C.** A full-term fetus with head pressed against the cervix. Note the placenta, the umbilical cord, and the fetal position.

3. Examine Figure 29.7B, noting the components of the umbilical cord, the placenta, and the amniotic fluid enveloping the fetus.

4. *Complete items 5a and 5b on the laboratory report.*

5. Examine the pregnant human torso model and compare it with Figure 29.7C. Note how the amnion and chorion are pressed together, forming the amniochorion in late stages of pregnancy. Does the fetus fill all available space?

6. Examine the pregnant cat or pig uterus, noting the amniochorion, placenta, and umbilical cord.
7. Examine the series of models showing human development. Note the progression of development with the age of the fetus.
8. Examine the series of preserved human fetuses, noting the degree of development and the age of each fetus.
9. *Complete the laboratory report.*

FERTILIZATION AND DEVELOPMENT

Student _____

Lab Instructor _____

1. INTRODUCTION

Write the term that matches each phrase.

a. Penetration of egg by a sperm _____

b. Mitotic division of fertilized egg _____

c. Fusion of egg and sperm nuclei _____

d. Solid ball of cells (embryonic stage) _____

e. Hollow ball of cells (embryonic stage) _____

f. Cell formed by fertilization _____

g. Embryonic stage formed by gastrulation _____

2. ACTIVATION AND EARLY DEVELOPMENT IN THE SEA URCHIN

a. Write the term that matches each phrase.

1. Chemicals that attract sperm to eggs _____

2. Hemisphere of egg containing most yolk _____

3. Hemisphere of egg penetrated by sperm _____

4. Extension of egg engulfing sperm head _____

5. Prevents penetration by additional sperm _____

6. Time until most eggs are activated _____

b. From your slide, draw a few nonactivated and activated eggs. Show the distribution of sperm around the activated eggs. Label pertinent parts.

Nonactivated Eggs   **Activated Eggs**

c. Explain the distribution of the sperm. _____

_____

d. List the labels for Figure 29.2.

1. _____   4. _____   7. _____

2. _____   5. _____   8. _____

3. _____   6. _____   9. _____

e. Record the time required to reach each stage.

Two-cell stage _____   Four-cell stage _____   Eight-cell stage _____

f. Indicate the most abundant stage after each time period.

12 hr _____   24 hr _____   48 hr _____

3. CHORDATE DEVELOPMENT

a. Indicate the plane (polar or equatorial) of the cleavage divisions.

1. First division   _____

2. Second division   _____

3. Third division   _____

b. Are the planes the same as in the sea urchin? _____

c. Why are the cells formed in the vegetal hemisphere larger than those of the animal hemisphere? _____

_____

d. In which hemisphere does gastrulation occur? _____

Is this the same site as in the sea urchin? _____

e. Indicate the embryonic tissue that forms these structures.

1. Embryonic gut   _____

2. Mesodermal pouches   _____

3. Notochord   _____

4. Neural tube   _____

5. Lining of the coelom   _____

f. From your slide, draw a frog neurula in cross-section. Label the pertinent parts. Color your drawing as follows: ectoderm—blue; neural tube—green; mesoderm—red; endoderm—yellow.

g. What similarities have you observed in embryos of the sea urchin, amphioxus, and frog?

_____

_____

_____

4. HUMAN DEVELOPMENT

a. By what process are materials exchanged between embryonic and maternal bloods? _____

_____

b. Describe the function of the placenta and umbilical cord in humans.

Placenta _____

_____

Umbilical cord _____

_____

c. List the ages and distinctive visible developmental features of the fetuses observed.

Age	Characteristics

    d. At what age is an embryo implanted in the uterus? _____

    e. At what age are germ layers formed? _____

    f. At what age does the fetus resemble a human? _____

    g. At 8 weeks, what percentage of the fetus's length consists of the head? _____

    _____

    h. At 8 weeks, are arms, legs, fingers, and toes evident? _____

    i. What trend occurs in the head–body proportions during development? _____

    _____

    _____

    _____

    j. What is the duration of pregnancy in humans? _____

    k. Physicians usually advise pregnant women to avoid all drugs during pregnancy. What is the rationale for such

    avoidance? _____

    _____

    _____

    _____

    l. Why does a fetus assume a "fetal position" in the uterus? _____

    _____

    _____

    _____

    m. What is the stimulus that causes a newborn baby to start breathing? _____

    Explain. _____

    _____

    _____

    _____

EARLY EMBRYOLOGY OF THE CHICK

OBJECTIVES

After completing the laboratory session, you should be able to:
1. Identify and give the function of each structure of a bird egg.
2. Compare the bird egg with a sea urchin egg.
3. Compare blastoderm formation in a chick embryo with blastula formation in a sea urchin embryo.
4. Describe how the pattern of bird embryo development is affected by the large amounts of yolk.
5. Locate the following structures in a 48-hr and 96-hr chick embryo:

a. Brain	e. Foregut	h. Vitelline arteries
b. Optic cup	f. Heart	i. Neural tube
c. Otic vesicle	g. Vitelline veins	j. Somites
d. Headfold		

6. Identify three factors in the external environment that influence embryonic development.
7. Define all terms in bold print.

The amniote eggs of reptiles and birds contain a large amount of yolk that serves as a nutrient for the embryo. The yolk never divides, but is eventually surrounded by a yolk sac and absorbed. Because of the large amount of yolk, the pattern of development in reptilian and bird eggs is quite different from that of the sea urchin, in which the entire fertilized egg cell divides. See Figure 30.1.

STRUCTURE OF A BIRD EGG

The structure of a bird egg is uniquely adapted for the development of the embryo and chick. It consists of the **true egg** and accessory structures. The basic structure is shown in Figure 30.2. The true egg is surrounded by the **vitelline membrane** and is an extremely large cell containing the fat-rich **yolk,** which serves as a nutrient

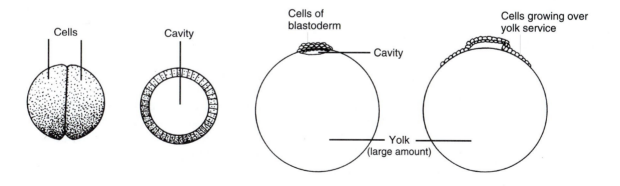

A Division of entire egg (sea urchin)   **B** Early division of the blastodisc (bird or reptile)

Figure 30.1 Contrasting patterns of early development.

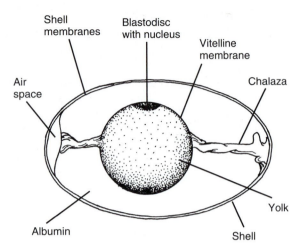

Figure 30.2 Structure of a bird's egg.

for the embryo. The **blastodisc** is a small, whitish, circular area located at the animal pole, and it contains the egg nucleus. All other components are accessory structures.

The **calcareous shell** protects the embryo from mechanical injury and allows a free exchange of gases with the environment. The inner and outer **shell membranes** allow the passage of gases but prevent evaporative water loss. An air space exists at the rounded end of the egg between the separated inner and outer shell membranes. **Albumin,** the white of the egg, is a clear, viscous protein that serves as a nutrient for the developing embryo. The **chalaza** is a strand of dense protein that suspends the true egg inside the shell and keeps the blastodisc upward.

 Assignment 1

Materials

Dissecting instruments
Finger bowls
Paper towels
Scissors, fine-tipped
Unfertilized chicken eggs

1. Obtain an unfertilized chicken egg from the stock table. *Keeping the same surface of the shell upward,* place it in a finger bowl partially filled with paper toweling to support the egg.
2. Draw a circle with a 2-cm diameter on the upper surface of the shell over the anticipated location of the blastodisc. Use a dissecting needle and forceps to pick away bits of the shell within the circle. Try not to rupture the shell membranes. When the shell has been removed, observe the appearance of the

shell membranes. Then use fine-tipped scissors to cut away the shell membranes and remove them.
3. Observe the true egg. The whitish blastodisc should be visible. If necessary, enlarge the opening to see the attachment points of the chalaza.
4. When you have completed your observations, gently break the shell and spill the contents into a finger bowl of water. Observe the attachment of the chalaza to the true egg. Locate the air space at the rounded end of the shell.
5. *Complete item 1 on Laboratory Report 30 that begins on page 387.*

DEVELOPMENT WHILE IN THE OVIDUCT

After **copulation,** sperm swim to the upper end of the oviduct and contact the ovulated egg. The egg nucleus undergoes its second meiotic division after a sperm has penetrated the blastodisc. As the fertile egg (zygote) descends the oviduct, the albumin, shell membranes, and a shell are secreted around it. Concurrently, embryonic cell divisions occur in the blastodisc. When an egg is laid (about 20 hr after fertilization), the embryo is already at the early gastrula stage.

Figure 30.3A shows the enlarged blastodisc in surface and sagittal views. The zygote nucleus has divided for the first time, and a cleavage furrow separates the two resultant nuclei. This stage is **homologous** to the two-cell stage of the sea urchin embryo. The eight-cell stage is shown in Figure 30.3B. In the sagittal view, note how the furrows between nuclei extend only partially through the blastodisc.

Continued cell divisions change the central part of the blastodisc into a cellular **blastoderm** several cells in thickness (Figure 30.4). A cavity forms between the blastoderm and the underlying yolk. The peripheral ring of noncellular blastodisc is now called the **periblast.**

The blastoderm and segmentation cavity are considered by some authorities to be homologous to the blastula and blastocoele of sea urchins. The blastoderm cannot form a hollow ball of cells like the sea urchin blastula, however, because the blastoderm develops on the surface of the yolk.

Figure 30.5 shows the stage of embryonic development at the time an egg is laid. The upper layer of small cells (sagittal view) has become **ectoderm** tissue; large granular cells have formed a lower layer of **endoderm** tissue toward the posterior end of the embryo. The endodermal tissue at this stage is crescent shaped in the surface view and represents the partly formed roof of an incomplete **embryonic gut.** The former segmentation cavity is now the cavity of the gut. After the egg is laid, the embryo does not develop any further until it is incubated.

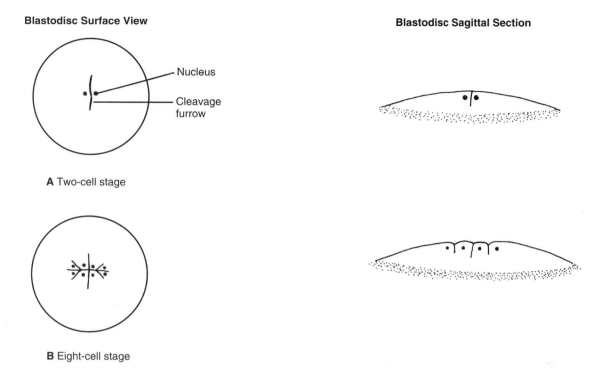

Blastodisc Surface View

Nucleus

Cleavage furrow

A Two-cell stage

B Eight-cell stage

Blastodisc Sagittal Section

Figure 30.3 Early divisions of the blastodisc.

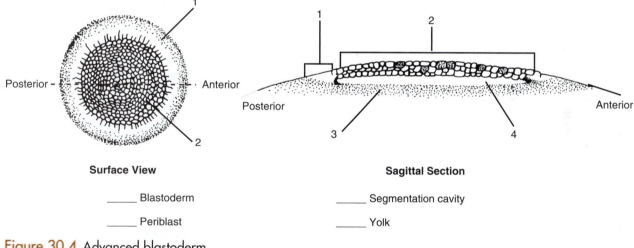

1

2

Posterior — — Anterior

2

Surface View

_____ Blastoderm

_____ Periblast

1

2

Posterior

3

4

Anterior

Sagittal Section

_____ Segmentation cavity

_____ Yolk

Figure 30.4 Advanced blastoderm.

A tubular gut is eventually achieved in the chick embryo, but unlike the rapid gut formation by invagination in the sea urchin, the process is delayed in the chick embryo because of its flattened form.

 Assignment 2

1. Study Figure 30.3 and note the incomplete cleavage. Why does this occur?
2. Study and label Figures 30.4 and 30.5.
3. **_Complete item 2 on the laboratory report._**

DEVELOPMENT AFTER INCUBATION BEGINS

Identify in Figure 30.6 the structures described here. A few hours after incubation is started, a **primitive streak** begins to form along the anterior–posterior axis of the embryo by the anterior migration of endoderm cells. After 16 hr of incubation, the primitive streak is complete and extends entirely across the blastoderm. Note how cells from the blastoderm have spread out over the yolk in an oval pattern (Figure 30.6). The primitive streak consists of an elongated depression, the **primitive**

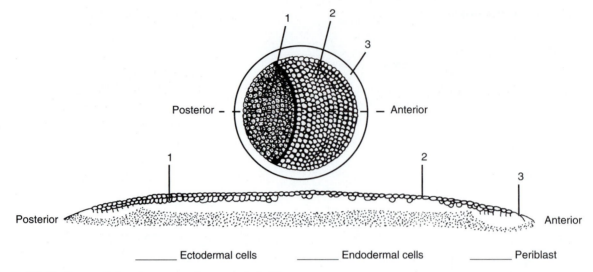

Figure 30.5 Stage of development when egg is laid.

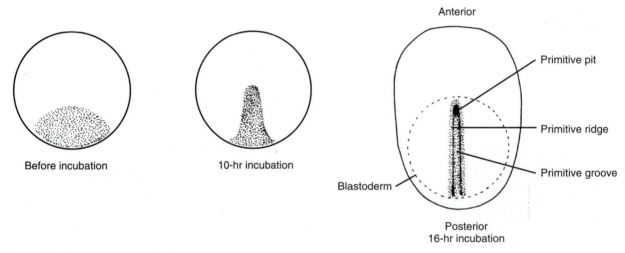

Figure 30.6 Primitive streak formation.

groove, bounded on either side by **primitive ridges.** At the anterior end of the groove is the **primitive pit.** The primitive streak and pit together are homologous to the blastopore of the sea urchin embryo.

Cells from the surface of the blastoderm migrate into the primitive groove and pit and spread out in a layer between the ectoderm and endoderm. These migrating cells form **mesoderm** tissue and the **notochord.** Subsequent embryonic development occurs anteriorly from the primitive pit.

The primitive streak and primitive pit will become the anus of the chick embryo; the embryo therefore develops anterior to the streak. As ectoderm, endoderm, and mesoderm tissues spread anteriorly and laterally over the yolk, they rise up and form a **headfold** in the head region (Figure 30.7). The headfold contains an endoderm-lined cavity that represents the beginning of

a **tubular foregut.** Note how relatively small the primitive streak now appears compared with the rest of the embryo.

In Figure 30.8, the neural tube has been modified to form four parts of the developing **brain**—the **optic cups,** which will become the eyes, and the **otic cups,** which will become the ears. Posterior to the brain, the neural tube will form the spinal cord. The tubular **heart** is starting to twist, a step toward its development into four chambers. The segmentally arranged **somites** will form the musculature of the body. The **vitelline arteries** carry blood from the embryo to the yolk surface, which facilitates gas exchange and nutrient absorption. Blood is returned to the embryo via the **vitelline veins.**

Refer to Figure 30.9 to see the continuing development in a 96-hr embryo. The more rapid anterior

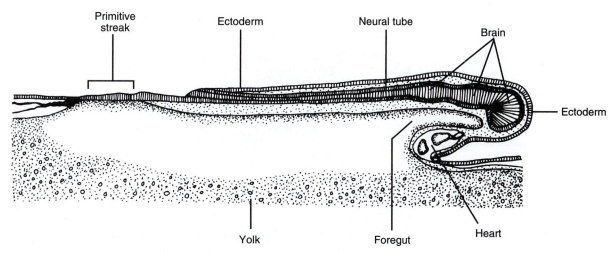

Figure 30.7 Embryo after 48-hr incubation (sagittal section).

development has caused the brain and head to be much larger than the rest of the body. The **limb buds** and **tail bud** are evident, and the liver and kidneys are starting to form. Note the presence of nonfunctional **pharyngeal slits.**

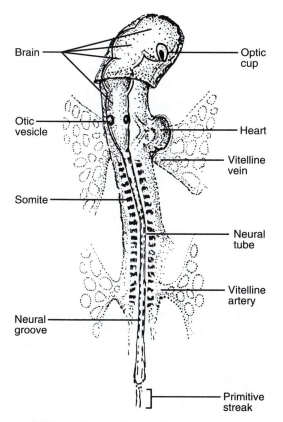

Figure 30.8 Embryo after 48-hr incubation (surface view).

Assignment 3

Materials

Colored pencils
Compound microscope
Models of chick embryonic development
Prepared slide of chick embryo, w.m., 18, 24, 36, 48, and 96 hr
Prepared slides of chick embryo, x.s., 24 hr

1. Examine a model of the primitive streak stage, and compare it with Figure 30.6.
2. Examine a prepared slide of chick embryo, w.m., after 18 hr of incubation. Locate the structures shown in Figure 30.6.
3. Examine the demonstration slide of chick embryo, x.s., 24 hr, which shows mesodermal cells migrating from the primitive groove laterally between ectoderm and mesoderm.
4. Study the models of chick development between 18 and 96 hr of development, and examine the demonstration slides.
5. Study Figures 30.7 and 30.8 to understand the structure of a 48-hr chick embryo. Correlate the sagittal section with the whole mount. Color the structures as follows:
   ectoderm—blue
   neural tube—green
   mesoderm—red
   endoderm—yellow
6. Examine prepared slides of 48-hr chick embryo, w.m., and sagittal section. Locate the structures shown in Figures 30.7 and 30.8.
7. Study Figure 30.9, and note the developmental changes between 48 and 96 hr. Color the structures as in step 5.

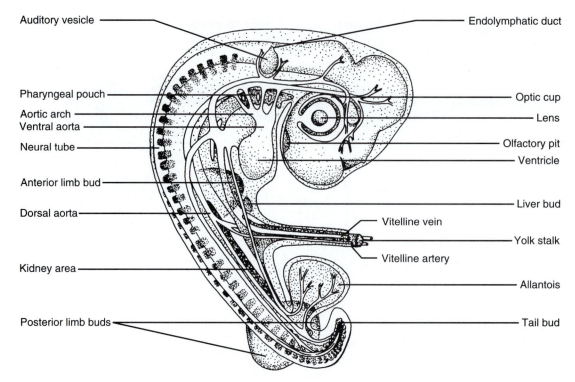

Figure 30.9 Chick embryo, 96 hr, showing four nonfunctional pharyngeal slits.

8. Examine a slide of chick embryo, w.m., 96 hr. Locate the structures shown in Figure 30.9.
9. *Complete item 3 on the laboratory report.*

STUDY OF A 48-HR EMBRYO

After you understand the structure of a 48-hr chick embryo, observe a living 48-hr embryo. Work in pairs and follow the directions carefully.

 Assignment 4

Materials

Desk lamp
Incubator, 39°C
Warming oven, 39°C
Depression slides
Dissecting needles
Filter paper rings with an opening 4 mm × 6 mm
Finger bowls
Forceps, label type, square-nosed
Forceps, sharp-tipped
Scissors, fine-tipped
Ringer's solution in dropping bottles
Fertile eggs, 48 hr

1. Obtain a fertilized egg that has been incubated 48 hr. *Keeping the same surface of the shell upward,* place it in a finger bowl on paper toweling so that it will not roll. Place a lighted lamp over the egg to keep it warm.
2. Draw a circle with a 2-cm diameter on the shell over the presumed location of the embryo. Using sharp-tipped forceps and a dissecting needle, chip away the shell. Then, using fine-tipped scissors, cut the shell membranes within the circle to expose the embryo. Make the opening larger, if necessary.
3. Locate the embryo, which will appear as a faint white streak in the center of an oval surface. When the large end of the egg shell is to the left, the head of the embryo will be directed away from the observer.
4. Using label forceps, place a filter paper ring over the embryo so that the embryo is visible within the hole in the ring. See Figure 30.10.
5. Allow time for the ring to become wet so it will adhere to the embryonic membranes. Then use fine-tipped scissors to cut through the embryonic membranes around the outside of the ring. Do not cut deeply into the yolk, as this will tend to contaminate your preparation with excessive yolk.
6. Obtain a depression slide from the warming oven and place 2 drops of warm Ringer's solution in the depression. Using the label forceps, lift the paper ring with adhered embryo and place it in the depression on the slide.

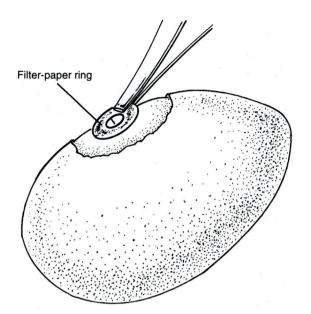

Filter-paper ring

Figure 30.10 Removal of a 48-hr chick embryo.

7. Examine the embryo at 40× and 100× with a compound microscope, and identify the structures shown in Figure 30.8.
8. *Complete item 4 on the laboratory report.*

GENERALIZATIONS

The chemical and structural organization (polarity) of the bird egg influences its pattern of early embryonic development, as was the case in sea urchin development. When incubation begins, certain factors in the external environment (outside the eggshell) exert an influence on the internal environment of the cells of the chick embryo. In turn, the internal environment of these cells selectively activates the genes that control the pattern of development.

EARLY EMBRYOLOGY
OF THE CHICK

Student _____

Lab Instructor _____

1. EGG STRUCTURE

a. Write the term that matches each phrase.

1. Envelops the true egg   _____

2. Contains the egg nucleus   _____

3. Fat-rich nutrient in true egg   _____

4. Protein-rich nutrient exterior to true egg   _____

5. Suspends true egg   _____

6. Prevent evaporative water loss   _____

7. Protects true egg from mechanical injury   _____

b. Describe the egg structure that enables the blastodisc to remain upward. _____

_____

_____

_____

c. Describe the advantage provided by the greater amount of yolk in a bird egg in contrast to the sea urchin egg.

_____

_____

2. DEVELOPMENT BEFORE INCUBATION

a. Write the term that matches each phrase.

1. Site of blastodisc penetration by sperm   _____

2. Embryonic stage at time egg is laid   _____

3. Formed by divisions of the blastodisc   _____

4. Homologous with the sea urchin blastula   _____

5. Cavity between yolk and blastoderm   _____

6. Outer layer of cells of blastoderm   _____

7. Inner layer of cells of blastoderm   _____

b. Explain why cleavage of a bird egg is so different from that of the sea urchin. _____

_____

_____

c. List the labels for Figures 30.4 and 30.5.

Figure 30.4	*Figure 30.5*
1. _____	1. _____
2. _____	2. _____
3. _____	3. _____
4. _____	

3. DEVELOPMENT AFTER INCUBATION BEGINS

a. Write the term that matches each phrase.

1. Direction of endoderm growth _____

2. Elongated depression bordered by ridges _____

3. Homologous with blastopore of sea urchin _____

4. Fate of primitive streak and primitive pit _____

5. First part of tubular gut to form _____

6. Carry blood from the yolk to the embryo _____

7. Form body muscles _____

8. Most rapidly developing part of 96-hr chick _____

9. Fate of otic vesicle _____

10. Fate of optic cup _____

b. Describe the formation of the mesoderm in the 24-hr chick. _____

_____

_____

_____

4. LIVING 48-HR EMBRYO

a. In the living 48-hr embryo, does the:

head turn to the right or left? _____

heart bulge to the right or left? _____

same pattern seem to be observed on the prepared slide? _____

pattern seem random or determined? _____

b. In the living 48-hr embryo:

was blood present? _____

was the heart beating? _____

c. Indicate the optimum temperature for:

chick development _____

sea urchin development _____

d. What is the ultimate control of embryonic development? _____

Part V

Plant Biology

STRUCTURE OF FLOWERING PLANTS

OBJECTIVES

After completing the laboratory session, you should be able to:
1. Identify the external structure of a flowering plant.
2. Identify monocots and dicots.
3. Identify types of root systems and leaves.
4. Identify tissues and their arrangements composing roots, stems, and leaves.
5. Define all terms in bold print.

Flowering plants are the most advanced vascular plants. They have well-developed vascular tissues providing support as well as transport of materials, and, as you learned in Exercise 13, their life cycle involves flowers, pollination, and seeds enclosed in fruits. In this exercise, you will study the external and internal structure of roots, stems, and leaves.

Flowering plants are subdivided into two major classes: **monocotyledonous** (monocots) **plants** and **dicotyledonous** (dicots) **plants.** Their distinguishing characteristics are shown in Figure 31.1.

Monocots include grasses, palms, and lilies. Most flowering plants are dicots, and they fall into one of two major categories: herbaceous or woody. **Herbaceous** (not woody) **dicots** are usually annuals or biennials. Annuals complete their life cycle in a single growing season. They include most wildflowers, beans, and tomatoes. Biennials require two growing seasons; flowers are produced only in the second season. **Woody dicots** are usually perennials that live several years and produce flowers each year, such as oaks and roses.

GENERAL EXTERNAL STRUCTURE

Figure 31.2 shows the basic parts of a flowering plant. The shoot system consists of a **stem** that supports the **leaves, flowers,** and **fruits.** Leaves branch from the stem at sites called **nodes.** A section of stem between nodes is an **internode.** Leaves are the primary photosynthetic organs of the plant, and they exhibit two types of venation: net venation and parallel venation.

Leaves with **net venation** have a central vascular bundle (vein) called a midrib from which smaller lateral veins branch. Such leaves consist of a thin, expanded portion called a **blade** and a leaf stalk called a **petiole.** Leaves with **parallel venation** lack a midrib and have veins running parallel to each other for the length of the blade. Such leaves usually lack a petiole.

The root system is located in the ground, and it may be more highly branched than the shoot system. Roots not only anchor the plant but also absorb water and nutrients.

 Assignment 1

Materials

Dissecting microscope
Coleus seedlings
Coleus and corn stems in solution of red food coloring
Corn seedlings
Representative dicots and monocots
Razor blades, single-edged

1. Label Figure 31.2. ***Complete items 1a and 1b on Laboratory Report 31 that begins on page 399.***
2. Obtain corn and *Coleus* seedlings. Gently wash the soil from their roots and lay them on a paper towel for observation. Compare the roots, stems, and leaves with Figures 31.1 and 31.2.
3. Stems of several *Coleus* and corn seedlings have been placed in a water-soluble dye that stains vascular tissue as it is carried up the stem. Use a

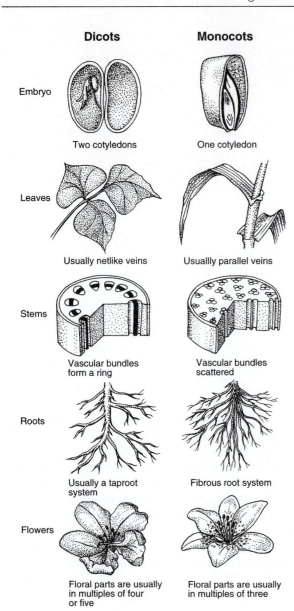

Figure 31.1 Distinguishing characteristics of monocots and dicots.

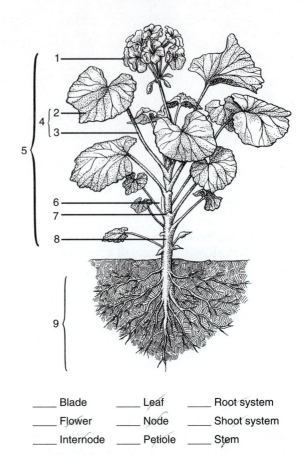

____ Blade   ____ Leaf   ____ Root system

____ Flower   ____ Node   ____ Shoot system

____ Internode ____ Petiole   ____ Stem

Figure 31.2 External structure of a flowering plant.

razor blade to cut thin cross-sections from the *Coleus* and corn stems, and examine them with a dissecting microscope. Observe the arrangement of the vascular bundles and compare them with Figure 31.1.

4. Identify the "unknown" plants as dicots or monocots. ***Complete item 1 on the laboratory report.***

ROOTS

Roots perform three important functions: (1) anchorage and support, (2) absorption and transport of water and minerals, and (3) storage and transport of organic nutrients.

The Root Tip

The basic structure of a root tip is shown in Figure 31.3. Cells formed in the **region of cell division** either become part of the root proper or form the root cap. The **root cap** is composed of rather large cells that protect the region of cell division. It also provides a sort of lubrication as its cells are eroded by the root growing through the soil. Cells of the root are enlarged in the **region of elongation,** and this accounts for the greatest increase in the linear growth of the root. As the cells become older, they develop their specialized characteristics in the **region of differentiation,** where the **primary root tissues** are formed. Note that the cells of the root are arranged in columns. The column in which a cell is located determines the type of cell it will become. For example, cells in the outermost columns become epidermal cells, and those in the innermost columns become xylem cells. **Root hairs** are extensions of epidermal cells in the region of differentiation. They greatly increase the surface area of the root tip and are the primary sites of water and mineral absorption.

Growth in length of both roots and stems results from the formation of new cells in the region of cell

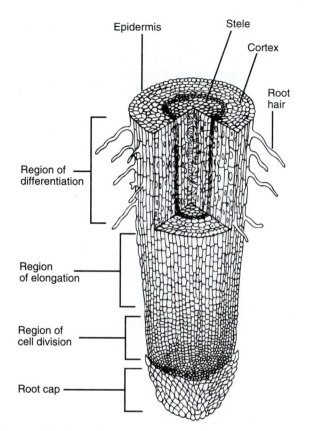

Figure 31.3 A dicot root tip.

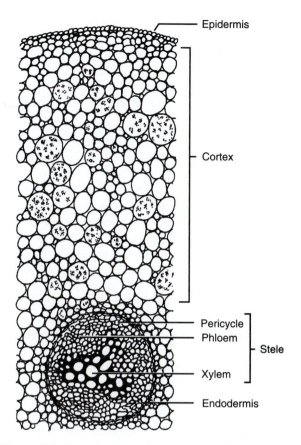

Figure 31.4 Herbaceous dicot (*Ranunculus*) root, x.s.

division and their subsequent enlargement in the region of elongation. Thus, growth of both roots and stems occurs at their tips, and this growth is continuous throughout the life of the plant.

Root Tissues

Figure 31.4 shows the tissues found in a root of a herbaceous dicot (*Ranunculus*) as viewed in cross-section. Note the three major divisions: epidermis, cortex, and stele.

The **epidermis** is the outermost layer of cells. It provides protection for the underlying tissues and reduces water loss.

The **cortex** composes the bulk of the root and consists mostly of large, thin-walled cells used for food storage.

The **endodermis,** the innermost layer of the cortex, is composed of thick-walled, water-impermeable cells and a few water-permeable cells with thinner walls. This ring of cells controls the movement of water and minerals into and out of the xylem.

The **stele** is that portion of the root within the endodermis. It is sometimes called the central cylinder. The outer layer of the stele is composed of thin-walled cells, the **pericycle,** from which branch roots originate. The large, thick-walled cells in the center of the stele compose the **xylem.** Between the rays of the xylem are are smaller cells composing the **phloem.**

Root Types

There are three types of roots. **Taproot systems** have a single dominant root from which branch roots arise. **Fibrous root systems** consist of a number of similar-sized roots that branch repeatedly. Fibrous roots are characteristic of monocots. Both types of root systems may be shallow or deep, depending on the species of plant, but taproots are capable of the deepest penetration. **Adventitious roots** are unique in that, unlike other roots, they do not grow from the primary root of the embryo. They originate from stems or leaves and are typically fibrous in nature.

 ## Assignment 2

Materials

Colored pencils
Compound microscope
Dissecting microscope
Dissecting instruments
Dropping bottles of methylene blue, 0.01%

Microscope slides and cover glasses
Examples of adventitious, fibrous, and taproots
Germinated grass (or radish) seeds
Prepared slides of:
  Allium (onion) root tip, l.s.
  Ranunculus (buttercup) root, x.s.

1. Color-code the root cap and regions of cell division, cell elongation, and cell differentiation in Figure 31.3 and the xylem and phloem in Figure 31.4.

2. Obtain a germinated grass seed. Place it on a microscope slide in a drop of water. Examine the young root with a dissecting microscope and at 40× with a compound microscope. Note the root hairs. Locate the oldest (longest) and youngest root hairs. Use a scalpel to cut the seed from the root. Add a drop of methylene blue and a cover glass. Examine the root at 100×. Note the attachment of a root hair to an epidermal cell. Try to locate a cell nucleus in a root hair. **Complete items 2a–2c on the laboratory report.**

3. Examine a prepared slide of *Allium* root tip, l.s., with a compound microscope at 40×. Locate the root cap and the regions shown in Figure 31.3, except for the region of differentiation, which is not shown on your slide. Compare the size of the cells in each region. **Complete item 2d on the laboratory report.**

4. Examine a prepared slide of *Ranunculus* root, x.s., with a compound microscope at 40×. Locate the tissues shown in Figure 31.4. Note the starch granules in the cells of the cortex and the arrangement of xylem and phloem.

5. Examine the examples of root types. Note their characteristics. **Complete item 2 on the laboratory report.**

STEMS

The stem serves as a connecting link between the roots and the leaves and reproductive organs. It also may serve as a site for food storage. The arrangement of vascular tissue in stems varies among the subgroups of vascular plants.

Monocot Stems

Figure 31.5 illustrates the cross-sectional structure of part of a corn stem. Note the **scattered vascular bundles** surrounded by large, thin-walled cells, a characteristic of monocots. Each vascular bundle has thick-walled, fibrous cells around the edges surrounding the large xylem vessels and the smaller sieve tubes and companion cells of phloem. Most of the support for the stem is provided by the xylem and fibrous cells.

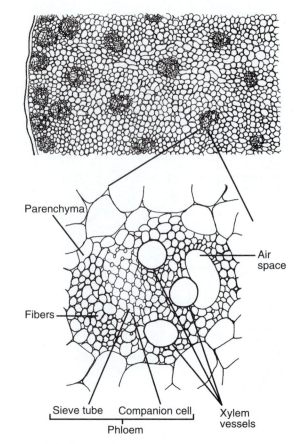

Figure 31.5 Corn (*Zea*) stem, x.s.

Herbaceous Dicot Stems

Figure 31.6 shows the structure of a portion of an alfalfa stem in cross-section. Note that the vascular bundles are arranged in a *broken ring* just interior to the **cortex.** This pattern is characteristic of herbaceous

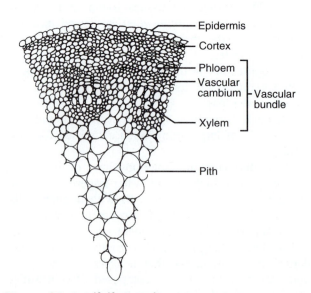

Figure 31.6 Alfalfa (*Medicago*) stem, x.s.

dicots. Each vascular bundle is composed of three tissues (from outside): **phloem, vascular cambium,** and **xylem.** The central portion of the stem, the **pith,** is composed of large, thin-walled cells.

Woody Dicot Stems

The structure of a young woody stem in cross-section is shown, in part, in Figure 31.7. Note the characteristic *continuous ring* of primary vascular tissue and the arrangement of the tissues within the stem. The **wood** of woody plants is actually xylem tissue.

In each growing season, the vascular cambium forms new (secondary) xylem and phloem. Each growing season is identifiable by an **annual ring** of xylem, which is composed of large-celled **spring wood** (lighter color) and small-celled **summer wood** (darker color). No cells are formed in fall or winter in temperate climates. The stem grows in diameter by the formation and enlargement of new xylem and phloem. Because the growth is actually from the inside of the stem and because the epidermis and cortex cannot grow, they tend to fracture and slough off. The **cork cambium** forms in the cortex, and it produces cork cells that assume the function of protecting the underlying tissues and preventing water loss. At this stage, the stem is subdivided into (1) bark, (2) vascular cambium, and (3) wood. The **bark** is everything exterior to the vascular cambium: phloem, cortex, cork cambium, and cork. Figure 31.8 illustrates the structure of an older woody stem.

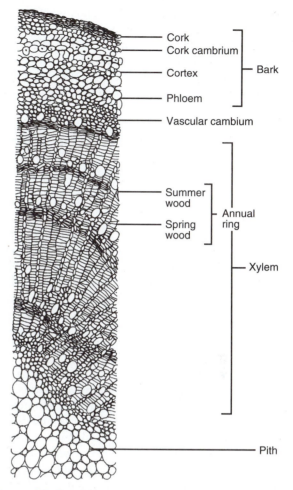

Figure 31.8 Older oak (*Quercus*) stem, x.s.

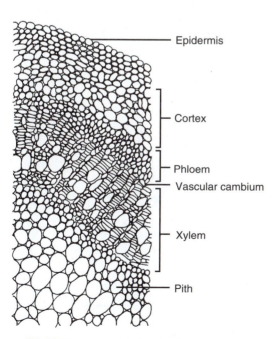

Figure 31.7 Young oak (*Quercus*) stem, x.s.

 Assignment 3

Materials

Colored pencils
Compound microscope
Tree stem, x.s.
Prepared slides of:
   Quercus (oak) stem, 1 yr, x.s.
   Quercus stem, 4 yr, x.s.
   Medicago (alfalfa) stem, x.s.
   Zea (corn) stem, x.s.

1. Color-code the xylem and phloem in Figures 31.5, 31.6, and 31.7. In Figure 31.8, color-code the spring wood, summer wood, vascular cambium, phloem, cork, and cork cambium.
2. Examine a prepared slide of *Zea* stem, x.s., and locate the parts shown in Figure 31.5. Note the arrangement of the vascular bundles in the stem. ***Complete items 3a and 3b on the laboratory report.***

3. Examine a prepared slide of *Medicago* stem, x.s., and locate the parts shown in Figure 31.6. Note the arrangement of the vascular bundles. ***Complete item 3c on the laboratory report.***

4. Examine prepared slides of *Quercus* stems, 1 yr and 4 yr, x.s. Locate the parts shown in Figures 31.7 and 31.8.

5. Determine the age of the demonstration section of tree stem. ***Complete item 3 on the laboratory report.***

LEAVES

Leaves are the primary organs of photosynthesis and are sites of gas exchange, including water loss by evaporation. The major part of a leaf is the broad, thin **blade,** which is attached to a stem by a leaf stalk, the **petiole,** in most dicots. Some leaves, especially in monocots, lack a petiole, so the blade is attached directly to a stem.

Venation

The blade is supported by many **veins** that are composed of vascular tissue plus a fibrous vascular bundle sheath. Leaves of monocots have **parallel venation** in which veins are arranged in parallel rows running the length of the leaf. See Figure 31.9E. Leaves for dicots have **net venation** in which a large central vein, the **midrib,** extends the length of the leaf. Veins branching from the midrib branch repeatedly into smaller and smaller veins that form a network throughout the leaf. See Figure 31.9A.

The pattern of veins in net-veined leaves follows one of two major types. In **pinnate** leaves, there is a single midrib with several lateral branches forming a pattern resembling the structure of a feather. In **palmate** leaves, there are several major veins branching from the petiole, resembling the fingers extending from the hand. If a leaf is divided into a number of leaflets (parts), it is said to be **compound.** If not, it is said to be **simple.** The combination of these patterns determines the type of venation, and it is used in classifying plants. See Figure 31.9.

Internal Structure

The internal structure of a lilac (*Syringa*) leaf is shown in Figure 31.10. An upper and lower **epidermis** form the surfaces of the leaf, and each is coated with a waxy **cuticle** that retards evaporative water loss through the epidermis. Epidermal cells do not contain chlorophyll, except for the scattered **guard cells** that surround tiny openings called **stomata** (*stoma* is singular). Gas exchange between leaf tissues and the atmosphere occurs through the stomata.

Stomata are often found only on the lower epidermis. This location prevents direct exposure to sunlight and helps to reduce evaporative water loss. When guard cells are photosynthesizing, the accumulation of sugars causes them to swell with water that enters by osmosis, and the swelling bends the guard cells, opening the stomata, which increases gas exchange. At night, water exits the guard cells, so they straighten out and close the stomata.

The **mesophyll** is the main photosynthesizing tissue in a leaf. It consists of two types. **Palisade mesophyll** consists of closely packed, upright cells located just under the upper epidermis, and it carries out most of the photosynthesis. **Spongy meosphyll** consists of a meshwork of cells with many spaces between them, and it occurs in the lower half of a leaf. This arrangement aids the movement of gases within the leaf.

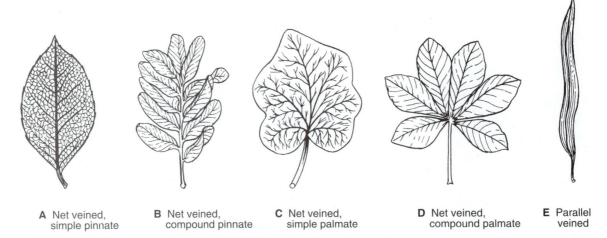

A Net veined, simple pinnate   **B** Net veined, compound pinnate   **C** Net veined, simple palmate   **D** Net veined, compound palmate   **E** Parallel veined

Figure 31.9 Examples of leaf venation in flowering plants. **A–D** are dicots; **E** is a monocot.

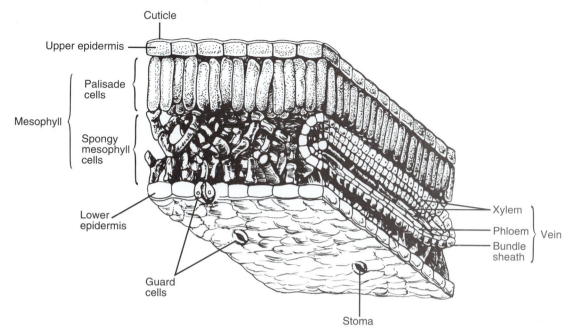

Figure 31.10 Lilac (*Syringa*) leaf, x.s.

The leaf veins provide not only support but also movement of materials. Xylem brings water and minerals from roots to the leaf cells, and phloem carries away sugar to stem and root cells for use or storage.

 Assignment 4

Materials

Colored pencils
Compound microscope
Leaves with different types of venation
Leaves with "unknown" venation
Prepared slides of:
    Syringa (lilac) leaf, x.s.
    Syringa leaf epidermis with stomata
    Zea (corn) leaf, x.s.

1. Examine the leaves with different types of venation. Identify the types of venation for the "unknowns." *Complete items 4a and 4b on the laboratory report.*
2. Study Figure 31.10. Color the mesophyll and guard cells green.
3. Examine a prepared slide of *Syringa* leaf, x.s. Locate the cells and tissues shown in Figure 31.10. *Complete items 4c–4e on the laboratory report.*
4. Examine a prepared slide of *Syringa* epidermis with stomata. Note the density of the stomata and the arrangement of the guard cells. *Complete items 4f–4i on the laboratory report.*
5. Examine a prepared slide of *Zea* (corn) leaf, x.s. Compare it with the *Syringa* leaf, noting differences and similarities. *Complete the laboratory report.*

STRUCTURE OF FLOWERING PLANTS

Student _____

Lab Instructor _____

1. EXTERNAL STRUCTURE

a. List the labels for Figure 31.2.

1. _____ 4. _____ 7. _____

2. _____ 5. _____ 8. _____

3. _____ 6. _____ 9. _____

b. Write the term that matches the phrase.

1. Site of leaf attachment to a stem _____

2. Type of leaf venation in dicots _____

3. A leaf stalk _____

4. Type of leaf venation in monocots _____

5. Arrangement of vascular bundles in:

   dicots _____

   monocots _____

6. Number of cotyledons in seeds of:

   dicots _____

   monocots _____

7. Arrangement of flower parts in:

   dicots _____

   monocots _____

8. Type of root systems in:

   dicots _____

   monocots _____

c. Compare the *Coleus* and corn seedlings.

Structure	Coleus	Corn
Root type		
Leaf venation		
Arrangement of vascular bundles		

d. Identify as a dicot or monocot: *Coleus* _____ Corn _____

e. Identify the "unknowns" as dicots or monocots.

1. _____ 3. _____ 5. _____

2. _____ 4. _____ 6. _____

2. ROOTS

a. Write the term that matches the phrase.

   1. Root tip region forming new cells   _____

   2. Root tip region of greatest growth   _____

   3. Root tip region of root hairs   _____

   4. Tissue transporting water and minerals   _____

   5. Tissue transporting organic nutrients   _____

   6. Tissue forming branch roots   _____

   7. Tissue of water-impermeable cells   _____

b. List the three functions of roots. _____

_____

c. Draw a young root of a germinated seed with root hairs at 40× and an epidermal cell with its root hair at 100×. Label the oldest and youngest root hairs.

            Young Root   **Root Hair**

d. Are the cells of an *Allium* root tip arranged in an orderly pattern or in a nonordered mass? Explain.

_____

e. Draw the arrangement of roots in taproot and fibrous root systems.

            Taproot System   **Fibrous Root System**

f. What are adventitious roots? _____

_____

g. Which type of root system is best for:

   preventing erosion of surface soil? _____

   reaching deep water sources? _____

3. STEMS

a. Describe the function of stems. _____

_____

_____

b. Describe the arrangement of vascular bundles in a cross-section of a *Zea* stem.

_____

Are they more abundant near the periphery of the stem? _____

Explain any mechanical advantage in their arrangement. _____

_____

_____

c. In a *Medicago* vascular bundle, is xylem closer to the epidermis or pith? _____

Which vascular tissue has larger cells? _____

Has thicker walls? _____ Provides greater support? _____

d. In a young *Quercus* stem, what tissue produces new (secondary) xylem and phloem? _____

Is phloem formed interior or exterior to this tissue? _____

What is the outermost tissue in the young stem? _____

e. In an old *Quercus* stem, is the newest xylem near the pith or vascular cambium? _____

Is the oldest phloem next to the cortex or vascular cambium? _____

List the tissues forming the bark. _____

What tissue forms wood in a tree stem? _____

f. How can you distinguish spring wood and summer wood in an annual ring? _____

_____

Does spring or summer wood provide the greatest growth? _____

g. What forms the cork cells of the bark? _____

What is the function of cork cells? _____

h. Explain why removing a strip of bark completely around a tree stem causes the tree to die.

_____

_____

_____

_____

i. What is the age of the demonstration section of tree stem? _____

4. LEAVES

a. Write the term that matches the phrase.

1. The broad, flat part of a leaf   _____

2. A leaf stalk   _____

3. Venation in monocots   _____

4. Basic venation in dicots   _____

b. Indicate the type of venation of the numbered "unknown" leaves.

1. _____   4. _____

2. _____   5. _____

3. _____   6. _____

   c. Write the term that matches the phrase.

     1. Photosynthetic tissue of a leaf   _____

     2. Tiny openings in the epidermis   _____

     3. Waterproof coating of epidermis   _____

     4. Compose a leaf vein   _____

                                      _____

                                      _____

     5. Control size of stomata   _____

   d. Explain the advantage of the palisade mesophyll cells being closely packed together._____

_____

_____

   e. Explain the advantage of the many spaces between spongy mesophyll cells._____

_____

_____

   f. Draw a stoma showing the arrangement of the guard cells and adjacent epidermal cells. Label the stoma, guard cells, and chloroplasts.

   g. Explain how stomata are opened and closed by the guard cells. _____

_____

_____

   h. What advantage is there in having a greater density of stomata in the lower epidermis? _____

_____

_____

   i. What advantage is there in having stomata open during the day? _____

_____

_____

   j. Draw a portion of a *Zea* (corn) leaf, x.s., from your slide. Label the epidermis, stomata, mesophyll, and vein.

TRANSPORT IN PLANTS

OBJECTIVES

After completing the laboratory session, you should be able to:
1. Describe the structure and function of vascular tissues in plants.
2. Describe the processes of root pressure, transpiration, and translocation.
3. Define all terms in bold print.

Each cell carries out its own metabolic processes whether it exists as a single cell or as part of a multicellular organism. Therefore, each cell must obtain the necessary raw materials from its environment and rid itself of the wastes produced by the metabolic activity.

In unicellular, colonial, and simple multicellular organisms, the cells are either in direct contact with the external environment or only a few cells away from it. The rate of diffusion is sufficient for the transport of substances for such cells. In more complex organisms, however, internal cells are located far from the external environment. In such organisms, special tissues, organs, or both are necessary to transport substances to and from the cells.

The successful colonization of the land by plants depended on the evolution of conducting tissues. This adaptation enabled the evolution of large plant bodies, typical of gymnosperms and angiosperms, with a greater specialization of parts than in the less advanced moss plants. Thus, leaves are specialized for photosynthesis and roots for the uptake of water and dissolved minerals, whereas the vascular tissue forms a continuous pathway for the transport of substances from one portion of the plant to another.

VASCULAR TISSUES IN PLANTS

Xylem and **phloem** are the vascular (conducting) tissues in plants. Water and dissolved minerals are absorbed by the roots, transported to the xylem, and conducted upward in the root and stem to the leaves, flowers, and fruits. Water and mineral transport in the xylem is always upward.

Organic nutrients produced in the leaves by photosynthesis are transported via the phloem to the various parts of the plant. The transport of organic nutrients is called **translocation.** The process is not clearly understood, although diffusion is involved. The translocation of organic nutrients may be either upward or downward within the phloem.

Xylem

Xylem tissue is composed of **vessel elements (cells)** and **tracheids,** plus supporting fibers. Vessel elements are tubelike cells that are stacked end-to-end to form **vessels,** tiny channels from root tip to shoot tip. The ends of vessel elements are perforated or lacking, which allows liquid to move freely from element to element. Tracheids are thin, tubelike cells that are also stacked end-to-end to form conduits for liquid transport. Their ends are sharply slanted and perforated. Both vessel elements and tracheids have perforated walls allowing liquid to move laterally between them. Vessels elements and tracheids ultimately die, leaving their cell walls to form the tiny pipelines. See Figure 32.1.

Xylem also contains elongated, thick-walled fiber cells that provide additional support. Xylem in most conifers contains only tracheids and fibers. Xylem in flowering plants contains vessels, tracheids and supporting fibers.

Phloem

Phloem tissue is composed of **sieve tube elements (cells)** and **companion cells** plus supporting fibers. See Figure 32.2. Sieve tube elements contain strands

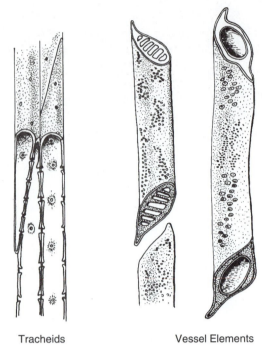

Tracheids   Vessel Elements

Figure 32.1 Tracheids and vessel elements occur in xylem.

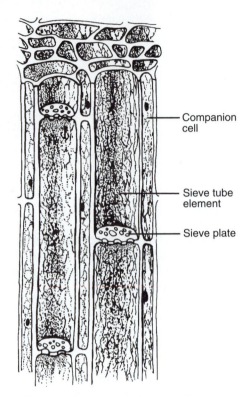

Companion cell

Sieve tube element

Sieve plate

Figure 32.2 Sieve tube elements and companion cells in phloem.

of cytoplasm but no nucleus. The cytoplasm of one cell is continuous with that of the adjacent cells in the series. This is possible because of the perforations in the sieve plate forming the end cell wall. Many sieve tube elements are joined end-to-end to form the long sieve tubes through which organic nutrients are transported upward or downward within phloem.

Companion cells are smaller cells that are adjacent to the sieve tube elements. They are also joined end-to-end but do not function in transport. Cytoplasmic strands extend between sieve tube elements and companion cells through tiny openings in their cell walls. The cytoplasm of sieve tube elements is maintained and nourished by companion cells.

Organization of Vascular Tissue

Both xylem and phloem form minute channels that are continuous from root tip to shoot tip, and they are always arranged in close association with each other. When observed in the cross-section of a stem, the vascular tissue may be arranged in (1) scattered vascular bundles in monocots (e.g., corn), (2) vascular bundles in a broken ring near the stem circumference in herbaceous dicots (e.g., alfalfa), or (3) a continuous vascular ring in woody dicots (e.g., oak).

Figure 32.3 illustrates patterns of vascular tissue distribution. Note that the xylem is always located toward the center of the stem and the phloem is always

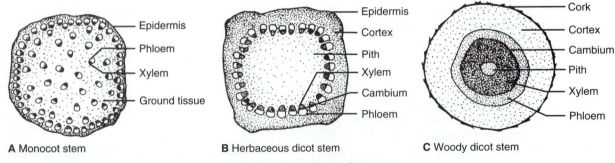

A Monocot stem — Epidermis, Phloem, Xylem, Ground tissue

B Herbaceous dicot stem — Epidermis, Cortex, Pith, Xylem, Cambium, Phloem

C Woody dicot stem — Cork, Cortex, Cambium, Pith, Xylem, Phloem

Figure 32.3 Arrangement of vascular tissue in various types of plant stems.

oriented toward the periphery. The cambium is located between the xylem and the phloem in stems capable of secondary growth. Consult Exercise 31 for the detailed cellular structure of stems.

 ## Assignment 1

Materials

Compound microscope
Net- and parallel-veined leaves
Prepared slides of:
    pine wood (*Pinus*), l.s. and x.s.
    basswood (*Tilia*), macerated wood
    pumpkin (*Curcubita*) stem, l.s.
    corn (*Zea*) stem, x.s.
    alfalfa (*Medicago*) stem, x.s.
    oak (*Quercus*) stem, 1 yr, x.s.
    privet (*Ligustrum*) and corn (*Zea*) leaves, x.s.

1. ***Complete item 1a on Laboratory Report 32 that begins on page 409.***
2. Examine a prepared slide of pine wood, x.s. and l.s., and locate tracheids similar to those in Figure 32.1.
3. Examine a prepared slide of macerated basswood and locate vessels like those in Figure 32.1.
4. Examine a prepared slide of pumpkin stem, l.s., and locate sieve tubes and companion cells like those in Figure 32.2.
5. Observe the location of vascular tissues and relative positions of xylem and phloem in prepared slides of:
    a. corn (*Zea*) stem, x.s.
    b. alfalfa (*Medicago*) stem, x.s.
    c. oak (*Quercus*) stem, x.s.
    d. privet (*Ligustrum*) leaf, x.s.
    e. corn (*Zea*) leaf, x.s.
6. ***Complete item 1 on the laboratory report.***

TRANSPORT OF WATER AND MINERALS

You will now investigate the upward movement of water and minerals in xylem from roots to the tips of the shoot system. Two processes are involved: root pressure and transpiration. You are to determine which process contributes more to the transport of water and minerals. In addition, you will determine how intrinsic and climatic factors affect the rate of transpiration.

Root Pressure

Water and dissolved minerals are absorbed by the root tips, especially the **root hairs,** which greatly increase the surface area of the root in contact with water in the soil. From the epidermal cells, water moves from cell to cell until it enters a xylem vessel. Water may also move along the cell walls until it reaches the endodermis. It must then pass through a living cell to enter the xylem because of the impermeable cell walls of the **endodermal cells.** These impermeable cell walls constitute the **Casparian strip** and allow some control of the amount of water absorbed. See Figure 32.4. Dissolved minerals may be absorbed by simple diffusion, but experiments have shown that active absorption is usually involved.

The absorptive force of the root cells produces a **root pressure** that forces water upward in the xylem vessels.

 ## Assignment 2

Materials

Compound microscope
Dissecting instruments
Microscope slides and cover glasses

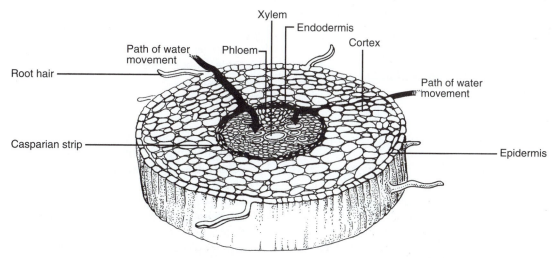

Figure 32.4 Path of water absorption in root.

Root pressure demonstration (see Figure 32.5)
Methylene blue, 0.01%, in dropping bottles
Germinated grass seeds

1. Examine the root pressure demonstration that has been prepared in accordance with Figure 32.5. Record the level of the fluid in the pipette at 30-min intervals. *Record your data in item 2a on the laboratory report.*

2. Obtain a germinated grass seed and place it on a microscope slide in a drop of water. Examine it with a dissecting microscope and at 40× and 100× with a compound microscope, using reduced light. Note the root hairs. The oldest root hairs are longest.

3. Use a scalpel to cut the seed from the root. Discard the seed. Add a drop of methylene blue, 0.01%, and a cover glass and observe a short root hair at 400×. Observe the junction of the root hair and its epidermal cell. Use reduced light to locate a nucleus.

4. ***Complete item 2 on the laboratory report.***

Transpiration

The evaporation of water from plant tissues is called **transpiration.** Most of the transpiration occurs from leaf tissues, especially the **mesophyll.** The water vapor escapes from the leaf via the open **stomata.** The evaporation of water exerts a "pull" on the columns of water in the xylem vessels. The cohesion of water molecules by hydrogen bonds enables the columns of water to be pulled upward by the evaporation of water molecules from the leaves.

The rate of transpiration varies with (1) intrinsic factors, such as the number and size of leaves and the density of stomata; and (2) extrinsic factors, such as available water and climatic and weather conditions.

Assignment 3

Materials

Desk lamps
Electric fan
Knife, sharp
Pipette, 1 ml
Potometer jar
Two-hole rubber stopper
Thermometers
Branches of woody plants, such as lilac, privet, or *Camellia*

1. Prepare a potometer setup as shown in Figure 32.6. Insert the pipette into the rubber stopper before cutting the branch from the plant. Do *not* get the leaves wet. Use a branch with a diameter slightly larger than the hole in the rubber stopper (about the diameter of the 1-ml pipette) and with at least four leaves. ***Complete item 3a on the laboratory report.***

2. Press the stopper firmly into the jar to cause the water to rise to near the top of the pipette.

1. With sharp blade cut off stem of a well-watered plant.

2. Snugly fit rubber tubing over stem and add a few drops of water.

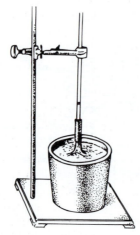

3. Insert a 1-ml pipette into tubing, forcing water up into pipette. Support with ring stand.

Figure 32.5 Root pressure setup.

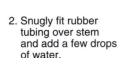

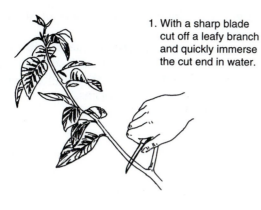

1. With a sharp blade cut off a leafy branch and quickly immerse the cut end in water.

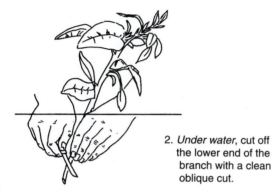

2. *Under water*, cut off the lower end of the branch with a clean oblique cut.

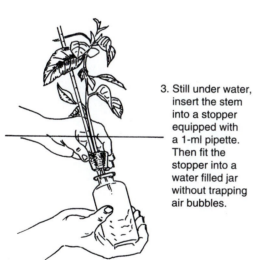

3. Still under water, insert the stem into a stopper equipped with a 1-ml pipette. Then fit the stopper into a water filled jar without trapping air bubbles.

Figure 32.6 Potometer setup.

3. Place the potometer on the table at your lab station and determine the rate of transpiration, in milliliters of water per minute (ml H_2O/min). Take readings at 2-min intervals. *Record your data in item 3b on the laboratory report.*

4. The rate of transpiration varies with climatic factors that affect evaporation of water from leaves.

What effect do you think wind has on transpiration rate? Place your potometer in the breeze created by an electric fan to test your hypothesis. Press the stopper in a bit to push water to near the 0 mark of the pipette. *Record your data in item 3b and complete items 3c–3g on the laboratory report.*

5. What other climatic factor might affect the rate of transpiration? Design an experiment to test your hypothesis. Refill your potometer, if necessary. *Complete item 3h on the laboratory report.*

6. Do you think the number of leaves affects the transpiration rate? In what way? Remove and save half the leaves on your branch, and determine the rate of transpiration at normal room conditions. *Complete item 3i on the laboratory report.*

Density of Stomata

Because water vapor passes from the leaf via the stomata, the density of stomata and the size of leaves affect the transpiration rate. In what way is the rate affected?

Assignment 4

Materials

Compound microscope
Forceps
Graph paper, mm and cm division
Microscope slides and cover glasses
Nail polish, clear
Leaves of woody plants such as lilac, privet, or *Camillia*

1. Obtain one of the leaves removed from your branch. Place it on a sheet of graph paper lined in millimeter and centimeter divisions and trace around it. From the tracing, estimate the surface area of the leaf in square centimeters (1 cm^2 = 100 mm^2) Consider the surface area to be the same for both upper and lower leaf surfaces. *Record your data in item 4a on the laboratory report.*

2. Coat part of both surfaces of the leaf with clear nail polish. After the nail polish has dried, use forceps to peel it off, one leaf surface at a time, and mount it dry on a slide under a cover glass. Use the 4× or 10× objective and *reduced light* to observe the stomata. Record the number of stomata in three fields and calculate the average density per field for each leaf surface. Are stomata on both surfaces of the leaf?

3. Calculate the area of the field in millimeters using the diameter of field determined in Exercise 2

$(A = \pi r^2)$. Convert the number of stomata per field to stomata per square centimeter.

$$\frac{\text{Number of stomata}}{\text{Area of field in mm}^2} \times \frac{100 \text{ mm}^2}{1 \text{ cm}^2} = \text{stomata/cm}^2$$

Multiply the area of leaf surface *containing stomata* by the stomata per square centimeter *on that surface* to determine the number of stomata on your leaf.

4. If time permits, determine the distribution and density of stomata on leaves of other plants, and correlate your findings with the normal environmental conditions of the plants.

5. **Complete the laboratory report.**

TRANSPORT IN PLANTS

Student _____

Lab Instructor _____

1. VASCULAR TISSUE

a. Write the term that matches the phrase.

1. Tissue transporting water and minerals _____

2. Tissue transporting organic nutrients _____

3. Composes xylem in conifers _____

4. Composes xylem in flowering plants _____

5. Larger cells of phloem _____

6. Smaller cells of phloem _____

7. Composes wood of woody plants _____

8. Vascular tissue formed by dead cells _____

9. Direction of fluid movement in xylem _____

10. Direction of nutrient movement in phloem _____

b. Draw a few of the following from your slides:

Vessels   **Tracheids**   **Sieve Tubes and Companion Cells**

c. Based on your microscopic study, write the term that matches the phrase.

1. Vascular tissue uppermost in a leaf vein _____

2. Vascular tissue innermost in alfalfa stem _____

3. Vascular tissue outermost in oak stem _____

4. Vascular tissue innermost in vascular bundles of corn stem _____

2. ROOT PRESSURE

a. Record the movement of fluid by root pressure in the following table:

	Time (min)				
	0	30	60	90	120
Reading (ml)					

b. Record the average rate of fluid movement by root pressure: _____ ml/min

c. Describe the importance of root hairs. _____

_____

d. Show by diagram the distribution of root hairs on the young root of a germinated grass seed. Draw a root hair showing its attachment to the root and its nucleus.

Root Hair Distribution (40×)   **Root Hair (400×)**

3. TRANSPIRATION

a. What process causes water loss from the leaves in transpiration? _____

b. Record the movement of fluid by transpiration in the following table:

	Time (min)					
	0	2	4	6	8	10
Reading (ml) (normal)						
Reading (ml) (wind)						

c. Record the average transpiration rate (ml/min).

No wind _____   Wind _____

d. State a conclusion from your results. _____

_____

_____

e. Place numbers in front of the following items to indicate the sequence of the water movement from potometer bottle to atmosphere:

_____ Bottle   _____ Stomata   _____ Atmosphere

_____ Substomatal space   _____ Xylem   _____ Mesophyll

f. Which process, root pressure or transpiration, contributes more to the movement of water up the stem? _____

g. What causes plants to wilt? _____

_____

_____

h. State your hypothesis to be tested. _____

_____

Describe your procedures. _____

_____

_____

_____

_____

Record the movement of fluid by transpiration in the following table:

	Time (min)					
	0	2	4	6	8	10
Reading (ml)						

Indicate your results (avg. ml H_2O/min). _____

What was your control? _____

_____

Do the results support your hypothesis? _____

State a conclusion from your results. _____

_____

_____

i. State the hypothesis to be tested. _____

_____

Record the movement of fluid by transpiration in the following table:

	Time (min)					
	0	2	4	6	8	10
Reading (ml)						

Indicate your results (avg. ml H_2O/min). _____

What was your control? _____

_____

Do the results support your hypothesis? _____

State a conclusion from your results. _____

_____

_____

What other characteristics of leaves do you think affects the rate of transpiration? _____

_____

Explain. _____

_____

4. DENSITY OF STOMATA

a. Calculate the density of stomata on your leaf based on your observations.

CALCULATION OF STOMATA DENSITY

Leaf Surface	Area of Field (mm²)	Stomata per Field	Stomata per cm²	Leaf Surface Area (cm²)	Stomata per Leaf Surface
Upper					
Lower					

b. Indicate the average density of stomata (stomata/cm²) on each surface.

Upper leaf surface _____   Lower leaf surface _____

Explain the advantage of this distribution. _____

_____

_____

How many stomata are on the leaf? _____

c. Does the density of stomata help explain your results recorded in item 3i? _____

Explain. _____

_____

_____

d. What adaptations would you expect in dicot leaves of desert plants in regard to:

size? _____

number of stomata per leaf? _____

Explain. _____

_____

_____

_____

e. What adaptations would you expect in dicot leaves of plants in a shaded, moist environment in regard to:

size? _____

number of stomata per leaf? _____

Explain. _____

_____

_____

_____

CHEMICAL CONTROL IN PLANTS

OBJECTIVES

After completing the laboratory session, you should be able to:
1. Describe the effect of IAA and gibberellins on plant growth.
2. Explain the mechanism of plant tropisms.
3. Describe the effect of gibberellins and abscisic acid on seed germination.
4. Define all terms in bold print.

Chemical control is widespread among organisms and is responsible for much of the orderly fashion in which living processes occur. Among vertebrates and arthropods, much of chemical control is carried out by chemical messengers called **hormones.** A hormone is produced by endocrine glands and causes a specific response in adjacent or distant cells.

Vascular plants lack endocrine glands, but they produce at least five types of chemical messengers collectively called **plant hormones.** Plant hormones are usually produced in actively metabolizing cells, and they are typically distributed by diffusion to target cells in which they produce a specific response.

In this exercise, you will investigate the effect of three plant hormones: indoleacetic acid (IAA), gibberellins, and abscisic acid.

INDOLEACETIC ACID AND PLANT GROWTH

Plant **tropisms** are positive or negative responses to specific environmental stimuli. Whenever a planted seed germinates, the shoot grows upward and the root grows downward into the soil. Have you ever wondered what mechanism is responsible for this growth pattern? This growth pattern is called **gravitropism,** a response to gravity. Shoots exhibit negative gravitropism because they grow away from the force of gravity. Roots exhibit positive gravitropism because they grow toward the force of gravity. Another tropism in plants is **phototropism,** a growth response either toward or away from light. Plant tropisms are regulated by a group of plant chemicals that are collectively called **auxins,** and the primary auxin is **indoleacetic acid.**

Oat Coleoptiles

The action of IAA was discovered in experiments using oat coleoptiles. A coleoptile is a hollow sheath that protects the young leaves in seedlings of cereal plants, such as oats, corn, and wheat, as they grow upward through the soil after germination of the seed. See Figure 33.1. Coleoptile growth slows after it breaks through the soil. Soon, the first leaves of the seedling break out of the coleoptile, expand, and begin photosynthesizing as the coleoptile disintegrates.

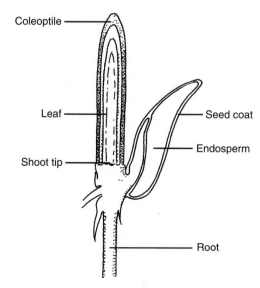

Figure 33.1 Oat seedling with a coleoptile.

The **region of cell division** in a coleoptile is at its base, where it is attached to the shoot tip. Most of its growth occurs in the **region of elongation** composing the central region of the coleoptile.

Because experimenting with oat coleoptiles is a bit tricky, you will examine results of experiments shown in Figure 33.2 to form your conclusions about the action of IAA on coleoptile growth.

Experiment 1

Two groups of oat seedlings were prepared as follows:
    Group A - The coleoptile tips were removed.
    Group B - Seedlings were left in normal state.
After 12 hr in darkness, the results were:
    Group A - No growth
    Group B - Average growth of 4.2 mm

Experiment 2

Three groups of oat seedlings were prepared as follows:
    Group A - Coleoptile tips were removed, and plain agar blocks were placed on the cut surface of the coleoptiles.
    Group B - Coleoptile tips were removed, and agar blocks containing an extract of coleoptile tips were placed on the cut surface of the coleoptiles.
    Group C - Seedlings were left in normal state.
The results after 12 hr darkness were:
    Group A - No growth
    Group B - Average growth of 3.9 mm
    Group C - Average growth of 4.0 mm

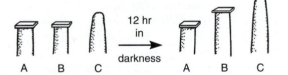

Experiment 3

Three groups of oat seedlings were prepared as follows:
    Group A - Coleoptile tips were removed, and agar blocks containing an extract of coleoptile tips were placed on the right side of the cut surface of the coleoptiles.
    Group B - Same as group A except that the agar blocks were placed on the left side of the cut surface of the coleoptiles.
    Group C - Seedlings were left in normal state.
The results after 12 hr darkness were:
    Group A - Growth causing bending to the left
    Group B - Growth causing bending to the right
    Group C - Normal growth

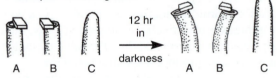

Experiment 4

Two groups of seedlings were placed in a horizontal position and prepared as follows:
    Group A - Coleoptile tips were removed.
    Group B - Seedlings were left in normal state.
After 12 hr in darkness the results were:
    Group A - No growth
    Group B - Growth causing tips to bend upwards (negative gravitropism)

Experiment 5

Three groups of seedlings were prepared as follows:
    Group A - Coleoptile tips were removed.
    Group B - Seedlings were left in normal state.
    Group C - Tinfoil "cap" was placed on coleoptile tip.
After 12 hr exposure to unilateral light, the results were:
    Group A - No growth
    Group B - Growth causing tips to bend towards the light (positive phototropism)
    Group C - Upward growth only, that is, no bending.

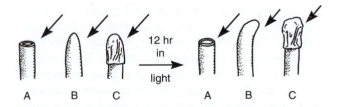

Figure 33.2 Control of plant tropisms by IAA.

Assignment 1

Materials

Oat seedlings with coleoptiles

1. Obtain an oat seedling and compare it with Figure 33.1. Use a single-edged razor blade to make a lengthwise cut through the wall of the coleoptile. Spread the coleoptile to observe the developing leaves within.
2. **Complete item 1 on Laboratory Report 33 that begins on page 417.**

Assignment 2

1. Study the results of a series of experiments with oat seedlings in Figure 33.2.
2. **Complete item 2 on the laboratory report.**

Tropisms in Dicots

You will now investigate the role of IAA in gravitropism and phototropism in herbaceous dicots. Do you think IAA has a similar action in dicots as in monocots, such as oats? If time or space is limited, your instructor may assign a division of labor in performing the experiments, or they may be presented as demonstrations.

Assignment 3

Materials

Applicator sticks
Dissecting instruments
Lanolin, plain
Lanolin with 0.5% IAA
Light-proof chamber
Protractors
Razor blades, single-edged
Tomato (*Lycopersicon*) or *Coleus* plants in small pots

1. Investigate the role of IAA in gravitropism by preparing three well-watered tomato or *Coleus* plants, as follows:
   Plant A: Cut off the terminal 1 in. of the shoot tip and apply a small globule of plain lanolin paste to the cut surface only.
   Plant B: Cut off the terminal 1 in. of the shoot tip and apply a small globule of lanolin-IAA paste to the cut surface only.
   Plant C: Do nothing.
2. Place the plants on their sides in the dark chamber until the next lab session.
3. At the next lab session, examine the plants and **complete item 3 on the laboratory report.**

Assignment 4

Materials

Applicator sticks
Desk lamps
Dissecting instruments
Lanolin, plain
Lanolin with 0.5% IAA
Protractors
Razor blades, single-edged
Tomato (*Lycopersicon*) or *Coleus* plants in small pots

1. Investigate the role of IAA in phototropism by preparing four well-watered tomato or *Coleus* plants as follows:
   Plant A: Cut off the terminal 1 in. of the shoot tip and apply a small globule of plain lanolin paste to the cut surface only.
   Plant B: Cut off the terminal 1 in. of the shoot tip and apply a small globule of lanolin-IAA paste to the cut surface only.
   Plant C: Cover the terminal 1 in. of the shoot tip with aluminum foil as a light shield. Enclose the young leaves around the shoot tip as well.
   Plant D: Do nothing.
2. Place the plants upright and expose them to unilateral light until the next lab session.
3. At the next lab session, examine the plants and **complete item 4 on the laboratory report.**

GIBBERELLINS

Gibberellins are formed in high-growth regions, such as buds, root tips, and germinating seeds. Over 50 different gibberellins are known. The best known is gibberellic acid (GA_3), which you will use in the following experiments to determine its effect on stem elongation and seed germination.

Assignment 5

Materials

Dropping bottles of:
  0.1% gibberellic acid
  distilled water
Metersticks
Dwarf bean (*Phaseolus*) plants, 2–3 wk old, in small pots
Labels, tie-on or stake

1. Obtain two dwarf bean plants in pots. Label one *C* for control and the other *GA* for gibberellic acid. Write your name on the label.

2. Measure the height of each plant, from the cotyledons to the terminal bud, and ***record it in item 5 on the laboratory report.***
3. Place the pots on the plant growth rack and water them well.
4. Place 1 drop of water to the stem tip of plant *C* and 1 drop of 0.1% gibberellic acid to the stem tip of plant *GA*. *Don't spill gibberellic acid solution on the soil because it may stop root growth.*
5. Each week, measure the height of the plants, reapply the drops as in step 4, and water the plants. Continue the experiment for 2–3 wk, as determined by your instructor.
6. ***Record your data in item 5. At the end of the experiment, complete item 5 on the laboratory report.***

 ## Assignment 6
Materials

Petri plates of:
 plain agar (labeled *C*)
 agar with 0.1% gibberellic acid (labeled *GA*)
Forceps
Glass-marking pen
Lettuce seeds

1. Obtain two petri dishes, one with plain agar (*C*) and one with 0.1% gibberellic acid agar (*GA*). Use a glass-marking pen to write your initials on each plate. Because agar is 98% water, it provides sufficient water for seed germination.
2. Use forceps to place ten lettuce seeds equidistant from each other on each agar plate. Press the seeds gently into the agar.
3. Replace the covers and place the plates in a location as directed by your instructor until the next laboratory session.
4. At the next lab session, examine the seeds for germination and ***complete item 6 on the laboratory report.***

ABSCISIC ACID

Abscisic acid (ABA) promotes aging in leaves and other plant parts, and it suppresses the growth of buds. Determine the effects of abscisic acid on seed germination in the following experiment.

 ## Assignment 7
Materials

Petri plates of:
 plain agar (labeled *C*)
 agar with 0.01% abscisic acid (labeled *ABA*)
Forceps
Glass-marking pen
Lettuce seeds

1. Obtain two petri dishes, one containing plain agar (*C*) and one containing agar with 0.01% abscisic acid (*ABA*). Use a glass-marking pen to write your initials on each side.
2. Use forceps to place ten lettuce seeds equidistant from each other on each agar plate. Press the seeds gently into the agar.
3. Replace the covers and place the plates in a location as directed by your instructor for 1 wk.
4. After 1 wk, examine the seeds and ***complete item 7 on the laboratory report.***

OTHER EXPERIMENTS

Your instructor may ask you to design and conduct an additional experiment using materials present in the laboratory. The following are suggested experiments:

1. Effect of IAA concentration on growth in roots (or stems)
2. Effect of GA_3 on germination of dormant seeds (or buds)
3. Effect of ABA on aging in leaves and leaf fall

 ## Assignment 8

1. Follow the directions of your instructor in designing and conducting your experiment.
2. ***Record your hypothesis, prediction statement, results, and conclusion in item 8 on the laboratory report.***

CHEMICAL CONTROL IN PLANTS

Student _____

Lab Instructor _____

1. INTRODUCTION

a. Write the term that matches the phrase.

1. Growth of stem or root toward or away from specific stimulus _____

2. Plant hormone involved in tropisms _____

3. Growth of root toward gravity _____

4. Growth of stem toward light _____

5. Primary auxin hormone _____

b. Where does most of the growth in length occur in a coleoptile?

_____

2. IAA AND OAT COLEOPTILES

a. Considering experiments 1 and 2 (Figure 33.2) and that IAA promotes growth:

1. Where is IAA produced in the coleoptile? _____

2. What portion of the coleoptile responds to IAA? _____

_____

3. How does IAA move from the site of production to the site of action? _____

4. Is IAA soluble in water?_____ Why do you think so? _____

b. When an agar block containing coleoptile-tip extract was placed on the right side of a tipless coleoptile in experiment 3:

1. Was the IAA concentration in the region of elongation greater on the right or left side of the coleoptile? _____

2. Why do you think so? _____

3. Which side of the coleoptile grew more rapidly? _____

4. What is the action of IAA? _____

c. In experiment 4:

1. What is the role of group A? _____

2. How do you explain the growth of group B? _____

_____

d. In experiment 5:

1. What is the role of groups A and C? _____

2. On which side of the coleoptile in group B was the IAA concentration:

higher? _____   lower? _____

3. Why do you think so? _____

4. State two possible explanations for the effect of light on IAA.

a. _____

_____

b. _____

_____

3. GRAVITROPISM IN A DICOT

a. State a hypothesis about the action of IAA being tested in this experiment. _____

_____

b. State an "if . . . then" prediction of what you think will happen based on the hypothesis. If

_____

_____

c. What is the purpose of plant A? _____

d. What is the purpose of plant C? _____

e. Estimate the degree of negative gravitropism in the plants as +++ (strong), ++ (moderate), + (weak), and 0 (none):

Plant A _____   Plant B _____   Plant C _____

f. Explain the results.

Plant A _____

Plant B _____

Plant C _____

g. Was your hypothesis supported? _____ State a conclusion from your results.

_____

_____

_____

h. In plant C, where is IAA produced? _____ Did the greatest cell elongation occur on the upper or lower

side of the stem? _____ Was the IAA concentration higher or lower in this region?

_____ Explain. _____

_____

4. PHOTOTROPISM IN A DICOT

a. State the hypothesis about the action of IAA being tested in this experiment. _____

_____

b. State an "if . . . then" prediction of what you think will happen based on the hypothesis. If _____

_____

_____

c. What is the purpose of plants A, C, and D? _____

d. Estimate the degree of positive phototropism in the plants as +++ (strong), ++ (moderate), + (weak), and 0 (none):

Plant A _____   Plant B _____   Plant C _____   Plant D _____

e. Explain the results.

Plant A _____

Plant B _____

Plant C _____

Plant D _____

f. Was your hypothesis supported? _____ State a conclusion from the results. _____

_____

g. In plant D, did the greatest cell elongation occur on the light or shade side of the stem? _____

Was the IAA concentration higher or lower in this region? _____ Develop two hypotheses to possibly

explain this observation.

1. _____

_____

2. _____

_____

5. GIBBERELLINS AND STEM GROWTH

a. State a hypothesis to be tested in this experiment.

_____

b. State an "if . . . then" prediction of what you think will happen based on the hypothesis. If _____

_____

_____

c. Record the height of the plants in centimeters.

	Start	*Week 1*	*Week 2*	*Week 3*
Plate *C*				
Plate *GA*				

d. Was your hypothesis supported? _____ State a conclusion from your results. _____

_____

e. Do your results suggest that gibberellins may interact with IAA in regulating growth in stems? _____

6. GIBBERELLIC ACID AND SEED GERMINATION

a. State a hypothesis to be tested in this experiment.

_____

b. State an "if . . . then" prediction of what you think will happen based on the hypothesis. If _____

_____

_____

c. Record the results as the percentage of seeds germinated.

Plate *C*:   Your results _____   Entire class _____

Plate *GA*:   Your results _____   Entire class _____

d. Was your hypothesis supported? _____ State a conclusion from your

results. _____

_____

e. Do the results suggest that gibberellins may play a role in breaking dormancy in seeds? _____

Explain. _____

_____

7. ABSCISIC ACID AND GERMINATION

   a. State a hypothesis to be tested in this experiment. _____

   _____

   b. State an "if . . . then" prediction of what you think will happen based on the hypothesis. If _____

   _____

   c. Record the results as the percentage of seeds germinated.

   Plate *C*:   Your results _____   Entire class _____

   Plate *ABA*:   Your results _____   Entire class _____

   d. Was your hypothesis supported? _____ State a conclusion from your results. _____

   _____

   _____

   e. Do your results suggest that abscisic acid may play a role in promoting dormancy in seeds? _____

   Explain your response. _____

   _____

8. FURTHER INVESTIGATIONS

   a. State a hypothesis to be tested in this experiment. _____

   _____

   b. State an "if . . . then" prediction of what you think will happen based on the hypothesis. If _____

   _____

   c. Describe your methods. _____

   _____

   _____

   _____

   _____

   _____

   d. Describe your results. _____

   _____

   _____

   _____

   _____

   _____

   _____

   e. Was your hypothesis supported? _____ State a conclusion from your results. _____

   _____

   _____

   _____

   _____

   _____

   f. Do your results suggest an additional experiment? _____ If so, what?

   _____

   _____

Part **VI**

Heredity
and Evolution

HEREDITY

OBJECTIVES

After completing the laboratory session, you should be able to:
1. Explain Mendel's principle of segregation and principle of independent assortment, and give examples of each.
2. Solve simple genetic problems involving dominance, recessiveness, codominance, sex linkage, and polygenes.
3. Determine gametes from genotypes where genes are linked or nonlinked.
4. Perform a chi-square analysis.
5. Define all terms in bold print.

The **inherited characteristics** of a diploid organism are determined at the moment of sperm and egg fusion. The zygote (2n) receives one member of each chromosome pair from each parent. The genetic information that determines the hereditary traits is found in the structure of the **DNA molecules** in the **chromosomes.** A short segment of DNA that codes for a particular protein constitutes a **gene,** a hereditary unit. Both genes and chromosomes occur in homologous pairs in diploid organisms. See Figure 34.1.

In the simplest situation, an inherited trait, such as flower color, is determined by a single pair of genes. The members of a **gene pair** may be identical (e.g., each codes for purple flowers) or may code for a different variation of the trait (e.g., one codes for purple flowers and the other codes for white flowers). Again, in the simplest case, only two forms of a gene exist. Alternate forms of a gene are called **alleles.**

A b c D E

a B C d e

Figure 34.1 Diagrammatic representation of homologous chromosomes and genes.

Geneticists use symbols (usually letters such as "P" or "p") to represent alleles when solving genetic problems. When both members of a gene pair consist of the same allele, such as PP or pp, the individual is **homozygous** for the expressed trait. When the members of the gene pair consist of unlike alleles, such as Pp, the individual is **heterozygous** (hybrid) for the expressed trait.

The genetic composition of the gene pair (e.g., PP, Pp, or pp) is known as the **genotype** of the individual. The observable (expressed) form of a trait (e.g., purple flowers or white flowers) is called the **phenotype.**

An understanding of inheritance patterns enables the prediction of an **expected ratio** for the occurrence of a trait in the progeny (offspring) of parents of known genotypes.

MENDEL'S PRINCIPLES

Gregor Mendel, an Austrian monk, worked out the basic patterns of simple inheritance in 1860, long before chromosomes or genes were associated with inheritance. Mendel's work correctly identified the existence of the units of inheritance now known as genes.

Mendel proposed two principles that form the basis for the study of inheritance. Look for evidence of these principles as you work through this exercise. In modern terms, these principles may be stated as follows:

1. The **principle of segregation** states that (a) genes occur in pairs and exist unchanged in the heterozygous state and (b) members of a gene pair are segregated (separated) from each other during gametogenesis, ending up in separate gametes.

2. The **principle of independent assortment** states that genes for one trait are assorted (segregated into the gametes) independently from genes for other traits. This principle applies *only* to traits whose genes are located on different chromosome pairs (i.e., the genes are not linked).

DOMINANT/RECESSIVE TRAITS

When a gene pair consists of two alleles, and one is expressed and the other is not, the expressed allele is **dominant.** The unexpressed allele is **recessive.** Many traits are inherited in this manner. For example, the following traits in garden peas exhibit a dominant/recessive pattern of inheritance. In each case, the dominant trait is in italics, and the dominant allele is capitalized in the genotype.

Flower color: *Purple flowers* (PP, Pp) or white flowers (pp)
Seed color: *Yellow seeds* (YY, Yy) or green seeds (yy)
Seed shape: *Round seeds* (RR, Rr) or wrinkled seeds (rr)
Plant height: *Tall plants* (TT, Tt) or dwarf plants (tt)

Table 34.1 shows the genotypes and phenotypes that are possible for purple or white flowers in peas. Note that the dominant allele is assigned an uppercase P, whereas the recessive allele is represented by a lowercase p. Only one dominant allele is required for the expression of purple flowers. In contrast, both recessive alleles must be present for white flowers to be expressed in the phenotype. *This relationship is constant for all dominant and recessive alleles.*

TABLE 34.1	GENOTYPES AND PHENOTYPES FOR FLOWER COLOR IN GARDEN PEAS	
Genotypes		*Phenotypes*
PP		Purple
Pp		Purple
pp		White

Solving Genetic Problems

Consider this genetic problem: What are the expected genotype and phenotype ratios (probabilities) in the progeny of purple-flowering and white-flowering pea plants when each parent is homozygous? Steps used to solve genetic problems such as this are listed in Table 34.2. Note how the steps are used in Figure 34.2 to solve this problem.

The determination of the gametes is a critical step. Recall that meiosis separates homologous chromosomes (and genes) into different gametes. Thus, the members of the gene pair are separated into different gametes. In Figure 34.2, each parent forms only one class of gametes because each parent is homozygous for flower color.

In setting up the Punnett square, the gametes of one parent are placed on the vertical axis, and the gametes of the other parent are placed on the horizontal axis. The number of squares composing a Punnett square depends on the number of gamete classes formed by the parents. Four squares are used in Figure 34.2 to enable you to understand the setup, although only one square is actually needed because each parent produces only one class (type) of gamete.

All possible combinations of gametes are simulated by recording the gametes on the vertical axis into each square to their right and those on the horizontal axis into each square below them. Note that uppercase letters (dominant alleles) always compose the first letter in each gene pair. When the Punnett square is complete, the individual squares contain the expected genotypes of progeny in the F_1 (first filial) generation. The genotype and phenotype ratios may then be determined.

TABLE 34.2	STEPS USED TO SOLVE STANDARD GENETICS PROBLEMS

1. Be sure you understand what you are to solve. Write down what is known.

2. Write out the cross using genotypes of the parents.

3. Determine the possible gametes that may be formed.

4. Use a Punnett square to establish the genotypes of all possible progeny.

5. Determine the genotype ratio of the progeny. Count the number of identical genotypes and express them as a ratio of the total genotypes (e.g., 1/4 PP : 2/4 Pp : 1/4 pp).

6. Use the information obtained in step 1 to determine the phenotype ratio from the genotypes (e.g., 3/4 purple flowers to 1/4 white flowers).

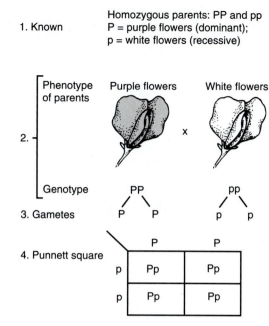

1. Known   Homozygous parents: PP and pp
            P = purple flowers (dominant);
            p = white flowers (recessive)

2. ┌ Phenotype   Purple flowers   White flowers
   │ of parents
   │
   │   x
   │
   │ Genotype   PP   pp
   └   / \   / \
3. Gametes   P   P   p   p

4. Punnett square
                    P   P
              ┌────────┬────────┐
           p │   Pp   │   Pp   │
              ├────────┼────────┤
           p │   Pp   │   Pp   │
              └────────┴────────┘

5. Genotype ratio: All or 1/1 Pp

6. Phenotype ratio: All or 1/1 purple flowers

Figure 34.2 Method of solving genetic crosses.

Test Cross

It is usually not possible to distinguish between homozygous and heterozygous phenotypes exhibiting a dominant trait, but they may be determined by using a **test cross.** In a test cross, the individual exhibiting the dominant phenotype is crossed with an individual exhibiting the recessive phenotype. Recall that an individual exhibiting a recessive phenotype is *always* homozygous for that trait. If all progeny exhibit the dominant trait, the parent with the dominant phenotype is homozygous. If half the progeny exhibit the dominant trait and half exhibit the recessive trait, the parent with the dominant phenotype is heterozygous.

Assignment 1

Complete item 1 on Laboratory Report 34 that begins on page 433.

Assignment 2

Materials

A tray of tall-dwarf corn seedlings from a monohybrid cross

A tray of green-albino corn seedlings from a monohybrid cross

1. Using Figure 34.2 as a guide and completing the Punnet square in Figure 34.3, determine the expected progeny in the F_2 generation by crossing two members of the F_1 generation, both of which are monohybrids (Pp). A monohybrid is heterozygous for one trait.

2. Once you have completed the Punnett square, determine the genotypes by counting the identical genotypes and recording them as fractions of the total number of genotypes. The different types of genotypes are then expressed as a proportion to establish the expected ratios:

$$1/4\ PP : 2/4\ Pp : 1/4\ pp$$

or

$$1\ PP : 2\ Pp : 1\ pp$$

Determine the phenotype ratio from the genotype ratio. Remember that the presence of a single dominant allele in a genotype produces a dominant phenotype. The phenotype ratio is:

3/4 purple-flowering plants : 1/4 white-flowering plants

These ratios are always obtained in a monohybrid cross where the gene consists of only two alleles and one allele is dominant.

3. **Complete items 2a–2d on the laboratory report.**

4. Examine the trays of corn seedlings from monohybrid crosses. **Complete items 2e–2j on the laboratory report.**

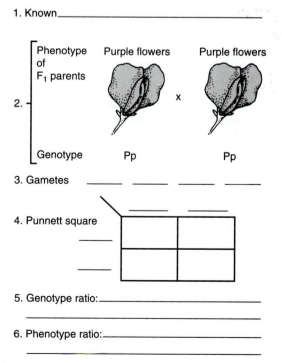

1. Known_____

2. ┌ Phenotype   Purple flowers   Purple flowers
   │ of
   │ F₁ parents
   │
   │   x
   │
   │ Genotype   Pp   Pp
   └

3. Gametes   ____ ____   ____ ____

4. Punnett square
                   ____   ____
              ┌────────┬────────┐
         ____ │   │   │
              ├────────┼────────┤
         ____ │   │   │
              └────────┴────────┘

5. Genotype ratio:_____
   _____

6. Phenotype ratio:_____
   _____

Figure 34.3 Determination of the progeny of monohybrid parents.

TABLE 34.3	COMMON DOMINANT/RECESSIVE TRAITS IN HUMANS DETERMINED BY A SINGLE GENE WITH TWO ALLELES	
Trait	Dominant Phenotype	Recessive Phenotype
Ear lobes	Free	Attached
Pigment distribution	Freckles	No freckles
Hairline	Widow's peak	Straight
Little finger	Bent	Straight
Tongue roller	Yes	No

5. Examine Table 34.3, which illustrates several easily detectable human traits exhibiting a simple dominant/recessive mode of inheritance. **Complete item 2 on the laboratory report.** *Set up your Punnett squares on a separate sheet of paper, as necessary.*

Incomplete Dominance and Codominance

Inheritance in Mendel's pea plants exhibited complete dominance because both heterozygotes and homozygous dominants had the same phenotypes. In contrast, **incomplete dominance** is characterized by the heterozygote's phenotype appearing as a blended intermediate between the phenotypes of the two homozygous parents. This type of inheritance occurs in flower color in snapdragons. A homozygous red-flowering snapdragon (RR) crossed with a homozygous white-flowering snapdragon (rr) yields all pink-flowering heterozygous snapdragons (Rr).

In spite of the blended appearance of the heterozygous phenotype, the alleles are not altered and segregate unchanged in the next generation.

In **codominance,** both alleles are expressed in the heterozygote, but no apparent blending occurs in the heterozygote's phenotype. Sickle-cell anemia, a disorder affecting some Black Americans, is inherited in this manner. Persons heterozygous for sickle cell (Hb^{A}Hb^S) produce both normal and abnormal hemoglobin in their red blood cells. They rarely experience illness, because sufficient normal hemoglobin is present to carry oxygen. However, persons homozygous for sickle cell (Hb^{S}Hb^S) exhibit sickle-cell anemia and die prematurely without medical intervention.

Assignment 3

Complete item 3 on the laboratory report.

TABLE 34.4	PHENOTYPES AND GENOTYPES OF THE A, B, O BLOOD TYPES	
Blood Type		Genotype
O		Ii
A		$I^A I^A$ or $I^A i$
B		$I^B I^B$ or $I^B i$
A		$I^A I^B$

MULTIPLE ALLELES

Some traits are controlled by genes with more than two alleles. The inheritance of ABO blood groups in humans is an example. Three alleles are involved: I^A codes for type A blood, I^B codes for type B blood, and i codes for type O blood. Table 34.4 shows the relationship between genotypes and phenotypes. Note that alleles I^A and I^B are both dominant over i, but that they are codominant with respect to each other.

 ## Assignment 4

Complete item 4 on the laboratory report.

DIHYBRID CROSS

In this section, you will solve genetic problems involving two traits whose genes are located on separate chromosomes. In garden peas, yellow seed color is dominant over green seed color, and round seed shape is dominant over wrinkled seed shape. The genes for seed shapes and seed color are located on separate chromosomes. Consider a cross of a plant producing yellow-round seeds with a plant producing green-wrinkled seeds. Both plants are homozygous for both traits (seed color and seed shape). What kinds of seeds will be produced by the progeny?

Based on what is known, you can establish the genotype of the parents and write out the cross:

$$\underset{\textbf{YYRR}}{\text{yellow-round}} \times \underset{\text{yyrr}}{\text{green-wrinkled}}$$

Recalling that the members of a gene pair are separated into different gametes and recognizing that one member of each gene pair must be in each gamete, you can determine the gametes. Because both parents are homozygous for both traits, each can produce only one type of genotype in the gametes. Now, you can set up and complete a Punnett square:

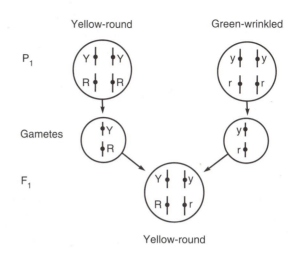

Figure 34.4 Formation of a dihybrid.

$$\begin{array}{c|c} & \text{YR} \\ \hline \text{yr} & \boxed{\text{YyRr}} \end{array}$$

All the progeny are yellow-round and heterozygous for both traits. They are **dihybrids.** Figure 34.4 shows the cross in a way that depicts the chromosomes and genes involved.

What are the expected genotype and phenotype ratios in the F2 generation? To answer this question, a cross must be made of two F_1 dihybrids with the same phenotype (yellow-round seeds) and genotype (YyRr). A dihybrid cross is solved the same way as a monohybrid cross by using a Punnett square. The most difficult and key part of the solution is the determination of gamete genotypes. Figure 34.5 and Table 34.5 show two different ways to determine the gamete genotypes. Note that there are four classes of gametes for each parent: YR, Yr, yR, and yr. Thus, you must use a Punnett square with 16 squares and place these gamete genotypes along the horizontal and vertical axes. It helps to place them in the same order along each axis, starting from the upper left corner of the Punnett square.

TABLE 34.5	GAMETE DETERMINATION IN A DIHYBRID WITH NONLINKED GENES		
Parent Genotype	First Gene Pair	Second Gene Pair	Possible Gametes

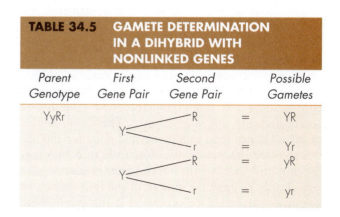

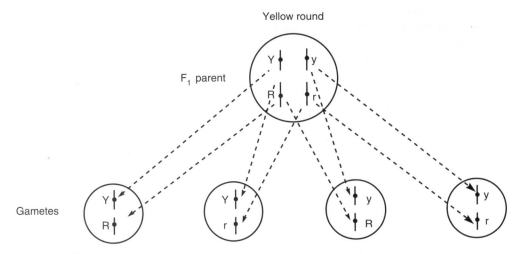

Figure 34.5 Gamete formation in a dihybrid with nonlinked genes.

Assignment 5

Materials

Trays of corn seedlings from a dihybrid cross: tall-green/dwarf-albino

1. **Complete item 5a on the laboratory report.** The expected phenotype ratio in dihybrid crosses when each trait is determined by two alleles, one of which is dominant, and when the genes are located on separate chromosomes is always 9 : 3 : 3 : 1.
2. **Complete item 5b and 5c on the laboratory report.** Set up your Punnett squares on a separate sheet of paper.
3. Examine the trays of corn seedlings produced by a cross of two tall-green dihybrids. **Complete item 5d on the laboratory report.**

LINKED GENES

Homologous genes occur in a linear sequence on homologous chromosomes, as shown in Figure 34.1. When homologous chromosomes are segregated into different gametes during gametogenesis, members of each gene pair are also segregated, as shown in Figure 34.6. Thus, alleles on each homologous chromosome are linked together and tend to be segregated into the same gamete. Therefore, alleles of linked genes tend to be inherited together and are not assorted independently. Note in Figure 34.6 that the dihybrid parent with linked genes produces only two classes of gametes instead of the four classes of gametes that would be produced if the genes were nonlinked genes. Thus, a dihybrid cross with linked genes requires a Punnett square with only four squares.

X-Linked Traits

In humans, sex is determined by a single pair of sex chromosomes. Females possess two X chromosomes (XX), and males possess an X and a Y (XY). Sex in humans is inherited as shown in Figure 34.7.

The Y chromosome is shorter than the X and lacks some of the genes present on the X chromosome. Those genes that are present on the X chromosome but absent on the Y chromosome are the **X-linked genes** that control inheritance of X-linked traits. See Figure 34.8.

For the X-linked genes, a female is diploid and a male is haploid. Therefore, a recessive allele on the

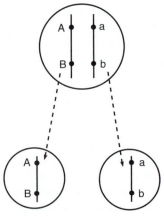

Figure 34.6 Gametogenesis involving linked genes.

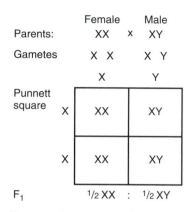

Figure 34.7 Sex inheritance in humans.

X chromosome of a male will be expressed, whereas the recessive allele must be present on both X chromosomes of a female to be expressed. Red-green color blindness and hemophilia are common recessive X-linked traits in humans. X-linked traits are expressed as a superscript in this manner: $X^C X^C$ (color-blind female); XX^C (carrier female); $X^C Y$ (color-blind male).

Assignment 6

1. Contrast the number of classes of gametes formed by dihybrids when the genes are linked and non-linked.
2. **Complete item 6 on the laboratory report.** Set up your Punnett squares on a separate sheet of paper.

PEDIGREE ANALYSIS

Now that you understand the fundamentals of simple inheritance patterns, it is possible to trace a trait in a pedigree (family tree) to determine if it is inherited in a simple dominant/recessive or an X-linked pattern of inheritance. Keep in mind that dominant traits appear in each generation, recessive traits skip one or more generations, and recessive X-linked traits affect more males than females.

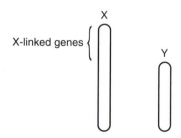

Figure 34.8 The sex chromosomes.

Assignment 7

1. **Complete item 7 on the laboratory report.**
2. Examine your family photos to see if you can trace in your family some of the traits noted in Table 34.3.

POLYGENIC INHERITANCE

Traits inherited as dominants or recessives are qualitative traits. For example, seeds of pea plants are either round or wrinkled, yellow or green, and the flowers are either purple or white. General observation of plants and animals suggests that some traits are not inherited in this manner; that is, some are quantitative in nature. For example, people are not either short or tall but show a gradation of heights typical of a normal (bell-shaped) curve. Such traits exhibit **polygenic inheritance** in which (1) several genes control the same trait and (2) incomplete dominance is evident among the alleles. What other human traits seem to be determined by polygenes?

In wheat, seed color is determined by two pairs of genes, each with two alleles that code for either red or white seed color. In a cross between a wheat plant homozygous for red seeds ($R_1R_1R_2R_2$) and a wheat plant homozygous for white seeds ($r_1r_1r_2r_2$), the F_1 heterozygote ($R_1r_1R_2r_2$) produces pink seeds (Figure 34.9). The blended appearance of the seed color in the F_1 heterozygote indicates a polygenic inheritance pattern of incomplete dominance. Crossing two F_1 heterozygotes yields five classes of progeny, as shown in Figure 34.9.

Assignment 8

1. Work out the cross of F_1 wheat plants in Figure 34.9 to determine the genotype ratio and to understand the basis of the phenotype ratio.
2. **Complete item 8 on the laboratory report.** Set up your Punnett square on a separate sheet of paper.

CHI-SQUARE ANALYSIS

To this point in the exercise, you have learned how to predict the expected genotype and phenotype ratios of progeny. However, biologists must verify the expected ratio of a cross to establish the pattern of inheritance. This is done by using the **chi-square** (χ^2) test. This statistical test indicates the probability (p) that differences between the expected ratio and the actual ratio are due to chance alone or whether use of a different

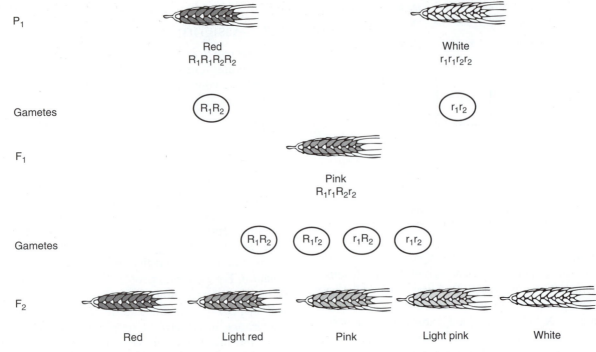

Figure 34.9 Polygenic inheritance in wheat.

hypothesis (expected ratio) would be more appropriate to explain the results (observed ratio). The formula for the chi-square test is $\chi^2 = \Sigma(d^2/e)$, where

χ = chi square

Σ = sum of

d = deviation (difference) between expected and observed results

e = expected results.

Consider a monohybrid cross involving flower color in garden peas. The predicted phenotype ratio is 3 purple-flowering plants to 1 white-flowering plant. Thus, if 100 plants were produced from the cross, 75 should have purple flowers and 25 should have white flowers. Table 34.6 shows the results of such a cross and the calculation of chi square.

Comparing the calculated value of chi square (χ^2) with the values in Table 34.7 is necessary to determine the probability (p) that the deviation from the expected ratio is either (1) by chance and verifies the predicted ratio or (2) greater than chance and does not support the predicted ratio. Note that the chi-square values are arranged in columns headed by probability values and in horizontal rows by phenotype classes minus 1 (C−1).

The two classes of progeny in the example are purple flowers and white flowers. Thus, C = 2, and (C−1) = 1, you must look for the calculated chi-square value in the first horizontal row of values. A χ^2 value of 0.48 falls between the columns of 0.50 and 0.20 probability. This means that by random chance the deviation between the expected and actual results will occur between 20% and 50% of the time. Thus, the predicted ratio for the progeny is supported. Probabilities greater than 5% ($p \geq 0.05$) are generally accepted as supporting the hypothesis (expected ratio), whereas those of 5% or less indicate that the results could not be due to chance.

TABLE 34.6	CHI-SQUARE DETERMINATION				
Phenotype	Actual Results	Expected Results	Deviation (d)	(d^2)	(d^2/e)
Purple flowers	78	75	3	9	9/75 = 0.12
White flowers	22	25	3	9	9/25 = 0.36
					$\Sigma(d^2/e) = 0.48$
					$\chi^2 = 0.48$

TABLE 34.7   CHI-SQUARE VALUES

	Probability (p)						
	Deviation Insignificant Hypothesis Supported					Deviation Significant Hypothesis Not Supported	
C–1	.99	.80	.50	.20	.10	0.5	.01
1	0.00016	0.064	0.455	1.642	2.706	3.841	6.635
2	0.0201	0.446	1.386	3.219	4.605	5.991	9.210
3	0.115	1.005	2.366	4.642	6.251	7.815	11.341
4	0.297	1.649	3.357	5.989	7.779	9.488	13.277

 Assignment 9

Materials

Corn ears with kernels showing a ratio of 3 purple to 1 white

Corn ears with kernels showing a ratio of:
   9 smooth-yellow to 3 smooth-white to 3 shrunken-yellow to 1 shrunken-white

Corn ears formed from experimental crosses

1. Examine a corn ear with both purple and white kernels that have resulted from a monohybrid cross. The predicted ratio is 3 purple to 1 white. Count the purple and white kernels to determine the actual ratio. Mark the row of kernels where you start counting with a pin stuck into the cob under the first kernel. Then do a chi-square analysis of the results in item 9b on the laboratory report.

2. Examine a corn ear formed by an F_2 dihybrid cross so a 9 : 3 : 3 : 1 phenotype ratio of the kernels is expected. The two independent phenotypes involved are smooth-shrunken and yellow-white.

Smooth and yellow are dominant. Determine the actual results, and do a chi-square analysis in item 9c on the laboratory report.

3. **Complete items 9a–9d on the laboratory report.**

4. Examine the corn ears from experimental crosses to determine the phenotype ratios of the kernels (progeny) and to determine whether the cross was a monohybrid cross, monohybrid test cross, dihybrid cross, or dihybrid test cross. For each ear, use the following steps.

   a. Determine the number of phenotypic classes in the progeny. Construct a chi-square analysis table.

   b. Count and record the number of progeny in each class.

   c. Inspect the number of progeny in each class and form a hypothesis as to which type of class is probably involved. Record the expected numbers of progeny in each class, as predicted by the hypothesis, in the chi-square table.

   d. Perform a chi-square analysis to determine if the data support your hypothesis.

5. **Complete the laboratory report.**

HEREDITY

Student _____

Lab Instructor _____

1. FUNDAMENTALS

Write the term that matches each phrase.

a. Traits passed from parents to progeny   _____

b. Part of DNA coding for a specific protein   _____

c. Contain homologous genes   _____

d. Alternate forms of a gene   _____

e. Alleles of a gene pair are identical   _____

f. Alleles of a gene pair are different   _____

g. Observable form of a trait   _____

h. Genetic composition determining a trait   _____

i. Allele expressed in heterozygote   _____

j. Allele not expressed in heterozygote   _____

2. MONOHYBRID CROSSES WITH DOMINANCE

a. In corn plants, plant height is controlled by one gene with two alleles. How many alleles that control the trait for height are present in each cell nucleus of a corn plant? _____

How many alleles that control the trait for height are present in the nucleus of each gamete formed by a corn plant? _____

b. Indicate the genotypes of these corn plants:

homozygous tall _____

heterozygous tall _____

dwarf _____

c. Indicate the genotypes of possible gametes of these corn plants:

homozygous tall _____   heterozygous tall _____   dwarf _____

d. Determine the predicted phenotype ratio in progeny of a cross between two heterozygous tall corn plants.

Parent phenotypes _____ × _____

Parent genotypes _____   _____

Gametes _____ _____   _____ _____

Punnett square _____ _____

Genotype ratio _____

Phenotype ratio _____

e. Examine the corn seedlings in tray 1 that are progeny of a cross of two monohybrid tall corn plants. Record the number of:

tall plants _____   dwarf plants _____

Determine the observed phenotype ratio of tall plants to dwarf plants by dividing the number of plants of each type by the number of dwarf plants. Round your answers to whole numbers.

_____ tall ÷ _____ dwarf = _____   _____ dwarf ÷ _____ dwarf =1

Record the observed phenotype ratio: _____ tall : 1 dwarf

Is the observed ratio different from the predicted ratio? _____

If so, explain why this may have happened. _____

_____

f. Examine the corn seedlings in tray 2. Record the number of:

green plants _____   albino plants _____

Determine the phenotype ratio of green plants to albino plants by dividing the number of plants of each type by the number of albino plants. Round your answers to whole numbers.

_____ green ÷ _____ albino = _____   _____ albino ÷ _____ albino =1

Record the observed phenotype ratio:_____ green : 1 albino

Indicate the type of simple dominant/recessive cross that produces this kind of ratio: _____

_____

Based on this observed ratio and using "G" to represent the allele for green leaves and "g" to represent the allele for albino leaves, record the phenotypes and genotypes of the parent plants:

Phenotypes: _____ ✕ _____

Genotypes: _____   _____

g. Determine the expected progeny of a cross between:

1. heterozygous tall and dwarf corn plants

Parent phenotypes   _____ ✕ _____

Parent genotypes   _____   _____

Gametes   _____ _____   _____ _____

Punnett square   _____ _____

_____

_____

Genotype ratio _____

Phenotype ratio _____

2. heterozygous green and albino corn plants (set up your own Punnett square on a separate sheet of paper)

Genotype ratio _____

Phenotype ratio _____

h. Indicate the possible genotypes of parent pea plants in crosses yielding the following ratios:

Phenotype Ratio	*Parent Genotypes*
1. 3 round seeds : 1 wrinkled seed	_____ ✕ _____
2. all white flowers	_____ ✕ _____
3. all purple flowers (3 possibilities)	_____ ✕ _____
	_____ ✕ _____
	_____ ✕ _____
4. 1 round seed : 1 wrinkled seed	_____ ✕ _____

i. Is it possible to observe differences between homozygous and heterozygous individuals possessing the dominant allele? _____

j. To determine the genotype of a tall pea plant by a test cross, it should be crossed with a _____ pea plant. In such a test cross, indicate the genotype of the tall pea plant if:

all the progeny are tall _____

half the progeny are tall _____

k. The following table lists several human traits that are determined by a simple dominant/recessive mode of inheritance. Record your phenotypes and the frequency of the phenotypes among members of your class.

HUMAN DOMINANT/RECESSIVE TRAITS DETERMINED BY A SINGLE GENE*

Trait	Phenotype	Your Phenotype	Number in Class with Phenotype
Handedness	Right-handed*		
	Left-handed		
Ear lobes	Free*		
	Attached		
Freckles	Freckled*		
	Nonfreckled		
Hairline	Widow's peak*		
	Straight		
Little finger	Bent*		
	Straight		
Tongue roll	Yes*		
	No		
Rh factor	Rh^+*		
	Rh^-		

*Indicates dominant traits

l. For each trait, is the dominant phenotype always more abundant among class members?

_____

m. Assuming your class to be representative of the general population, what can you conclude about the relationship between dominant and recessive phenotypes and their frequencies in the population? _____

_____

n. Newlyweds Bill and Sue are nonfreckled. Because each had one parent who had freckles, they wonder what the probability is of their children having freckles. What would you tell them? _____

_____

o. Mary has freckles, but her husband Dick does not. Mary's father has freckles, but her mother does not. What is the probability that Mary and Dick's child will have freckles? _____

p. Rh blood type is controlled by two alleles. The allele for Rh^+ blood is dominant; the allele for Rh^- blood is recessive. Linda is pregnant and Rh^-. Her husband Tom is Rh^+, as are both of his parents. What is Linda's genotype? _____ What are the two possible genotypes for Tom? _____

What is the probability of their baby being Rh^-:

if Tom is homozygous for Rh^+? _____

if Tom is heterozygous for Rh^+? _____

3. INCOMPLETE DOMINANCE AND CODOMINANCE

a. Contrast inheritance patterns of incomplete dominance and codominance.

Incomplete dominance _____

_____

Codominance _____

_____

b. Indicate the genotype and phenotype ratios of this cross in snapdragons: Pink flowers × pink flowers.

Genotype ratio: _____

Phenotype ratio: _____

c. What is the probability that parents who are both heterozygous for sickle-cell hemoglobin would have a child with sickle-cell anemia? _____

4. MULTIPLE ALLELES: ABO BLOOD TYPES

a. Indicate the expected genotype and phenotype ratios for these matings:

1. $I^A I^B \times ii$

Genotype ratio: _____

Phenotype ratio: _____

2. $I^A i \times I^B i$

Genotype ratio: _____

Phenotype ratio: _____

b. True/False

_____ A type O child may have two type A parents.

_____ A type O child may have one type AB parent.

_____ A type B child may have two type AB parents.

_____ A type AB child may have one type O parent.

c. Ann (type A, Rh^+ blood) is suing Joe (type AB, Rh^- blood) for child support, claiming that he is the father of her child (type B, Rh^+ blood). On the basis of blood type, is it possible that Joe is the father? _____
What ABO blood type(s) would Joe have to possess to *prove* that he is *not* the father? _____

5. DIHYBRID CROSSES WITH DOMINANCE

a. Determine the phenotype ratio in progeny of a cross between two yellow-round dihybrid pea plants.

Parent phenotypes _____ × _____

Parent genotypes _____ _____

Gametes ____ ____ ____ ____ ____ ____ ____ ____

Punnett square _____ _____ _____ _____

Phenotype ratio   _____ yellow-round   _____ green-round

                          _____ yellow-wrinkled   _____ green-wrinkled

b. List the possible gametes of a dihybrid with a genotype of AaBb.

Possible gametes   _____   _____   _____   _____

c. If two tall-green dihybrid corn plants were crossed, each with a genotype of TtGg, what is the *expected phenotype ratio* in the progeny (T = tall, t = dwarf, G = green, g = albino)? Recall that tall and green are dominant.

Phenotype ratio   _____ tall-green   _____ dwarf-green

                          _____ tall-albino   _____ dwarf-albino

d. In the lab, there are several trays of corn seedlings that are progeny of a cross between tall-green dihybrids. Each tray is labeled tray 3. Count and record the phenotypes in *all* these trays. Determine the *observed phenotype ratio* by dividing the number of each phenotype by the number of dwarf-albino plants and rounding to the nearest whole number.

Phenotype ratio   _____ tall-green   _____ dwarf-green

                          _____ tall-albino   _____ dwarf-albino

Explain any difference between the expected and observed phenotype ratios.

_____

_____

_____

6. LINKED GENES

a. A dihybrid was crossed with a homozygous recessive for each trait, and the cross produced only two classes of progeny. How do you explain this? _____

_____

b. Why are more males color-blind than females? _____

_____

c. Determine the expected phenotype ratio of children from these matings:

1. Normal female (XX) × color-blind male (X^CY)

_____

2. Carrier female (XX^C) × normal male (XY)

_____

7. PEDIGREE ANALYSIS

Examine the pedigrees shown here, and determine whether the inherited trait (■ or ●) is dominant, recessive, or X-linked recessive; □ = male; ○ = female.

a.

b.

c.

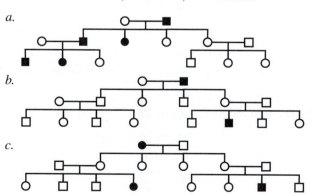

8. POLYGENIC INHERITANCE

Determine the expected phenotype ratio in progeny produced by the cross of a wheat plant with pink seeds ($R_1r_1R_2r_2$) and a wheat plant with white seeds ($r_1r_1r_2r_2$).

_____

_____

9. CHI-SQUARE ANALYSIS

a. In the example of chi-square determination in Table 34.6, what is the hypothesis being tested?

_____

_____

Would the hypothesis be supported by a chi square of 3.952? _____

b. Record your results of the monohybrid cross and do a chi-square analysis of your results in the following table.

CHI-SQUARE ANALYSIS OF PROGENY FROM A MONOHYBRID CROSS

Phenotype	Actual Results	Expected Results	Deviation (d)	(d^2)	(d^2/e)
Purple flowers	_____	_____	_____	_____	_____ = _____
White flowers	_____	_____	_____	_____	_____ = _____
					$\Sigma(d^2/e) =$ _____
					$\chi^2 =$ _____
p falls between _____ and _____					

Is the hypothesis supported? _____

c. Record your results of the dihybrid cross and do a chi-square analysis of your results in the following table.

CHI-SQUARE ANALYSIS OF PROGENY FROM A DIHYBRID CROSS

Kernel Phenotype	Actual Results	Expected Results	Deviation (d)	(d^2)	(d^2/e)
Smooth-yellow	_____	_____	_____	_____	_____ = _____
Smooth-white	_____	_____	_____	_____	_____ = _____
Shrunken-yellow	_____	_____	_____	_____	_____ = _____
Shrunken-white	_____	_____	_____	_____	_____ = _____
					$\Sigma(d^2/e) =$ _____
					$\chi^2 =$ _____
p falls between _____ and _____					

Is the hypothesis supported? _____

d. Biologists like to have a large number of progeny to analyze when doing genetic crosses. Why is this of value?

_____

e. Indicate the chi-square values and parental genotypes for the corn ears from experimental crosses.

Ear No.	Chi Square	Parental Genotypes
_____	_____	_____
_____	_____	_____
_____	_____	_____

MOLECULAR AND CHROMOSOMAL GENETICS

OBJECTIVES

After completing the laboratory session, you should be able to:
1. Describe the basic structure of DNA and RNA.
2. Describe the process of information transfer in (a) DNA replication, (b) RNA synthesis, and (c) protein synthesis.
3. Explain how mutations involving base substitution, addition, or deletion affect protein synthesis.
4. Prepare a karyotype from a metaphase smear of human chromosomes.
5. Describe the basis of the chromosomal abnormalities studied.
6. Define all terms in bold print.

Chromosomes are responsible for transmitting the hereditary material from cell to cell in cell division and from organism to progeny in reproduction. This is why the distribution of replicated chromosomes in mitotic and meiotic cell divisions is so important in eukaryotic cells. The genetic information is contained in the structure of **deoxyribonucleic acid** (**DNA**), which forms the hereditary portion of the chromosomes.

DNA AND THE GENETIC CODE

DNA is a long, thin molecule consisting of two strands twisted in a spiral arrangement to form a double helix, somewhat like a twisted ladder. See Figure 35.1. The sides of the ladder are formed of sugar and phosphate molecules, and the rungs are formed by nitrogenous bases joined together by hydrogen bonds.

Each strand of DNA consists of a series of **nucleotides** joined together to form a polymer of nucleotides. Each nucleotide of DNA is formed of three parts: (1) a deoxyribose (C_5) sugar, (2) a phosphate group, and (3) a nitrogenous base. Four kinds of nitrogenous bases are in DNA. The purine bases (double-ring structure) are **adenine** (A) and **guanine** (G). The pyrimidine bases (single-ring structure) are **thymine** (T) and **cytosine** (C). Note the **complementary pairing** of the bases in Figure 35.1. Can you discover a pattern to their pairing? It is the sequence of nucleotides with their respective purine or pyrimidine bases that contains the genetic information of the DNA molecule.

DNA Replication

Each DNA molecule is able to replicate itself during interphase of the cell cycle and thereby maintain the constancy of the genetic information in new cells that are formed. **Replication** begins with the breaking of the weak hydrogen bonds that join the nitrogen bases of the nucleotides. This results in the separation of the DNA molecule into two strands of nucleotides. See Figure 35.2. Each strand then serves as a **template** for the synthesis of a complementary strand of nucleotides that is formed from nucleotides available in the nucleus. The complementary pairing of nitrogen bases determines the sequence of nucleotides in the new strands and results in the formation of two DNA molecules that are identical. Because each new DNA molecule contains one "old" strand and one "new" strand, replication is said to be **semiconservative.** Occasionally, errors are made during replication, and such errors are a type of **mutation.** Of course, replication is controlled by a series of enzymes that catalyze the process.

 Assignment 1

Materials

Colored pencils
DNA synthesis kits

1. Color-code each of the four nitrogenous bases in Figure 35.1 and circle one nucleotide.

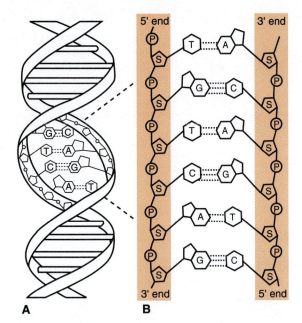

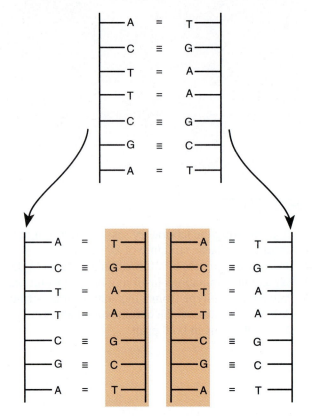

Figure 35.1 DNA structure. **A.** The double helix of a DNA molecule. Complementary pairing of the nitrogenous bases joins the sides like rungs of a twisted ladder. A = adenine, G = guanine, C = cytosine, and T = thymine. **B.** If a DNA molecule is untwisted, it would resemble a ladder in shape. The sides of the ladder are formed of deoxyribose sugar and phosphate, and the rungs consist of nitrogenous bases.

Figure 35.2 DNA replication. Shading indicates the new strand of nucleotides. Each replicated DNA molecule consists of an "old" strand of nucleotides and a "new" strand of nucleotides. A = adenine, G = guanine, C = cytosine, and T = thymine.

2. ***Complete items 1a–1c on Laboratory Report 35 that begins on page 447.***
3. Use a DNA kit to construct a segment of a DNA molecule that matches the base sequence of the DNA segment in item 1c on the laboratory report.
4. ***Complete item 1d on the laboratory report.***
5. Use a DNA kit to construct a segment of a DNA molecule that matches the "old," nonreplicated DNA segment in item 1d on the laboratory report. Then, separate the strands and construct the replicated strands as shown in item 1d.

RNA Synthesis

DNA serves as the template for the synthesis of **ribonucleic acid (RNA)**. RNA differs from DNA in three important ways: (1) It consists of a single strand of nucleotides, (2) its nucleotides contain ribose sugar instead of deoxyribose sugar, and (3) **uracil** (U) is substituted for thymine as one of the four nitrogenous bases.

To synthesize RNA, a segment of a DNA molecule untwists, and the hydrogen bonds between the nucleotides are broken. The nucleotides of one strand pair with complementary RNA nucleotides in the nucleus. When the RNA nucleotides are joined by

sugar–phosphate bonds, the RNA strand is complete and separates from the DNA strand. Few or many RNA molecules may be formed before the DNA strands reunite.

 ## Assignment 2

Materials

DNA-RNA synthesis kit

1. ***Complete item 2 on the laboratory report.***
2. Use a DNA-RNA kit to synthesize an RNA molecule with a base sequence identical to the hypothetical RNA molecule in item 2 on the laboratory report.

Protein Synthesis

The genetic information of DNA functions by determining the kinds of protein molecules that are synthesized in the cell. A sequence of three bases—a base triplet—in a DNA molecule has been shown to code indirectly for an amino acid. By controlling the sequence of amino acids, DNA determines the kind of protein produced. Recall that enzymes are proteins and that the chemical reactions in a cell are controlled by enzymes. Thus, DNA indirectly controls cellular functions by controlling enzyme production.

Each of the three types of RNA molecules plays an important role in protein synthesis. **Messenger RNA (mRNA)** is a complement of the genetic information of DNA. A **transcription** of the genetic information in DNA is made when mRNA is synthesized. A base triplet of mRNA is called a **codon,** and the codons for the 20 amino acids have been determined. The genetic information is carried by mRNA as it passes from the nucleus to the **ribosomes,** sites of protein synthesis in the cytoplasm.

Transfer RNA (tRNA) carries amino acids to the ribosomes. At one end of a tRNA molecule are three nitrogenous bases, an **anticodon,** which is complementary to a codon of the mRNA. At the other end is an attachment site for 1 of the 20 types of amino acids.

The codon of mRNA and anticodon of tRNA briefly join to place a specific amino acid in its position in a forming polypeptide chain. This interaction takes place on the surface of a ribosome containing the necessary enzymes for the reaction. **Ribosomal RNA (rRNA),** the third type of RNA, is an integral component of ribosomes, and it plays an important role in decoding the mRNA message to enable protein synthesis. The formation of an amino acid chain is the **translation** of the genetic information.

Figure 35.3 depicts the interaction of mRNA, tRNA, and rRNA in the formation of a polypeptide. Note how the sequence of the amino acids is controlled by the pairing of the codons and anticodons. It may be simplified as follows:

$$\begin{array}{ccc} \textbf{Transcription} & & \textbf{Translation} \\ \downarrow & & \downarrow \\ \text{Base triplets} \longrightarrow \text{Codons} & \longrightarrow & \text{Polypeptide} \\ \text{of DNA} \quad \text{of RNA} \end{array}$$

The Genetic Code

In protein synthesis, the genetic information inherent in the sequence of base triplets in DNA is transcribed into the sequence of codons in mRNA, which in turn are translated into the sequence of amino acids in a polypeptide chain. In this way, DNA determines both

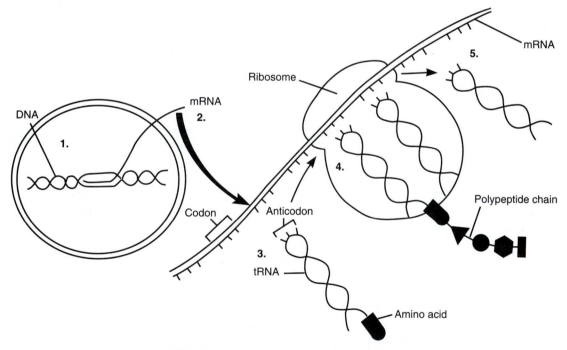

Figure 35.3 A summary of protein synthesis. 1. Chromosomal DNA is the template for the synthesis of mRNA. 2. mRNA carries the genetic code in its codons and moves out of the nucleus to a ribosome, where amino acids are joined together to form a polypeptide (protein). 3. Each tRNA transports a specific amino acid. 4. Temporary pairing of mRNA codons and tRNA anticodons at a ribosome determines the position of the transported amino acid in the forming polypeptide chain. 5. After adding its amino acid and transferring the polypeptide chain to the next tRNA, the tRNA molecule departs the ribosome.

the kinds of amino acids and their sequence in proteins that are synthesized.

There are 64 possible combinations of nucleotide bases in mRNA codons. Their translation is shown in Table 35.1. Note the AUG specifies the amino acid methionine and is also the start signal for protein synthesis. Three codons, UAA, UAG, and UGA, do not specify an amino acid, but they signal the ribosomes to stop assembling the polypeptide chain. Most amino acids are specified by more than one codon, but no codon specifies more than one amino acid. Thus, the code is **redundant** but not **ambiguous.**

Probes

If the sequence of nucleotides in the parent DNA strand is known, you can deduce the sequence of nucleotides in the complementary DNA strand or in the complementary mRNA strand. Because the sequence of nucleotides in mRNA is complementary to the sequence of nucleotides in DNA, it is possible to use a small mRNA molecule with a known sequence of nucleotides to locate a particular segment of DNA that is of interest among thousands of DNA segments. Such a segment of mRNA is called a **probe.** Probes are valuable tools in nucleic acid technology.

Mutations

Base-pair substitution, deletion, or addition constitutes a mutation in a gene. The effect of such mutations is variable, depending on how the mutation is translated via the genetic code.

For example, if a base substitution mutation resulted in a codon change from GCU to GCC, there would be no effect because both specify the amino acid alanine. But if the change was from GCU to GUU, valine would be substituted for alanine in the polypeptide chain and may have a marked effect on the protein. Similarly, if UAU mutated to UAA, it would terminate the polypeptide chain at that point instead of adding tyrosine to the chain. This likely would form a nonfunctional protein.

The mutation involving the addition or deletion of one or more base pairs will cause a **frameshift** in the reading of the codons that may either terminate the polypeptide chain or insert different amino acids into the chain. Usually, addition or deletion mutations have a more disastrous effect than substitution mutations.

Assignment 3

Materials

DNA-RNA-protein synthesis kits

1. Add the bases of the DNA template and the anticodons of tRNA in Figure 35.4.
2. ***Complete items 3a–3c on the laboratory report.***
3. Use a DNA-RNA-protein synthesis kit to synthesize an amino acid sequence as shown in item 3c on the laboratory report starting with the DNA template.
4. ***Complete item 3 on the laboratory report.***
5. Use a DNA-RNA-protein synthesis kit to synthesize the amino acid sequences determined in items 3e and 3f on the laboratory report.

TABLE 35.1	THE CODONS OF mRNA AND THE AMINO ACIDS THAT THEY SPECIFY						
AAU, AAC	Asparagine	CAU, CAC	Histidine	GAU, GAC	Aspartic acid	UAU, UAC	Tyrosine
AAA, AAG	Lysine	CAA, CAG	Glutamine	GAA, GAG	Glutamic acid	UAA, UAG	(Stop)*
ACU, ACC, ACA, ACG	Threonine	CCU, CCC, CCA, CCG	Proline	GCU, GCC, GCA, GCG	Alanine	UCU, UCC, UCA, UCG	Serine
AGU, AGC	Serine	CGU, CGC	Arginine	GGU, GGC	Glycine	UGU, UGC	Cysteine
AGA, AGG	Arginine	CGA, CGG		GGA, GGG		UGA	(Stop)*
						UGG	Tryptophan
AUU, AUC, AUA	Isoleucine	CUU, CUC, CUA, CUG	Leucine	GUU, GUC, GUA, GUG	Valine	UUU, UUC	Phenylalanine
AUG	Methionine and start					UUA, UUG	Leucine

*Signals the termination of the polypeptide chain.

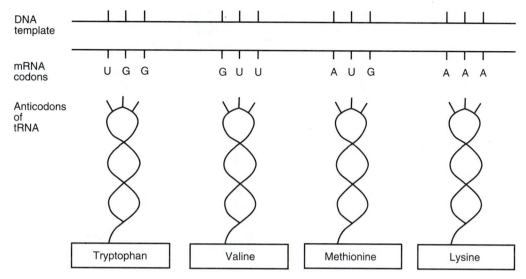

Figure 35.4 Interaction of DNA, mRNA, and tRNA in protein synthesis. Complete the figure by adding the bases in the anticodons of tRNA and the base triplets of DNA.

HUMAN CHROMOSOMAL DISORDERS

In the study of cell division, you learned that (1) chromosomes occur in pairs in diploid cells, (2) chromosomes are faithfully replicated and equally distributed in mitotic cell division, and (3) cells formed by meiotic cell division receive only one member of each chromosome pair. In both types of cell division, the distribution of the chromosomes is systematically controlled, but errors sometimes occur. In this section of the exercise, you will consider human genetic defects, **chromosomal aberrations,** in which whole chromosomes or large parts thereof are missing or added. Because you now understand the role of DNA, you can appreciate the effect of the deletion or addition of large amounts of DNA on normal cellular functions.

The 46 chromosomes of human body cells are classified as 22 pairs of **autosomes,** nonsex chromosomes, and 1 pair of **sex chromosomes,** XX in females and XY in males. Normally, the separation of chromosomes in meiosis of gametogenesis places 22 autosomes and 1 sex chromosome in each gamete, but occasionally errors place both members of a chromosome pair in the same gamete. Such errors are caused by nondisjunction. **Nondisjunction** is the failure of homologous chromosomes or sister chromatids to separate during gametogenesis. If it occurs during the first meiotic division, all the gametes contain an abnormal chromosome number. If it occurs during the second meiotic division, half the gametes will have an abnormal chromosome number. See Figure 35.5. If a gamete with an abnormal number of chromosomes is involved in fertilization, the resulting zygote will have an abnormal chromosome number and will be minimally or severely affected, depending on the chromosome involved. Severe defects or death usually occur. For example, if the zygote contains an extra copy of chromosome 21, the presence of three 21 chromosomes (trisomy 21) results in Down syndrome. See Table 35.2. In contrast, the loss of a chromosome 21, like the loss of any autosome, is lethal.

Chromosomal abnormalities also may stem from the **translocation** of a portion of one chromosome to a member of a different chromosome pair, resulting in reduced or extra chromosomal material in a gamete. Table 35.2 indicates a few disorders caused by chromosomal abnormalities.

Cytogeneticists are able to determine some of the abnormalities among chromosomes by examining them at the metaphase stage of mitosis. A photograph of the spread chromosomes is taken and enlarged. Then the chromosomes are cut out one at a time from the photo and sorted on an analysis sheet to form a **karyotype,** an arrangement of chromosome pairs by size that allows determination of the chromosome number and any abnormalities that can be visually identified.

Examine the normal male karyotype in Figure 35.6. Note that the chromosomes are sorted into seven groups, A through G, on the basis of size and the location of the centromere. This process separates the sex chromosomes, X and Y, from each other in the karyotype.

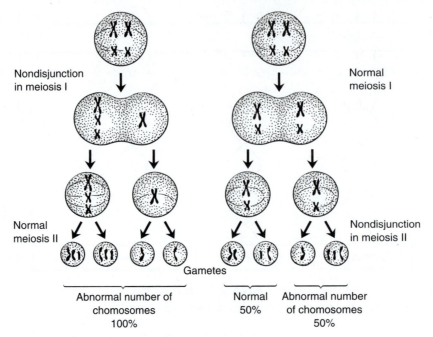

Figure 35.5 Nondisjunction yields gametes with an abnormal number of chromosomes. From Krogh, *Biology, A Guide to the Natural World,* Figure 12.6, 2002, p. 237.

TABLE 35.2	EXAMPLES OF CHROMOSOMAL ABNORMALITIES
Chromosomal Abnormality	*Effect*
Trisomy 18	E syndrome: usually fatal within 3 mo due to multiple congenital defects
Trisomy 21	Down syndrome: mental retardation; short and incurved fifth finger; marked creases in palm; characteristic facial appearance
Deletion of chromosome 5 from short arm	Cri-du-chat syndrome: mental and physical retardation; round face; plaintive catlike cry; death by early childhood
XXY	Klinefelter syndrome (male): underdeveloped testes; breasts enlarged; usually sterile; mentally retarded
XO	Turner syndrome (female): underdeveloped ovaries; no ovulation or menstruation

 Assignment 4

Materials

Scissors
Forceps
Human karyotype analysis set (chromosome spreads)
Prepared slide of human chromosomes
Rubber cement or glue stick

1. Examine a prepared slide of human chromosomes. Note their small size and replicated state.
2. Study Figure 35.6 to understand how chromosomes are sorted in preparing a karyotype.
3. Obtain an enlarged photocopy of a human chromosome spread, as directed by your instructor. Cut out the chromosomes one by one and place them on the karyotype analysis form on the laboratory report with the centromere on the dashed line. Do not glue them in place until you are sure of their correct positions. Use the size of the chromosome and the position of the centromere to determine the correct position of each chromosome. Refer to Figure 35.6 as needed.
4. After your karyotype is completed, determine the sex of the individual and any abnormality that may be indicated. Compare your karyotype and analysis with classmates who have prepared karyotypes from different individuals.
5. ***Complete item 4 on the laboratory report.***

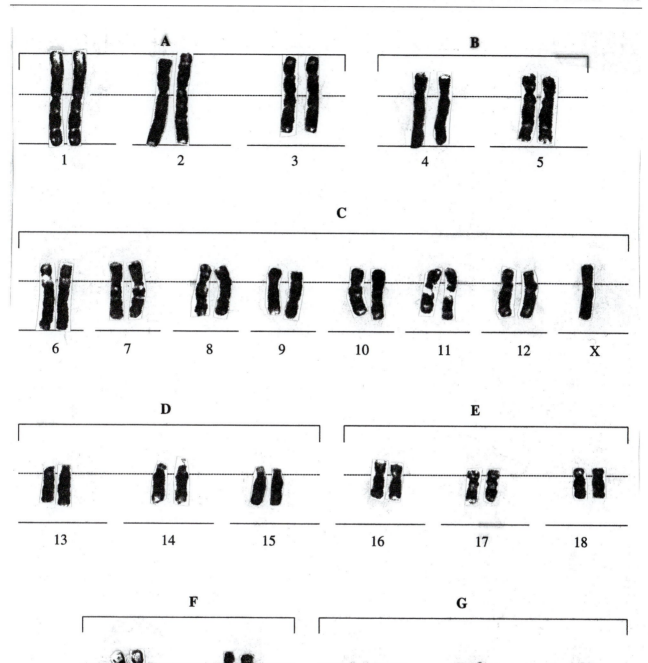

Figure 35.6 Karyotype of a normal male. (Chromosomes from a chromosome spread used with permission of Carolina Biological Supply Company.)

MOLECULAR AND CHROMOSOMAL GENETICS

Student _____

Lab Instructor _____

1. DNA

a. Write the term that matches each phrase.

1. Molecule containing genetic information _____

2. Sugar in DNA nucleotides _____

3. Pyrimidine that pairs with adenine _____

4. Purine that pairs with cytosine _____

5. Chemical bonds joining complementary nitrogen bases _____

6. Number of nucleotide strands in DNA _____

7. Two molecules forming sides of the DNA "ladder" _____

b. In DNA replication, what determines the sequence of nucleotides in the new strands of nucleotides that are formed? _____

_____

c. After determining the pairing pattern of the nitrogenous bases in Figure 35.2, add the missing bases to this hypothetical strand of DNA:

A = adenine, C = cytosine, G = guanine, T = thymine

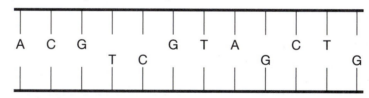

d. At the left is a hypothetical segment of DNA. After replication, it forms two strands, as shown at the right. Add the missing bases in the replicated strands.

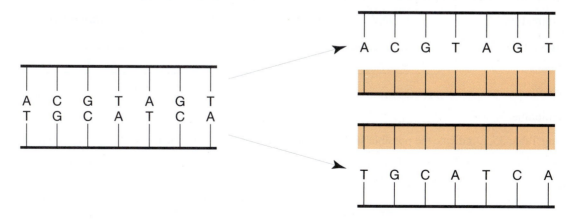

447

2. RNA SYNTHESIS

a. Write the term that matches each meaning.

1. Sugar in an RNA nucleotide _____

2. Number of nucleotide strands in RNA _____

3. RNA base that pairs with adenine _____

4. Template for RNA synthesis _____

b. The following figure shows a portion of a DNA molecule whose strands have separated to synthesize mRNA. The bases are shown for the strand that is *not* serving as the template. Add the bases of the DNA strand serving as the template and the bases of the newly formed RNA molecule.

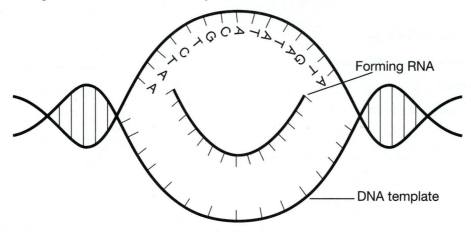

Forming RNA

DNA template

3. PROTEIN SYNTHESIS

a. Write the term that matches each meaning.

1. Carries the genetic code to ribosomes _____

2. Formed of an mRNA base triplet _____

3. Specifies a particular amino acid _____

4. Sites of protein synthesis _____

5. Carries amino acids to ribosome _____

6. tRNA triplet that pairs with codon _____

7. Codon that starts polypeptide synthesis _____

8. Codons that stop polypeptide synthesis _____

b. Indicate the DNA base triplet and tRNA anticodons for the mRNA codons in Figure 35.4.

Amino Acid	*DNA Base Triplet*	*Anticodon*
Tryptophan	_____	_____
Lysine	_____	_____
Valine	_____	_____
Methionine	_____	_____

c. Using Table 35.1, indicate the possible sequences of mRNA codons that code for the following polypeptide:

methionine - lysine - histidine - tryptophan - glutamine - tryptophan - (stop)

_____   _____   _____   _____   _____   _____   _____

_____   _____   _____

d. Molecular biologists use probes of mRNA to locate specific gene loci on DNA segments. Use the codon sequences determinesd 3c as probes to locate the gene (complementary base sequence) in the segment of template (single-strand) DNA shown here. Circle the located gene, and indicate the probe that was successful in locating it.

-A-G-T|T-C-T|T-A-C|C-C-T|G-A-A|C-G-G|C-A-T|T-C-A|

-G-A-C|A-T-T|C-C-G|C-T-A|G-A-C|C-A-T|T-A-C|T-T-C|

-G-T-A|A-C-C|G-T-T|A-C-C|A-T-C|G-T-A|T-C-A|

Successful probe _____

e. Consider the following hypothetical gene. Indicate the mRNA codons that it forms and the sequence of amino acids produced in the polypeptide chain.

-T-A-C|T-T-A|G-A-A|A-T-A|C-C-G|A-A-G|A-C-T

mRNA codons

_____

Amino acid sequence

_____

f. For the following mutations, show the effect by indicating the mRNA codons and the amino acid sequence in the polypeptide.

1. Base substitution

-T-A-C|T-T-A|G-A-G|A-T-A|C-C-G|A-A-G|A-C-T-

mRNA codons

_____

Amino acid sequence

_____

2. Triplet addition

-T-A-C|T-T-A|G-G-T|G-A-A|A-T-A|C-C-G|A-A-G|A-C-T-

mRNA codons

_____

Amino acid sequence

_____

3. Base addition

-T-A-C|T-T-A|G-A-A|A-T-C-A|C-C-G|A-A-G|A-C-T-

mRNA codons

_____

Amino acid sequence

_____

4. Base deletion (site indicated by arrow)

-T-A-C|T-T-A|G-A-A|A-T|C-C-G|A-A-G|A-C-T-

mRNA codons

_____

Amino acid sequence

_____

g. What seems to determine the magnitude of a mutation's effect?

_____

_____

_____

_____

h. Summarize how DNA controls cellular functions.

_____

_____

_____

_____

_____

4. CHROMOSOMAL ABNORMALITIES

a. Prepare a karyotype analysis of your assigned spread of human chromosomes on the next page.

Sex _____ Number of chromosomes _____ Abnormality, if any _____

b. Indicate chromosomal abnormalities detected in the karyotypes analyzed in your class.

Abnormality	*Cause*
_____	_____
_____	_____
_____	_____
_____	_____

Human Karyotype Analysis Form

A
B

_____   _____   _____   _____   _____
1   2   3   4   5

C

_____   _____   _____   _____   _____   _____   _____   _____
6   7   8   9   10   11   12   X

D
E

_____   _____   _____   _____   _____   _____
13   14   15   16   17   18

F
G

_____   _____   _____   _____   _____
19   20   21   22   Y

Sex of subject _____   Number of chromosomes _____

Chromosomal disorder _____

Human Karyotype Analysis Form

A

1	2	3

B

4	5

C

6	7	8	9	10	11	12	X

D

13	14	15

E

16	17	18

F

19	20

G

21	22	Y

Sex of subject _____   Number of chromosomes _____

Chromosomal disorder _____

DNA FINGERPRINTING

OBJECTIVES

After completing the laboratory session, you should be able to:
1. Describe the use of restriction endonucleases in DNA technology.
2. Describe how electrophoresis separates DNA restriction fragments.
3. Describe the procedures used in electrophoresis of DNA restriction fragments.
4. Describe the basis of DNA fingerprinting.
5. Recognize simple identical DNA fingerprints.
6. Define all terms in bold print.

DNA technology is based on the ability of **restriction endonucleases,** or **restriction enzymes,** to cut DNA molecules at specific points. The resulting DNA fragments are called **restriction fragments.** Restriction endonucleases are naturally produced by bacteria as protection against foreign DNA, such as bacteriophage DNA. Over 2,500 restriction endonucleases have been identified, and many are mass-produced for commercial and scientific use.

Each restriction endonuclease recognizes a specific sequence of nucleotides, usually a palindrome—a sequence of four to eight nucleotides that reads the same when reading forward in one strand and in the opposite direction in the complementary strand. For example, the commonly used restriction endonuclease Eco R1 recognizes the sequence GAATTC. The cut is made at specific points within the palindrome. See Figure 36.1.

DNA cleavage by most restriction endonucleases leaves single-stranded, complementary ends ("sticky ends") on the restriction fragments (Figure 36.1). If two different DNA molecules are cut by the same restriction endonuclease, the "sticky ends" enable DNA from one source to combine with DNA from the other source. This is how a DNA fragment (a gene) from one organism is inserted into DNA of a different organism to form **recombinant DNA.** For example, human genes for the production of insulin and growth hormone have been inserted into certain bacteria, and these bacteria are used to mass-produce these human hormones for medical use.

The size (length) of the restriction fragments produced by a restriction endonuclease depends on the frequency of the recognition sites and the distances between them. Thus, the longer a DNA molecule is, the more recognition sites are likely to be present, and the fewer the nucleotides composing a recognition site, the more frequently the site is likely to occur. Thus, a four-nucleotide recognition site will occur more frequently than a six-nucleotide recognition site.

A large portion of mammalian DNA consists of **tandemly arranged repeats,** repetitious nucleotide sequences that occur between genes. Tandemly arranged repeats have no known function, but they

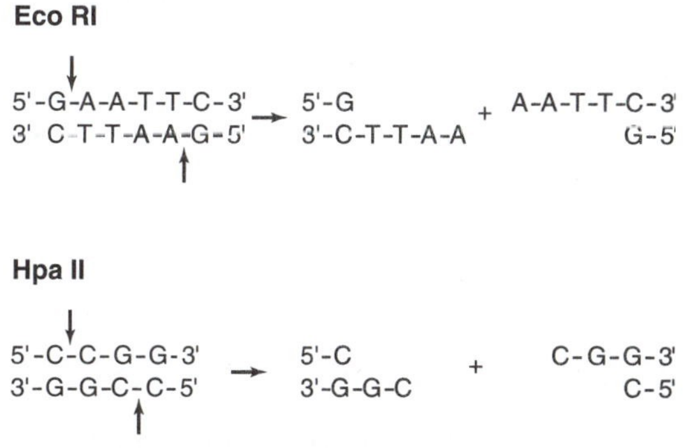

Eco RI

```
      ↓
5'-G-A-A-T-T-C-3'        5'-G           A-A-T-T-C-3'
3' C-T-T-A-A-G-5'   →    3'-C-T-T-A-A   +        G-5'
      ↑
```

Hpa II

```
      ↓
5'-C-C-G-G-3'        5'-C          C-G-G-3'
3'-G-G-C-C-5'   →   3'-G-G-C   +        C-5'
      ↑
```

Figure 36.1 Recognition palindromes and cutting sites of two restriction endonucleases. Nucleotide sequences are read from the 5' to the 3' direction. Both of these endonucleases produce restriction fragments with "sticky ends."

are genetically determined and vary in number from individual to individual. Therefore, when mammalian DNA is cut with a restriction endonuclease, the resulting restriction fragments are of different lengths as determined by the distances between recognition sites due to the number of tandemly arranged repeats.

DNA from the same individual, cut with the same restriction endonuclease, always yields the same pattern of restriction fragment lengths. Similarly, DNA from the same individual, cut with two or more restriction endonucleases, will yield a distinctive pattern of restriction fragment lengths for each endonuclease. Because these patterns depend on the sequence of nucleotides, which is genetically determined, they constitute the unique **DNA fingerprint** for the individual.

If the same endonucleases are used to cut DNA from two different individuals, a different pattern of restriction fragment lengths is produced from each individual's DNA. The distinctive patterns are known as **RFLPs** (restriction fragment length polymorphisms). RFLP differences reflect the genetic differences in the sequence of nucleotides in DNA of different individuals. Because each individual's genetic composition is unique (except for identical twins), a person's DNA fingerprint is different from the DNA fingerprint of all other persons, and it can be used to identify an individual with utmost precision.

After DNA is cut by a restriction endonuclease, the restriction fragments can be separated by **agarose gel electrophoresis.** An agarose gel provides a thin meshwork or sieve through which DNA fragments migrate when exposed to an electrical field. Because DNA has a negative charge, the restriction fragments migrate toward the positive pole, and their rate of movement varies with their size. Smaller fragments migrate faster than larger fragments. Thus, restriction fragments separate according to their molecular weights. The separated restriction fragments may be used for DNA fingerprinting, producing recombinant DNA, or mapping DNA.

 Assignment 1

Complete item 1 on Laboratory report 36, which begins on page 461.

OVERVIEW OF THE EXERCISE

In this exercise, a DNA fingerprinting simulation will be used to solve a hypothetical crime. DNA fingerprinting has become important evidence in criminal cases because it can positively identify a suspect as being present at a crime scene. You will carry out agarose gel electrophoresis to separate restriction fragments of

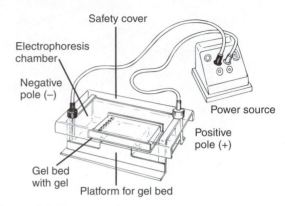

Figure 36.2 Electrophoresis apparatus.

DNA simulating samples collected at a crime scene and from two suspects. The DNA samples have been cut with two different endonucleases.

The electrophoresis apparatus (Figure 36.2) consists of a gel bed, an electrophoresis chamber, a safety cover, and a direct current power source. The sequence of steps that you will perform is as follows:

1. Prepare an agarose gel on the gel bed.
2. Place the gel bed in the electrophoresis chamber.
3. Add buffer solution to the electrophoresis chamber to cover the gel.
4. Transfer samples of restriction fragments to the gel using an automatic micropipetter.
5. Run the gel (expose it to an electric current) to separate the restriction fragments.
6. Destain the gel.
7. Read and interpret the results.

PREPARING THE GEL

Your instructor has prepared the agarose gel solution and the electrophoresis buffer solution. Your task is to form a gel on the gel bed and set up the electrophoresis chamber for loading the gel. See Figure 36.3. Your instructor will describe and demonstrate how to prepare the gel. Follow the steps listed next unless directed otherwise by your instructor. *Wear the safety gloves provided throughout the exercise.*

 Assignment 2

Materials

(Materials based on Edvotek kit 109)
Flasks of sterile:
   distilled or deionized water
   agarose gel solution (0.8%), containing methylene
     blue, melted and cooled to 55°C
   buffer solution, containing methylene blue

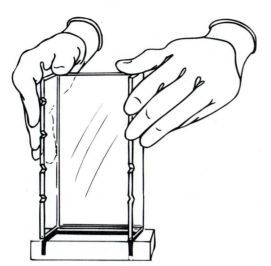

1A. If your gel bed has rubber dams, press rubber dams firmly onto ends of gel bed.

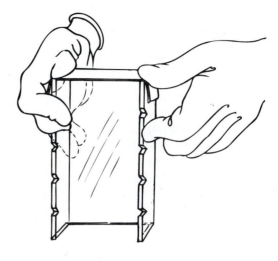

1B. If your gel bed lacks rubber dams, tape the ends securely with tape.

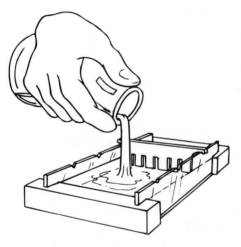

2. Install the comb at a notch near one end and fill gel bed with agarose gel solution.

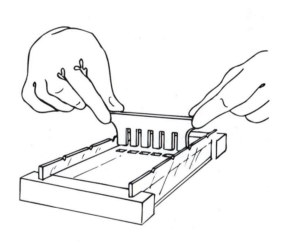

3. After the gel has solidified, gently remove the rubber dams (or tape) and comb.

Figure 36.3 Preparation of the gel.

Gloves, protective and disposable
Horizontal electrophoresis setups with DC power source
Tape, labeling or masking
Water bath, 55°C

1. Rinse the gel bed with distilled water and dry with paper towels.
2. If your gel bed has rubber dams, install a rubber dam at each end, making sure they are firmly attached to the bottom and sides of the gel bed. If your gel bed does not have rubber dams, use $\frac{3}{4}$-in. masking tape or labeling tape to close off the open ends. Fold the ends (about 1 in.) of the tape back on the sides of the gel bed and be sure that they are firmly secured. Make sure that the tape is firmly adhered to the edges of the bottom and sides by running your fingertip over the tape on these surfaces.
3. Place a six-tooth comb in the set of notches near one end of the gel bed. Note that there is a small space between the bottom of the teeth and the gel bed. Be sure the comb is positioned evenly across the gel bed. The comb is used to form wells (depressions) in the gel.

4. Obtain one of the small flasks of 0.8% agarose gel solution containing methylene blue stain from the water bath at 55°C. Place the gel bed on a level surface and pour the gel solution into the gel bed until it is filled. The gel bed must remain motionless while the gel is solidifying, which takes 15–20 min. The gel will become translucent when solidified. *While the gel is solidifying, skip to the next section and practice loading wells in a practice gel. Then return here to complete the gel preparation.*

5. After the gel has solidified, gently remove the rubber dams or tape, being careful not to damage the gel. Running a plastic knife between the gel and the rubber dams or tape helps to prevent the gel from tearing.

6. Gently remove the comb by lifting it straight up while keeping it level. This will prevent damage to the wells.

7. Place the gel bed in the electrophoresis chamber, centered on the platform, with the wells near the negative pole (black).

8. Obtain a flask of buffer plus methylene blue stain. The flask contains the amount of buffer needed for the electrophoresis chamber. Pour the buffer into the electrophoresis chamber until the gel is covered by about 2 mm of buffer.

GEL LOADING PRACTICE

Loading the wells of a gel can be a bit tricky, so a little practice is helpful. Figure 36.4 shows how to hold a micropipetter with your thumb on the plunger at

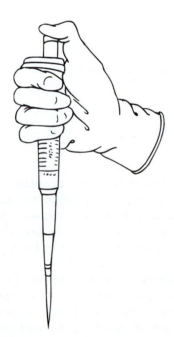

Figure 36.4 The correct way to hold a micropipetter.

the top. A removable micropipette tip is attached at the bottom of the micropipetter. When filling and dispensing fluid, fluid moves into and out of the tip only. Micropipetters differ in the volume of fluid that they can handle. Your micropipetter may be adjusted to automatically pipette small volumes from 5 μl to 50 μl, but it is set at 40 μl because that is the volume needed to fill the wells of the agarose gel. If you need to change the volume setting, see your instructor.

Here's how the micropipetter works. The plunger at the top is used to withdraw and expel fluid.

1. To withdraw fluid, the plunger is depressed to the first stop, the removable tip is inserted into the fluid, and then the plunger is slowly released to draw the fluid into the tip.

2. Dispensing fluid is just the reverse. The tip is inserted into the receiving chamber, and the plunger is slowly depressed to the first stop to expel the fluid. When pipetting fluid into microcentrifuge tubes, the plunger is depressed to the second stop, which ejects air to blow out any remaining fluid. ***Do not do this when filling gel wells.***

3. Pressing the plunger to the third stop ejects the tip from the micropipetter. Some micropipetters have a separate plunger to eject the tip.

Your instructor will demonstrate the correct use of a micropipetter to get you started. Here are some general rules to keep in mind.

1. When not in use, always keep the micropipetter on its stand. Never lay it down on a table or countertop.

2. Always hold the micropipetter with the tip pointing down. This will prevent fluid from running into the micropipetter and contaminating or damaging it.

3. Always use a new sterile tip for each solution that you transfer.

 Assignment 3

Materials

(Materials based on Edvotek kit 109)
Beakers, tip-collection, 250 ml
Gels for practice loading
Gloves, protective and disposable
Microcentrifuge tube racks
Micropipetters, 5–50 μl size
Micropipetter tips, sterile, 50 μl
Tubes of practice loading dye

Picking up a sample:

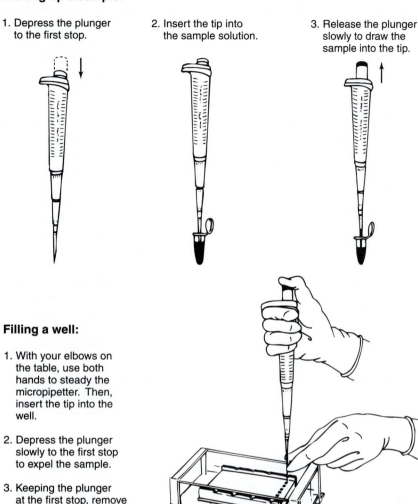

1. Depress the plunger to the first stop.

2. Insert the tip into the sample solution.

3. Release the plunger slowly to draw the sample into the tip.

Filling a well:

1. With your elbows on the table, use both hands to steady the micropipetter. Then, insert the tip into the well.

2. Depress the plunger slowly to the first stop to expel the sample.

3. Keeping the plunger at the first stop, remove the tip from the well.

Figure 36.5 How to use a micropipetter to obtain a sample and fill a gel well.

Practice filling the wells of the practice gel with the practice loading dye using these steps unless directed otherwise by your instructor. See Figure 36.5.

1. Obtain a practice gel to practice loading and a tube of practice loading dye. Add water to cover the gel to a depth of 2–4 mm. Note the wells for practice loading. Placing the gel on a piece of black construction paper will make the wells more visible.

2. Depress the plunger of the micropipetter to the first stop, insert the tip into the practice loading dye, and *slowly* release the plunger allowing it to move up to its highest position. This will draw 40 μl of dye into the micropipette tip.

3. Filling the gel wells requires a steady hand, precision, and care. Placing your elbows on the table will help you control the micropipetter. Hold the micropipetter with the tip down in your right hand (if right-handed), and hold the base of the tip between thumb and forefinger of your other hand to guide it with precision, as shown in Figure 36.5. *Remember, keep your elbows on the table.*

4. Place the pipette tip in the water over a gel well and insert the tip *just barely below* the top of the well, being careful not to touch the sides or bottom of the well. *Slowly* press the plunger to the first stop and note how the dense dye sinks into the well. Keeping the plunger at the first stop position, remove the tip from the well and water.

5. Repeat steps 2–4 several times until you feel comfortable that you can load a gel well correctly without damaging the gel or losing the dye.

6. When finished practicing, hold the micropipetter over the tip collection beaker and depress the plunger to the third position, ejecting the tip into the beaker.

7. Return the micropipetter to its rack.

Now, return to your gel to see if it's solidified. Complete the preparation of the gel by following steps 5 through 8 in the previous section, **Preparing the Gel,** on page 456.

LOADING THE GEL

There are six microcentrifuge tubes (A through F) containing restriction fragments to be loaded into the gel. The hypothetical sources of these tubes are as follows:

Tube A: DNA from the crime scene cut with endonuclease 1

Tube B: DNA from the crime scene cut with endonuclease 2

Tube C: DNA from suspect 1 cut with endonuclease 1

Tube D: DNA from suspect 1 cut with endonuclease 2

Tube E: DNA from suspect 2 cut with endonuclease 1

Tube F: DNA from suspect 2 cut with endonuclease 2

Separating the restriction fragments by agarose gel electrophoresis will enable you to determine if DNA from either suspect matches the DNA from the crime scene.

 Assignment 4

Materials

(Materials based on Edvotek kit 109)
Beakers, tip-collection, 250 ml
Gloves, protective and disposable
Horizontal electrophoresis setups with DC power source and gels in gel beds
Microcentrifuge tubes of restriction fragments labeled A to F
Microcentrifuge tube racks
Micropipetters, 5–50 μl size
Micropipetter tips, sterile, 50 μl

1. Be sure the wells of the gel are nearer the negative pole (black) of the electrophoresis chamber and that the leads will reach from the chamber to the power source. *Once the wells are filled, the chamber must not be moved until electrophoresis is completed.*

2. When viewing a gel from above with the wells farthest from you, the wells are read left to right, either 1 through 6 or, in this case, A through F.

3. Place a sterile tip on the micropipetter.

4. Depress the plunger of the micropipetter to the first stop, insert the tip into the fluid in tube A, and *slowly* release the plunger drawing a sample into the pipette tip.

5. With your elbows on the table, insert the tip just below the top of well A, being careful not to touch the sides or bottom of the well. *Slowly* press the plunger to the first stop to fill the well with the sample. Keeping the plunger at the first stop position, remove the tip from the well and buffer.

6. Hold the micropipetter over the tip collection beaker and eject the tip into the beaker.

7. Place a new sterile tip on the micropipetter, and use the same procedure to place a sample from tube B into well B of the gel. Then eject the tip into the tip collection beaker.

8. Use the same procedure to place samples from tubes C through F into wells C through F, respectively, *being certain to use a new sterile pipette tip for each sample.*

9. When you have loaded the last well, eject the pipette tip into the tip collection beaker and return the micropipetter to its rack.

RUNNING THE GEL

Once the gel is loaded, it is ready to be run. Complete the setup and hook up to the power source as described below.

 Assignment 5

Materials

(Materials based on Edvotek kit 109)
Gloves, protective and disposable
Horizontal electrophoresis setups with DC power source and loaded gels in place

1. Place the safety cover on the electrophoresis chamber, lining up the electrode terminals correctly so the (+) and (–) terminal indicators on the cover and chamber match.

2. Insert the plug of the black (–) wire into the black (–) jack of the power source.

3. Insert the plug of the red (+) wire into the red (+) jack of the power source.

4. Set the voltage indicator of the power source on 50 V unless directed otherwise by your instructor.

5. Turn on the switch of the power source and run the gel for 1.5–2.0 hr as directed by your instructor.

6. If you have not completed Assignment 1, do so while the gel is running.

READING THE GEL

After the gel has run and the restriction fragments have migrated along the gel, you can read the gel to determine the results.

 Assignment 6

Materials

(Materials based on Edvotek kit 109)
Destaining trays
Gel viewing box, white light
Gloves, protective and disposable
Horizontal electrophoresis setups with DC power source with loaded gels that have been run
Spatula, plastic gel-removing

1. Because both the gel and buffer contain methylene blue stain, you will be able to see the faint bands of the restriction fragments as they move down the gel.

2. After running the gel, turn off the power source and unplug the wires from the jacks of the power source.

3. Remove the safety cover from the electrophoresis chamber.

4. Grasp each end of the gel bed to prevent the gel from sliding off, and gently lift the gel bed and gel from the electrophoresis chamber.

5. Place the gel bed and gel on a white-light gel viewing box and read the positions of the bands of restriction fragments in the lanes below each well. If the bands are not easily visible, destaining will be necessary, as described in step 6.

6. Using a plastic spatula, gently slide the gel into a destaining tray containing distilled water. Use the plastic spatula to nip off the upper left corner of the gel to allow easy recognition of the orientation of the gel wells. Destain for 20 min with frequent stirring of the water, and change the distilled water every 5 min or so.

7. Use a plastic spatula to remove the gel to a white-light gel viewing box, and read the positions of the bands of restriction fragments in the lanes below the wells.

8. Clean your apparatus as directed by your instructor.

9. ***Complete item 2 on the laboratory report.***

DNA FINGERPRINTING

Student _____

Lab Instructor _____

1. INTRODUCTION

a. Write the term that matches the phrase.

1. DNA formed of DNA fragments from two different organisms _____

2. Enzymes cutting DNA at specific sites to form restriction fragments _____

3. Technique used to separate restriction fragments in this exercise _____

4. A person's unique RFLP pattern _____

b. What is the source of restriction endonucleases? _____

c. Consider three restriction endonucleases: 1 recognizes a four-nucleotide site; 2 recognizes a five-nucleotide site; and 3 recognizes a six-nucleotide site. Indicate which restriction endonuclease will produce:

1. the greatest number of restriction fragments _____

2. the fewest number of restriction fragments _____

d. What determines the length of restriction fragments? _____

_____

e. What are tandemly arranged repeats? _____

_____

f. How do tandemly arranged repeats affect the lengths of restriction fragments? _____

_____

_____

g. A single-stranded sticky end with a nucleotide sequence of -A-G-C-T can combine with a complementary sticky end with a nucleotide sequence of _____

h. In agarose gel electrophoresis of DNA restriction fragments, the restriction fragments:

1. migrate toward which pole? _____

2. possess, at a neutral pH, an electrical charge that is _____

3. are separated according to their _____

i. If DNA fingerprinting can positively identify an individual, why is it better than regular fingerprints?

_____

_____

j. What are some possible uses of DNA fingerprinting? _____

_____

k. Some persons advocate determining and recording the DNA fingerprint of each newborn baby. What are some pros and cons of such a practice?

Pro _____

_____

_____

Con _____

_____

_____

l. DNA fingerprints reflect genetic variability. List a few genetic mechanisms that increase genetic variability. ____

_____

2. READING THE GEL

a. Draw the bands of restriction fragments as they appear on your gel. Circle the band(s) of the smallest restriction fragments.

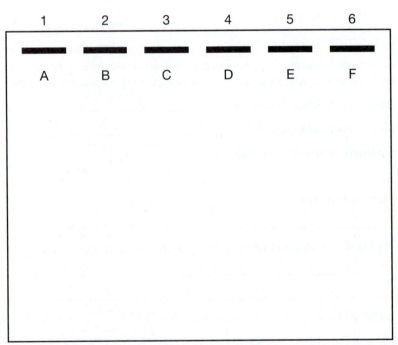

b. According to your data, which suspect was present at the crime scene? _____

c. Why was it important to use two restriction endonucleases in determining the DNA fingerprints of the samples?

_____

_____

d. Does the use of two or more restriction endonucleases increase the validity of DNA fingerprinting?

_____ Explain. _____

_____

_____

EVOLUTION

OBJECTIVES

After completing the laboratory session, you should be able to:
1. Identify the era and period of first appearance for the major organismic groups.
2. Identify the major trends in the history of life.
3. Briefly describe evidence supporting the theory of evolution.
4. Identify and compare the hominid fossils that precede the appearance of modern humans.
5. Perform a comparative skull analysis of fossil and modern hominids.
6. Define all terms in bold print.

The **theory of evolution** postulates (1) that modern species of organisms are descendants of a common ancestor and (2) that their present characteristics have resulted from genetic variation and natural selection. Genetic variation provides the raw material of evolution, and natural selection by the environment is the driving force.

Here's a brief summary of the evolutionary process. Members of a **species** share common genes and can interbreed, producing fertile offspring. When a beneficial mutation enables one or more individuals to obtain more resources and produce more offspring than other individuals, individuals with the mutated gene increase within the species' population. After many generations, most surviving members of the species will have the mutated gene, which provides a better adaptation to the environment. Over long periods of time, this interaction of genetic variation (mutations and other genetic mechanisms) and natural selection produces evolutionary adaptations that make each species distinct from its ancestral species and well suited to its environment. Note that selection occurs in individuals, but that evolutionary change occurs in a population of a species.

The theory of evolution is supported by evidence from diverse fields of inquiry, such as paleontology, biogeography, comparative anatomy, comparative biochemistry, and genetics. In this exercise, you will (1) note the appearance of major organismic groups in the fossil record, (2) examine some of the evidence supporting evolution, (3) consider a possible

pathway of human evolution, and (4) perform a comparative analysis of chimpanzee, hominid, and human skulls.

THE FOSSIL RECORD

A **fossil** is a preserved remnant or impression of an organism. A remnant, such as a leaf or an exoskeleton of an invertebrate, is often replaced by minerals, preserving the detailed structure of the organism. An impression (or mold) results from hardening of material around an organism, which decays and is removed by groundwater. Sometimes an impression is later filled with mineral deposits, forming a cast that maintains the external features of the organism.

Fossils are usually formed when an organism dies and is quickly covered by sand or silt, preventing its destruction by scavengers and decay processes. When subsequent geological processes convert the surrounding soil into sedimentary rock, the remnant of the organism, now called a fossil, is trapped within the rock layers. Some fossils are found in natural asphalt pits and peat bogs, which also preserve organisms.

Our knowledge of the history of life is based on the **fossil record**—the sequence of fossils from the oldest to the most recent. Although incomplete, the fossil record is amazingly good, considering that both the formation and the discovery of fossils occur by chance. Table 37.1 provides a brief summary of the

463

TABLE 37.1   HISTORY OF LIFE

Position of Land Masses	Era	Period	Epoch	Millions of Years Ago (mya)
	CENOZOIC (Age of Mammals)	Quaternary	Recent	
				0.01
Present			Pleistocene	
				2
		Tertiary		
				65
Cretaceous 90 million years ago Pangea breaking up	MESOZOIC (Age of Reptiles)	Cretaceous		
				144
		Jurassic		
				213
Pangea Triassic 240 million years ago		Triassic		
				248
	PALEOZOIC	Permian		
				285
Carboniferous 340 million years ago Pangea forming		Carboniferous		
				380
		Devonian		
				408
Silurian 425 million years ago Gondwana		Silurian		
				438
		Ordovician		
				505
Laurentia Cambrian 520 million years ago Gondwana		Cambrian		
				590
	PROTEROZOIC			
				2500
	ARCHEAN			

TABLE 37.1   CONT.

Major Geologic Events and Extinctions	Major Origins and Radiations of Organisms
Glaciers retreat and climate warms. Extinction of large ice-age mammals.	Radiation of herbaceous plants. Development of human civilization and culture.
The Ice Age. Extinction of many plants and animals. Continents move to present positions.	Giant ice-age mammals dominant. Modern humans emerge.
Large climatic shifts as continents rupture, separate, and collide. Mountain building and cooling at end of the period.	Grasslands appear. Major radiation of flowering plants, mammals, birds, and insects. First hominids appear and later first humans appear.
Breakup of Pangea continues. Modern continents are separated. Mass extinction of dinosaurs and other giant reptiles at end of period (asteroid impact?).	Radiation of marine invertebrates, fishes, insects, and dinosaurs early in period. First flowering plants appear. Gymnosperms decline. First primates appear.
Breakup of Pangea continues.	Gymnosperms (cycads and conifers) are dominant. Ferns, horsetails, and club mosses decline. Dinosaurs radiate and dominate land. First birds appear.
Breakup of Pangea begins.	Gymnosperms and ferns dominant. Radiation of surviving marine invertebrates, fishes, and reptiles. First dinosaurs appear. First mammals appear.
Formation of Pangea is completed. Largest extinction at end of period of most marine invertebrates and many land species.	Radiation of gymnosperms; conifers appear. Radiation of reptiles. Decline of amphibians. First mosses and liverworts appear at beginning of period.
Continents shifting and colliding as Pangea is being formed; mountain building. Extinction of many marine invertebrates.	Age of amphibians. Immense swamp forests of club mosses, ferns, and horsetails. Radiation of amphibians and insects. First reptiles and gymnosperms appear.
Continents move closer together, some collide in early stages of forming Pangea. Mountain building.	First forests of club mosses and horsetails. First ferns and seed ferns appear. Age of fishes. Radiation of fishes. First amphibians and insects appear.
Gondwana at south pole. Mass extinction of many marine invertebrates.	Arthropods and mollusks abundant in seas. First jawed fish appear. Modern algae present. Plants invade land; first vascular plants appear. Arthropods invade land.
Gondwana moves south.	Marine algae and invertebrates dominant. First jawless fishes. Radiation of fishes and marine invertebrates.
Land masses at tropical latitudes; Gondwana wraps around Earth.	Marine algae abundant. Representatives of most marine invertebrate phyla "suddenly" appear and diversify.
Oxygen present in atmosphere.	Bacteria, including cyanobacteria, abundant. First protists appear (1,400 mya).
Origin of Earth (about 4,600 mya).	First bacteria and cyanobacteria (3,500 mya).

major events in the history of life. It also includes positions of land masses at selected times and other geologic events that influenced the history of life.

Assignment 1

Materials

Representative fossils

1. Study Table 37.1 Read it from bottom (oldest) to top. Note when major organismic groups appeared, the sequence of their appearances, and trends evident in the history of life.
2. ***Complete items 1a–1g on Laboratory Report 37, which begins on page 475.***
3. Examine the fossils, which are labeled as to the type of organism fossilized and the period in which the fossil was formed. Record the organismic group of each fossil in the mya column of Table 37.1.
4. ***Complete item 1 on the laboratory report.***

EVIDENCE FROM VERTEBRATE EMBRYOLOGY

The early embryos of all vertebrates are remarkably similar (Figure 37.1). They possess pharyngeal pouches (which form gills and gill slits in fish), aortic arches (arteries that serve gills in fish), and a tubular, two-chambered heart. These structures persist and are functional in adult fish, but they are modified to form other structures in the development of other vertebrates. For example, in humans the first pair of pharyngeal pouches forms the cavity of the middle ear and the auditory tubes, the second pair forms tonsils, the third and fourth pairs form thymus and parathyroid glands. Apparently, all vertebrates have similar genes directing early embryonic development. How do you explain this?

Assignment 2
Materials

Models of vertebrate embryos at comparable stages
Compound microscope
Stereomicroscope
Prepared slides of:
  chick embryo, w.m., 96 hr
  pig embryo, w.m., 10 mm

1. Compare the models of vertebrate embryos with each other.
2. Examine the prepared slides of chick and pig embryos with a stereomicroscope and with the 4× objective of a compound microscope. Compare them with Figure 37.1. Locate the pharyngeal pouches and tail buds.
3. ***Complete item 2 on the laboratory report.***

EVIDENCE FROM VERTEBRATE ANATOMY

Modern vertebrates are adapted for a variety of lifestyles. For example, the forelimbs of reptiles, mammals, and birds are specialized for particular functions, such as running, swimming, or flying. Figure 37.2 illustrates the forelimb bones of selected vertebrates. Although the forelimbs are specialized for different functions, the bones of each species' forelimbs exhibit a similar basic body plan. How do you explain the common set of bones in the same relationship to each other even though their size and relative proportion varies in each specialized forelimb?

Biologists view the forelimb bones as examples of **homologous structures,** structures whose similarity is due to descent from a common ancestor. Homologous structures result from **divergent evolution** as related organisms expand into new environments.

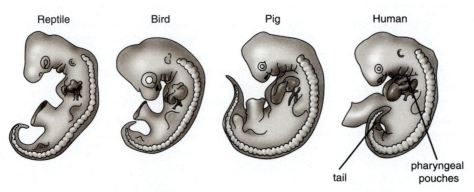

Figure 37.1 Comparison of early vertebrate embryos.

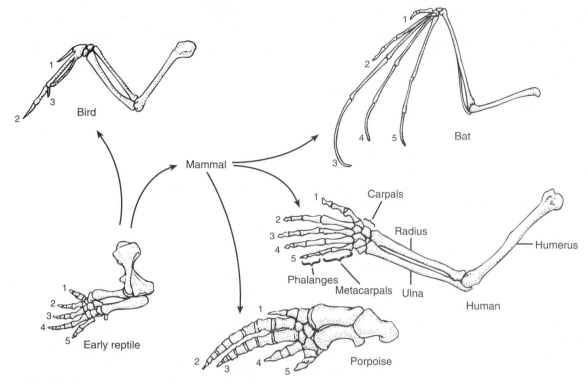

Figure 37.2 Forelimbs of terrestrial vertebrates are homologous structures.

In contrast, **analogous structures** have similar functions in unrelated organisms, but they are not homologous. Examples of analogous structures are wings of birds and insects, lungs of mammals and tracheae of insects, and streamlined bodies of tuna, penguins, and porpoises. Analogous structures result from **convergent evolution** when unrelated organisms are subjected to similar environmental pressures.

 ## Assignment 3

Materials

Colored pencils
Articulated skeletons of representative vertebrates
Examples of analogous structures
Model showing homology in forelimbs

1. Study Figure 37.2. The basic skeletal structure of a vertebrate forelimb, from proximal to distal, consists of humerus, ulna and radius, several small carpals, five metacarpals, and five phalanges (digits). Note this structure in the early reptile forelimb and how it has been modified in the other forelimbs shown. In each forelimb, color-code the humerus red, ulna blue, radius green, metacarpals orange, and phalanges brown.
2. Examine the forelimbs of the skeletons and models provided and locate the homologous bones. Observe

the structural changes to the basic vertebrate plan exhibited by each specimen. Note how the bat and bird have evolved different wing structures.
3. Examine the hindlimbs of the vertebrate skeletons. Is there a basic vertebrate plan for the hindlimbs? Are there other similarities among the skeletons suggesting modification of a basic plan with descent? *Complete items 3a and 3b on the laboratory report.*
4. Examine the examples of analogous structures.
5. *Complete item 3 on the laboratory report.*

EVIDENCE FROM BIOCHEMISTRY

Basic chemical similarities, such as deoxyribonucleic acid (DNA), ribonucleic acid (RNA), the genetic code, and adenosine triphosphate (ATP), are found in all living things and are evidence of the unity of organisms. Do these chemical similarities support the theory of descent from a common ancestor?

Laboratory mice are commonly used in medical research. For example, drugs that alter metabolic processes are first tested in mice before being used in human trials. If the drug is successful in mice, it is likely to be successful in humans, because the biochemistry of mice and humans is quite similar. Does this similarity support the theory of common descent?

The more closely two species are related, the greater are their structural similarities, which, in turn, are determined by their biochemical similarities. Because each protein is a product of a specific gene, a comparative analysis of proteins may be used to determine the degree of relationship between two species. The more identical proteins two species possess, the more genes (DNA) they have in common, and the more closely they are related.

In this section, you will perform an experiment that simulates the antigen–antibody reactions that are used to indicate the degree of similarity among blood serum proteins of vertebrates. These reactions are based on the fact that the immune system produces specific **antibodies** (proteins) against specific **antigens,** foreign proteins that enter the body. An antibody produced against a specific antigen will combine with that antigen and will react less well with a closely related antigen.

In the research laboratory, the procedure is performed as follows:

1. Human serum proteins (antigens) are injected into a rabbit. In a few weeks, the rabbit's immune system produces large numbers of antibodies against the injected human serum proteins; that is, it forms antihuman antibodies. The rabbit's immune system is now sensitized to human serum proteins.

2. A sample of rabbit blood is drawn, and the serum is separated out. The serum contains the rabbit-produced antihuman antibodies. If a sample of human serum (which contains serum proteins) is added to it, an antigen–antibody reaction occurs. In this reaction, the rabbit-produced antihuman antibodies attach to the human serum proteins, forming an antigen–antibody complex that precipitates out and that can be measured by visual inspection or special instruments.

3. If a sample of nonhuman vertebrate serum is added to the serum from the sensitized rabbit, the amount of precipitation of the antigen–antibody complex indicates the degree of similarity between human and nonhuman serum proteins, which in turn indicates how closely that vertebrate is related to humans. The results of a series of such tests are shown in Table 37.2.

 Assignment 4

Materials

Ward's immunology and evolution experiment kit
Toothpicks

1. Obtain a small plastic tray with several wells (depressions) and a dropping bottle each of synthetic: rabbit serum, human sensitizing serum, human serum proteins, and "unknown" test sera numbered, I, II, III, IV, and V.

TABLE 37.2	POSITIVE REACTIONS TO ANTIHUMAN ANTIBODIES
Serum Source	% Positive Reactions
Human	100
Chimpanzee	98
Gorilla	96
Orangutan	92
Old World monkeys	80
New World monkeys	77
Marmosets	50
Lemurs	20

2. Place 2 drops of synthetic rabbit serum in each well numbered 1 through 6.

3. Add 2 drops of synthetic human serum to each well and mix with a toothpick to "sensitize" the rabbit serum to human serum proteins. *Discard the toothpick.*

4. Add 4 drops of human serum to well 6. Mix with a toothpick and note the reactions. *Discard the toothpick.* The amount of precipitation (white particle formation) of human serum proteins in well 6 serves as a standard against which the precipitations of the "unknown" test sera are to be compared.

5. Add 4 drops of test serum I to well 1 and mix with a toothpick. *Discard the toothpick* and note the reaction.

6. Add 4 drops of test serum II to well 2 and mix with a toothpick. *Discard the toothpick* and note the reaction.

7. In a similar manner, add drops of test sera III, IV, and V to wells 3, 4, and 5, respectively. Mix each with a *separate* toothpick and note the reactions. *Discard each toothpick.* Compare the amount of precipitation in wells 1–5 with the amount of precipitation in well 6. The test sera represent human, chimpanzee, orangutan, monkey, and cow. You are to determine, on the basis of the degree of precipitation, the animal from which each serum sample was obtained.

8. ***Complete item 4 on the laboratory report.***

HUMAN EVOLUTION

The pattern of human evolution is not clear, and the fossil record is sparse. Considerable disagreement exists among paleoanthropologists regarding the species to which specific fossils belong and the evolutionary relationships among the fossils. However, they do agree on the following points:

1. Modern human's closest living relative is the chimpanzee.

2. The great apes and hominids (humans and human-like forms) evolved from a common ancestor.

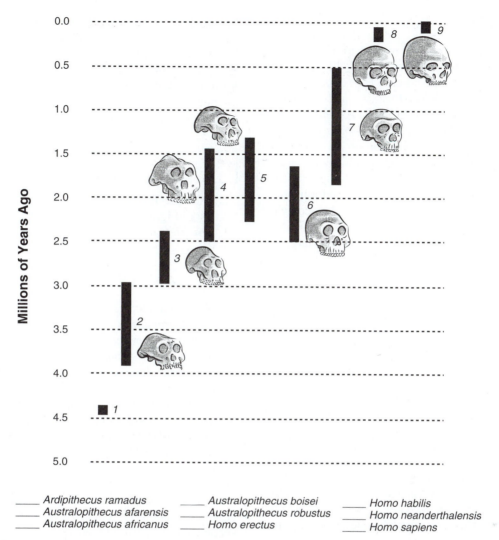

Millions of Years Ago

_____ Ardipithecus ramadus	_____ Australopithecus boisei	_____ Homo habilis
_____ Australopithecus afarensis	_____ Australopithecus robustus	_____ Homo neanderthalensis
_____ Australopithecus africanus	_____ Homo erectus	_____ Homo sapiens

Figure 37.3 Time line for the major hominid fossils.

Figure 37.3 shows the time line of the major hominids in the fossil record. Note their location in the time line as you read the following discussion of these hominids.

Hominids

The hominid lineage diverged from the ape lineage in Africa between 8 and 5 mya. All hominid fossils older than 1.5 mya have been found in eastern and southern Africa. The earliest hominid, *Ardipithecus ramidus,* is known from only a few teeth and bone fragments dated at 4.4 mya.

Australopithecus

The oldest australopithecine is *Australopithecus ana-mensis,* which was discovered in Kenya and lived about 4 mya. The second-oldest australopithecine is *A. afarensis,* which was discovered in Ethiopia and lived about 3.9 to 2.9 mya. This species was slender and small—only about 1 meter tall. It was bipedal,

although its short legs, long arms, and skull were rather apelike. Its brain was about the size of the brain of a modern chimpanzee. It is believed that *A. afarensis* is ancestral to *A. africanus,* another slender (gracile) species, which lived about 3 to 2.3 mya. Two later-appearing species, *A. boisei* (2.5–1.4 mya) and *A. robustus* (2.3–1.3 mya) were distinguished by much heavier bone structure, including massive jaws and large teeth. These "robust" species are a dead-end offshoot and not in the human lineage. Most paleoanthropologists think that the ancestor of the genus *Homo* is probably *A. afarensis.*

Homo habilis

The first human, *H. habilis,* lived 2.5 to 1.6 mya in Africa and coexisted with *A. boisei* and *A. robustus.* Most paleoanthropologists think *H. habilis* is ancestral to *H. erectus.* The teeth and cranium of *H. habilis* are more humanlike than those of australopithecines. *H. habilis* is believed to have used crude stone tools.

Homo erectus

Homo erectus lived from 1.8 million to 500,000 years ago. Some paleoanthropologists identify the early forms of *H. erectus* as a separate species, *H. ergaster,* and consider it an ancestor to both *H. erectus* and *H. sapiens. H. erectus* migrated out of Africa to Asia and Europe about 1.5 mya. Numerous fossils of this species have been found in Africa, Asia, and Europe. *H. erectus* was taller (5 ft), larger brained, and more efficiently bipedal than was *Australopithecus.* Evidence indicates construction and use of stone tools, use of fire, and a hunting-and-gathering lifestyle. The skull is characterized by (1) heavy brow ridges; (2) sloping forehead; (3) long, low-crowned cranium; (4) no chin; and (5) protruding face.

Homo neanderthalensis

Neanderthals appeared about 200,000 years ago and survived until 35,000 years ago. They are named for the valley in Germany where many early specimens were found. Most paleoanthropologists think that Neanderthals are an evolutionary dead end arising from *H. erectus* and that they were replaced by *H. sapiens.* Others believe that they are a subspecies of *H. sapiens* and that they became "extinct" through interbreeding with *H. sapiens.*

Neanderthals have many characteristics intermediate between *H. erectus* and *H. sapiens.* They were stocky with heavy muscles and about 5 ft tall. Skulls exhibit (1) heavy brow ridges; (2) large, protruding faces; (3) no chin; and (4) a sloping, low-crowned but large cranium. They constructed stone, bone, and stick tools, used fire, wore clothing, were skilled hunters, and were able to cope with the frigid climate of the glacial periods. They also buried flowers with their dead. They coexisted briefly with modern humans before their extinction. The cause of their extinction is not understood.

Homo sapiens

There are two major hypotheses regarding the origin of *Homo sapiens.* The **out-of-Africa hypothesis** proposes that early *H. sapiens* evolved from *H. erectus* in Africa about 150,000 years ago and migrated to other parts of the world, replacing earlier *Homo* species. The **multiregional hypothesis** holds that localized populations of early *H. sapiens* evolved independently from *H. erectus* in several regions of the world. Supporters of this view may refer to early *H. sapiens* as *H. heidelbergensis,* or "archaic" *H. sapiens.* The debate about the origin of *H. sapiens* is ongoing. In any event, the radiation and interbreeding of *H. sapiens* has produced the modern human populations of today.

Much of our knowledge about the appearance of modern *H. sapiens* comes from fossils found in the Middle East and Europe. This population, often called Cro-Magnons, appeared about 40,000 years ago. Their skeletal features are essentially identical to modern-day humans. They were (1) excellent tool and weapon makers, using stone, sticks, bone, and antlers as raw materials; (2) skilled hunters; and (3) fine artists, as evidenced by cave paintings and sculptures.

Assignment 5

1. Examine Table 37.3.
2. Label Figure 37.4.
3. ***Complete item 5 on the laboratory report.***

SKULL ANALYSIS

Figure 37.4 indicates some of the major differences between chimpanzee and human skulls. Although this comparison is between two modern species, it does suggest the types of characteristics that should be present in intermediate forms when a common ancestry is assumed.

An understanding of hominid evolution is based on the study of fossil bones, teeth, and artifacts, such as tools, associated with the specimens. Analysis of the fossils involves detailed measurements and careful comparisons. In this section, you will analyze certain skeletal materials and fossil replicas by making measurements and calculating indexes, as well as conduct qualitative comparisons of the available specimens.

Indexes are useful for comparative purposes because they overcome the problems caused by differences in the size of the specimens. An index is a ratio calculated by dividing one measurement by another, and it indicates the proportional relationship of the two measurements. Calipers and metersticks are to be used in making the measurements, and all measurements are to be in millimeters. Figures 37.5 and 37.6 show how to make the measurements. The indexes are calculated by dividing one measurement by another (carried to two decimal places) and multiplying by 100.

TABLE 37.3	AVERAGE CRANIAL CAPACITY (CC) OF HOMINIDS
Australopithecus afarensis	414
Australopithecus africanus	441
Australopithecus robustus	530
Homo habilis	640
Homo erectus	990
Homo neanderthalensis	1,465
Homo sapiens sapiens	1,350

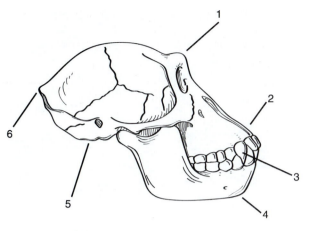 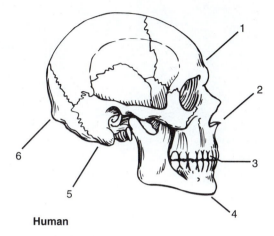

Chimpanzee

1. Heavy brow ridges; low-crowned, sloping cranium
2. Muzzlelike faces
3. Large canines
4. No chin
5. Posterior skull attachment
6. Occipital crest for attachment of heavy neck muscles

Human

1. Almost no brow ridges; high forehead and high-crowned cranium
2. Flattened face
3. Small canines
4. Well-developed chin
5. Central skull attachment
6. No occipital crest, neck muscles not as heavy and attached lower

Figure 37.4 Comparison of chimpanzee and human skulls.

Although your measurements will be less precise than those in a professional analysis, you will gain an understanding of the process and recognize the difficulty in making conclusions from analysis of a single specimen.

Handle the bones and fossil replicas carefully because they can be easily damaged. Avoid making marks or scratches on the specimens. The measurements are best made by working in groups of two to four students.

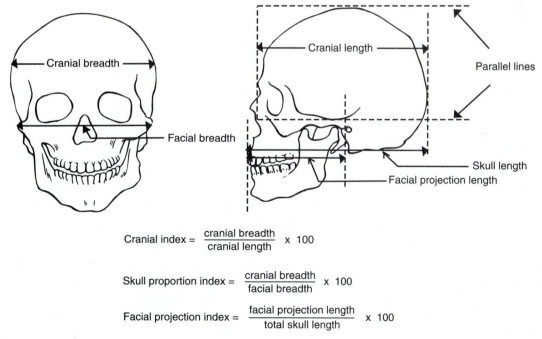

$$\text{Cranial index} = \frac{\text{cranial breadth}}{\text{cranial length}} \times 100$$

$$\text{Skull proportion index} = \frac{\text{cranial breadth}}{\text{facial breadth}} \times 100$$

$$\text{Facial projection index} = \frac{\text{facial projection length}}{\text{total skull length}} \times 100$$

Figure 37.5 Measurements to be made for skull analysis.

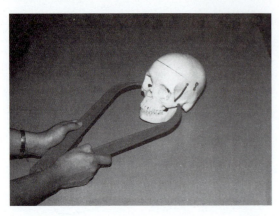
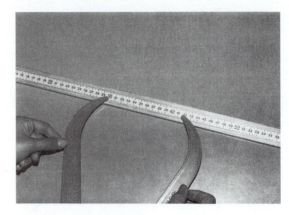

Figure 37.6 Procedures for making measurements for skeletal analysis. **A.** Use the tips of the calipers to establish the distance between two points on the skull. **B.** Then place the tips of the calipers on a meterstick to determine the distance (mm) between the two points.

Cranial (Cephalic) Index

When this index is determined for living forms, it is called a **cephalic index**. The term **cranial index** is used in reference to nonliving specimens.

$$\text{Cranial index} = \frac{\text{cranial breadth}}{\text{cranial length}} \times 100$$

Cranial breadth: maximum width of the cranium
Cranial length: maximum distance between the posterior surface and the small prominence (glabella) between the brow ridges

Skull Proportion Index

The skull is composed of the face and the cranium, and the skull proportion index identifies the proportional relationship between these two components. The greater the value of the index, the larger the cranium is in relation to the face.

$$\text{Skull proportion index} = \frac{\text{cranial breadth}}{\text{facial breadth}} \times 100$$

Cranial breadth: maximum breadth of the cranium
Facial breadth: maximum distance between the lateral surfaces of the cheekbones (zygomatic arches)

Facial Projection Index

A projecting, muzzlelike face is a primitive condition among primates. This index identifies the degree of facial projection in a specimen. The greater the value of the index, the greater is the degree of facial projection.

$$\text{Facial proportion index} = \frac{\text{facial projection length}}{\text{total skull length}} \times 100$$

Facial projection length: distance between the anterior margins of the auditory canal and upper jaw (maxilla)
Total skull length: maximum distance between the posterior surface of the cranium and the anterior margin of the upper jaw (maxilla)

 ## Assignment 6

Materials

Calipers
Metersticks
Skull replicas of:
   Australopithecus africanus
   Homo erectus
   Homo neanderthalensis
   Homo sapiens sapiens (Cro-Magnon)
Skull of recent human

1. Determine the cephalic index for each member of your laboratory group. Exchange data with all groups in the class, and list the indexes for each person in Table 37.4. Determine the range and average index value for the class.
2. ***Complete items 6a–6d on the laboratory report.***
3. Determine the (a) cranial index, (b) skull proportion index, and (c) facial projection index for each of the skull replicas and skulls available. ***Record your data in item 6e on the laboratory report.***

TABLE 37.4 CEPHALIC INDEXES OF CLASS MEMBERS

1. _____	6. _____	11. _____	16. _____	21. _____	26. _____
2. _____	7. _____	12. _____	17. _____	22. _____	27. _____
3. _____	8. _____	13. _____	18. _____	23. _____	28. _____
4. _____	9. _____	14. _____	19. _____	24. _____	29. _____
5. _____	10. _____	15. _____	20. _____	25. _____	30. _____

Range _____   Average _____

4. Compare the available skulls as to the following characteristics and rate each on a 1–5 scale. *Record your responses in item 6e on the laboratory report.*

Skull and vertebra attachment
1 = most posterior
5 = most central

Brow ridges
1 = most pronounced
5 = least pronounced

Length of canines
1 = longest
5 = no longer than incisors

Forehead
1 = most sloping
5 = best developed

Chin
1 = no chin
5 = best developed

5. *Complete the laboratory report.*

EVOLUTION

Student _____

Lab Instructor _____

1. THE FOSSIL RECORD

a. In Table 37.1, indicate the era/period when these organismic groups first appeared.

Flowering plants _____   Humans _____

Conifers _____   Mammals _____

Gymnosperms _____   Birds _____

Ferns _____   Reptiles _____

Horsetails _____   Amphibians _____

Vascular plants _____   Jawed fish _____

Mosses and liverworts _____ Jawless fish _____

Marine algae _____   Marine invertebrates _____

Cyanobacteria _____   Protists _____

Prokaryotic cells _____   Eukaryotic cells _____

b. Considering Table 37.1, what general trend is evident in organismic:

diversity? _____

_____

complexity? _____

_____

Are there exceptions to this trend? _____ Explain. _____

_____

_____

c. How did cyanobacteria prepare the way for the origin of aerobic respiration? _____

_____

d. Indicate the period when these groups invaded the land.

Vertebrates _____

Vascular plants _____

Arthropods _____

e. Considering the drifting of continents, how do you explain the presence of fossils of tropical organisms in

Canada? _____

_____

f. Considering the drifting of continents, how do you explain the abundance of marsupials and the absence of

native placental mammals in Australia? _____

_____

g. For a group of organisms to radiate, there must be niches (ways of making a living) available for the taking. What made niches available for the radiation of:

mammals _____

amphibians _____

h. Extinction is a common occurrence in the fossil record. Why do you think species become extinct? _____

_____

_____

i. Examine the fossils available and complete the following table:

Number	Organismic Group	Period Formed	Description
1			
2			
3			
4			
5			
6			
7			
8			
9			
10			

2. EVIDENCE FROM VERTEBRATE EMBRYOLOGY

a. Do early chick and pig embryos have a similar structure and appearance? _____

b. Would you expect embryos of closely related species or distantly related species to have the greater similarity?

Explain. _____

_____

c. How do you explain the pharyngeal pouches in chick and pig embryos? _____

_____

_____

d. Does the presence of pharyngeal pouches in all vertebrate early embryos support the theory of descent from a

common ancestor? _____ Explain. _____

_____

3. EVIDENCE FROM VERTEBRATE ANATOMY

a. Compare specializations of the forelimbs by completing the table below:

Vertebrate	Describe the Forelimb Specializations

b. Do other features of the skeletons examined indicate modification from a common vertebrate plan? _____

Explain. _____

_____

_____

_____

c. Whales (mammals) possess a rudimentary (vestigial) pelvic girdle and limb bones. How do you explain this?

_____

Great apes and humans possess a coccyx, a rudimentary tail bone. How do you explain this? _____

_____

d. Birds and bats illustrate two different wing designs. Are these the only designs possible? _____

Does evolution of a body part produce the single best structural design or a functional modification of the basic

structure available? _____

Give two examples to illustrate your response. _____

_____

_____

e. Based on the fossil record and embryological and anatomical data, the presumed generalized phylogeny for vertebrates is as shown here. Add labels to identify the geologic periods and organismic groups.

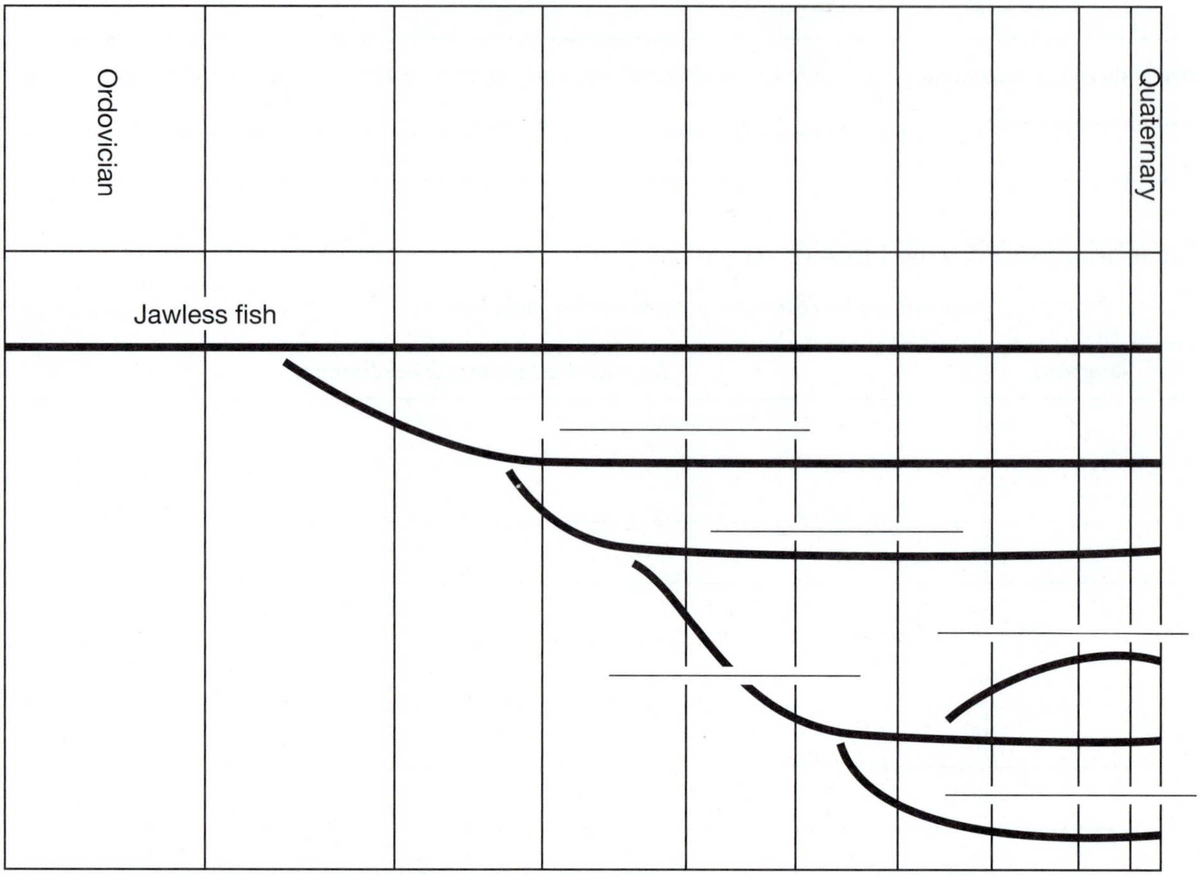

4. EVIDENCE FROM BIOCHEMISTRY

a. All organisms possess ATP, RNA, and DNA. What can you infer from this? _____

_____

b. Would you expect plasma proteins of closely related species or distantly related species to have the greater

similarity? _____

Explain. _____

_____

c. Record your results and conclusions in the following table. Rate the degree of precipitation on a 1-to-5 scale, 5 being the greatest amount of precipitation.

Test Serum	Degree of Precipitation	Animal Source of the Serum
I		
II		
III		
IV		
V		
VI	5	Human

d. Do your results agree with the data in Table 37.1? _____

   If not, explain the differences. _____

   _____

5. HUMAN EVOLUTION

List the labels for Figure 37.3.

1. _____   6. _____

2. _____   7. _____

3. _____   8. _____

4. _____   9. _____

5. _____

6. SKULL ANALYSIS

a. Plot the cephalic indexes of members of the class.

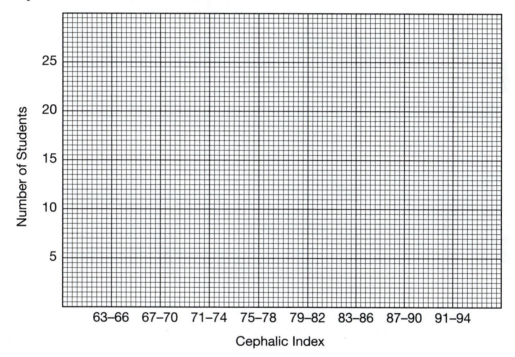

b. Indicate the following data from your class:

   Average cephalic index _____ Ranges of indexes _____

c. Considering the data in item 6a and 6b, what difficulties are encountered in trying to make firm conclusions about a species from a single fossil specimen?

   _____

   _____

d. Why are indexes better than simple measurements for comparing fossil specimens?

   _____

   _____

e. Record the results of your skull analyses in the following table.

SKULL COMPARISON

Characteristics	Chimpanzee	A. africanus	H. erectus	H. neanderthalensis	H. sapiens sapiens (Cro-Magnon)	H. sapiens sapiens (recent)
Cranial breadth						
Cranial length						
Cranial index						
Facial breadth						
Skull proportion index						
Facial projection length						
Skull length						
Facial projection index						
Brow ridges						
Forehead						
Canine length						
Skull and vertebra attachment						
Chin						

f. Which of the hominid skulls has a cranial (cephalic) index that falls within the range of indexes for the class? _____

_____

g. Have there been any major skull changes in humans during the last 40,000 years? _____
Explain your response. _____

_____

_____

EVOLUTIONARY MECHANISMS

OBJECTIVES

After completing the laboratory session, you should be able to:
1. Identify the factors that produce evolutionary change.
2. Use the Castle-Hardy-Weinberg law for determining genetic stability or change in hypothetical populations.
3. Describe the effect of mutation, gene flow, natural selection, and genetic drift on the gene pool of populations.
4. Define all terms in bold print.

Simply defined, **evolution** is a change in the genetic composition of a population. The genetic composition of a population is usually called the **gene pool,** and it refers to all the alleles in the interbreeding population. Evolution results from a change in the frequency of the alleles in the gene pool produced by one or more of the following factors:

Mutation. The naturally occurring mutations of genes and chromosomes produce variation in the gene pool. These heritable changes serve as the raw material of evolution. Evolution can occur only in populations where preexisting genetic variation exists. Genetic recombination tends to distribute the mutations throughout the gene pool.

Gene Flow. The frequency of alleles in the gene pool may change due to the immigration or emigration of organisms.

Natural Selection. The impact of the environment on the survival and reproduction of genotypes within the population is a major force in changing the frequency of alleles. Selection is always on the phenotypes, but the genetic effect is on the genotypes. The interaction of genetic variability and natural selection is the primary mechanism causing evolution.

Genetic Drift. The frequency of alleles in the gene pool may change by pure chance. It is a significant evolutionary mechanism in small populations only.

Nonrandom Mating. If there is a mating preference for a certain combination of alleles, the frequency of alleles in a population will change.

CASTLE-HARDY-WEINBERG LAW

Castle, Hardy, and Weinberg independently determined a simple mathematical expression that establishes a point of reference in evaluating genetic changes in a population. This is known as the **Castle-Hardy-Weinberg (CHW) law** or **equilibrium,** and it is operative whenever these conditions are met:

1. Mutations do not occur.
2. The population size is large.
3. The population is isolated from other populations of the same species (i.e., no immigration or emigration occurs).
4. No selection takes place (i.e., the genotypes are equally viable and fertile).
5. Mating is random.

Contrast this list of conditions with the factors involved in the process of evolution noted previously.

The CHW equilibrium states that *in the absence of forces that change gene frequencies, the frequencies of the alleles in a population will remain constant from generation to generation.* The frequency (percentage) of alleles in a population may be mathematically expressed as

$$p + q = 1$$

where

p = the frequency of one allele of a gene

q = the frequency of the alternate allele of the gene

For example, consider the gene controlling the ability to taste PTC (phenylthiocarbamide) in a human population meeting the requirements noted previously. The ability to taste PTC is dominant, and the inability to taste PTC is recessive. Therefore, the alleles may be expressed as T (taster) and t (nontaster). Considering $p + q = 1$, if the frequency (percentage) of the q (t) allele is 50% (0.5), the frequency of the p (T) allele is expressed as

$$p = 1 - q$$

By substitution,

$$p = 1 - 0.5$$

$$p = 0.5$$

If the allele frequencies are known, a Punnett square may be used to predict the genotype and phenotype frequencies in the next generation, when the CHW equilibrium is in effect. Figure 38.1 shows how this is done and also the general equation expressing the genotype frequencies. The **CHW equation** is expressed as

$$p^2 + 2pq + q^2 = 1$$

where

$p^2 =$ the frequency of TT individuals (homozygous dominants)

$2pq =$ the frequency of Tt individuals (heterozygotes)

$q^2 =$ the frequency of tt individuals (homozygous recessives)

For a gene with only two alleles, the CHW equation may be used to calculate the genotype frequencies in a population if the frequency of one allele is known, or to calculate the frequencies of the alleles if the genotypes are known. Once the allele frequencies have been established, you can predict the genotype and phenotype frequencies in the next generation, if the conditions of the CHW equilibrium are met, by substituting the allele frequencies into the CHW equation instead of using a Punnett square. See Figure 38.2.

Consider a hypothetical population of 75% tasters and 25% nontasters. If you want to determine the percentage of homozygous tasters (TT) and heterozygous tasters (Tt) in the population, you can use the CHW equation. Because nontasters are homozygous recessives, their genotype is known to be tt (q^2). It is used to determine the frequencies of both alleles, as shown in Figure 38.3. Once the allele frequencies are known, you can determine the genotype and phenotype frequencies in the population by substituting the allele frequencies into the CHW equation.

Allele frequencies (known) $p =$ frequency of T = 0.5
$q =$ frequency of t = 0.5

Set up and complete a Punnett square.

	Sperm	
	0.5 T (p)	0.5 t (q)
0.5 T (p)	0.25 TT (p^2)	0.25 Tt (pq)
0.5 t (q)	0.25 Tt (pq)	0.25 tt (q^2)

Eggs

The sum of the predicted genotype frequencies equals 1, or 100%.

0.25 TT + 0.50 Tt + 0.25 tt = 1
25% TT + 50% Tt + 25% tt = 100%

When expressed in terms of p and q, we get the equation:

$$p^2 + 2pq + q^2 = 1$$

Figure 38.1 Using a Punnett square to predict the genotype and phenotype frequencies in the next generation of a population when the frequencies of the alleles are known. The genotype frequency expressed in terms of p and q gives us the CHW equation.

Allele frequencies (known) $p =$ frequency of T = 0.5
$q =$ frequency of t = 0.5

CHW equation	p^2	+	$2pq$	+	q^2	= 1
By substitution	$(0.5)^2$	+	$2(0.5)(0.5)$	+	$(0.5)^2$	= 1
	0.25	+	0.50	+	0.25	= 1

Genotype frequency in next generation   25% TT : 50% Tt : 25% tt

Phenotype frequency in next generation   75% tasters : 25% nontasters

Figure 38.2 Using the CHW equation to predict the genotype and phenotype frequencies in the next generation.

Assignment 1

1. Assuming that the conditions of the CHW equilibrium are met, determine the genotype and phenotype frequencies in the two successive generations of a population containing 64% tasters and 36% nontasters. How long will these frequencies persist?

2. Try a piece of PTC paper to see if you are a taster or a nontaster.

Phenotypes (known)	75% tasters	25% nontasters
Genotypes	Tt and Tt	tt
Percentage of nontasters in population	tt = 25% = 0.25	
Frequency of t	q = frequency of t = $\sqrt{q^2}$ by substitution, $q = \sqrt{0.25} = 0.5$	
Freqency of T	p = frequency of T = $1 - q$ by substitution, $p = 1 - 0.5 = 0.5$	
CHW equation	$p^2 + 2pq + q^2 = 1$ substitute allele frequencies, $(0.5)^2 + 2(0.5)(0.5) + (0.5)^2 = 1$ $0.25 + 0.50 + 0.25 = 1$	

Genotype 25% TT : 50% Tt : 25%

Figure 38.3 Using the CHW equation to determine the allele frequencies in a population and then determining the genotype frequencies in that population.

3. Tabulate the number of tasters and nontasters in your class, and calculate the percentage of each. Assuming that the conditions of the CHW law are met, calculate the frequencies of the T and t alleles in your class.
4. *Complete item 1 on Laboratory Report 38 that begins on page 487.*

NATURAL SELECTION

The theory of natural selection was independently developed by Darwin and Wallace. The basic tenets of natural selection are as follows:

1. All organisms are capable of producing more off-spring than the environment can support. Therefore, many die before they reach reproductive age.
2. Within each population, considerable variability exists among phenotypes, and most of this variation is genetically based. Because of this variability, some phenotypes are better adapted to the environment than others, resulting in differences in the viability and survival of the phenotypes.
3. Individuals whose traits are more adaptive have a greater chance of reproducing, and their offspring tend to form the largest portion of the population.

Natural selection is the selection *for* those phenotypes that are better adapted, with the result that such phenotypes contribute more surviving offspring to the population than other phenotypes. This means that some phenotypes are selected *against* as well. Note that the environment does the selecting and that natural selection really hinges on **differences in reproductive success.**

Selection Against Dominant Alleles

Suppose that the allele that allows the ability to taste PTC suddenly mutated so it also caused sterility. Selection would be *against* tasters and *for* nontasters. How would this affect succeeding generations? Using our previous population of 64% tasters and 36% nontasters, the allele frequencies are $p = 0.4$ and $q = 0.6$. Substituting into the CHW equation, we get

$$(0.4)^2 + 2(0.4)(0.6) + (0.6)^2 = 1$$

$$\cancel{0.16} + \cancel{0.48} + 0.36 = 1$$

Because all tasters are now sterile, the 36% nontasters of the original population are now the total, 100%, of the reproductive population. Will there be any tasters in the next generation?

Selection is rarely 100%. Usually it provides a slight advantage or disadvantage. However, selection is rather rapid on dominant alleles. For example, assume that the 64% tasters in the population have a lowered reproductive success so they produce only 80% of their expected progeny. What will be the effect on succeeding generations? The contribution (percentage) of tasters and nontasters to the next generation would be

$$\text{Tasters} = 0.64 \times 0.80 = 0.51$$

$$\text{Nontasters} = 1 - 0.51 = 0.49$$

The genotype frequencies may be predicted by calculating the frequency of the alleles and substituting into the CHW equation. Note that, as the frequency of one allele decreases, the frequency of the alternate allele increases.

$$\text{Frequency of t} = \sqrt{0.49} = 0.7$$

$$\text{Frequency of T} = 1 - 0.7 = 0.3$$

$$(0.3)^2 + 2(0.3)(0.7) + (0.7)^2 = 1$$

$$0.09 + 0.42 + 0.49 = 1$$

Selection Against Recessive Alleles

Now let's reverse the previous hypothetical selective pressure. In the population of 64% tasters and 36% nontasters, assume that the nontaster allele has mutated so it produces sterility in the homozygous state. The effect of this 100% selection against nontasters may be determined for the first generation as follows:

$$(0.4)^2 + 2(0.4)(0.6) + (0.6)^2 = 1$$

$$0.16 + 0.48 + 0.36 = 1$$

Because nontasters will not contribute to the next generation, new allele frequencies must be calculated to determine the genotype and phenotype percentages in the next generation. The 64% tasters constitute the total, 100%, of the reproductive population. The only source for nontaster alleles is the heterozygous tasters. Half of their alleles for this locus are nontaster alleles, which may be expressed as $0.5(2pq)$. Therefore, the frequency of q in the reduced population equals the nontaster alleles divided by the total alleles ($p^2 + 2pq$) in the population. Thus,

$$q = \frac{0.5(2pq)}{p^2 + 2pq} = \frac{0.5(0.48)}{0.16 + 0.48} = \frac{0.24}{0.64} = 0.375$$

$$p = 1 - q = 0.625$$

Substituting into the CHW equation, we get

$$(0.625)^2 + 2(0.625)(0.375) + (0.375)^2 = 1$$

$$0.39 \quad + \quad 0.47 \quad + \quad 0.14 \quad = 1$$

Phenotype percentages in the next generation would be

86% tasters : 14% nontasters

 ## Assignment 2

1. Determine the effect of 20% selection against tasters for five generations. If continued, would tasters ever become extinct? Use the data in Table 38.1 in your calculations.
2. If a computer and an appropriate program are available, determine the effect of this selective pressure to see if the dominant allele can be eliminated from the population.
3. *Complete items 2a–2e on the laboratory report.*
4. Determine the effect of 100% selection against nontasters for the next five generations. Will nontasters ever become extinct?
5. If a computer and an appropriate program are available, determine if nontasters will persist in the population after 10, 100, and 1,000 generations.
6. *Complete item 2 on the laboratory report.*

GENETIC DRIFT

In a large, random-mating population, allele frequencies tend to remain constant in the next generation if other requirements for the CHW principle are met. However, in a small population, random mating can change allele frequencies in the next generation by pure chance. Such a change is called **genetic drift.** In this section, you will simulate the effect of genetic drift on the taster/nontaster alleles in a small population.

 ## Assignment 3
Materials

Vials of beads, 100 red and 100 white
Beaker, 250 ml

1. Obtain a vial of red beads and a vial of white beads. These beads will represent the alleles. A red bead represents a T allele, and a white bead represents a t allele.
2. Place 50 red and 50 white beads in a beaker and mix them up thoroughly to form the gene pool of the original population. Note that the frequency of T is 0.5 (50%) and the frequency of t is 0.5 (50%). *Calculate the genotype and phenotype frequencies of the individuals in this population, and record these frequencies in the table in item 3a on the laboratory report.*
3. To determine the alleles in the next generation, simulate random mating by closing your eyes and drawing two beads from the gene pool. *Record the alleles of this "individual" in the tabulation square in the table for Generation 1 in item 3a.* Replace the beads in the beaker and shake the beaker to reshuffle the beads. Draw two beads as before and record the genotype of the "individual." Repeat this process until you have drawn 50 pairs of beads representing 50 individuals composing the next generation of the small population. From these results, determine the frequency of the genotypes and phenotypes and the frequencies of the T and t alleles in the first (F_1) generation.
4. Using the calculated frequencies of alleles T (red beads) and t (white beads) that you have determined for the first generation, create a new gene pool of 100 beads that reflects these percentages. For example, if T = 0.6 and t = 0.4, place 60 red beads and 40 white beads in the beaker to form the new gene pool.
5. Using the previously described procedures for (a) drawing 50 pairs of beads from the gene pool to simulate random mating in each generation, (b) calculating the genotype and phenotype ratios and the frequencies of the alleles, and (c) reforming the gene pool in accordance with the frequencies of the alleles in the preceding generation, determine the allele frequencies for five generations. *Record your tabulations and calculated frequencies in a table in item 3a on the laboratory report.*
6. If a computer and an appropriate program are available, determine the allele frequencies after genetic drift for 10, 50, and 100 generations.
7. *Complete the laboratory report.*

TABLE 38.1   TABLE OF SQUARES AND SQUARE ROOTS

n	n^2	$\sqrt{n}$	n	n^2	$\sqrt{n}$	n	n^2	$\sqrt{n}$
1	1	1.000	34	1,156	5.831	68	4,624	8.246
2	4	1.414	35	1,225	5.916	69	4,761	8.307
3	9	1.732	36	1,296	6.000	70	4,900	8.367
4	16	2.000	37	1,369	6.083	71	5,041	8.426
5	25	2.236	38	1,444	6.164	72	5,184	8.485
6	36	2.450	39	1,521	6.245	73	5,329	8.544
7	49	2.646	40	1,600	6.325	74	5,476	8.602
8	64	2.828	41	1,681	6.403	75	5,625	8.660
9	81	3.000	42	1,764	6.481	76	5,776	8.718
10	100	3.162	43	1,849	6.557	77	5,929	8.775
11	121	3.317	44	1,936	6.633	78	6,084	8.832
12	144	3.464	45	2,025	6.708	79	6,241	8.888
13	169	3.605	46	2,116	6.782	80	6,400	8.944
14	196	3.742	47	2,209	6.856	81	6,561	9.000
15	225	3.873	48	2,304	6.928	82	6,724	9.055
16	256	4.000	49	2,401	7.000	83	6,889	9.110
17	289	4.123	50	2,500	7.071	84	7,056	9.165
18	324	4.243	51	2,601	7.141	85	7,225	9.220
19	361	4.359	52	2,704	7.211	86	7,396	9.274
20	400	4.472	53	2,809	7.280	87	7,569	9.327
21	441	4.583	54	2,916	7.348	88	7,704	9.381
22	484	4.690	55	3,025	7.416	89	7,921	9.434
23	529	4.796	56	3,136	7.483	90	8,100	9.487
24	576	4.899	57	3,249	7.550	91	8,281	9.539
25	625	5.000	58	3,364	7.616	92	8,464	9.592
26	676	5.099	59	3,481	7.681	93	8,649	9.644
27	729	5.196	60	3,600	7.746	94	8,836	9.695
28	784	5.292	61	3,721	7.810	95	9,025	9.747
29	841	5.385	62	3,844	7.874	96	9,216	9.798
30	900	5.477	63	3,969	7.937	97	9,409	9.849
31	961	5.568	64	4,096	8.000	98	9,604	9.899
32	1,024	5.657	65	4,225	8.062	99	9,801	9.950
33	1,089	5.745	66	4,356	8.124	100	10,000	10.000
			67	4,489	8.185			

EVOLUTIONARY MECHANISMS

Student _____

Lab Instructor _____

1. CHW EQUILIBRIUM

a. Define:

Population _____

Gene pool _____

Evolution _____

Gene mutation _____

Natural selection _____

_____

Gene flow _____

_____

Genetic drift _____

_____

b. Genetic variability in a population may be increased by three factors: _____,

_____, and _____.

c. Under what conditions does the CHW equilibrium operate? _____

_____

_____

Do these exist in natural populations? _____

d. Indicate the allele and genotype frequencies in these populations:

1. 91% tasters and 9% nontasters

Allele frequencies: T _____ t _____

Genotype frequencies: TT _____ Tt _____ tt _____

2. 36% tasters and 64% nontasters

Allele frequencies: T _____ t _____

Genotype frequencies: TT _____ Tt _____ tt _____

e. If the conditions of the CHW law are met, what will be the allele frequencies in a population of 75% tasters and 25% nontasters after 10 generations? T _____ t _____

f. Indicate the number and frequency (%) of tasters and nontasters in the class.

Tasters: number _____; % _____   Nontasters: number _____; % _____

Indicate the frequency of the following alleles and genotypes for the class.

t _____ T _____   TT _____ Tt _____ tt _____

2. NATURAL SELECTION

a. In the selective breeding of horses, who or what does the selecting? _____

   In natural selection, who or what does the selecting? _____

b. With 100% selection against a dominant phenotype, how long does it take to remove the dominant allele from the population? _____

   How will the dominant allele return to the population, if ever? _____

c. In a population of 64% tasters and 36% nontasters, complete the following chart to show the effect of 20% selection against tasters in each generation. The T allele for tasting PTC is dominant.

	% Tasters	% Nontasters	Frequency of q	Frequency of p
Original Generation	64	36	0.60	0.40
First Generation				
Second Generation				
Third Generation				
Fourth Generation				
Fifth Generation				

d. In a population of 64% tasters and 36% nontasters, complete the following chart to show the effect of 100% selection against nontasters in each generation. The T allele for tasting PTC is dominant.

	0.5(2pq)	$p^2 + 2pq$	Frequency of q	Frequency of p	% Nontasters	% Tasters
Original Generation			0.60	0.40	36	64
First Generation						
Second Generation						
Third Generation						
Fourth Generation						
Fifth Generation						

e. With 100% selection against nontasters, how do you explain the presence of nontasters in the fifth generation?

   _____

   _____

f. Are dominant or recessive alleles easier to remove by selection? _____

   Explain. _____

   _____

g. Is selection against the genotype or phenotype? _____

h. Is selection more effective against recessive alleles in haploid or diploid organisms? _____

   Explain your response. _____

   _____

i. Would a lethal recessive mutant tend to spread through a population of:

haploid organisms? _____ diploid organisms? _____

Explain your responses. _____

_____

_____

3. GENETIC DRIFT

a. Record your tabulations, genotype frequency, phenotype frequency, and the frequency of T and t in the following tables for five generations of a population subjected to genetic drift. The allele frequencies in the original population are T = 0.5 and t = 0.5. A red bead = T, and a white bead = t.

Original Generation

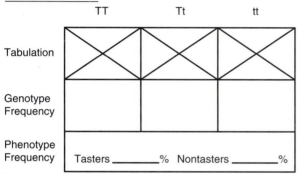

New Allele Frequencies: T = 0.5; t = 0.5

Generation 1

New Allele Frequencies: T = _____; t = _____

Generation 2

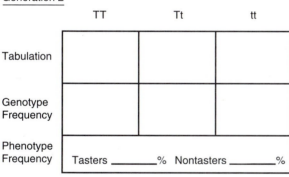

New Allele Frequencies: T = _____; t = _____

Generation 3

New Allele Frequencies: T = _____; t = _____

Generation 4

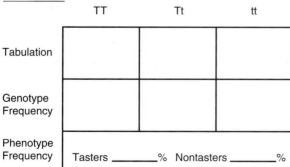

New Allele Frequencies: T = _____; t = _____

Generation 5

New Allele Frequencies: T = _____; t = _____

b. If a computer program is available, determine the allele frequencies after 10, 50, and 100 generations.

Generation	Frequency of T	Frequency of t
10	_____	_____
50	_____	_____
100	_____	_____

c. Is there a pattern to the changes in allele frequencies? _____

What is the cause of the changes in allele frequency? _____

d. Why is genetic drift unimportant in large populations? _____

_____

_____

_____

Part VII

Ecology
and Behavior

ECOLOGICAL RELATIONSHIPS

OBJECTIVES

After completing the laboratory session, you should be able to:

1. Describe the interrelationships between producers, consumers, decomposers, and abiotic components of an ecosystem.
2. Diagram a biomass pyramid, and explain the decrease in biomass at each trophic level.
3. Describe the flow of energy through a community.
4. Diagram the carbon and nitrogen cycles, and explain the role of organisms in each.
5. Describe the effect of acid and organic pollution on producers and consumers in aquatic ecosystems.
6. Demonstrate how to determine the LC_{50} for a specific pollutant affecting *Daphnia*.
7. Define all terms in bold print.

Life on Earth is maintained in a delicate balance by ecological interrelationships. All organisms, including humans, are subject to the ecological principles that govern the relationships of organisms with their environment. Humans, more than other organisms, have the ability to modify the environment to their own choosing. Such power must be used with care because its misuse can yield disastrous results.

Ecology is the study of the interrelationships between organisms and their environment, including both biotic (living) and abiotic (nonliving) components. It is primarily concerned with the study of populations, communities, ecosystems, and the biosphere.

A **population** is a group of interbreeding members of the same species living in a defined area.

A **community** is composed of all populations within a defined area.

An **ecosystem** consists of both the community and the abiotic components of a defined area. Soil, air, water, temperature, and organic debris are examples of abiotic components.

The **biosphere** consists of all ecosystems of the world, that is, all regions where organisms exist.

ENERGY FLOW

Members of a community may be classified according to their functional roles. **Producers** consist of photosynthetic and chemosynthetic **autotrophs** that convert inorganic nutrients into the organic chemicals of protoplasm. In most ecosystems, photosynthesis is the most important process in this conversion. Producers provide all the organic molecules and chemical energy for the entire community of organisms.

Consumers are **heterotrophs** that depend on the organic compounds formed by producers for their nutrient needs. The primary (first-level) consumers are **herbivores** that feed directly on the producers. Secondary and tertiary consumers are **carnivores** that feed on other animals. **Omnivores** feed on both plants (producers) and animals. The **top carnivore** of a community is not preyed upon by any other organism.

Waste products of organisms and dead organisms are decayed (decomposed) by the **decomposers,** primarily fungi and bacteria. They extract the last bit of energy from the organic molecules and convert them into inorganic molecules that can be used once again by the producers.

Chemical energy of organic nutrients flows through the community from producers to consumers to decomposers. The transfer of energy follows the laws of thermodynamics. The **first law of thermodynamics** states that energy is neither created nor destroyed but

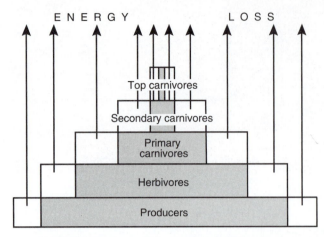

Figure 39.1 Pyramid of biomass. The biomass in each layer of the pyramid decreases from producers to the top carnivore.

only changed in form. The **second law of thermodynamics** states that usable energy is reduced with each energy transfer.

The quantitative relationships of producers and consumers based on energy flow may be depicted in an **ecological pyramid.** The three types of ecological pyramids are (1) a pyramid of numbers, (2) a pyramid of biomass (dry weight), and (3) a pyramid of calories (energy). Generally, each pyramid shows a decrease in quantity from the base (producers) to the apex (top carnivore). In unique situations, pyramids of numbers and biomass may be inverted, but the pyramid of calories never is inverted. See Figure 39.1. In general, only about 10% of the energy in one trophic level (layer in the pyramid) is transferred to the next trophic level. The remaining 90% is respired or unassimilated in the process. Thus, herbivores contain about 10% of the calories in the plants consumed. Primary carnivores are reduced to 1% of the energy stored in producers.

 Assignment 1

1. Study Figure 39.2 and note the pathway of energy flow. Determine the producers, the herbivores, and the primary, secondary, and tertiary carnivores. Label the figure.
2. ***Complete item 1 on Laboratory Report 39 that begins on page 501.***

BIOGEOCHEMICAL CYCLES

Although energy flows through the community only once, the chemicals composing organisms cycle repeatedly between the abiotic components and the community. For example, the molecules composing your body were previously composing bodies of plants and other animals. The molecules pass along the food web and ultimately reach the decomposers, which break them down into inorganic molecules that, in turn, can be used again by the producers.

The **carbon cycle** shown in Figure 39.3 illustrates the cycling of carbon from the inorganic carbon in the atmosphere to the organic carbon of organisms and its return to the atmosphere. The cycle is continuous.

Nitrogen is an indispensable part of amino acids and proteins. The **nitrogen cycle** is shown in Figure 39.4. Plants are able to use inorganic nitrogen in the form of nitrates, but they depend on certain bacteria to convert organic debris into this usable form. The most common pathway is the conversion of organic nitrogen first into ammonia, then into nitrites (NO_2^-), and finally into nitrates (NO_3^-) that can be used by plants. Certain bacteria are able to fix (convert) atmospheric nitrogen into organic nitrogen. *Rhizobium* is a nitrogen-fixing bacterium found in nodules on the roots of legumes (e.g., beans, peas, alfalfa). The bacteria and plants live in a **mutualistic symbiosis** because both organisms benefit from the association. *Azotobacter* is a bacterium that fixes nitrogen nonsymbiotically.

 Assignment 2

Materials

Beaker, 400 ml
Microscope slides
Slide holder
Bunsen burner
Methylene blue in dropping bottle
Clover and grass roots
Prepared slide of *Rhizobium* in a clover nodule

1. Study Figures 39.3 and 39.4 and identify the processes involved. Label Figure 39.3.
2. Examine the grass and clover roots. Locate the nodules on the clover roots.
3. Prepare a slide of *Rhizobium,* following these steps:
   a. Cut a nodule in half and smear the cut surface on a clean microscope slide.
   b. After the smear has dried, pass the slide quickly through a flame several times.
   c. Add 3–5 drops of methylene blue to the slide and let it stand for 3 min. Rinse the slide gently in a beaker of water and observe it with your microscope at 400×.
4. Examine a prepared slide of *Rhizobium* in a clover nodule and compare it with your slide.
5. ***Complete item 2 on the laboratory report.***

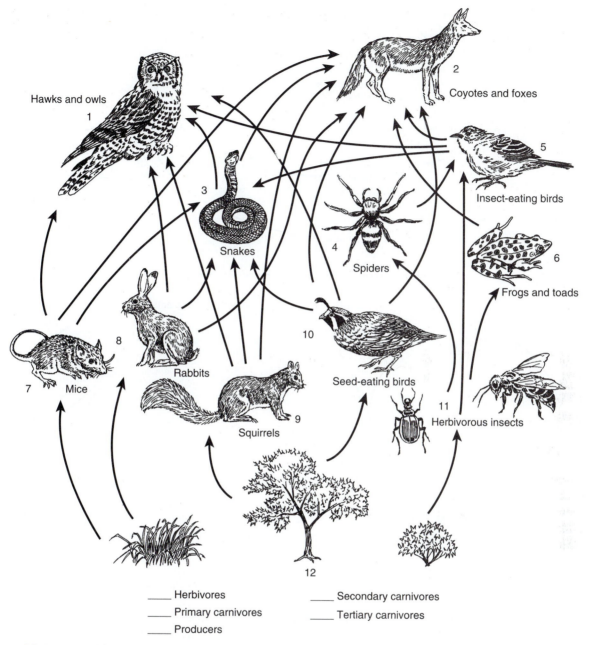

Figure 39.2 A simplified food web.

ENVIRONMENTAL POLLUTION

Pollution often results from the by-products of human activities accumulating in the environment at levels that are harmful to living organisms. Sources of pollution are varied. For example, air pollution is usually caused by submicroscopic particles, sulfur oxides, and nitrogen oxides produced by the burning of coal or petroleum products. Lakes and rivers may be polluted by pesticides and fertilizers in water runoff from agricultural land, contaminated waste-water from mining operations or chemical plants, and sewage from cities. Whatever the type and source of

pollution, it overloads the natural processes of an ecosystem and damages its delicate balance by altering one or more populations within the community of organisms. Its effect may be sudden and dramatic or gradual and subtle.

Aquatic Microecosystems

In this section, you will analyze aquatic microecosystems set up in the laboratory to simulate three lake environments: (1) normal, (2) polluted with acid rain, and (3) polluted with organic materials. Your task is to compare these ecosystems to ascertain the effect of

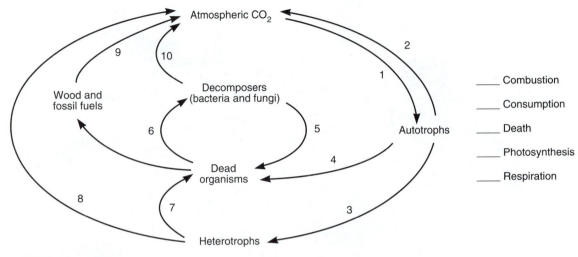

Figure 39.3 The carbon cycle.

_____ Combustion

_____ Consumption

_____ Death

_____ Photosynthesis

_____ Respiration

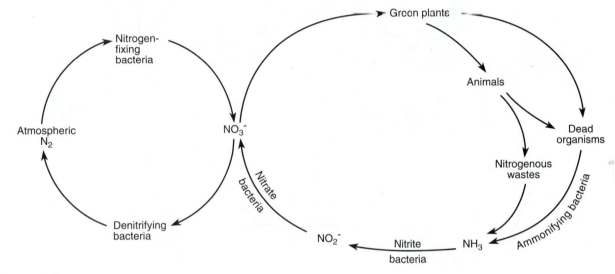

Figure 39.4 The nitrogen cycle.

acid and organic pollution on the diversity and density of organisms.

Acid rain is a side effect of air pollution. Sulfur and nitrogen oxides react with water in the atmosphere to form sulfuric acid and nitric acid, respectively, which are returned to earth in rain. Rainwater with a pH less than 5.6 is considered to be acid rain. Lakes and ponds exposed to acid rain have a lowered pH, which adversely alters their biogeochemical cycles, decreasing the availability of nutrients.

Lakes and ponds normally undergo a process called **eutrophication,** which gradually increases the available nutrients. Eutrophication, over many years, leads to changes in the community of organisms within an ecosystem and, ultimately, leads to the conversion of a pond to a meadow. However, the process is so slow that it is seldom noticed within a person's lifetime. In contrast to the slow process of eutrophication, **organic**

pollution of lakes and ponds by fertilizers or sewage causes a sudden increase in the available nutrients, which alters the ecosystem by favoring some members of the community while harming others. A polluted aquatic ecosystem tends to have a reduced concentration of oxygen, which may kill off desirable heterotrophic species, *especially game fish.*

 Assignment 3

Materials

Aquatic ecosystems in small aquaria or 600-ml beakers simulating (1) normal, (2) acid pollution, and (3) organic pollution

Beaker, small

Keys to algae (Ward's)

Medicine droppers
Microscope slides and cover glasses
Pipettes, 5 ml, with rubber bulbs

1. Observe the three microecosystems macroscopically to determine the relative density of autotrophs. This is easily done by noting the degree of clarity or greenish color of the water. *Record your observations in item 3a on the laboratory report.*

2. Learn the recognition characteristics of the three types of crustaceans in Figure 39.5. *Gammarus* is the largest form and tends to stay near the bottom of the container. *Daphnia* is intermediate in size and swims in the water by pulsating beats of the antennae as it engulfs microscopic autotrophs. *Cyclops* is the smallest and will be either swimming in the water or attached to filamentous algae or to the sides of the container. All these crustaceans can be located with the unaided eye or a hand lens.

3. Locate the crustaceans in the microecosystems. Note the type and relative density of the crustaceans in each one. *Record your observations in items 3a and 3b on the laboratory report.*

4. Learn the recognition characteristics of the autotrophs in the microecosystems by studying Figure 39.6 and the specimens under demonstration microscopes set up by your instructor.

5. Determine the autotrophs in each microecosystem by making slides of water from each microecosystem and observing them with your microscope. If filamentous algae are present, remove a small amount with forceps and mount it on a slide for observation. Most of the autotrophs are *very small.* Use the 10× objective to locate a specimen, then switch to the 40× objective to identify it. Use reduced illumination for best results. It may be necessary to make many slides to locate specimens in the microecosystem with the lowest density of autotrophs. *Record your results in item 3b and complete item 3 on the laboratory report.*

Bioassay

One way biologists assess the potential damage a pollutant may exert on an ecosystem is to determine its effect on an **indicator organism,** a process known as a **bioassay.** By exposing the indicator organism to a series of concentrations of a pollutant, it is possible to determine the concentration that will produce a harmful effect. One method is to determine the LC_{50} (lethal concentration$_{50}$), the concentration that kills 50% of the indicator organisms within a specific time of exposure. Figure 39.7 illustrates how to determine the LC_{50}.

In this section, you will use *Daphnia* (Figure 39.5), a small crustacean found in ponds and lakes, as an indicator organism to investigate the effect of acid,

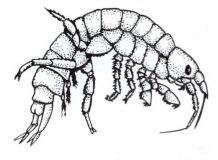

Gammarus

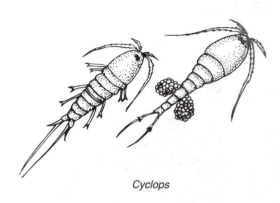

Cyclops

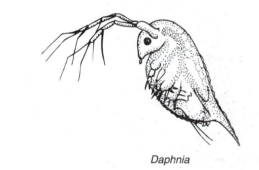

Daphnia

Figure 39.5 Crustaceans in the aquatic ecosystems.

pesticide, and thermal pollution. *Daphnia* feed on microscopic algae and other protists and, in turn, serve as a prime food source for small fish. They live suspended in the water, alternately moving upward in a series of "hops" propelled by quick downward thrusts of their antennae and slowly sinking until the next antennal thrust. All the while, their legs sweep food organisms into their mouths.

In the following experiments, you will use modified glass or plastic medicine droppers to transfer *Daphnia* from the stock culture to solutions of pond water in vials. *Caution:* Be sure to immerse the tip of the dropper below the surface of the pond water before squeezing the bulb to force *Daphnia* out of the dropper. This prevents air being trapped under the carapace, causing *Daphnia* to float at the surface.

Oscillatoria
4–6 μm
Blue-green

Euglena
30–55 μm
Englenoid

Peridinium
45 μm
Dinoflagellate

Pandorina
20–250 μm

Closterium
30–70 μm
Green (Desmid)

Chlorella
Green 50 μm

Microspora
20–30 μm
Green

Spirogyra
15–100 μm

Figure 39.6 Photosynthetic autotrophs that may be in the simulated ecosystems. Organisms are not drawn to scale.

Your instructor has prepared flasks of filtered pond water for your use in the experiments. Some have an acid pH, and others contain a small amount of pesticide. *Caution:* Handle the acid and pesticide solutions with care. If you spill any on your hands, quickly wash your hands with soap and lots of water. Notify your instructor of any spills.

The experiments are best done by groups of three or four students. Work out a division of labor among members of your group.

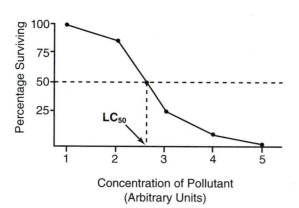

Figure 39.7 Determination of the LC_{50}. 1. Plot the percentage surviving at each concentration of pollutant. 2. Connect the dots with straight lines to form a survivor curve. 3. Draw a perpendicular line from the point where the survivor curve intersects the 50% surviving line. 4. The point where the perpendicular line intersects the X axis indicates the LC_{50}.

 ## Assignment 4

Materials

Beakers, 250 ml
Glass-marking pens
Glass vials, 1 in. × 3 in.
Hot plates or water baths
Ice, crushed
Medicine droppers modified to transfer *Daphnia*
Thermometers, Celsius
Vial holders
Cultures of *Daphnia*
Flasks of pond water
Flasks of pond water adjusted to pH 7, pH 6, pH 5, pH 4, and pH 3
Flasks of pond water containing these percentages of Diazinon Plus: 0.001%, 0.002%, 0.003%, and 0.004%.

1. Perform the following experiments and pool the results of the entire class.
2. ***Record the results in item 4 on the laboratory report.***
3. After completing the experiments, ***complete item 4 on the laboratory report.***

Experiment 1: Acid Pollution

1. Label five vials, 1 to 5, and fill each vial about two-thirds full with pond water adjusted to these pH values:
   Vial 1 = pH 7
   Vial 2 = pH 6
   Vial 3 = pH 5
   Vial 4 = pH 4
   Vial 5 = pH 3
   Place the vials in a rack or holder at your work station so they will not spill.
2. Starting with Vial 1, transfer four healthy *Daphnia* to each vial. Discard any floating *Daphnia* and replace them with healthy *Daphnia*. Rinse the transfer pipette in tap water between each transfer. (Why?) Record the time.
3. ***Record the number of Daphnia surviving after 1 hour on the laboratory report.***

Experiment 2: Pesticide Pollution

1. Label five vials, 1 to 5, and fill each vial about two-thirds full with pond water containing these percentages of Diazinon Plus:
   Vial 1 = 0.000%
   Vial 2 = 0.001%
   Vial 3 = 0.002%
   Vial 4 = 0.003%
   Vial 5 = 0.004%
   Place the vials in a rack or holder at your work station so they will not spill.
2. Starting with Vial 1, transfer four healthy *Daphnia* to each vial. Discard any floating *Daphnia* and replace them with healthy *Daphnia*. Rinse the transfer pipette in tap water between each transfer. (Why?) Record the time.
3. ***Record the number of Daphnia surviving after 1 hour on the laboratory report.***

Experiment 3: Thermal Pollution

1. Fill two 250-ml beakers half full with tap water to make small water baths. Use ice to adjust the temperature of one to 20°C and leave the other one at room temperature. Large water baths are available in the lab at 30°C and 35°C. Record the temperature of each water bath.
2. Label four vials, 1 through 4, and fill each about two-thirds full with pond water.
3. Place one vial of pond water in each water bath.
   Vial 1 = 20°C
   Vial 2 = room temperature
   Vial 3 = 30°C
   Vial 4 = 35°C
   Allow 5 minutes for temperature equilibration.
4. Transfer four healthy *Daphnia* to each vial. Discard any floating *Daphnia* and replace them with healthy *Daphnia*. Record the time.
5. ***Record the number of Daphnia surviving after 1 hour on the laboratory report.***

ECOLOGICAL RELATIONSHIPS

Student _____

Lab Instructor _____

1. ENERGY FLOW

a. Write the term that matches each meaning.

1. Sum of all ecosystems in the world _____

2. All populations within an ecosystem _____

3. Interbreeding members of a species
in an ecosystem _____

4. A defined area consisting of biotic and
abiotic components _____

5. All parts of Earth where organisms exist _____

6. Provide energy and organic nutrients for
all members of the community _____

7. Extract last energy from organic matter _____

8. Animals feeding on plants only _____

9. Composed mainly of fungi and bacteria _____

10. Animals feeding on other animals _____

11. Animals feeding on both plants and animals _____

12. Animal group not preyed upon by other animals _____

13. Convert light energy into chemical energy _____

14. Convert organic matter into inorganic matter _____

15. Ultimate source of energy for the Earth _____

b. Show the flow of energy through a community consisting of decomposers, herbivores, producers, and carnivores.

_____ → _____ → _____ → _____

c. Explain why only about 10% of the energy stored in one trophic level is transferred to the next higher level. ____

_____

_____

_____

d. In dry years, the biomass of producers is significantly decreased. What would be the effect on:

a rodent population? _____

a hawk population? _____

e. What group of organisms is not shown in Figure 39.1? _____

f. *Circle* the simplified food chain that would support a larger human population.

Cereals and legumes → beef → humans

Cereals and legumes → humans

Explain your answer. _____

_____

g. Categorize the organisms (by number) in Figure 39.2 according to their ecological roles.

Producers _____   Herbivores _____

Primary carnivores _____   Secondary carnivores _____

Tertiary carnivores _____

h. In which categories would humans be placed? _____

_____

i. What would result from a significant reduction of the insect population by insecticides?

_____

_____

j. Coyotes feed primarily on rodents but may occasionally kill a lamb in sheep-ranching areas. Ranchers sometimes conduct coyote-killing campaigns to protect their flocks. Evaluate the advisability of such practices.

_____

k. Evaluate the effect on the ecosystem of farmers killing rodents with poison. _____

2. BIOGEOCHEMICAL CYCLES

a. List the processes (by number) involved in the carbon cycle shown in Figure 39.3.

_____ Combustion   _____ Consumption   _____ Death

_____ Photosynthesis   _____ Respiration

Which of these processes is a major contributor to air pollution in our industrialized society?

_____

b. Scientists are concerned that an increase of CO_2 in the atmosphere may result in an increase in the average global temperature. What would be the source of this additional CO_2?

_____

c. How is the carbon in dead organisms returned to circulation? _____

_____

d. What is the origin of the energy available in fossil fuels? _____

_____

e. What is the global abiotic reservoir for nitrogen? _____

What organisms are able to use this nitrogen source? _____

f. Indicate the form of nitrogen required by:

animals _____   plants _____

g. What advantage does symbiotic nitrogen-fixing bacteria provide legumes? _____

_____

h. Diagram (1) a portion of a clover root nodule showing the *Rhizobium* bacteria and (2) *Rhizobium* bacteria in a smear on your slide.

In a Root Nodule   **In a Smear of a Root Nodule**

3. AQUATIC MICROECOSYSTEMS

a. Based on macroscopic observations, which microecosystem seems to have:

the greatest density of autotrophs? _____

the lowest density of autotrophs? _____

the greatest density of heterotrophs? _____

the lowest density of heterotrophs? _____

b. Based on macroscopic observations, record in the following table the relative abundance of each kind of crustacean in the microecosystems. Use this scale: 0 = none; 5 = most abundant. Based on examination of your slides, record the presence of autotrophs in the microecosystems by placing an "X" in the appropriate spaces in the table. Compare your data with that of the entire class.

Organism	Acid Pollution	Normal	Organic Pollution
Cyanobacteria Oscillatoria			
Protista Euglena			
Peridinium			
Green Algae Microspora			
Pandorina			
Spirogyra			
Crustacea Cyclops			
Daphnia			
Gammarus			

c. Indicate the autotroph and heterotroph most tolerant of:

	Autotroph	Heterotroph
acid pollution	_____	_____
organic pollution	_____	_____

d. Indicate the ecosystem with the:

greatest diversity of autotrophs. _____

least diversity of autotrophs. _____

e. Indicate the ecosystem with the:

greatest diversity of heterotrophs. _____

least diversity of heterotrophs. _____

f. Which ecosystem (in nature) would have the:

greatest density of decomposers? _____

least density of decomposers? _____

g. If the action of decomposers is curtailed, what would be the effect on other organisms in the community? _____

Explain your answer. _____

_____

_____

4. BIOASSAY

a. Record here the number of surviving *Daphnia* from your experiments and the total class. Calculate the percentage of *Daphnia* surviving for the total class.

1. Acid Pollution

Your group: Number surviving pH7 ____ pH6 ____ pH5 ____ pH4 ____ pH3 ____
Total class: Number surviving pH7 ____ pH6 ____ pH5 ____ pH4 ____ pH3 ____
Number exposed pH7 ____ pH6 ____ pH5 ____ pH4 ____ pH3 ____
% surviving pH7 ____ pH6 ____ pH5 ____ pH4 ____ pH3 ____

2. Pesticide Pollution

Your group: Number surviving 0.000% ____ 0.001% ____ 0.002% ____ 0.003% ____ 0.004% ____
Total class: Number surviving 0.000% ____ 0.001% ____ 0.002% ____ 0.003% ____ 0.004% ____
Number exposed 0.000% ____ 0.001% ____ 0.002% ____ 0.003% ____ 0.004% ____
% surviving 0.000% ____ 0.001% ____ 0.002% ____ 0.003% ____ 0.004% ____

3. Thermal Pollution

Your group: Number surviving 20°C ____ 25°C ____ 30°C ____ 35°C ____
Total class: Number surviving 20°C ____ 25°C ____ 30°C ____ 35°C ____
Number exposed 20°C ____ 25°C ____ 30°C ____ 35°C ____
% surviving 20°C ____ 25°C ____ 30°C ____ 35°C ____

b. Using data from the entire class, plot the percentage of surviving *Daphnia* here and determine the LC_{50} for each pollutant. Use Figure 39.7 as a guide.

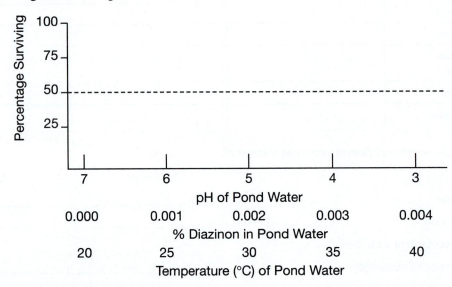

c. Record the estimated LC_{50} for each pollutant.

Acid pollution _____ Pesticide pollution _____ Thermal pollution _____

d. If only the *Daphnia* population of a lake is reduced by pollution, what would you predict to be the effect of *Daphnia* reduction on populations of:

microscopic algae? _____

small fish? _____

larger fish? _____

e. Is it possible that *Daphnia* that survived exposure to these pollutants for 1 hr may:

not survive exposure for hours, days, or weeks? _____

not survive exposure to two or more pollutants at the same time? _____

POPULATION GROWTH

OBJECTIVES

After completing the laboratory session, you should be able to:

1. Distinguish between population growth and population growth rate.
2. Compare theoretical and realized population growth curves.
3. Describe the role of the following in population growth:
   a. Biotic potential
   b. Environmental resistance
   c. Density-dependent factors
   d. Density-independent factors
   e. Environmental carrying capacity
4. Compare the growth of human and natural animal populations.
5. Compare the growth of human populations in developed and developing countries.
6. Define all terms in bold print.

A **population** is a group of interbreeding members of the same species within a defined area. Populations possess several unique characteristics that are not found in individuals: **density, birthrate, death rate, age distribution, biotic potential, dispersion,** and **growth form.** One of the central concerns in ecology is the study of population growth and factors that control it.

Natural populations are maintained in a state of dynamic equilibrium with the environment by two opposing factors: biotic potential and environmental resistance. **Biotic potential** is the maximum reproductive capacity of a population that is theoretically possible in an unlimiting environment. It is never realized except for brief periods of time. **Environmental resistance** includes all **limiting factors** that prevent the biotic potential from being attained.

The limiting factors may be categorized as **density-dependent factors,** the effects of which increase as the population increases. Space, food, water, waste accumulation, and disease are examples of such factors. In contrast, **density-independent factors** exert the same effect regardless of the population size. Climatic factors and the kill of individual predators generally function in this manner.

 Assignment 1

Complete item 1 on Laboratory Report 40 that begins on page 509.

GROWTH CURVES

There are two basic types of population growth patterns. **Theoretical population growth** is the growth that would occur in a population if the biotic potential of the species were realized, that is, if all limiting factors were eliminated. This type of growth does not occur in nature, except for very short periods of time. **Realized population growth** is the growth of a population that actually occurs in nature. Let's consider both types of population growth in bacteria.

Theoretical Growth Curves

In the absence of limiting factors, bacteria exhibit tremendous theoretical population growth potential. Bacterial population size is measured as the number of bacteria in a milliliter (ml) of nutrient broth, a culture medium. Assume that bacterial cells divide at half-hour intervals. If the initial population density were 10,000 bacteria per milliliter of nutrient broth (10×10^3 bacteria/ml), a half-hour later the population would be 20,000 bacteria per milliliter (20×10^3 bacteria ml), and the population would have doubled. See Table 40.1. In another half-hour, the population would double again to 40×10^3 bacteria/ml. In an unlimiting environment, this population would double every half-hour. If the population size is calculated for each half-hour interval and plotted on a graph, a line joining the points on the graph yields a **theoretical growth curve.**

A **theoretical growth rate curve** may be produced by determining the growth rate for each half-hour interval, plotting these values on a graph, and drawing a line to join the points. The **growth rate** is the change in population size (ΔN) per change in time (Δt), or $\Delta N/\Delta t$. In our example, the growth rate for the first half-hour is

$$\frac{\Delta N}{\Delta t} = (20 \times 10^3) - (10 \times 10^3)$$

$$= (10 \times 10^3) \text{ bacteria/ml/0.5 hr}$$

And for the second half-hour, it is

$$\frac{\Delta N}{\Delta t} = (40 \times 10^3) - (20 \times 10^3)$$

$$= (20 \times 10^3) \text{ bacteria/ml/0.5 hr}$$

Realized Growth Curves

Natural populations exhibit realized growth that results from the opposing effects of biotic potential and environmental resistance. Table 40.2 indicates the growth of a bacterial population in a tube of nutrient broth—a limited environment that eventually results in the death of the entire population.

TABLE 40.1	THEORETICAL GROWTH OF A BACTERIAL POPULATION	
Time (hr)	Population Density (10^3 bacteria/ml)	Growth Rate (10^3 bacteria/ml)
0.0	10	
0.5	20	
1.0	_____	_____
1.5	_____	_____
2.0	_____	_____
2.5	_____	_____
3.0	_____	_____
3.5	_____	_____
4.0	_____	_____

TABLE 40.2	REALIZED GROWTH OF A BACTERIAL POPULATION	
Time (hr)	Population Density (10^3 bacteria/ml)	Growth Rate (10^3 bacteria/ml)
0.0	10	
0.5	20	_____
1.0	40	_____
1.5	80	_____
2.0	150	_____
2.5	290	_____
3.0	450	_____
3.5	520	_____
4.0	520	_____
4.5	515	_____
5.0	260	_____
5.5	80	_____
6.0	0	_____

The population was sampled at half-hour intervals, and no additional nutrients were added.

Assignment 2

1. In Table 40.1, determine and record the theoretical growth of a bacterial population that doubles in size at half-hour intervals.
2. Determine the theoretical growth rate of this bacterial population at half-hour intervals and record the data in Table 40.1.
3. *Plot the theoretical growth and growth rate curves on the appropriate graphs in item 2a on the laboratory report.*
4. In Table 40.2, determine and record the realized growth rate of the bacterial population at half-hour intervals.
5. *Plot the realized growth and growth rate curves on the appropriate graphs in item 2a on the laboratory report.* Use a different symbol or color for these curves to distinguish them from the theoretical curves.
6. Compare the theoretical and realized population growth curves.

   The theoretical curve is a J-shaped curve. The realized growth curve starts off in the same manner, but it is soon changed by environmental resistance. As the limiting factors take effect, the curve becomes S shaped. In the artificial environment of a

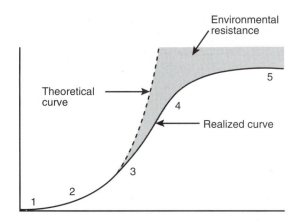

Figure 40.1 Theoretical and realized growth curves.
1 = lag phase, 2 = exponential growth phase,
3 = inflection point, 4 = decreasing growth phase,
5 = carrying capacity.

test tube, the bacterial population dies out. In natural populations where materials can cycle and energy is continuously available, the population size levels off where it is at equilibrium with the environment.

7. Study Figure 40.1, which compares generalized theoretical and realized population growth curves. The theoretical growth curve exhibits exponential growth, with the slope of the curve becoming ever steeper. In contrast, the effect of environmental resistance causes the realized growth curve to become S shaped.

Several parts of the realized growth curve are recognized. Growth is slow, initially, in the **lag phase** due to the small number of organisms present. This is followed by the **exponential growth phase** when the periodic doubling of the population yields explosive growth. As environmental resistance takes effect, the growth gradually slows, leading to the **inflection point** that indicates the start of the **decreasing growth phase** as the curve turns to the right. Continued environmental resistance causes the curve to level off at the **carrying capacity** of the environment where the population size is in balance with the environment.

The carrying capacity is the maximum population that can be supported by the environment—a balance between biotic potential and environmental resistance.

8. *Complete item 2 on the laboratory report.*

HUMAN POPULATION GROWTH

Over six billion four hundred million and counting! The continued growth of the human population is one of the major concerns of ecologists and modern society. Although a widespread catastrophe has not occurred,

localized famines have devastated some populations in developing countries, where the population temporarily approaches the carrying capacity of the environment. When a long-term drought or other temporary environmental disaster disrupts food production and distribution, this increase in environmental resistance increases the death rate. With over 80% of the world population located in developing countries, these societies would be the first to feel the impact of a population overload.

Just how large a human population Earth can support is unknown. The increased production of food and improved medical care as a result of the industrial, agricultural, and technological revolutions have enabled the rapid growth of the human population. Advances in science and technology will undoubtedly provide continued support for the population in the future. However, the human population cannot grow *ad infinitum* without negative consequences. If not controlled voluntarily, the limiting factors that control natural populations will ultimately kick in with unpleasant consequences.

Table 40.3 shows the estimated size of the world population at various times in history and projects a rather conservative growth in the near future.

TABLE 40.3	ESTIMATED WORLD HUMAN POPULATION	
Year	Population Size (millions)	Growth Rate (millions/year)
8000 B.C.	5	_____
4000	86	_____
1 A.D.	133	_____
1650	545	_____
1750	728	_____
1800	906	_____
1850	1,130	_____
1900	1,610	_____
1950	2,520	_____
1960	3,021	_____
1970	3,697	_____
1980	4,444	_____
1990	5,279	_____
2000	6,085	_____
2010	6,842*	_____
2020	7,577*	_____
2030	8,199*	_____
2040	8,701*	_____
2050	9,075*	_____

* Projection.

Growth Rate

Ultimately, a population is limited by a decrease in **birthrate,** an increase in **death rate,** or both. In natural nonhuman populations, an increase in death rate is the usual method. More individuals are produced than can survive, and the environment selects the better-adapted individuals who will contribute the greater share of the genes to the next generation. Although this mode of selection occurred in preindustrialized human societies, it has been minimized by technology in modern societies, especially in developed countries.

Concern over the growth rate of the human population gained worldwide attention in the 1960s and persists today. The growth rate is the difference between the number of persons born (birthrate) and the number who die (death rate) per year. For humans, the growth rate usually is expressed per 1,000 persons. For example, if a human population shows 25 births and 10 deaths per 1,000 persons in a year, the growth rate may be determined as follows:

$$\frac{25 - 10}{1,000} = \frac{15}{1,000} = \frac{1.5}{100}$$

$$= 1.5\% \text{ growth rate}$$

The growth rate reached a peak of 2% in 1965 and has gradually declined to 1.1% in 2005. This decline is encouraging but does not mean that the population growth is no longer a concern.

The growth rate is often expressed as the **doubling time,** that is, the time required for the population to double in size. In 1850, the population required 135 yr to double, but this was reduced to only 35 yr in 1965. All of the efforts to reduce population growth had only extended the doubling time to 41 yr by 1984. Continued efforts have increased the doubling time to 58.3 years in 2000 and to 63.6 in 2005.

The doubling time of a population is determined by dividing 70 yr (a demographic constant) by the growth rate. For example, the world population growth rate was 1.1% in 2005, so the doubling time (d) would be determined as follows:

$$d = \frac{70 \text{ yr}}{\text{growth rate}} = \frac{70 \text{ yr}}{1.1} = 63.6 \text{ yr}$$

This means that the entire world population would double in only 63.6 years. To maintain the same standard of living, the available resources must also double in 63.6 years. Is this likely? Are Earth's resources infinite?

Population growth has essentially stabilized in developed countries, where societies enjoy the benefits of technology, modern medicine including contraceptives, educational and employment opportunities for both men and women, and a relatively high standard of living. These benefits are not available to societies in developing countries, where the death rates are high and birthrates are substantially higher. In the foreseeable future, most of the growth in the human population will occur in developing countries.

Assignment 3

1. Use the data in Table 40.3 to *plot the human population growth curve in item 3a on the laboratory report*.
2. Calculate and record in Table 40.3 the growth rate per year for the time intervals shown.
3. *Complete items 3b–3i on the laboratory report*.
4. Calculate the growth rate and doubling time for the countries or regions shown in Table 40.4.
5. *Complete the laboratory report*.

TABLE 40.4	ESTIMATED 2005 HUMAN POPULATION DATA				
Country or Region	Population Size (10^6)	Birthrate (per 10^3)	Death Rate (per 10^3)	Growth Rate (% inc./yr)	Doubling Time (yr)
World	6,465	20.3	8.9	1.1	63.6
Africa	906	36.3	14.8	_____	_____
Asia	3,905	19.0	7.5	_____	_____
Europe	728	10.1	11.9	_____	_____
Latin America	561	20.1	6.0	_____	_____
North America	331	13.8	8.8	_____	_____
Oceania	33	16.2	7.4	_____	_____

Source: UN Population Division, The 2004 Revision Population Database.

POPULATION GROWTH

Student _____

Lab Instructor _____

1. INTRODUCTION

Write the term that matches each meaning.

1. Sum of all limiting factors

   _____

2. Maximum theoretical reproductive capacity

   _____

3. Limiting factors whose effect is unchanged
   by population size

   _____

4. Limiting factors whose effect changes
   proportionately with population size

   _____

2. GROWTH CURVES

a. Plot the theoretical and realized growth and growth rate curves here.

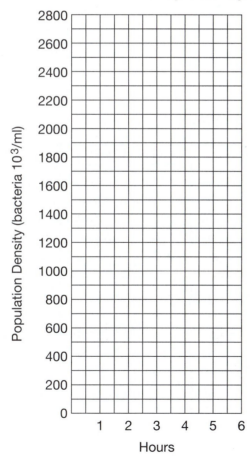

Theoretical and realized growth
curves of bacterial populations

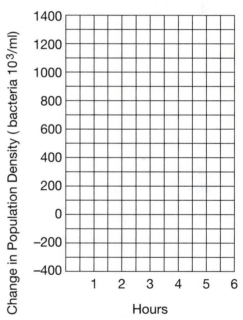

Theoretical and realized
growth rate curves of
bacterial populations

b. Indicate the time interval when the growth rate was greatest.

Theoretical growth _____   Realized growth _____

c. In the realized growth curve:

When did cell deaths most nearly equal cell "births"? _____

Why did the realized bacterial population die out? _____

_____

What limiting factors determined the realized population growth pattern? _____

_____

Were density-dependent or density-independent factors involved? _____

_____

d. In item 2a, show the shape the realized growth curve would have taken if additional nutrient broth was added at the fifth hour. Use a red pencil to draw the new curve.

e. Write the term that matches each meaning regarding Figure 40.1.

1. Growth curve with a J shape _____

2. Growth curve with an S shape _____

3. Phase of explosive growth _____

4. Point where limiting factors take effect _____

5. Phase of equilibrium between biotic potential
   and environmental resistance _____

6. Where birth equals death _____

7. Where resources per capita are least _____

3. HUMAN POPULATION GROWTH

a. Plot the human population growth curve here using data in Table 40.3.

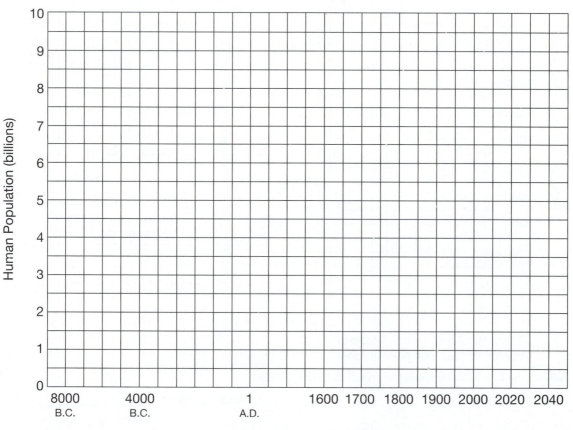

Human Population Growth Curve

b. Does the human population growth curve resemble the theoretical growth curve or the realized growth curve? _____

_____

c. In what phase of the realized growth curve does the human population curve appear to be in? _____

_____

_____

d. What events in human history have allowed the rapid growth of the human population? _____

_____

e. Are human populations subject to the interaction of biotic potential and environmental resistance like natural

populations? _____

f. List any limiting factors controlling the size of natural populations that do not affect human populations. _____

_____

g. Humans are subject to certain limiting factors that are not observed in other animal populations. Name two of

these. _____

h. If nothing is done to check the growth of the human population, will deaths ultimately equal births? _____

What will be the shape of the population growth curve? _____

i. Ultimately, there are only two ways to decrease population growth. Name them. _____

Which is preferable? _____

j. Record the growth rate and doubling time as calculated in Table 40.4.

	Growth Rate (%/yr)	*Doubling Time (yr)*
Africa	_____	_____
Asia	_____	_____
Europe	_____	_____
Latin America	_____	_____
North America	_____	_____
Oceania	_____	_____

k. What region has the:

fastest population growth? _____

slowest population growth? _____

l. The rapid growth rate in developing countries has resulted from a decreased death rate without a proportionate

decrease in birthrate. What has brought this about? _____

_____

m. What relationship exists between population size and resources per person? _____

n. Are Earth's resources infinite? _____ Does it seem likely that resources can increase to

keep pace with the projected population growth? _____

_____

o. The standard of living is based on available resources per person. If a country's population is increasing at a rate

of 2% per year, what rate of increase in productivity is required to maintain the same standard of living? _____

_____

p. At the carrying capacity, deaths equal births and resources are just adequate to maintain the existing population size. Would it be advantageous to curtail the human population (zero growth) before it levels off at the carrying capacity due to natural causes? _____

Explain your response. _____

_____

_____

_____

_____

_____

q. Are developing countries likely to improve their standard of living without decreasing their population growth? _____ Explain your answer. _____

_____

_____

_____

r. Famine has been a serious problem in Africa in recent years. What has caused the famine?

_____

_____

_____

Food was provided by developed countries to help the people faced with starvation. Was the donated food a real or artificial increase in the carrying capacity of the environment?

_____

What is the long-range solution to this problem? _____

_____

_____

_____

_____

_____

_____

s. If you were a policymaker in the U.S. government, what policies would you propose to stimulate a decrease in population growth:

in developing countries receiving foreign aid? _____

_____

_____

_____

_____

_____

in the United States? _____

_____

_____

_____

ANIMAL BEHAVIOR

OBJECTIVES

After completing the laboratory session, you should be able to:
1. Contrast innate and learned behavior.
2. Describe the four types of innate behavior.
3. Describe the responses of planaria to light, gravity, current, and food.
4. Describe the responses of sowbugs to being touched and to light and moisture.
5. Describe the responses of fruit flies to light and gravity.
6. Define all terms in bold print.

Animal behaviors may be divided into two fundamentally distinct types: innate behavior and learned behavior. **Innate behavior** is inherited behavior. It is an automatic and consistent response to a specific stimulus. Innate behavior is performed correctly the first time a specific stimulus is perceived. **Learned behavior** is not inherited. It tends to change or improve with experience, that is, with repeated exposure to the specific stimulus.

Simple animals exhibit only innate behavior. As more complex nervous systems evolved, learned behavior became possible and increasingly important. The more complex the nervous system, the more learned behavior is exhibited by the animal. Learned behavior reaches its peak in humans, who exhibit learning and reasoning in complex behavioral responses to stimuli. However, humans also exhibit innate behavior such as reflexes.

There are four types of innate behavior: (1) kineses, (2) taxes, (3) reflexes, and (4) fixed action patterns. A **kinesis** is an orientation behavior in which the animal changes its speed of movement, but not direction, in response to a stimulus. If the animal slows down in response to a stimulus, the response is *positive kinesis*. If it speeds up, the response is *negative kinesis*. Positive kinesis enables animals to stay close to the stimulus, for example, favorable environmental conditions, although they are not actually attracted by it.

Taxes are orientation behaviors in which an animal is either attracted to (positive) or repelled by (negative)

the stimulus. In each case, the animal changes direction, although the rate of movement may or may not change. Light, heat, cold, gravity, chemicals, and sound are known to elicit such responses in animals. For example, an animal attracted to light exhibits *positive phototaxis,* and an animal repelled by a chemical exhibits *negative chemotaxis.*

Reflexes involve a rapid, stereotyped movement of the body or a body part rather than the orientation of the entire body as in kineses and taxes. Common reflexes in humans include the knee-jerk reflex, blinking an eye, and coughing.

Fixed action patterns are predictable, stereotyped movements that may be rather complex. The stimulus that causes a fixed action pattern is called a *releaser.* For example, a newly hatched seagull chick will peck at the red spot on the parent bird's bill causing the parent to regurgitate food for the young chick. The red spot is the releaser stimulating the pecking by the chick, and the pecking of the chick is the releaser stimulating the parent bird to regurgitate food. Fixed action patterns may be so complex and so appropriate that they may seem to be learned upon casual observation.

In this exercise, you will investigate innate behavior in simple animals. Because a variety of stimuli affects animal behavior, conditions except for the stimulus being tested should remain constant in your experiments.

Assignment 1

Complete item 1 on Laboratory Report 41 that begins on page 517.

TAXES IN PLANARIA

Recall that planaria are bilaterally symmetrical animals with special sense organs located at the anterior end. Eyespots are sensitive to light, and the auricles contain receptors for tactile (touch) and chemical stimuli. In this section, you will investigate hypotheses that planaria do not respond to light, water current, chemicals, or gravity.

1. **Phototaxis:** movement toward or away from light.
2. **Rheotaxis:** movement into or away from a current.
3. **Chemotaxis:** movement toward or away from a chemical source.
4. **Gravitaxis:** movement toward or away from the force of gravity.

Assignment 2

Materials

Brown planaria
Black construction paper
Camel's hair brushes
Finger bowls
Glass-marking pen
Laboratory lamp
Magnetic stirrer
Masking tape
Medicine droppers
Petri dishes
Pond water
Test tubes
Test-tube racks
Cork stoppers for test tubes
Chopped liver

1. Obtain a petri dish. Fill it half full with pond water. Place five planaria in the dish using small camel's hair brushes or the modified medicine droppers with enlarged openings. Keep the dish motionless at your lab station and wait 5 min for the planarians to adapt. Observe the movement of the planaria in the dish under normal classroom lighting. Do the planaria disperse randomly over the bottom of the dish? Does the direction of their movement seem to be random? Do they change direction often? **Complete item 2a on the laboratory report.**

2. To test the hypothesis that planaria do not exhibit phototaxis, make a light shield of black construction paper and masking tape and place it over one-half of the dish. Expose half the dish to normal classroom light while the other half is shaded by the light shield. Observe the movements of the planaria. Do they move directly into the shaded half or into the light? Do they tend to stay in the light or shade? Do they remain motionless in light or shade? After 10 min, record the number of planaria in the lighted half and in the shaded half of the dish. Record your results and those of the entire class. **Complete item 2b on the laboratory report.**

3. Remove the light shield and allow 5 min for the planaria to adapt. To test the hypothesis that planaria do not exhibit chemotaxis, gently place a small piece of liver in the center of the dish without disturbing the water. Observe the planaria to see if they exhibit chemotaxis. Record your results and those of the entire class. **Complete item 2c on the laboratory report.**

4. Transfer your planaria to a finger bowl half filled with pond water and allow 5 min for the planaria to adapt. To test the hypothesis that planaria do not exhibit rheotaxis, place your petri dish of planaria on a magnetic stirrer and place the stirring piece in the center of the dish. Turn on the stirrer at the *slowest possible speed* to create a gentle, circular flow of water in the dish. If a magnetic stirrer is not available, slowly stir the water with a pencil. Observe the flatworms to see if they exhibit rheotaxis. Record your results and those of the entire class. **Complete item 2d on the laboratory report.**

5. To test the hypothesis that planaria do not exhibit gravitropism, fill a vial with pond water and transfer one planarian into the vial. Insert a cork stopper. Use a marking pen to mark the midpoint of the vial. Make a cylindrical light shield from black construction paper and masking tape that will loosely slip over the vial. One end of the light shield must be closed to exclude light.

   Lay the vial horizontally on the table. When the planarian moves to near the midpoint of the vial, gently stand the vial upright and slip the light shield over the vial. At 2-min intervals for 6 min, lift the light shield briefly to determine the location of the planarian. Record whether the planarian is located in the lower half or upper half of the vial. If the planarian tends to stay at one end of the vial after 6 min, gently invert the vial and repeat the experiment. Record you results and those of the entire class. **Complete item 2 on the laboratory report.**

BEHAVIOR IN SOWBUGS

Sowbugs (isopods) or wood lice are typically found under rotting logs and other organic debris in a habitat that is dark and moist. In this section, you will investigate the orientation behavior that causes sowbugs to reside in such habitats. Two possible stimuli for this orientation behavior are moisture and light. You will investigate the reactions of sowbugs to these stimuli by testing two hypotheses:

1. Sowbug locomotion is not affected by light.
2. Sowbug locomotion is not affected by moisture.

 Assignment 3

Materials

Sowbugs
Black construction paper
Dropping bottles of water
Masking tape
Plastic petri dishes, 14 cm diameter
Filter paper, 14.5 cm diameter

1. *Complete items 3a and 3b on the laboratory report.*
2. Obtain a large petri dish and fit a circle of filter paper into the bottom of it.
3. Place five sowbugs into the petri dish and replace the cover. How did they respond to being touched? Place the dish on the table at your work station under normal classroom lighting. Observe their behavior for several minutes. What is their response to vibrations? Tap the dish and see. Do they seem to move rapidly or slowly? Are they equally or randomly distributed over the dish? *Complete items 3c–3e on the laboratory report.*
4. To test the hypothesis that sowbug behavior is unaffected by light, make a low, rooflike light shield of black construction paper to shade one-half of the petri dish. The light shield should be low but allow you to look under it to make your observations. Shake the dish to activate the sowbugs, and position them in one-half of the dish. Place the light shield over the other half of the dish to shade it, and observe their responses. Do their orientation movements seem random? Do they make quick directional movements toward or away from the light? Once in the light or shade do they stay there? Observe the movement of the sowbugs and record their locations at 3-min intervals for 15 min. Record your results and those of the total class. *Complete items 3f and 3g on the laboratory report.*

5. Remove the light shield. To test the hypothesis that sowbug behavior is not affected by moisture, add 5 drops of water at a spot near the edge of the filter paper. This will provide a choice for the sowbugs: dry and moist. Shake the dish to get the sowbugs moving, and expose the dish to normal classroom lighting. Observe their movements and locations at 3-min intervals for 15 min. Watch to see if their movements are faster or slower in the moist area compared with the dry area. Record your results and those of the total class. *Complete items 3h and 3i on the laboratory report.*
6. To determine which stimulus (moisture or light) is stronger, you will provide a choice of either moisture or dim light and enable the sowbugs to choose their preferred location. Shake the dish to activate the sowbugs, and place the light shield over the dish to shade the *dry half* of the dish opposite the moist area. Observe the movements and locations of the sowbugs at 3-min intervals for 15 min. Record your results and those of the total class. *Complete item 3 on the laboratory report.*

TAXES IN FRUIT FLIES

Drosophila (fruit flies) are much more advanced than flatworms and sowbugs, so their nervous system is more complex. They perceive images with their compound eyes, and tactile and chemical receptors are especially abundant on their antennae and feet.

 Assignment 4

Materials

Vial (3 in. or longer) of fruit flies
Black construction paper
Glass-marking pen
Masking tape
Vials (3 in. or longer) with cork stoppers

1. Obtain a vial of five fruit flies. Mark the midpoint of the vial with a glass-marking pen. Stand the vial up vertically, stopper up, on the table at your lab station. After 3 min, record the location of the flies. Invert the vial and after 3 min record the location of the flies. *Complete item 4a on the laboratory report.*
2. Using the materials available, design and conduct experiments to determine if the distribution of fruit flies in the vial is due to phototaxis and/or gravitaxis. *Complete item 4 on the laboratory report.*

ANIMAL BEHAVIOR

Student _____

Lab Instructor _____

1. INTRODUCTION

a. Define:

Innate behavior _____

_____

Learned behavior _____

_____

b. Provide the terms that match the statements.

1. A rapid, stereotyped movement of the body
    or body part in response to a stimulus   _____

2. Movement toward or away from a stimulus   _____

3. Predictable, stereotyped, complex movements
    triggered by a releaser   _____

4. Slower or faster movement of an animal in
    response to a stimulus   _____

2. TAXES IN PLANARIA

a. Are the planaria randomly dispersed over the bottom of the dish?

Describe the distribution. _____

How do you explain their distribution? _____

_____

b. When the planaria moved into the shaded area did they:

stay there? _____ become motionless? _____

After 10 min, indicate the number of planaria in the:

lighted half _____ shaded half _____

Do the results support the hypothesis?_____

Explain. _____

_____

How is this behavior beneficial to planaria? _____

_____

c. What physical process disperses the chemicals in the water? _____

How long did it take to get a response from the planaria? _____

Explain any time lag. _____

Do your observations support the hypothesis? _____

Explain. _____

_____

How is this response beneficial to planaria? _____

_____

d. Did the planaria exhibit rheotaxis? _____   What type? _____

   How is this behavior beneficial to planaria? _____

   _____

e. Why was the light shield necessary? _____

   _____

f. Record your results and place the class totals in parentheses.

Time	Location of Planarian	
	Bottom Half	Top Half

g. Did the results support the hypothesis? _____

   Explain. _____

   _____

   How is this response beneficial to planaria? _____

   _____

3. BEHAVIOR IN SOWBUGS

a. What behavior would you expect to observe if sowbugs exhibited negative kinesis to light?

   _____

   _____

   What would indicate negative phototaxis? _____

   _____

b. What behavior would you expect to observe if sowbugs exhibited positive kinesis to moisture?

   _____

   _____

c. How did sowbugs respond to being touched?

   _____

   What type of innate response is this? _____

   How is this response beneficial to sowbugs? _____

   _____

d. How do moving sowbugs respond when you tap the dish?

_____

What type of innate response is this? _____

How is this response beneficial to sowbugs? _____

_____

_____

e. Do the sowbugs move randomly over the bottom of the dish? _____

If not, describe your observations. _____

_____

Do they seem to use their antennae to establish their position? _____

Explain. _____

_____

_____

f. Record your observations and place the class totals in parentheses.

Time	Number in the Shade		Number in the Light	
	Moving	Not Moving	Moving	Not Moving

g. Do the results support the hypothesis? _____

State a conclusion from the results. _____

_____

_____

h. Record your observations and place the class totals in parentheses.

Time	Number in Moist Area		Number in Dry Area	
	Moving	Not Moving	Moving	Not Moving

i. Do the results support the hypothesis? _____

State a conclusion from the results. _____

_____

_____

_____

j. Record your observations and place the class totals in parentheses.

Time	Number in the Dry Shade		Number in the Moist Light	
	Moving	Not Moving	Moving	Not Moving

k. Which stimulus produced a stronger response? _____

l. Is there a tendency for sowbugs to form aggregations or to be isolated when inactive?

_____

Can you explain this behavior solely on the basis of preferences for darkness and moisture?

_____

If not, what other factors may be involved? _____

_____

4. TAXES IN FRUIT FLIES

a. Record the distribution of fruit flies in the vial.

Condition	Number of Flies	
	Top Half	Bottom Half
After 3 min with stopper up		
After 3 min with bottom up		

b. State three hypotheses to explain these observations.

1. _____

_____

2. _____

_____

3. _____

_____

c. Use a separate sheet of paper to explain how you determined the response of fruit flies to light and gravity. For each hypothesis tested, provide the following:

1. State the hypothesis.

2. Describe your methods.

3. Record your results.

4. State your conclusion.

Common Prefixes, Suffixes, and Root Words

▶ Prefixes of Quantity and Size

amphi-, diplo-	both, double, two
bi, di-	two
centi-	one hundredth
equi-, iso-	equal
haplo-	single, simple
hemi-, semi-	one-half
hex-	six
holo-	whole
macro-	large
micro-	small
milli-	one thousandths
mono-, uni-	one
multi-, poly-	many
oligo-	few
omni-	all
pento-	five
quadri-	four
tri-	three

▶ Prefixes of Direction or Position

ab-, de-, ef-, ex-	away, away from
acro-, apici-	top, highest
ad-, af-	to, toward
antero-, proto-	front
anti-, contra-	against, opposite
archi-, primi-, proto-	first
circum-, peri-	around
deutero-	second
dia-, trans-	across
ecto-, exo-, extra-	outside
endo-, ento-, intra-	inner, within
epi-	upon, over
hyper-, supra-	above, over
hypo-, infra-, sub-	under, below
inter-, meta-	between
intra-	within
medi-, meso-	middle
post-, postero-	behind
pre-, pro-	before, in front of
retro-	backward
ultra-	beyond

▶ Miscellaneous Prefixes

a-, an-, e-	without, lack of
chloro-	green
con-	together, with
contra-	against
dis-	apart, away
dys-	difficult, painful
erythro-	red
eu-	true, good
hetero-	different, other
homo-, homeo-	same, similar
leuco-, leuko-	white
meta-	change, after
necro-	dead, corpse
neo-	new
pseudo-	false
re-	again
sym-, syn-	together, with
tachy-	quick, rapid

▶ Miscellaneous Suffixes

-ac, -al, -alis, -an, -ar, -ary	pertaining to
-asis, -asia, -esis	condition of
-blast	bud, sprout
-cide	killer
-clast	break down, broken
-elle, -il	little, small
-emia	condition of blood
-fer	to bear
-ia, -ism	condition of
-ic, -ical, -ine, -ous	pertaining to
-id	member of group
-itis	inflammation
-logy	study of
-lysis	loosening, split apart
-old	like, similar to
-oma	tumor
-osis	a condition, disease
-pathy	disease
-phore, -phora	bearer
-sect, -tome, -tomy	cut
-some, -soma	body
-stat	stationary, placed
-tropic	change, influence
-vor	to eat

▶ Miscellaneous Root Words and Combining Vowels

andro	man, male
arthr, -i, -o	joint
aut, -o	self
bio	life
blast, -i, -o	bud, sprout
brachi, -o	arm

branch, -i	gill	neph, -i, -o	kidney
bronch, -i	windpipe	neur, -i, -o	nerve
carcin, -o	cancer	oculo	eye
cardi, -a, -o	heart	odont	tooth
carn, -i, -o	flesh	oo, ovi	egg
cephal, -i, -o	head	oss, -eo, osteo	bone
chole	bile	oto	ear
chondr, -i, -o	cartilage	path, -i, -o	disease
chrom, -at, -ato, -o	color	phag, -o	to eat
coel, -o	hollow	phyll	leaf
crani, -o	skull	phyte	plant
cuti	skin	plasm	formative substance
cyst, -i, -o	bag, sac, bladder	pneumo, -n	lung
cyt, -e, -o	cell	pod, -ia	foot
derm, -a, -ato	skin	proct, -o	anus
entero	intestine	soma, -to	body
gastr, -i, po	stomach	sperm, -a, -ato	seed
gen	to produce	stasis, stat, -i, -o	stationary, standing still
gyn, -o, gyneco	female	stoma, -e, -to	mouth, opening
hem, -e, -ato	blood	therm, -o	heat
hist, -io, -o	tissue	troph	food, nourish
hydro	water	ur, -ia	urine
lip, -o	fat	uro, uran	tail
mere	segment, body section	viscer	internal organ
morph, -i, -o	shape, form	vita	life
myo	muscle	zoo, zoa	animal

Common Metric Units and Temperature Conversions

COMMON METRIC UNITS				
Category	Symbol	Unit	Value	English Equivalent
Length	km	kilometer	1000 m	0.62 mi
	m	meter*	1 m	39.37 in.
	dm	decimeter	0.1 m	3.94 in.
	cm	centimeter	0.01 m	0.39 in.
	mm	millimeter	0.001 m	0.04 in.
	μm	micrometer	0.000001 m	0.00004 in.
Mass	kg	kilogram	1000 g	2.2 lb
	g	gram*	1 g	0.04 oz
	dg	decigram	0.1 g	0.004 oz
	cg	centigram	0.01 g	0.0004 oz
	mg	milligram	0.001 g	
	μg	microgram	0.000001 g	
Volume	l	liter*	1 l	1.06 qt
	ml	milliliter	0.001 l	0.03 oz
	μl	microliter	0.000001 l	

* Denotes the base unit.

▶ Temperature Conversions

Fahrenheit to Celsius

$$^\circ C = \frac{5(^\circ F - 32)}{9}$$

Celsius to Fahrenheit

$$^\circ F = \frac{9 \times {}^\circ C}{5} + 32$$

Oil-Immersion Technique

The oil-immersion objective allows a magnification of 95× to 100× because the oil prevents loss of light rays and permits the resolution of two points as close as 0.2 μm (1/100,000 in.).

The working distance of the oil-immersion objective is extremely small, and care must be exercised to avoid damage to the slide or the objective or both. Although focusing with the low-power and high-dry objectives before using the oil-immersion objective is not essential, it is the preferred technique.

Use the following procedure:

1. Bring the object into focus with the low-power objective.

2. Center the object in the field and rotate the high-dry objective into position.

3. Refocus using the fine-adjustment knob and readjust the iris diaphragm. Also, recenter the object if necessary.

4. Rotate the revolving nosepiece halfway between the high-dry and oil-immersion objectives to allow sufficient room to add the immersion oil.

5. Place a drop of immersion oil on the slide over the center of the stage aperture. Move quickly from bottle to slide to avoid dripping the oil on the stage or table.

6. Rotate the oil-immersion objective into position. Note that the tip of the objective is immersed in the oil.

7. A slight adjustment with the fine-adjustment knob may be needed to bring the object into focus. Remember that the distance between the slide and objective cannot be decreased much without damaging the slide or objective or both.

8. After using the oil-immersion objective, wipe the oil with lens paper before using the low- and high-dry objectives to avoid getting oil on these objectives.

9. When finished with the oil-immersion objective, wipe the oil from the objective, slide, and stage.

The Classification of Organisms

The classification system used here is a three-domain system. This system divides prokaryotes into domains Archaea and Bacteria, based on fundamental differences in cell biochemistry, and places all eukaryotes into domain Eukarya. It requires at least six kingdoms and probably more, but the details have not yet been worked out. The classification system used in your textbook may be a bit different because biologists do not all agree on any one classification system. Keep in mind that the classification of organisms is a continually evolving process.

DOMAIN ARCHAEA	Bacterialike prokaryotes that live in harsh environments, such as hot springs, deep sea vents, and salt marshes. Their biochemistry differs from true bacteria.
DOMAIN BACTERIA	The true bacteria. Prokaryotes include chemosynthetic, photosynthetic, saprotrophic, and parasitic bacteria.

DOMAIN EUKARYA

▶ Kingdom PROTISTA — Mostly unicellular, and some colonial, eukaryotic organisms that live in marine, freshwater, or moist terrestrial environments. Consists of a diverse group of simple organisms that do not "fit" into other kingdoms.

Animal-like Protists

Phylum ZOOMASTIGOPHORA	Flagellated protozoans
Phylum SARCODINA	Amoeboid protozoans
Phylum CILIOPHORA	Ciliated protozoans
Phylum SPOROZOA	Nonmotile, parasitic protozoans

Plantlike Protists

Phylum BACILLARIOPHYTA	Diatoms
Phylum EUGLENOPHYTA	Flagellated algae lacking cell walls
Phylum DINOFLAGELLATA	Fire algae (dinoflagellates)
Phylum RHODOPHYTA	Red algae
Phylum PHAEOPHYTA	Brown algae
Phylum CHLOROPHYTA	Green algae

Funguslike Protists

Phylum MYXOMYCOTA	Plasmodial slime molds
Phylum ACRASIOMYCOTA	Cellular slime molds
Phylum OOMYCOTA	Water molds

▶ Kingdom FUNGI — Plantlike saprotrophic heterotrophs, typically with multinucleate cells. Body composed of hyphae.

Phylum ZYGOMYCOTA	Conjugating or algal fungi
Phylum ASCOMYCOTA	Sac fungi
Phylum BASIDIOMYCOTA	Club fungi
Phylum DEUTEROMYCOTA	Imperfect fungi

▶ Kingdom PLANTAE — Multicellular eukaryotes with rigid cell walls; primary photosynthetic autotrophs.

Nonvascular Land Plants

Phylum BRYOPHYTA	Mosses, liverworts, and hornworts

Vascular Land Plants Without Seeds

Phylum PSILOPHYTA	Psilopsids
Phylum LYCOPHYTA	Club mosses

Phylum SPHENOPHYTA	Horsetails
Phylum PTEROPHYTA	Ferns

Vascular Plants with Seeds

Gymnosperms

Phylum CONIFEROPHYTA	Conifers
Phylum CYCADOPHYTA	Cycads
Phylum GINKGOPHYTA	Ginkgos
Phylum GNETOPHYTA	Gnetophytes

Angiosperms

Phylum ANTHOPHYTA	Flowering plants
Class MONOCOTYLEDONES	Monocots (e.g., grasses, lilies)
Class DICOTYLEDONES	Dicots (e.g., beans, oaks, snapdragons)

▶ **Kingdom ANIMALIA**	Multicellular eukaryotic heterotrophs; cells without cell walls or chlorophyll; primarily motile.
Phylum PORIFERA	Sponges

Radial Protostomes

Phylum CNIDARIA	Cnidarians (diploblastic)
Class HYDROZOA	Hydroids
Class SCYPHOZOA	True jellyfish
Class ANTHOZOA	Sea anemones and corals

Bilateral Protostomes

Phylum PLATYHELMINTHES	Flatworms (triploblastic)
Class TURBELLARIA	Free-living flatworms
Class TREMATODS	Flukes (parasitic)
Class CESTODA	Tapeworms (parasitic)
Phylum NEMATODA	Roundworms
Phylum ROTIFERA	Rotifers
Phylum MOLLUSCA	Mollusks, soft bodies and usually with a shell
Class POLYPLACOPHORA	Chitons
Class MONOPLACOPHORA	*Neopilina* (remnants of segmentation)
Class SCAPHOPODA	Tooth shells
Class GASTROPODA	Snails and slugs
Class BIVALVIA	Clams and mussels
Class CEPHALOPODA	Squids and octopi
Phylum ANNELIDA	Segmented worms
Class POLYCHAETA	Sandworms
Class OLIGOCHAETA	Earthworms
Class HIRUDINEA	Leeches
Phylum ONYCHOPHORA	Peripatus
Phylum ARTHROPODA	Animals with an exoskeleton and jointed appendages
Class CRUSTACEA	Crustaceans, crabs, crayfish, barnacles
Class ARACHNIDA	Scorpions, spiders, ticks, mites
Class CHILOPODA	Centipedes
Class DIPLOPIDA	Millipedes
Class INSECTA	Insects

Deuterostomes

Phylum ECHINODERMATA	Spiny-skinned, radially symmetrical animals
Class CRINOIDEA	Sea lilies, feather stars
Class ASTEROIDEA	Sea stars

Class OPHIUROIDEA	Brittle stars
Class ECHINOIDEA	Sea urchins
Class HOLOTHUROIDEA	Sea cucumbers
Phylum HEMICHORDATA	Acorn worms
Phylum CHORDATA	Chordates
Subphylum UROCHORDATA	Tunicates
Subphylum CEPHALOCHORDATA	Lancelets
Subphylum VERTEBRATA	Vertebrates
Class AGNATHA	Jawless fishes
Class CHONDRICHTHYES	Cartilaginous fishes
Class OSTEICHTHYES	Bony fishes
Class AMPHIBIA	Salamanders, frogs, toads
Class REPTILIA	Lizards, snakes, crocodiles, turtles
Class AVES	Birds
Class MAMMALIA	Mammals
Subclass PROTOTHERIA	Egg-laying mammals
Subclass METATHERIA	Marsupial mammals
Subclass EUTHERIA	Placental mammals